In Dodgson's era, mathematicians at times used the original Arabic term for algebra—"al-jabr w'al muquabala," or "restoration and opposition." This interesting bit of history is discussed in Chapter 13—the *Concepts and History of Calculus*.

Alice unsuccessfully tries to remember her multiplication tables, saying "Let me see: four times five is twelve, and four times six is thirteen, and four times seven is—oh dear!" Mathematicians know that 4 × 5 = 12, if we use base 18 rather than the familiar base 10, and that 4 × 6 = 13, in base 21. This is discussed in Chapter 7—Number Systems and Number Theory.

Mathematics: A Practical Odyssey

is a brief survey of many different branches of mathematics. Paralleling *Alice's* journeys to Wonderland, the authors hope to take students on an odyssey throughout the amazing world of mathematics where they may encounter strange, wonderful, practical, and sometimes whimsical topics. The first chapter of the book is Logic, symbolized by the image of *Alice*. In fact, every image on this book's cover is a reference to one of the book's topics. We hope you have fun finding some of them.

Mathematics: A Practical Odyssey

University of Rhode Island

David B. Johnson | Thomas A. Mowry

CENGAGE Learning™

Australia • Brazil • Japan • Korea • Mexico • Singapore • Spain • United Kingdom • United States

CENGAGE Learning™

Mathematics: A Practical Odyssey, Univeristy of Rhode Island

Executive Editors:
Maureen Staudt
Michael Stranz

Senior Project Development Manager:
Linda deStefano

Marketing Specialist:
Courtney Sheldon

Senior Production/Manufacturing Manager:
Donna M. Brown

PreMedia Manager:
Joel Brennecke

Sr. Rights Acquisition Account Manager:
Todd Osborne

Cover Image:
Getty Images*

*Unless otherwise noted, all cover images used by Custom Solutions, a part of Cengage Learning, have been supplied courtesy of Getty Images with the exception of the Earthview cover image, which has been supplied by the National Aeronautics and Space Administration (NASA).

Printed in the United States of America

Mathematics: A Practical Odyssey, University of Rhode Island
Johnson | Mowry

For product information and technology assistance, contact us at
Cengage Learning Customer & Sales Support, 1-800-354-9706
For permission to use material from this text or product,
submit all requests online at **cengage.com/permissions**
Further permissions questions can be emailed to
permissionrequest@cengage.com

This book contains select works from existing Cengage Learning resources and was produced by Cengage Learning Custom Solutions for collegiate use. As such, those adopting and/or contributing to this work are responsible for editorial content accuracy, continuity and completeness.

Compilation © 2011 Cengage Learning
ISBN-13: 978-1-133-44312-4

ISBN-10: 1-133-44312-5

Cengage Learning
5191 Natorp Boulevard
Mason, Ohio 45040
USA
Cengage Learning is a leading provider of customized learning solutions with office locations around the globe, including Singapore, the United Kingdom, Australia, Mexico, Brazil, and Japan. Locate your local office at:
international.cengage.com/region.

Cengage Learning products are represented in Canada by Nelson Education, Ltd.
For your lifelong learning solutions, visit **www.cengage.com/custom.**
Visit our corporate website at **www.cengage.com.**

Brief Contents

PREFACE

TO THE INSTRUCTOR ...

Course Prerequisite

Mathematics: A Practical Odyssey is written for the student who has successfully completed a course in intermediate algebra, not the student who excelled in it. It would be difficult for a student without background in intermediate algebra to succeed in a course using this book. However, some chapters are not algebra-based. These chapters require a level of critical thinking and mathematical maturity more commonly found in students who have taken intermediate algebra.

Topics

Algebra is the language of mathematics. A background in algebra allows you to learn mathematics that is usable and relevant to any educated person. In particular, it allows the student to do the following:

- Learn enough about *Logic* to analyze the validity of an argument.
- Learn enough about *Sets, Counting and Probability* to understand the risks of inherited diseases and to realize what an incredibly bad bet a lottery ticket is.
- Learn enough about *Statistics* to understand the accuracy and validity of a public opinion poll.
- Learn enough about *Finance* to calculate the monthly payment required by a car of home loan and to understand loans well enough to make an educated decision when selecting one.
- Learn enough about *Voting and Appointment* to understand that there is no perfect voting system or method of appointment; all methods have inherent flaws.
- Learn enough about *Number Systems and Number Theory* to understand why our commonly used base ten and base two number systems work, and to appreciate the prevalence of the Fibonacci numbers in nature and the use of the golden ratio in art.
- Learn enough about *Geometry* to understand its place in the history of Western civilization.
- Learn enough about *Graph Theory* to be able to use it in scheduling and to create networks.
- Learn enough about *Exponential and Logarithmic Functions* to understand how populations grow, how radiocarbon dating works, how the Richter scale measures earthquakes, and how sound is measured in decibels.
- Learn enough about *Matrices and Markov Chains* to understand how manufacturers can predict their products' success or failure in the marketplace.
- Learn enough about *Linear Programming* to understand how a small business can determine how to utilize its limited resources in order to maximize its profit.
- Learn enough about *Calculus* to understand just what the subject is and why it is so important.

New in the Seventh Edition

▶ THE NEXT LEVEL
These special exercises are designed to help prepare the student for admissions examinations such as the GRE (required for graduate school) or the GMAT (required for graduate study in business).

express the following in words.
a. $p \wedge q$ b. $p \to q$
c. $\sim q \to \sim p$ d. $q \vee \sim p$

37. Using the symbolic representations
p: I am an environmentalist.
q: I recycle my aluminum cans.
r: I recycle my newspapers.
express the following in words.
a. $(q \vee r) \to p$ b. $\sim p \to \sim(q \vee r)$
c. $(q \wedge r) \vee \sim p$ d. $(r \wedge \sim q) \to \sim p$

38. Using the symbolic representations
p: I am innocent.
q: I have an alibi.
r: I go to jail.
express the following in words.
a. $(p \vee q) \to \sim r$ b. $(p \wedge \sim q) \to r$
c. $(\sim p \wedge q) \vee r$ d. $(p \wedge r) \to \sim q$

39. Which statement, #1 or #2, is more appropriate? Explain why.
Statement #1: "Cold weather is necessary for it to snow."
Statement #2: "Cold weather is sufficient for it to snow."

● **HISTORY QUESTIONS**

51. In what academic field did Gottfried Leibniz receive his degrees? Why is the study of logic important in this field?

52. Who developed a formal system of logic based on syllogistic arguments?

53. What is meant by *characteristica universalis?* Who proposed this theory?

THE NEXT LEVEL

If a person wants to pursue an advanced degree (something beyond a bachelor's or four-year degree), chances are the person must take a standardized exam to gain admission to a graduate school or to be admitted into a specific program. These exams are intended to measure verbal, quantitative, and analytical skills that have developed throughout a person's life. Many classes and study guides are available to help people prepare for the exams. The following questions are typical of those found in the study guides.

Exercises 54–58 refer to the following: A culinary institute has a small restaurant in which the students prepare various dishes.

FEATURED IN THE NEWS

CHURCH CARVING MAY BE ORIGINAL 'CHESHIRE CAT'

London—Devotees of writer Lewis Carroll believe they have found what inspired his grinning Cheshire Cat, made famous in his book "Alice's Adventures in Wonderland."

Members of the Lewis Carroll Society made the discovery over the weekend in a church at which the author's father was once rector in the Yorkshire village of Croft in northern England.

It is a rough-hewn carving of a cat's head smiling near an altar, probably dating to the 10th century. Seen from below and from the perspective of a small boy, all that can be seen is the grinning mouth.

Carroll's Alice watched the Cheshire Cat disappear "ending with the grin, which remained for some time after the rest of the head had gone."

Alice mused: "I have often seen a cat without a grin, but not a grin without a cat. It is the most curious thing I have seen in all my life."

Reprinted with permission from Reuters

◀ FEATURED IN THE NEWS
Newspaper and magazine articles illustrate how the book's topics come up in the real world, in a way that might affect your students personally.

- **CHAPTER 1,** "Logic," now includes "necessary" and "sufficient" conditions as applies to conditional statements. Over 60 new exercises relating to necessary and sufficient conditions have been added.
- **CHAPTER 4,** "Statistics," now includes material on finding the minimum sample size needed to be approximately confident that the margin of error in a survey is at most a specified amount.
- **CHAPTER 5,** "Finance," now includes detailed descriptions of how to use a TI graphing calculator's "TVM" (time value of money) feature in a variety of financial situations.
- **CHAPTER 8,** "Geometry," now includes a new section on linear perspective. This application of geometry to art makes paintings seem more realistic by giving them a sense of depth.
- **THROUGHOUT THE BOOK:**
  - Chapter openers have been rewritten in a more engaging, student-oriented style.
  - Real world data has been updated.
  - Over 500 new exercises have been added.

Features

▶ ALGEBRA REVIEW

Many books review algebra. Usually, the algebra reviews are overly general and don't really help your students review the specific algebraic topics that arise in the book. Our algebra reviews occur at the very beginning of the chapters, and they only review the algebra that comes up in that chapter. there are many topics in algebra where students need to review and sharpen their skills.

10.0A Review of Exponentials and Logarithms

OBJECTIVES

- Define and graph an exponential function; define the natural exponential function
- Use a calculator to find values of the exponential function 10^x and the natural exponential function e^x
- Define and understand the meaning of a logarithm
- Rewrite a logarithm as an exponential equation and vice versa
- Use a calculator to find values of the common and natural logarithmic functions $\log x$ and $\ln x$

TOPIC X BLOOD TYPES:
SET THEORY IN THE REAL WORLD

Human blood types are a classic example of set theory. As you may know, there are four categories (or sets) of blood types: A, B, AB, and O. Knowing someone's blood type is extremely important in case a blood transfusion is required; if blood of two different types is combined, the blood cells may begin to clump together, with potentially fatal consequences! (Do you know your blood type?)

What exactly are "blood types"? In the early 1900s, the Austrian scientist Karl Landsteiner observed the presence (or absence) of two distinct chemical molecules on the surface of all red blood cells in numerous samples of human blood. Consequently, he labeled one molecule "A" and the other "B." The presence or absence of these specific molecules is the basis of the universal classification of blood types. Specifically, blood samples containing only the A molecule are labeled type A, whereas those containing only the B molecule are labeled type B. If a blood sample contains both molecules (A and B) it is labeled type AB; and if neither is present, the blood is typed as O. The presence (or absence) of these molecules can be depicted in a standard Venn diagram as shown in Figure 2.27. In the notation of set operations, type A blood is denoted $A \cap B'$, type B is $B \cap A'$, type AB is $A \cap B$, and type O is $A' \cap B'$.

If a specific blood sample is mixed with blood containing a blood molecule (A or B) that it does not already have, the presence of the foreign molecule may cause the mixture of blood to clump. For example, type A blood cannot be mixed with any blood containing the B molecule (type B or type AB). Therefore, a person with type A blood can receive a transfusion only of type A or type O blood. Consequently, a person with type AB blood may receive a transfusion of any blood type; type AB is referred to as the "universal receiver." Because type O blood contains neither the A nor the B molecule, all blood types are compatible with type O blood; type O is referred to as the "universal donor."

It is not uncommon for scientists to study rhesus monkeys in an effort to learn more about human physiology. In so doing, a certain blood protein was discovered in rhesus monkeys. Subsequently, scientists found that the blood of some people contained this protein, whereas the blood of others did not. The presence, or absence, of this protein in human blood is referred to as the *Rh factor*; blood containing the protein is labeled "Rh+", whereas "Rh−" indicates the absence of the protein. The Rh factor of human blood is especially important for expectant mothers; a fetus can develop problems if its parents have opposite Rh factors.

When a person's blood is typed, the designation includes both the regular blood type and the Rh factor. For instance, type AB− indicates the presence of both the A and B molecules (type AB), along with the absence of the rhesus protein; type O+ indicates the absence of both the A and B molecules (type O), along with the presence of the rhesus protein. Utilizing the Rh factor, there are eight possible blood types as shown in Figure 2.28.

We will investigate the occurrence and compatibility of the various blood types in Example 5 and in Exercises 35–43.

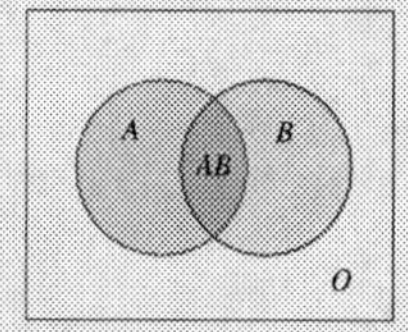

FIGURE 2.27 Blood types and the presence of the A and B molecules.

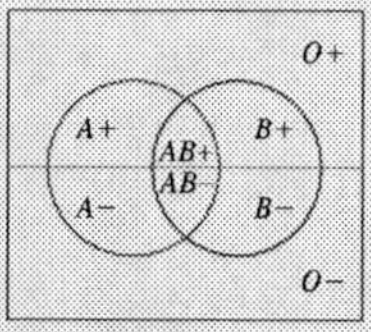

FIGURE 2.28 Blood types combined with the Rh factor.

◀ REAL-WORLD APPLICATIONS

"TOPIC X IN THE REAL WORLD"

Discussing current, powerful, real-world uses of the text's topics provides an opportunity to reinforce the practical emphasis of the text. To create the extended applications of Topic X in the Real World, we researched articles in professional journals, magazines, newspapers and the web, and distilled the information so that it is accessible to the liberal arts mathematics student. These highly practical applications are used as jumping-off points for both reality-based conceptual exercises. Below is a partial listing:

- Blood Type: *Set Theory in the Real World* (Section 2.2, page 88)
- The Business of Gambling: *Probabilities in the Real World* (Section 3.4, page 175)
- HIV/AIDS: *Probabilities in the Real World* (Section 3.6, page 198)
- NASA: *PERT Charts in the Real World* (Section 9.5, page 720)

For a complete listing and to access additional course materials and companion resources, please visit www.cengagebrain.com. At the CengageBrain.com home page, search for the ISBN of your title (from the back cover of this book) using the search box at the top of the page. This will take you to the product page where free companion resources can be found.

PROJECTS

Below is a partial listing:

CHAPTER 3: PROBABILITY

- Relative frequency of an odd phone number. (Section 3.2, Exercise 88, page 156)
- Compare relative frequency and theoretical probability with a coin toss. (Section 3.2, Exercise 89, page 156)
- Compare relative frequency and theoretical probability with a roll of a die. (Section 3.2, Exercise 90, page 157)
- Determine if the outcomes are equally likely with coin experiments. (Section 3.2, Exercises 91 and 92, page 157)
- Design a game of chance. Use probabilities and expected values to set house odds. (Section 3.5, Exercise 50, page 190)
- Investigate seemingly contradictory information regarding sex bias in graduate school and/or tuberculosis incidence in New York City and Richmond. (Section 3.6, Exercises 73 and 74, page 205)

CHAPTER 13: CALCULUS

- Write a research paper on the contributions of Apollonius, Oresme, Descartes and Fermat to analytic geometry. (Section 13.1, Exercise 36, page 13-22)
- Compare and contrast the algebra of Apollonius, al-Khowarizmi, and Decartes. (Section 13.1, Exercise 37, page 13-22)
- Use dimensional analysis to convert the distance and speed falling object formulas from the English system to the metric system. (Section 13.7, Exercise 1, page 13-81)
- Use calculus to sketch the graph of a given polynomial function. (Section 13.7, Exercise 2, page 13-81)
- Use calculus to find the areas of some complicated shapes. (Section 13.7, Exercises 3–6, pages 13-81 and 13-82)

For a complete listing, please visit www.cengagebrain.com.

Organization

CHAPTER OPENERS

The chapter openers briefly discuss "What We Will Do in this Chapter," so that your students can get an idea of what to expect.

EXAMPLES

All math texts have examples. Many skip just enough steps to be frustrating. Ours don't skip steps, and ours include a verbal summary to aid in your students understanding.

BRIDGES BETWEEN CHAPTERS

As much as possible, this book has been written so that its chapters are independent of each other. The instructor therefore has wide latitude in selecting topics to cover and can teach a course that is responsive to the needs of his or her students and institution. Sometimes,

Chapter is available online.

however, this independence is not desirable, because it does not allow connections to be made between seemingly unrelated topics. For this reason, the text features a number of bridges between chapters. These bridges feature both discussion and exercises. They are clearly labeled; that is, a bridge between an earlier Chapter X and a later Chapter Y is labeled "for those students who have completed Chapter X."

▶ BUILDING BRIDGES
There is a bridge between Chapter 1 (Logic) and Chapter 2 (Sets and Counting). At the end of Section 2.1, the similarities between the respective concepts and notation of logic and sets are discussed.

78 CHAPTER 2 Sets and Counting

Set Theory and Logic

If you have read Chapter 1, you have probably noticed that set theory and logic have many similarities. For instance, the union symbol ∪ and the disjunction symbol ∨ have the same meaning, but they are used in different circumstances; ∪ goes between sets, while ∨ goes between logical expressions. The ∪ and ∨ symbols are similar in appearance because their usages are similar. A comparison of the terms and symbols used in set theory and logic is given in Figure 2.13.

Set Theory		Logic		Common Wording
Term	Symbol	Term	Symbol	
union	$\cup$	disjunction	$\vee$	or
intersection	$\cap$	conjunction	$\wedge$	and
complement	$'$	negation	$\sim$	not
subset	$\subseteq$	conditional	$\rightarrow$	if . . . then . . .

FIGURE 2.13 Comparison of terms and symbols used in set theory and logic.

Applying the concepts and symbols of Chapter 1, we can define the basic operations of set theory in terms of logical biconditionals. The biconditionals in Figure 2.14 are tautologies (expressions that are always true); the first biconditional is read as "x is an element of the union of sets A and B if and only if x is an element of set A or x is an element of set B."

Basic Operations in Set Theory	Logical Biconditional
union	$[x \in (A \cup B)] \leftrightarrow [x \in A \vee x \in B]$
intersection	$[x \in (A \cap B)] \leftrightarrow [x \in A \wedge x \in B]$
complement	$(x \in A') \leftrightarrow \sim (x \in A)$
subset	$(A \subseteq B) \leftrightarrow (x \in A \rightarrow x \in B)$

FIGURE 2.14 Set theory operations as logical biconditionals.

- There is a bridge between Chapter 5 (Finance) and Chapter 10 (Exponential and Logarithmic Functions). That bridge discussed the relationship between the exponential growth model ($y = ae^{bt}$) and the common interest model [$FV = P(\mathrm{I} + i)^n$], and the circumstances under which model can be used in place of the other.
- There is a bridge between Chapter 8 (Geometry) and Chapter 13 (Calculus). That bridge discussed the use of the trigonometric functions to determine the equation of a trajectory when the initial angle of elevation other that 30°, 45°, or 60°.

TOPIC SELECTION AND PREREQUISITE MAPS

Obviously, the book contains more material that you could ever cover in a one-quarter course. It is written in such a way that most chapters are independent of each other. The Instructor's Manual includes a "prerequisite map" so that you can easily tell which earlier topics must be covered. It also includes sample course outlines, suggesting specific chapters to be used in forming a typical course.

FLEXIBILITY AND COURSE LEVEL

The book contains a wide range of topics, varying in level of sophistication and difficulty. Chapters 1 through 9 cover topics that are not uncommon to beginning algebra-abased liberal arts mathematics texts. However, because of the intermediate algebra prerequisite, the topics are covered more thoroughly, and the acquisition of problem-solving skills is emphasized. Chapters 10 through 13 are more sophisticated, covering topics not commonly found in liberal arts texts. However, the treatment is such that the material is accessible to the students who enroll in this course.

A chapter need not be covered in its entirety; topics can easily be left out. In most cases, a chapter has a suggested core of key sections as well as a selection of optional sections, which tend to be more sophisticated. They are not labeled "optional" in the text; rather, this distinction is made in the prerequisite map in the Instructor's Resource Manual.

The text is designed so that the instructor determines the difficulty of the course by selecting the chapters and sections to be covered. The instructor can thus create his or her own course—one that will fit the students' needs.

USABILITY

This book is user-friendly:

- The examples don't skip steps.
- Key points are boxed for emphasis.
- Step-by-step procedures are given.
- There is an abundance of exposition.

ANSWER CHECKING

Throughout the text, we emphasize the importance of checking ones answers. It's important that students learn to evaluate the reasonableness of their answers, rather than accepting them on face value. To this end, some exercises do not have answers in the back of the book when students are instructed to check answers for themselves.

▶ HISTORY

The history of the subject matter is interwoven throughout most chapters. In addition, Historical Notes give in-depth biographies of the prominent people involved. It is our hope that students will see the human side of mathematics. After all, mathematics was invented by real people for real purposes and is a part of our culture. Interesting research topics are given, and writing assignments are suggested.

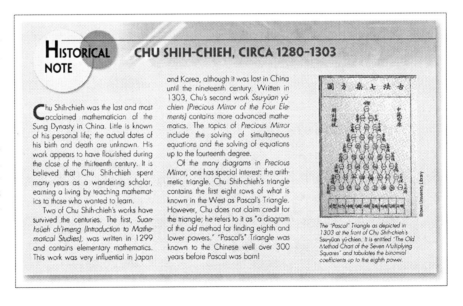

HISTORICAL NOTE **CHU SHIH-CHIEH, CIRCA 1280–1303**

Chu Shih-chieh was the last and most acclaimed mathematician of the Sung Dynasty in China. Little is known of his personal life; the actual dates of his birth and death are unknown. His work appears to have flourished during the close of the thirteenth century. It is believed that Chu Shih-chieh spent many years as a wandering scholar, earning a living by teaching mathematics to those who wanted to learn.

Two of Chu Shih-chieh's works have survived the centuries. The first, *Suan-hsüeh ch'i-meng (Introduction to Mathematical Studies)*, was written in 1299 and contains elementary mathematics. This work was very influential in Japan and Korea, although it was lost in China until the nineteenth century. Written in 1303, Chu's second work *Ssu-yüan yü-chien (Precious Mirror of the Four Elements)* contains more advanced mathematics. The topics of *Precious Mirror* include the solving of simultaneous equations and the solving of equations up to the fourteenth degree.

Of the many diagrams in *Precious Mirror*, one has special interest: the arithmetic triangle. Chu Shih-chieh's triangle contains the first eight rows of what is known in the West as Pascal's Triangle. However, Chu does not claim credit for the triangle; he refers to it as "a diagram of the *old* method for finding eighth and lower powers." "Pascal's" Triangle was known to the Chinese well over 300 years before Pascal was born!

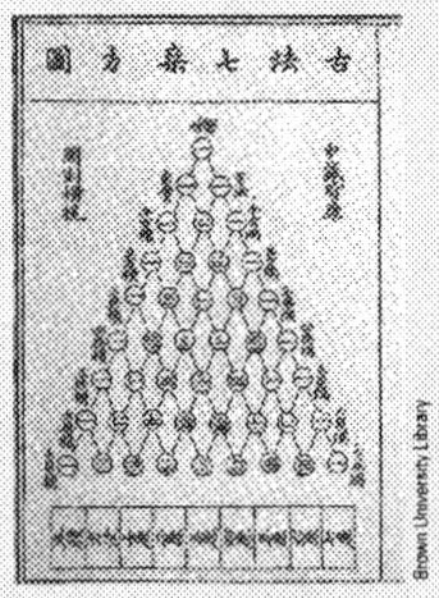

The "Pascal" Triangle as depicted in 1303 at the front of Chu Shih-chieh's Ssu-yüan yü-chien. It is entitled "The Old Method Chart of the Seven Multiplying Squares" and tabulates the binomial coefficients up to the eighth power.

Technology

WEB PROJECTS

These projects include links to web pages that the students can use as starting points in their research. Following is a partial listing:

CHAPTER 2: SETS AND COUNTING

- Determine the compatibility of the various blood types, including Rh factors. (Section 2.2, Exercise 53, Page 94)
- Write a research paper on an historic topic. (Section 2.4, Exercise 62, page 117; and section 2.5, Exercise 28, page 127)

CHAPTER 3: PROBABILITY

- Write a research paper on an historic topic. (Section 3.1, Exercise 46, page 139)
- Write a research paper on the successful screening of Jews for Tay-Sachs disease and the unsuccessful screening of blacks for sickle-cell anemia. (Section 3.2, Exercise 93, page 157)
- Determine the prevalence of the various blood types in the United States, the percentage of U.S. residents that can donate blood to people of each of each of the various blood types, and the percentage of U.S. residents that can receive blood from people of each of the various blood types. (Section 3.6, Exercise 71, page 204)
- Investigate automobile rollovers. Determine which type of vehicle—a sedan, and SUV or a van—is more prone to rollovers. (Section 3.6, Exercise 72, page 205)

CHAPTER 9: GRAPH THEORY

- Investigate some of the unsolved problems in graph theory. (Section 9.2, Exercise 33, page 683)
- Write an essay on the four-color map problem. (Section 9.2, Exercise 34, page 683)
- Investigate the traveling salesman problem on the web. Summarize their history, and describe specific problems that led to progress. (Section 9.3, Exercise 32, page 696)

> **WEB PROJECTS**
>
> 33. There are many interesting problems in graph theory. Some of these problems have been solved, and some remain unsolved. Many of these problems are discussed on the web. Visit several web sites and choose a specific problem. Describe the problem, its history, and its applications. If it has been solved, describe the method of solution if possible.

CHAPTER 13: CALCULUS

- Write a research paper on an historical topic, or on Fermat's last theorem. (Section 13.1, Exercise 38, page 13-22; and Section 13.2, Exercises 42 and 43, page 13-35)

For a complete listing, please visit www.cengagebrain.com.

GRAPHING AND SCIENTIFIC CALCULATORS

Calculator boxes give you all of the necessary keystrokes for both scientific calculators and graphing calculators. Calculator subsections help you learn how to use your calculator when a list of keystrokes is just not enough. The following graphing calculator topics are addressed:

- Section 3.3: Fractions on a Graphing Calculator (page 168)
- Section 5.2: Doubling Time with a TI's TVM Application (page 354)
- Section 5.5 Finding the APR with a TI's TVM Application (page 388)

- Section 11.0: Matrix Multiplication on a Graphing Calculator (page 819)
- Section 11.4: Solving Larger Systems of Linear Equations on a Graphing Calculator (page 847)

For a complete listing, please visit www.cengagebrain.com.

EXCEL AND COMPUTERS

Computers are ubiquitous at the workplace, and are becoming increasingly common in the classroom. However, many students have no mathematical experience using computer software such as Microsoft Excel. For this reason, we have included a number of subsections that give instruction on the use of Excel. We have also included a number of optional subsections that give instructions on the use of *Amortix*, custom text-specific software that is available on the text website (go to http://www.brookscole.com/math_d/resources/amortrix/) The following topics are addressed:

- Section 4.1: Histograms and Pie Charts on a Computerized Spreadsheet (page 241)
- Section 4.3: Measures of Central Tendency and Dispersion on Excel (page 274)
- Section 5.4: Amortization Schedules on Amortix (page 383)
- Section 5.4: Amortization Schedules on Excel (page 384)
- Section 12.3: The Row Operations on Amortix (page 12-25)

These subsections allow instructors to incorporate the computer into their class if they so desire, but they are entirely optional, and the book is in no way computer dependent. The subsections do not assume any previous experience with Excel.

WEB SITE

To access additional course materials and companion resources, please visit www.cengagebrain.com. At the CengageBrain.com home page, search for the ISBN of your title (from the back cover of your book) using the search box at the top of the page. This will take you to the product page where free companion resources can be found.

The site offers book-specific student and instructor resources, as well as discipline-specific links. Student resources at the web site include a web-based version of Amortix (the software used in Chapters 5, 7, and 8), downloadable partially completed Excel spreadsheets that are keyed to examples in the text, graphing and calculation tools, and Internet links for further research.

EXERCISES

The exercises in this text are designed to solidify the students' understanding of the material and make them proficient in the calculations involved. It is assumed that most students who complete this course will not continue in their formal study of mathematics. Consequently, neither the exposition nor the exercises are designed to expose the students to all aspects of the topic.

The exercises vary in difficulty. Some are exactly like the examples, and others demand more of the students. The exercises are not explicitly graded into A, B and C categories, nor are any marked "optional;" students in this audience tend to react negatively if asked to do anything labeled in this manner. The more difficult the exercises are indicated in the Instructor's Resource Manual.

The short-answer historical questions are meant to focus and reinforce the students' understanding of the historical material. They also serve to warn them that history questions may appear on exams. The essay questions can be used as an integral part of the students' grades, as background for classroom discussion, or for extra credit work. Most are research topics and are kept as open-ended as possible.

Answers to the odd-numbered exercises are given in the back of the book, with two exceptions:

- Answers to historical questions and essay questions are not given
- Answers are not given when the exercises instruct the student to check the answers themselves.

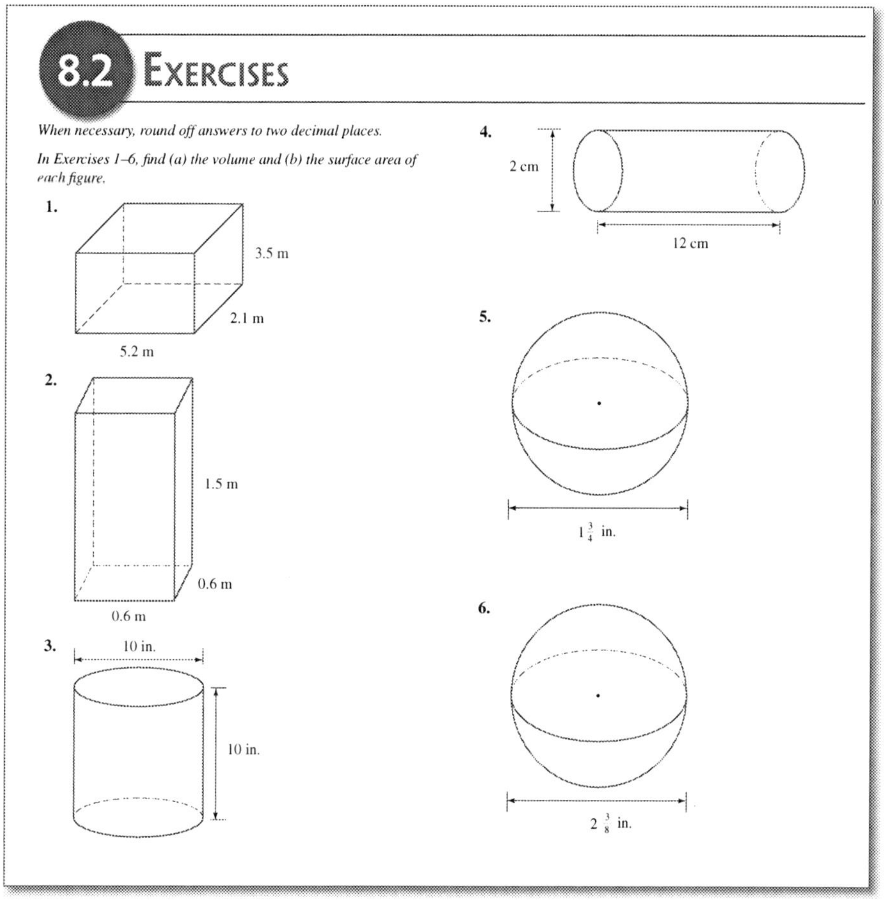

8.2 Exercises

When necessary, round off answers to two decimal places.

In Exercises 1–6, find (a) the volume and (b) the surface area of each figure.

Complete solutions to every other odd-numbered exercise are given in the Student Solutions Manual, with the above two exceptions.

Standards

This text is well-suited in addressing the AMATYC and NCTM standards.

Regarding *Standards for Intellectual Development*, the text contains a wide range of exercises designed to engage students in mathematical problem solving and modeling real-world situations. In addition, the exercises provide students an opportunity to expand their mathematical reasoning skills and to communicate their results effectively. Also, the text emphasizes the interrelationships between mathematics, human culture, and other disciplines. The use of appropriate technology is woven throughout the text, along with the material that encourages independent exploration and confidence building in mathematics.

Regarding *Standards for Content*, the text contains material covering all requisite topics including number sense, symbolism and algebra, geometry, function, combinatorics, probability and statistics, and deductive proof.

Regarding *Standards for Pedagogy*, the text provides ample opportunity for faculty to use technology in the teaching of mathematics, to foster interactive learning through collaborative activities and effective communication, to make connections between various branches of mathematics and between mathematics and the students' lives, and to use numerical, graphical, symbolic, and verbal approaches in the teaching of mathematics.

TO THE STUDENT, AS YOU EMBARK ON YOUR ODYSSEY ...

This textbook is designed for students in liberal arts programs and other fields that do not require a core of mathematics. The term *liberal arts* is a translation of a Latin phrase that means "studies befitting a free person." It was applied during the Middle Ages to seven branches of learning: arithmetic, geometry, logic, grammar, rhetoric, astronomy, and music. You might be surprised to learn that almost half of the original liberal arts are mathematics subjects.

In accordance with the tradition, handed down from the Middle Ages, that a broad-based education includes some mathematics, many institutions of higher education require their students to complete a college-level mathematics course. These schools award a bachelor's degree to a person who not only has acquired a detailed knowledge of his or her field but also has a broad background in the liberal arts.

The goal of this textbook is to expose you to topics in mathematics that are usable and relevant to any educated person. We hope that you will encounter topics that will be useful at some time during your life. In addition, you are encouraged to recognize the relevance of mathematics to a well-rounded education and to appreciate the creative, human aspect of mathematics.

This book is written for the student who has successfully completed a course in intermediate algebra, not the student who excelled in it. Your mathematical background doesn't have to be perfect, but algebra will come up all the time. It's also true that this book is written for a college-level math course, and that's a significant step up from your high school experience. You will have to work hard and put in a solid effort.

Your success in this course is important to us. To help you achieve that success, we have incorporated features in the textbook that promote learning and support various learning styles. Among these features are algebra review and instructions in using a calculator.

Our algebra reviews occur at the very beginning of the chapters, and they review only the algebra that comes up in that chapter. There are many topics in algebra in which students need to review and sharpen their skills. Calculator boxes give you all of the necessary keystrokes for scientific calculators and for graphing calculators. Calculator subsections help you learn how to use your calculator when a list of keystrokes is just not enough. We encourage you to examine these features and use them on your successful odyssey throughout this course.

SUPPLEMENTS

FOR THE STUDENT	FOR THE INSTRUCTOR
	Annotated Instructor's Edition (ISBN: 0840049137) The Annotated Instructor's Edition features an appendix containing the answers to all problems in the book as well as icons denoting which problems can be found in Enhanced WebAssign. (Print)
Student Solutions Manual for (ISBN: 0840053878) The Student Solutions Manual provides worked-out solutions to the odd-numbered problems in the text. Use of the solutions manual ensures that students learn the correct steps to arrive at an answer. (Print)	**Instructor's Resource Manual for** (ISBN: 0840053452) The Instructor's Resource Manual provides worked-out solutions to all of the problems in the text and includes suggestions for course syllabi and chapter summaries. (Print)
Text Specific DVDs (ISBN: 1111571570) Hosted by Dana, these professionally produced DVDs cover key topics of the text, offering a valuable alternative for classroom instruction or independent study and review. (Media)	**Text Specific DVDs** (ISBN: 1111571570) Hosted by Dana, these professionally produced DVDs cover key topics of the text, offering a valuable alternative for classroom instruction or independent study and review. (Media)
Enhanced WebAssign (ISBN: 0538738103) Enhanced WebAssign, used by over one million students at more than 1100 institutions, allows you to do homework assignments and get extra help and practice via the web. This proven and reliable homework system includes hundreds of algorithmically generated homework problems, and eBook, links to relevant textbook sections, video examples, problem specific tutorials, and more. (Online)	**Enhanced WebAssign** (ISBN: 0538738103) Enhanced WebAssign, used by over one million students at more than 1100 institutions, allows you to assign, collect, grade, and record homework assignments via the web. This proven and reliable homework system includes hundreds of algorithmically generated homework problems, and eBook, links to relevant textbook sections, video examples, and more. (Online) Note that the WebAssign problems for this text are highlighted by a ▶.
	PowerLecture with ExamView (ISBN: 0840054114) This CD-ROM provides the instructor with dynamic media tools for teaching. Create, deliver, and customize tests (both print and online) in minutes with *ExamView® Computerized Testing Featuring Algorithmic Equations*. Easily build solution sets for homework or exams using *Solution Builder's* online solutions manual. Microsoft® PowerPoint® lecture slides and figures from the book are also included on this CD-ROM. (CD)
	Solution Builder This online solutions manual allows instructors to create customizable solutions that they can print out to distribute or post as needed. This is a convenient and expedient way to deliver solutions to specific homework sets. Visit www.cengage.com/solutionbuilder. (Online)

Acknowledgements

The authors would like to thank Marc Bove, Kyle O'Loughlin, Meaghan Banks, Stefanie Beeck, and all the fine people at Cengage Learning: Anne Seitz and Gretchen Miller at Hearthside Publishing Services; Ann Ostberg, and Rhoda Oden.

Special thanks go to the users of the text and reviewers who evaluated the manuscript for this edition, as well as those who offered comments on previous editions.

Reviewers

Dennis Airey, *Rancho Santiago College*
Francisco E. Alarcon, *Indiana University of Pennsylvania*
Judith Arms, *University of Washington*
Bruce Atkinson, *Palm Beach Atlantic College*
Wayne C. Bell, *Murray State University*
Wayne Bishop, *California State University—Los Angeles*
David Boliver, *Trenton State College*
Stephen Brick, *University of South Alabama*
Barry Bronson, *Western Kentucky University*
Frank Burk, *California State University—Chico*
Laura Cameron, *University of New Mexico*
Jack Carter, *California State University—Hayward*
Timothy D. Cavanaugh, *University of Northern Colorado*
Joseph Chavez, *California State University—San Bernadino*
Eric Clarkson, *Murray State University*
Rebecca Conti, *State University of New York at Fredonia*
S.G. Crossley, *University of Southern Alabama*
Ben Divers, Jr., *Ferrum College*
Al Dixon, *College of the Ozarks*
Joe S. Evans, *Middle Tennessee State University*
Hajrudin Fejzie, *California State University—San Bernardino*
Lloyd Gavin, *California State University—Sacramento*
William Greiner, *McLennan Community College*
Martin Haines, *Olympic College*
Ray Hamlett, *East Central University*
Virginia Hanks, *Western Kentucky University*
Anne Herbst, *Santa Rosa Junior College*
Linda Hinzman, *Pasadena City College*
Thomas Hull, *University of Rhode Island*
Robert W. Hunt, *Humboldt State University*
Robert Jajcay, *Indiana State University*
Irja Kalantari, *Western Illinois University*
Daniel Katz, *University of Kansas*
Katalin Kolossa, *Arizona State University*
Donnald H. Lander, *Brevard College*
Lee LaRue, *Paris Junior College*
Thomas McCready, *California State University—Chico*
Vicki McMillian, *Stockton State University*
Narendra L. Maria, *California State University—Stanislaus*
John Martin, *Santa Rosa Junior College*
Gael Mericle, *Mankato State University*
Robert Morgan, *Pima Community College*
Pamela G. Nelson, *Panhandle State University*
Carol Oelkers, *Fullerton College*
Michael Olinick, *Middlebury College*
Matthew Pickard, *University of Puget Sound*
Joan D. Putnam, *University of Northern Colorado*
J. Doug Richey, *Northeast Texas Community College*
Stewart Robinson, *Cleveland State University*
Eugene P. Schlereth, *University of Tennessee at Chattanooga*
Lawrence Somer, *Catholic University of America*
Michael Trapuzzano, *Arizona State University*
Pat Velicky, *Mid-Plains Community College*
Karen M. Walters, *University of Northern California*
Dennis W. Watson, *Clark College*
Denielle Williams, *Eastern Washington University*
Charles Ziegenfus, *James Madison University*

LOGIC

1

© Bart Sadowski/iStockPhoto

When writer Lewis Carroll took Alice on her journeys down the rabbit hole to Wonderland and through the looking glass, she had many fantastic encounters with the tea-sipping Mad Hatter, a hookah-smoking Caterpillar, the White Rabbit, the Cheshire Cat, the Red and White Queens, and Tweedledum and Tweedledee. On the surface, Carroll's writings seem to be delightful nonsense and mere children's entertainment. Many people are quite surprised to learn that *Alice's Adventures in Wonderland* is as much an exercise in logic as it is a fantasy and that Lewis Carroll was actually Charles Dodgson, an Oxford mathematician. Dodgson's many writings include the whimsical *The Game of Logic* and the brilliant *Symbolic Logic,* in addition to *Alice's Adventures in Wonderland* and *Through the Looking Glass.*

continued

WHAT WE WILL DO IN THIS CHAPTER

WE'LL EXPLORE DIFFERENT TYPES OF LOGIC OR REASONING:

- Deductive reasoning involves the application of a general statement to a specific case; this type of logic is typified in the classic arguments of the renowned Greek logician Aristotle.
- Inductive reasoning involves generalizing after a pattern has been recognized and established; this type of logic is used in the solving of puzzles.

WE'LL ANALYZE AND EXPLORE VARIOUS TYPES OF STATEMENTS AND THE CONDITIONS UNDER WHICH THEY ARE TRUE:

- A statement is a simple sentence that is either true or false. Simple statements can be connected to form compound, or more complicated, statements.
- Symbolic representations reduce a compound statement to its basic form; phrases that appear to be different may actually have the same basic structure and meaning.

continued

WHAT WE WILL DO IN THIS CHAPTER — *continued*

WE'LL ANALYZE AND EXPLORE CONDITIONAL, OR "IF . . . THEN . . .," STATEMENTS:

- In everyday conversation, we often connect phrases by saying "if *this*, then *that*." However, does "*this*" actually guarantee "*that*"? Is "*this*" in fact necessary for "*that*"?
- How does "if" compare with "only if"? What does "if and only if" really mean?

WE'LL DETERMINE THE VALIDITY OF AN ARGUMENT:

- What constitutes a valid argument? Can a valid argument yield a false conclusion?
- You may have used Venn diagrams to depict a solution set in an algebra class. We will use Venn diagrams to visualize and analyze an argument.
- Some of Lewis Carroll's whimsical arguments are valid, and some are not. How can you tell?

Webster's New World College Dictionary defines **logic** as "the science of correct reasoning; science which describes relationships among propositions in terms of implication, contradiction, contrariety, conversion, etc." In addition to being flaunted in Mr. Spock's claim that "your human emotions have drawn you to an illogical conclusion" and in Sherlock Holmes's immortal phrase "elementary, my dear Watson," logic is fundamental both to critical thinking and to problem solving. In today's world of misleading commercial claims, innuendo, and political rhetoric, the ability to distinguish between valid and invalid arguments is important.

In this chapter, we will study the basic components of logic and its application. Mischievous, wild-eyed residents of Wonderland, eccentric, violin-playing detectives, and cold, emotionless Vulcans are not the only ones who can benefit from logic. Armed with the fundamentals of logic, we can surely join Spock and "live long and prosper!"

Vanni/Art Resource, NY

Logic is the science of correct reasoning. Auguste Rodin captured this ideal in his bronze sculpture *The Thinker*.

PARAMOUNT TELEVISION/THE KOBAL COLLECTION

In their quest for logical perfection, the Vulcans of *Star Trek* abandoned all emotion. Mr. Spock's frequent proclamation that "emotions are illogical" typified this attitude.

1.1 Deductive versus Inductive Reasoning

OBJECTIVES

- Use Venn diagrams to determine the validity of deductive arguments
- Use inductive reasoning to predict patterns

Logic is the science of correct reasoning. *Webster*'s *New World College Dictionary* defines **reasoning** as "the drawing of inferences or conclusions from known or assumed facts." Reasoning is an integral part of our daily lives; we take appropriate actions based on our perceptions and experiences. For instance, if the sky is heavily overcast this morning, you might assume that it will rain today and take your umbrella when you leave the house.

Problem Solving

Logic and reasoning are associated with the phrases *problem solving* and *critical thinking*. If we are faced with a problem, puzzle, or dilemma, we attempt to reason through it in hopes of arriving at a solution.

The first step in solving any problem is to define the problem in a thorough and accurate manner. Although this might sound like an obvious step, it is often overlooked. Always ask yourself, "What am I being asked to do?" Before you can

Using his extraordinary powers of logical deduction, Sherlock Holmes solves another mystery. "Finding the villain was elementary, my dear Watson."

solve a problem, you must understand the question. Once the problem has been defined, all known information that is relevant to it must be gathered, organized, and analyzed. This analysis should include a comparison of the present problem to previous ones. How is it similar? How is it different? Does a previous method of solution apply? If it seems appropriate, draw a picture of the problem; visual representations often provide insight into the interpretation of clues.

Before using any specific formula or method of solution, determine whether its use is valid for the situation at hand. A common error is to use a formula or method of solution when it does not apply. If a past formula or method of solution is appropriate, use it; if not, explore standard options and develop creative alternatives. Do not be afraid to try something different or out of the ordinary. "What if I try this . . . ?" may lead to a unique solution.

Deductive Reasoning

Once a problem has been defined and analyzed, it might fall into a known category of problems, so a common method of solution may be applied. For instance, when one is asked to solve the equation $x^2 = 2x + 1$, realizing that it is a second-degree equation (that is, a quadratic equation) leads one to put it into the standard form ($x^2 - 2x - 1 = 0$) and apply the Quadratic Formula.

EXAMPLE 1 **USING DEDUCTIVE REASONING TO SOLVE AN EQUATION** Solve the equation $x^2 = 2x + 1$.

SOLUTION The given equation is a second-degree equation in one variable. We know that all second-degree equations in one variable (in the form $ax^2 + bx + c = 0$) can be solved by applying the Quadratic Formula:

$$x = \frac{-b \pm \sqrt{b^2 - 4ac}}{2a}$$

Therefore, $x^2 = 2x + 1$ can be solved by applying the Quadratic Formula:

$$x^2 = 2x + 1$$

$$x^2 - 2x - 1 = 0$$

$$x = \frac{-(-2) \pm \sqrt{(-2)^2 - (4)1(-1)}}{2(1)}$$

$$x = \frac{2 \pm \sqrt{4 + 4}}{2}$$

$$x = \frac{2 \pm \sqrt{8}}{2}$$

$$x = \frac{2 \pm 2\sqrt{2}}{2}$$

$$x = \frac{2(1 \pm \sqrt{2})}{2}$$

$$x = 1 \pm \sqrt{2}$$

The solutions are $x = 1 + \sqrt{2}$ and $x = 1 - \sqrt{2}$.

In Example 1, we applied a general rule to a specific case; we reasoned that it was valid to apply the (general) Quadratic Formula to the (specific) equation $x^2 = 2x + 1$. This type of logic is known as **deductive reasoning**—that is, the application of a general statement to a specific instance.

Deductive reasoning and the formal structure of logic have been studied for thousands of years. One of the earliest logicians, and one of the most renowned, was Aristotle (384–322 B.C.). He was the student of the great philosopher Plato and the tutor of Alexander the Great, the conqueror of all the land from Greece to India. Aristotle's philosophy is pervasive; it influenced Roman Catholic theology through St. Thomas Aquinas and continues to influence modern philosophy. For centuries, Aristotelian logic was part of the education of lawyers and politicians and was used to distinguish valid arguments from invalid ones.

For Aristotle, logic was the necessary tool for any inquiry, and the syllogism was the sequence followed by all logical thought. A **syllogism** is an argument composed of two statements, or **premises** (the major and minor premises), followed by a **conclusion.** For any given set of premises, if the conclusion of an argument is guaranteed (that is, if it is inescapable in all instances), the argument is **valid.** If the conclusion is not guaranteed (that is, if there is at least one instance in which it does not follow), the argument is **invalid.**

Perhaps the best known of Aristotle's syllogisms is the following:

1. All men are mortal.	major premise
2. Socrates is a man.	minor premise
Therefore, Socrates is mortal.	conclusion

When the major premise is applied to the minor premise, the conclusion is inescapable; the argument is valid.

Notice that the deductive reasoning used in the analysis of Example 1 has exactly the same structure as Aristotle's syllogism concerning Socrates:

1. All second-degree equations in one variable can be solved by applying the Quadratic Formula.	major premise
2. $x^2 = 2x + 1$ is a second-degree equation in one variable.	minor premise
Therefore, $x^2 = 2x + 1$ can be solved by applying the Quadratic Formula.	conclusion

Each of these syllogisms is of the following general form:

1. If A, then B. All A are B. (major premise)
2. x is A. We have A. (minor premise)

Therefore, x is B. Therefore, we have B. (conclusion)

Historically, this valid pattern of deductive reasoning is known as *modus ponens.*

Deductive Reasoning and Venn Diagrams

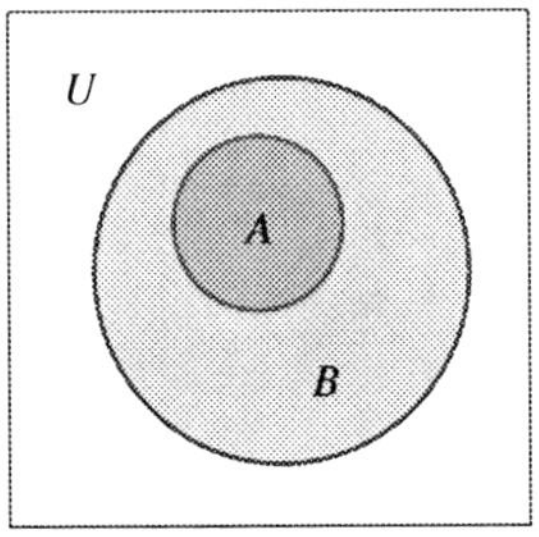

FIGURE 1.1
All A are B. (If A, then B.)

The validity of a deductive argument can be shown by use of a Venn diagram. A **Venn diagram** is a diagram consisting of various overlapping figures contained within a rectangle (called the "universe"). To depict a statement of the form "All A are B" (or, equivalently, "If A, then B"), we draw two circles, one inside the other; the inner circle represents A, and the outer circle represents B. This relationship is shown in Figure 1.1.

Venn diagrams depicting "No A are B" and "Some A are B" are shown in Figures 1.2 and 1.3, respectively.

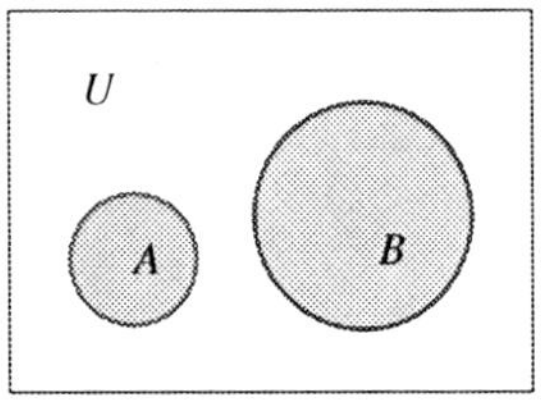

FIGURE 1.2 No A are B.

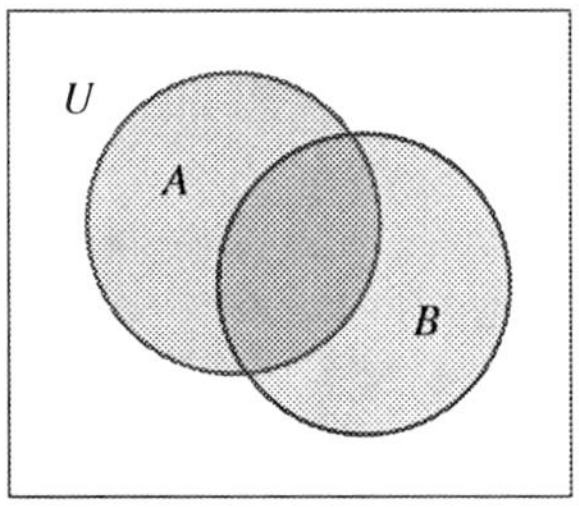

FIGURE 1.3 Some A are B. (At least one A is B.)

EXAMPLE 2

ANALYZING A DEDUCTIVE ARGUMENT Construct a Venn diagram to verify the validity of the following argument:

1. All men are mortal.
2. Socrates is a man.

Therefore, Socrates is mortal.

SOLUTION

Premise 1 is of the form "All A are B" and can be represented by a diagram like that shown in Figure 1.4.

Premise 2 refers to a specific man, namely, Socrates. If we let x = Socrates, the statement "Socrates is a man" can then be represented by placing x within the circle labeled "men," as shown in Figure 1.5. Because we placed x within the "men" circle, and all of the "men" circle is inside the "mortal" circle, the conclusion "Socrates is mortal" is inescapable; the argument is valid.

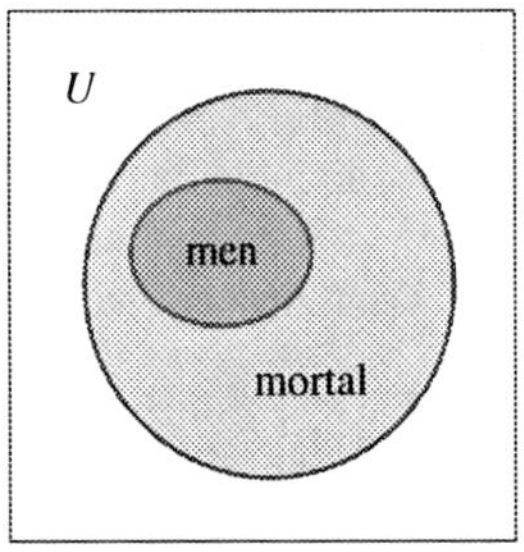

FIGURE 1.4 All men are mortal.

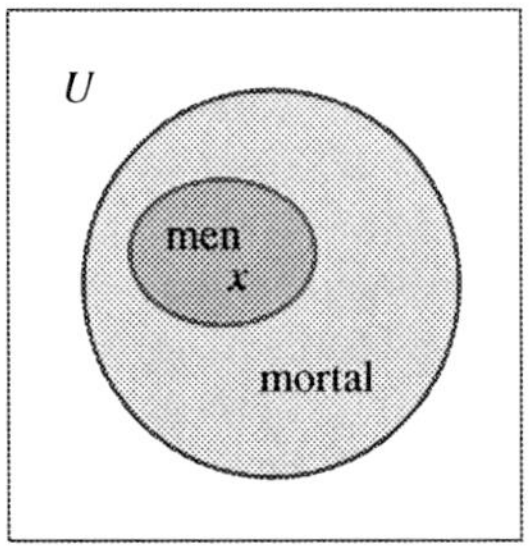

x = Socrates

FIGURE 1.5 Socrates is mortal.

HISTORICAL NOTE

ARISTOTLE 384–322 B.C.

Aristotle was born in 384 B.C. in the small Macedonian town of Stagira, 200 miles north of Athens, on the shore of the Aegean Sea. Aristotle's father was the personal physician of King Amyntas II, ruler of Macedonia. When he was seventeen, Aristotle enrolled at the Academy in Athens and became a student of the famed Plato.

Aristotle was one of Plato's brightest students; he frequently questioned Plato's teachings and openly disagreed with him. Whereas Plato emphasized the study of abstract ideas and mathematical truth, Aristotle was more interested in observing the "real world" around him. Plato often referred to Aristotle as "the brain" or "the mind of the school." Plato commented, "Where others need the spur, Aristotle needs the rein."

Aristotle stayed at the Academy for twenty years, until the death of Plato. Then the king of Macedonia invited Aristotle to supervise the education of his son Alexander, the future Alexander the Great. Aristotle accepted the invitation and taught Alexander until he succeeded his father as ruler. At that time, Aristotle founded a school known as the Lyceum, or Peripatetic School. The school had a large library with many maps, as well as botanical gardens containing an extensive collection of plants and animals. Aristotle and his students would walk about the grounds of the Lyceum while discussing various subjects (*peripatetic* is from the Greek word meaning "to walk").

Many consider Aristotle to be a founding father of the study of biology and of science in general; he observed and classified the behavior and anatomy of hundreds of living creatures. Alexander the Great, during his many military campaigns, had his troops gather specimens from distant places for Aristotle to study.

Aristotle was a prolific writer; some historians credit him with the writing of over 1,000 books. Most of his works have been lost or destroyed, but scholars have recreated some of his more influential works, including *Organon*.

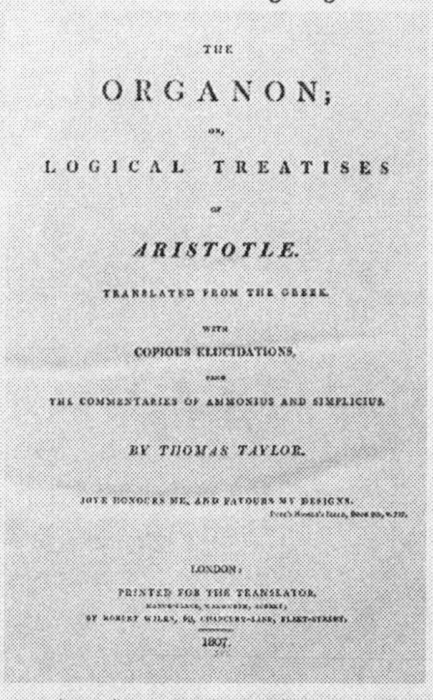

THE
ORGANON;
OR,
LOGICAL TREATISES
OF
ARISTOTLE.
TRANSLATED FROM THE GREEK.
WITH
COPIOUS ELUCIDATIONS,
FROM
THE COMMENTARIES OF AMMONIUS AND SIMPLICIUS.
BY THOMAS TAYLOR.
LONDON:
PRINTED FOR THE TRANSLATOR.
1807.

Aristotle's collective works on syllogisms and deductive logic are known as Organon, meaning "instrument," for logic is the instrument used in the acquisition of knowledge.

EXAMPLE 3

ANALYZING A DEDUCTIVE ARGUMENT Construct a Venn diagram to determine the validity of the following argument:

1. All doctors are men.
2. My mother is a doctor.

Therefore, my mother is a man.

SOLUTION

Premise 1 is of the form "All A are B"; the argument is depicted in Figure 1.6.

No matter where x is placed within the "doctors" circle, the conclusion "My mother is a man" is inescapable; the argument is valid.

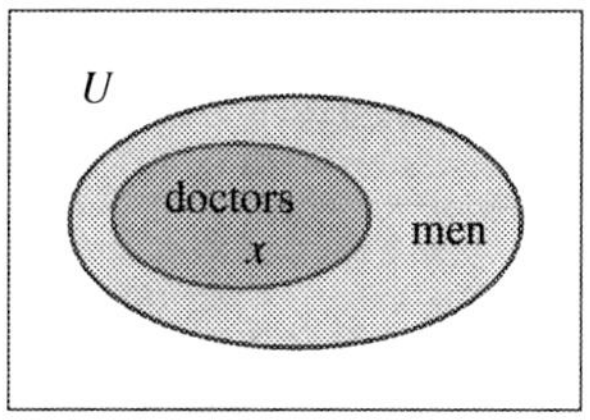

x = My mother

FIGURE 1.6

My mother is a man.

Saying that an argument is valid does not mean that the conclusion is true. The argument given in Example 3 *is* valid, but the conclusion is *false.* One's mother cannot be a man! Validity and truth do not mean the same thing. An argument is valid if the conclusion is inescapable, *given the premises.* Nothing is said about the truth of the premises. Thus, when examining the validity of an argument, we are not determining whether the conclusion is true or false. Saying that an argument is valid merely means that, *given the premises,* the reasoning used to obtain the conclusion is logical. However, if the premises of a valid argument are true, then the conclusion will also be true.

EXAMPLE **4**

ANALYZING A DEDUCTIVE ARGUMENT Construct a Venn diagram to determine the validity of the following argument:

1. All professional wrestlers are actors.
2. The Rock is an actor.

Therefore, The Rock is a professional wrestler.

SOLUTION

Premise 1 is of the form "All A are B"; the "circle of professional wrestlers" is contained within the "circle of actors." If we let x represent The Rock, premise 2 simply requires that we place x somewhere within the actor circle; x could be placed in either of the two locations shown in Figures 1.7 and 1.8.

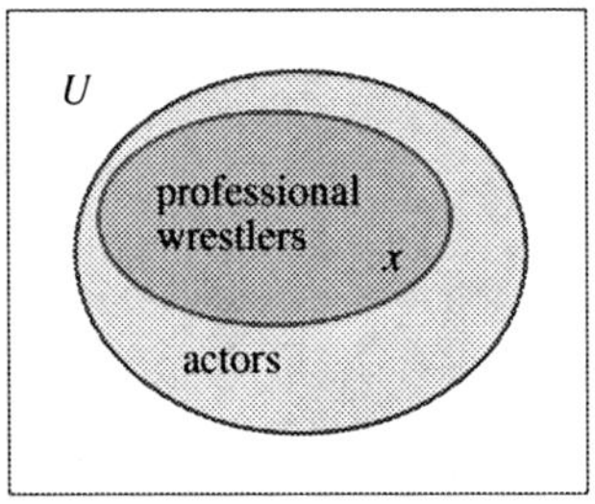

x = The Rock

FIGURE 1.7

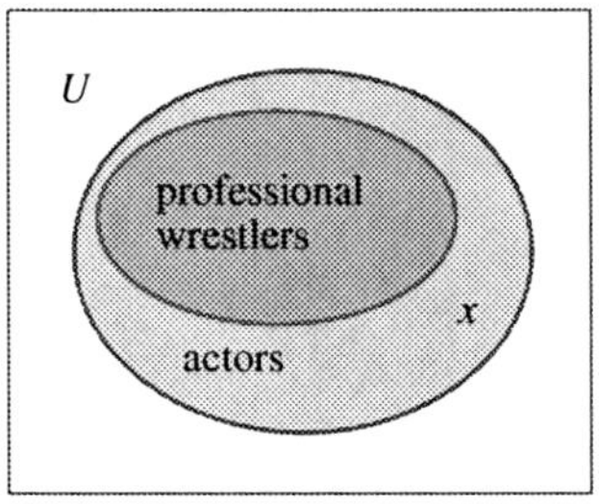

x = The Rock

FIGURE 1.8

If x is placed as in Figure 1.7, the argument would appear to be valid; the figure supports the conclusion "The Rock is a professional wrestler." However, the placement of x in Figure 1.8 does not support the conclusion; given the premises, we cannot *logically* deduce that "The Rock is a professional wrestler." Since the conclusion is *not* inescapable, the argument is invalid.

© Duomo/CORBIS

Even though The Rock *is* a professional wrestler, the argument used to obtain the conclusion is invalid.

Saying that an argument is invalid does not mean that the conclusion is false. Example 4 demonstrates that an invalid argument can have a true conclusion; even though The Rock is a professional wrestler, the argument used to obtain the conclusion is invalid. In logic, validity and truth do not have the same meaning. *Validity* refers to the process of reasoning used to obtain a conclusion; *truth* refers to conformity with fact or experience.

VENN DIAGRAMS AND INVALID ARGUMENTS

To show that an argument is invalid, you must construct a Venn diagram in which the premises are met yet the conclusion does not necessarily follow.

EXAMPLE **5**

ANALYZING A DEDUCTIVE ARGUMENT Construct a Venn diagram to determine the validity of the following argument:

1. Some plants are poisonous.
2. Broccoli is a plant.

Therefore, broccoli is poisonous.

SOLUTION

Premise 1 is of the form "Some *A* are *B*"; it can be represented by two overlapping circles (as in Figure 1.3). If we let *x* represent broccoli, premise 2 requires that we place *x* somewhere within the plant circle. If *x* is placed as in Figure 1.9, the argument would appear to be valid. However, if *x* is placed as in Figure 1.10, the conclusion does not follow. Because we can construct a Venn diagram in which the premises are met yet the conclusion does not follow (Figure 1.10), the argument is invalid.

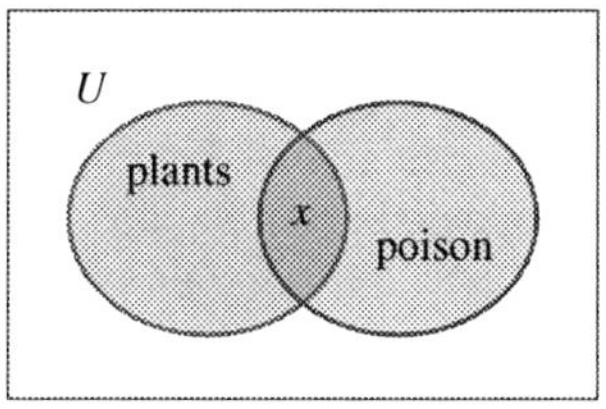

x = broccoli

FIGURE 1.9

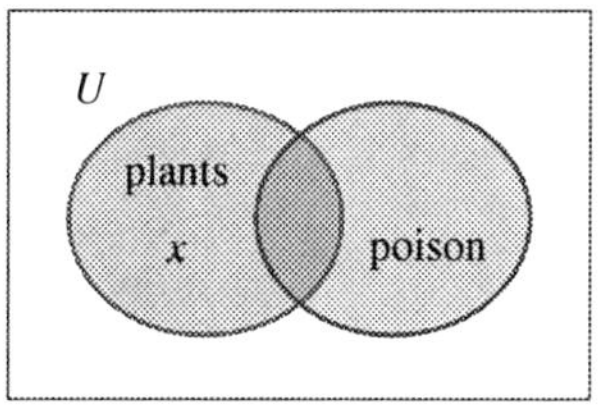

x = broccoli

FIGURE 1.10

When analyzing an argument via a Venn diagram, you might have to draw three or more circles, as in the next example.

EXAMPLE 6

ANALYZING A DEDUCTIVE ARGUMENT Construct a Venn diagram to determine the validity of the following argument:

1. No snake is warm-blooded.
2. All mammals are warm-blooded.

Therefore, snakes are not mammals.

SOLUTION

Premise 1 is of the form "No *A* are *B*"; it is depicted in Figure 1.11. Premise 2 is of the form "All *A* are *B*"; the "mammal circle" must be drawn within the "warm-blooded circle." Both premises are depicted in Figure 1.12.

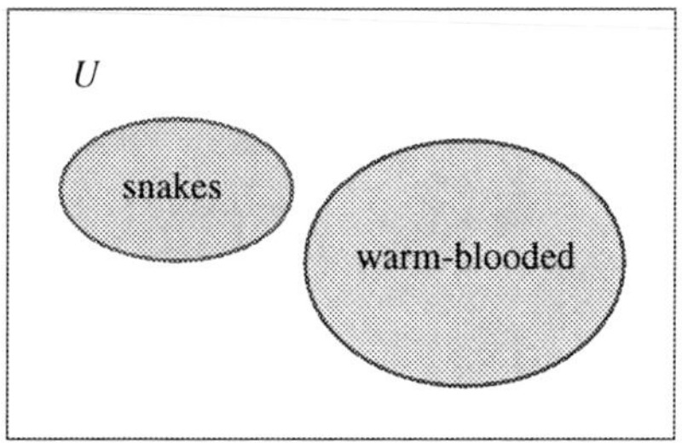

FIGURE 1.11 No snake is warm-blooded.

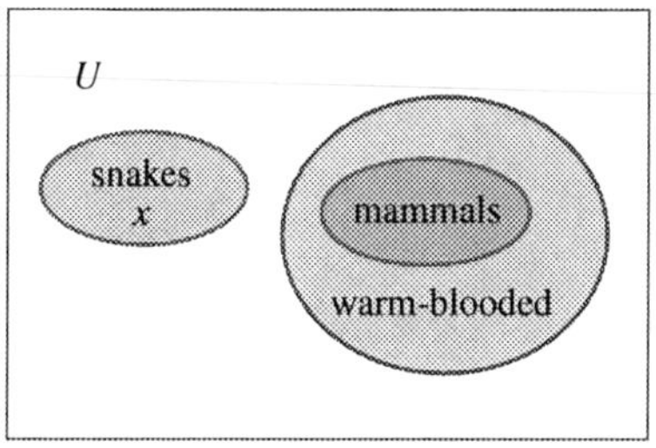

x = snake

FIGURE 1.12 All mammals are warm-blooded.

Because we placed *x* (= snake) within the "snake" circle, and the "snake" circle is outside the "warm-blooded" circle, *x* cannot be within the "mammal" circle (which is inside the "warm-blooded" circle). Given the premises, the conclusion "Snakes are not mammals" is inescapable; the argument is valid.

You might have encountered Venn diagrams when you studied sets in your algebra class. The academic fields of set theory and logic are historically intertwined; set theory was developed in the late nineteenth century as an aid in the study of logical arguments. Today, set theory and Venn diagrams are applied to areas other than the study of logical arguments; we will utilize Venn diagrams in our general study of set theory in Chapter 2.

Inductive Reasoning

The conclusion of a valid deductive argument (one that goes from general to specific) is guaranteed: Given true premises, a true conclusion must follow. However, there are arguments in which the conclusion is not guaranteed even though the premises are true. Consider the following:

1. Joe sneezed after petting Frako's cat.
2. Joe sneezed after petting Paulette's cat.

Therefore, Joe is allergic to cats.

Is the conclusion guaranteed? If the premises are true, they certainly *support* the conclusion, but we cannot say with 100% certainty that Joe is allergic to cats. The conclusion is *not* guaranteed. Maybe Joe is allergic to the flea powder that the cat owners used; maybe he is allergic to the dust that is trapped in the cats' fur; or maybe he has a cold!

Reasoning of this type is called inductive reasoning. **Inductive reasoning** involves going from a series of specific cases to a general statement (see Figure 1.13). Although it may seem to follow and may in fact be true, *the conclusion in an inductive argument is never guaranteed.*

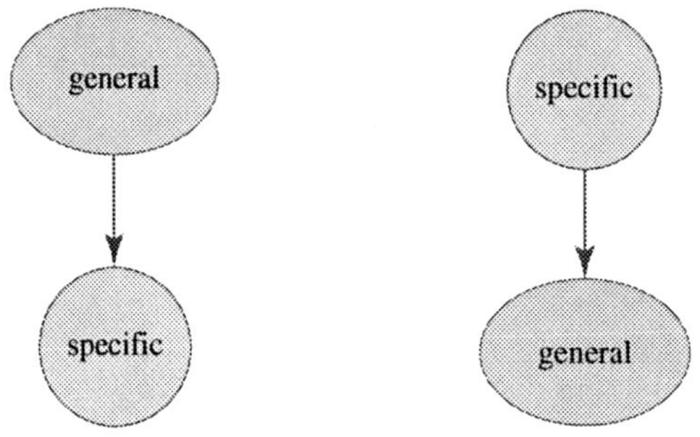

Deductive Reasoning (Conclusion is guaranteed.)

Inductive Reasoning (Conclusion may be probable but is not guaranteed.)

FIGURE 1.13

EXAMPLE 7

INDUCTIVE REASONING AND PATTERN RECOGNITION What is the next number in the sequence 1, 8, 15, 22, 29, . . . ?

SOLUTION

Noticing that the difference between consecutive numbers in the sequence is 7, we may be tempted to say that the next term is $29 + 7 = 36$. Is this conclusion guaranteed? No! Another sequence in which numbers differ by 7 are dates of a given day of the week. For instance, the dates of the Saturdays in the year 2011 are (January) 1, 8, 15, 22, 29, (February) 5, 12, 19, 26, ... Therefore, the next number in the sequence 1, 8, 15, 22, 29, . . . might be 5. Without further information, we cannot determine the next number in the given sequence. We can only use inductive reasoning and give one or more *possible* answers.

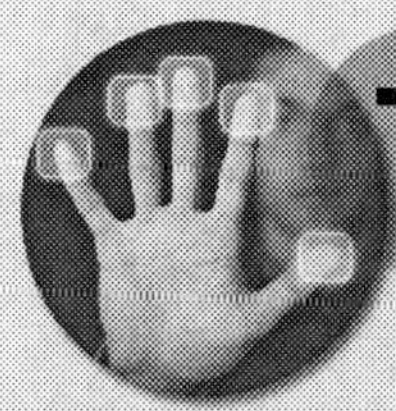

TOPIC X SUDOKU: *LOGIC IN THE REAL WORLD*

Throughout history, people have always been attracted to puzzles, mazes, and brainteasers. Who can deny the inherent satisfaction of solving a seemingly unsolvable or perplexing riddle? A popular new addition to the world of puzzle solving is ***sudoku,*** a numbers puzzle. Loosely translated from Japanese, *sudoku* means "single number"; a sudoku puzzle simply involves placing the digits 1 through 9 in a grid containing 9 rows and 9 columns. In addition, the 9 by 9 grid of squares is subdivided into nine 3 by 3 grids, or "boxes," as shown in Figure 1.14.

The rules of sudoku are quite simple: Each row, each column, and each box must contain the digits 1 through 9; and no row, column, or box can contain 2 squares with the same number. Consequently, sudoku does not require any arithmetic or mathematical skill; sudoku requires logic only. In solving a puzzle, a common thought is "What happens if I put this number here?"

Like crossword puzzles, sudoku puzzles are printed daily in many newspapers across the country and around the world. Web sites containing sudoku puzzles and strategies provide an endless source of new puzzles and help. See Exercise 62 to find links to popular sites.

FIGURE 1.14 A blank sudoku grid.

EXAMPLE **8**

SOLVING A SUDOKU PUZZLE Solve the sudoku puzzle given in Figure 1.15.

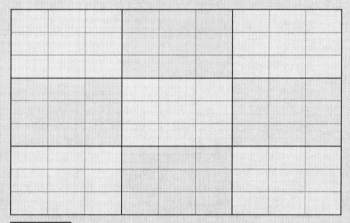

		2		6	4	8	5	
3	8						6	
			8	5			3	
			6		5	9		
6						1		5
7		8			1			
	6			7	9			
	4		1				9	8
	9	3	4	2		7		6

FIGURE 1.15 A sudoku puzzle.

SOLUTION

Recall that each 3 by 3 grid is referred to as a box. For convenience, the boxes are numbered 1 through 9, starting in the upper left-hand corner and moving from left to right, and each square can be assigned coordinates (x, y) based on its row number x and column number y as shown in Figure 1.16.

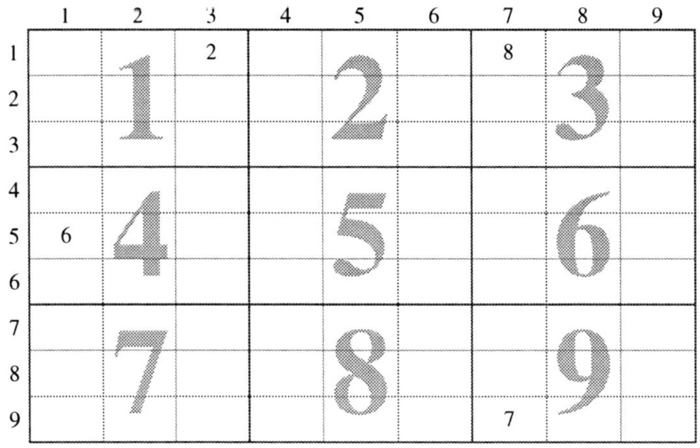

FIGURE 1.16 Box numbers and coordinate system in sudoku.

For example, the digit 2 in Figure 1.16 is in box 1 and has coordinates (1, 3), the digit 8 is in box 3 and has coordinates (1, 7), the digit 6 is in box 4 and has coordinates (5, 1) and the digit 7 is in box 9 and has coordinates (9, 7).

When you are first solving a sudoku puzzle, concentrate on only a few boxes rather than the puzzle as a whole. For instance, looking at boxes 1, 4, and 7, we see that boxes 4 and 7 each contain the digit 6, whereas box 1 does not. Consequently, the 6 in box 1 must be placed in column 3 because (shaded) columns 1 and 2 already have a 6. However, (shaded) row 2 already has a 6, so we can deduce that 6 must be placed in row 3, column 3, that is, in square (3, 3) as shown in Figure 1.17.

		2		6	4	8	5	
3	8						6	
		6	8	5			3	
			6		5	9		
6						1		5
7		8			1			
	6			7	9			
	4		1				9	8
	9	3	4	2		7		6

FIGURE 1.17 The 6 in box 1 must be placed in square (3, 3).

Examining boxes 1, 2, and 3, we see that boxes 2 and 3 each contain the digit 5, whereas box 1 does not. We deduce that 5 must be placed in square (2, 3) because rows 1 and 3 already have a 5. In a similar fashion, square (1, 4) must contain 3. See Figure 1.18.

		2	3	6	4	8	5	
3	8	5					6	
		6	8	5			3	
			6		5	9		
6						1		5
7		8			1			
	6			7	9			
	4		1				9	8
	9	3	4	2		7		6

FIGURE 1.18 Analyzing boxes 1, 2, and 3; placing the digits 3 and 5.

Because we have placed two new digits in box 1, we might wish to focus on the remainder of (shaded) box 1. Notice that the digit 4 can be placed only in square (3, 1), as row 1 and column 2 already have a 4 in each of them; likewise, the digit 9 can be placed only in square (1, 1) because column 2 already has a 9. Finally, either of the digits 1 or 7 can be placed in square (1, 2) or (3, 2) as shown in Figure 1.19. At some point later in the solution, we will be able to determine the exact values of squares (1, 2) and (3, 2), that is, which square receives a 1 and which receives a 7.

9	1,7	2	3	6	4	8	5	
3	8	5					6	
4	1,7	6	8	5			3	
			6		5	9		
6						1		5
7		8			1			
	6			7	9			
	4		1				9	8
	9	3	4	2		7		6

FIGURE 1.19 Focusing on box 1.

Using this strategy of analyzing the contents of three consecutive boxes, we deduce the following placement of digits: 1 must go in (2, 5), 6 must go in (6, 7), 6 must go in (8, 6), 7 must go in (8, 3), 3 must go in (8, 5), 8 must go in (9, 6), and 5 must go in (7, 4). At this point, box 8 is complete as shown in Figure 1.20. (Remember, each box must contain each of the digits 1 through 9.)

Once again, we use the three consecutive box strategy and deduce the following placement of digits: 5 must go in (8, 7), 5 must go in (9, 1), 8 must go in (7, 1), 2 must go in (8, 1), and 1 must go in (7, 3). At this point, box 7 is complete as shown in Figure 1.21.

9	1,7	2	3	6	4	8	5	
3	8	5		1			6	
4	1,7	6	8	5			3	
			6		5	9		
6						1		5
7		8			1	6		
	6		5	7	9			
	4	7	1	3	6		9	8
	9	3	4	2	8	7		6

FIGURE 1.20 Box 8 is complete.

9	1,7	2	3	6	4	8	5	
3	8	5		1			6	
4	1,7	6	8	5			3	
			6		5	9		
6						1		5
7		8			1	6		
8	6	1	5	7	9			
2	4	7	1	3	6	5	9	8
5	9	3	4	2	8	7		6

FIGURE 1.21 Box 7 is complete.

We now focus on box 4 and deduce the following placement of digits: 1 must go in (4, 1), 9 must go in (5, 3), 4 must go in (4, 3), 5 must go in (6, 2), 3 must go in (4, 2), and 2 must go in (5, 2). At this point, box 4 is complete. In addition, we deduce that 1 must go in (9, 8), and 3 must go in (5, 6) as shown in Figure 1.22.

9	1,7	2	3	6	4	8	5	
3	8	5		1			6	
4	1,7	6	8	5			3	
1	3	4	6		5	9		
6	2	9			3	1		5
7	5	8			1	6		
8	6	1	5	7	9			
2	4	7	1	3	6	5	9	8
5	9	3	4	2	8	7	1	6

FIGURE 1.22 Box 4 is complete.

Once again, we use the three consecutive box strategy and deduce the following placement of digits: 3 must go in (6, 9), 3 must go in (7, 7), 9 must go in (2, 4), and 9 must go in (6, 5). Now, to finish row 6, we place 4 in (6, 8) and 2 in (6, 4) as shown in Figure 1.23. (Remember, each row must contain each of the digits 1 through 9.)

9	1,7	2	3	6	4	8	5	
3	8	5	9	1			6	
4	1,7	6	8	5			3	
1	3	4	6		5	9		
6	2	9			3	1		5
7	5	8	2	9	1	6	4	3
8	6	1	5	7	9	3		
2	4	7	1	3	6	5	9	8
5	9	3	4	2	8	7	1	6

FIGURE 1.23 Row 6 is complete.

After we place 7 in (5, 4), column 4 is complete. (Remember, each column must contain each of the digits 1 through 9.) This leads to placing 4 in (5, 5) and 8 in (4, 5), thus completing box 5; row 5 is finalized by placing 8 in (5, 8) as shown in Figure 1.24.

9	1,7	2	3	6	4	8	5	
3	8	5	9	1			6	
4	1,7	6	8	5			3	
1	3	4	6	8	5	9		
6	2	9	7	4	3	1	8	5
7	5	8	2	9	1	6	4	3
8	6	1	5	7	9	3		
2	4	7	1	3	6	5	9	8
5	9	3	4	2	8	7	1	6

FIGURE 1.24 Column 4, box 5, and row 5 are complete.

Now column 7 is completed by placing 4 in (2, 7) and 2 in (3, 7); placing 2 in (2, 6) and 7 in (3, 6) completes column 6 as shown in Figure 1.25.

At this point, we deduce that the digit in (3, 2) must be 1 because row 3 cannot have two 7's. This in turn reveals that 7 must go in (1, 2), and box 1 is now complete. To complete row 7, we place 4 in (7, 9) and 2 in (7, 8); row 4 is finished with 2 in (4, 9) and 7 in (4, 8). See Figure 1.26.

9	1,7	2	3	6	4	8	5	
3	8	5	9	1	2	4	6	
4	1,7	6	8	5	7	2	3	
1	3	4	6	8	5	9		
6	2	9	7	4	3	1	8	5
7	5	8	2	9	1	6	4	3
8	6	1	5	7	9	3		
2	4	7	1	3	6	5	9	8
5	9	3	4	2	8	7	1	6

FIGURE 1.25 Columns 7 and 6 are complete.

9	7	2	3	6	4	8	5	
3	8	5	9	1	2	4	6	
4	1	6	8	5	7	2	3	
1	3	4	6	8	5	9	7	2
6	2	9	7	4	3	1	8	5
7	5	8	2	9	1	6	4	3
8	6	1	5	7	9	3	2	4
2	4	7	1	3	6	5	9	8
5	9	3	4	2	8	7	1	6

FIGURE 1.26 Box 1, row 7, and row 4 are complete.

To finish rows 1, 2, and 3, 1 must go in (1, 9), 7 must go in (2, 9), and 9 must go in (3, 9). The puzzle is now complete as shown in Figure 1.27.

9	7	2	3	6	4	8	5	1
3	8	5	9	1	2	4	6	7
4	1	6	8	5	7	2	3	9
1	3	4	6	8	5	9	7	2
6	2	9	7	4	3	1	8	5
7	5	8	2	9	1	6	4	3
8	6	1	5	7	9	3	2	4
2	4	7	1	3	6	5	9	8
5	9	3	4	2	8	7	1	6

FIGURE 1.27 A completed sudoku puzzle.

As a final check, we scrutinize each box, row, and column to verify that no box, row, or column contains the same digit twice. Congratulations, the puzzle has been solved!

1.1 EXERCISES

In Exercises 1–20, construct a Venn diagram to determine the validity of the given argument.

1. **a.** 1. All master photographers are artists.
2. Ansel Adams is a master photographer.

Therefore, Ansel Adams is an artist.

b. 1. All master photographers are artists.
2. Ansel Adams is an artist.

Therefore, Ansel Adams is a master photographer.

2. **a.** 1. All Olympic gold medal winners are role models.
2. Michael Phelps is an Olympic gold medal winner.

Therefore, Michael Phelps is a role model.

b. 1. All Olympic gold medal winners are role models.
2. Michael Phelps is a role model.

Therefore, Michael Phelps is an Olympic gold medal winner.

3. **a.** 1. All homeless people are unemployed.
2. Bill Gates is not a homeless person.

Therefore, Bill Gates is not unemployed.

b. 1. All homeless people are unemployed.
2. Bill Gates is not unemployed.

Therefore, Bill Gates is not a homeless person.

4. **a.** 1. All professional wrestlers are actors.
2. Ralph Nader is not an actor.

Therefore, Ralph Nader is not a professional wrestler.

b. 1. All professional wrestlers are actors.
2. Ralph Nader is not a professional wrestler.

Therefore, Ralph Nader is not an actor.

5. 1. All pesticides are harmful to the environment.
2. No fertilizer is a pesticide.

Therefore, no fertilizer is harmful to the environment.

6. 1. No one who can afford health insurance is unemployed.
2. All politicians can afford health insurance.

Therefore, no politician is unemployed.

7. 1. No vegetarian owns a gun.
2. All policemen own guns.

Therefore, no policeman is a vegetarian.

8. 1. No professor is a millionaire.
2. No millionaire is illiterate.

Therefore, no professor is illiterate.

9. 1. All poets are loners.
2. All loners are taxi drivers.

Therefore, all poets are taxi drivers.

10. 1. All forest rangers are environmentalists.
2. All forest rangers are storytellers.

Therefore, all environmentalists are storytellers.

11. 1. Real men don't eat quiche.
2. Clint Eastwood is a real man.

Therefore, Clint Eastwood doesn't eat quiche.

12. 1. Real men don't eat quiche.
2. Oscar Meyer eats quiche.

Therefore, Oscar Meyer isn't a real man.

13. 1. All roads lead to Rome.
2. Route 66 is a road.

Therefore, Route 66 leads to Rome.

14. 1. All smiling cats talk.
2. The Cheshire Cat smiles.

Therefore, the Cheshire Cat talks.

15. 1. Some animals are dangerous.
2. A tiger is an animal.

Therefore, a tiger is dangerous.

16. 1. Some professors wear glasses.
2. Mr. Einstein wears glasses.

Therefore, Mr. Einstein is a professor.

17. 1. Some women are police officers.
2. Some police officers ride motorcycles.

Therefore, some women ride motorcycles.

18. 1. All poets are eloquent.
2. Some poets are wine connoisseurs.

Therefore, some wine connoisseurs are eloquent.

19. 1. All squares are rectangles.
2. Some quadrilaterals are squares.

Therefore, some quadrilaterals are rectangles.

20. 1. All squares are rectangles.
2. Some quadrilaterals are rectangles.

Therefore, some quadrilaterals are squares.

21. Classify each argument as deductive or inductive.

a. 1. My television set did not work two nights ago.
2. My television set did not work last night.

Therefore, my television set is broken.

b. 1. All electronic devices give their owners grief.
2. My television set is an electronic device.

Therefore, my television set gives me grief.

22. Classify each argument as deductive or inductive.

a. 1. I ate a chili dog at Joe's and got indigestion.
2. I ate a chili dog at Ruby's and got indigestion.
Therefore, chili dogs give me indigestion.

b. 1. All spicy foods give me indigestion.
2. Chili dogs are spicy food.
Therefore, chili dogs give me indigestion.

In Exercises 23–32, fill in the blank with what is most likely to be the next number. Explain (using complete sentences) the pattern generated by your answer.

23. 3, 8, 13, 18, _____

24. 10, 11, 13, 16, _____

25. 0, 2, 6, 12, _____

26. 1, 2, 5, 10, _____

27. 1, 4, 9, 16, _____

28. 1, 8, 27, 64, _____

29. 2, 3, 5, 7, 11, _____

30. 1, 1, 2, 3, 5, _____

31. 5, 8, 11, 2, _____

32. 12, 5, 10, 3, _____

In Exercises 33–36, fill in the blanks with what are most likely to be the next letters. Explain (using complete sentences) the pattern generated by your answers.

33. O, T, T, F, _____, _____

34. T, F, S, E, _____, _____

35. F, S, S, M, _____, _____

36. J, F, M, A, _____, _____

In Exercises 37–42, explain the general rule or pattern used to assign the given letter to the given word. Fill in the blank with the letter that fits the pattern.

37.

circle	square	trapezoid	octagon	rectangle
c	s	t	o	_____

38.

circle	square	trapezoid	octagon	rectangle
i	u	a	o	_____

39.

circle	square	trapezoid	octagon	rectangle
j	v	b	p	_____

40.

circle	square	trapezoid	octagon	rectangle
c	r	p	g	_____

41.

banana	strawberry	asparagus	eggplant	orange
b	z	t	u	_____

42.

banana	strawberry	asparagus	eggplant	orange
y	r	g	p	_____

43. Find two different numbers that could be used to fill in the blank.

1, 4, 7, 10, _____

Explain the pattern generated by each of your answers.

44. Find five different numbers that could be used to fill in the blank.

7, 14, 21, 28, _____

Explain the pattern generated by each of your answers.

45. Example 1 utilized the Quadratic Formula. Verify that

$$x = \frac{-b + \sqrt{b^2 - 4ac}}{2a}$$

is a solution of the equation $ax^2 + bx + c = 0$.
HINT: Substitute the fraction for x in $ax^2 + bx + c$ and simplify.

46. Example 1 utilized the Quadratic Formula. Verify that

$$x = \frac{-b - \sqrt{b^2 - 4ac}}{2a}$$

is a solution of the equation $ax^2 + bx + c = 0$.
HINT: Substitute the fraction for x in $ax^2 + bx + c$ and simplify.

47. As a review of algebra, use the Quadratic Formula to solve

$$x^2 - 6x + 7 = 0$$

48. As a review of algebra, use the Quadratic Formula to solve

$$x^2 - 2x - 4 = 0$$

Solve the sudoku puzzles in Exercises 49–54.

49.

	5							9
		7				3	4	
					2	1	5	6
6		1	8	9				
	3						2	
				3		6		5
5	9	2	1		4			
	6	3				8		
4							1	

50.

4		7					3	
	2		5		1		7	
					4	5		
	7		9					2
2		9		4		6		3
3					6		9	
		2	4					
	8		6		5		4	
	6					2		9

51.

	5		3			2	8	
		3			7			9
	7		9		2			
	8	7						
5	9			6			1	3
						4	6	
			8		9		2	
2			5			6		
	3	9			6		7	

52.

		2				1		
6	8			2			5	4
		3	8		6	2		
7			6		4			5
2			3		9			7
		6	7		5	4		
3	5			1			8	6
		7				5		

53.

7	6		5					
		2	7					8
		4	6				5	
	7	1			3		2	
5								9
	9		8			7	6	
	8				1	3		
3					6	1		
					7		9	6

54.

7	1	2					8	6
				7				
		4					1	
4		3	5	8				
1			2		4			8
				9	1	2		5
	8					4		
				3				
6	3					8	2	9

Answer the following questions using complete sentences and your own words.

CONCEPT QUESTIONS

55. Explain the difference between deductive reasoning and inductive reasoning.

56. Explain the difference between truth and validity.

57. What is a syllogism? Give an example of a syllogism that relates to your life.

HISTORY QUESTIONS

58. From the days of the ancient Greeks, the study of logic has been mandatory in what two professions? Why?

59. Who developed a formal system of deductive logic based on arguments?

60. What was the name of the school Aristotle founded? What does it mean?

61. How did Aristotle's school of thought differ from Plato's?

WEB PROJECT

62. Obtain a sudoku puzzle and its solution from a popular web site. Some useful links for this web project are listed on the text web site:

www.cengage.com/math/johnson

1.2 Symbolic Logic

OBJECTIVES

- Identify simple statements
- Express a compound statement in symbolic form
- Create the negation of a statement
- Express a conditional statement in terms of necessary and sufficient conditions

The syllogism ruled the study of logic for nearly 2,000 years and was not supplanted until the development of symbolic logic in the late seventeenth century. As its name implies, symbolic logic involves the use of symbols and algebraic manipulations in logic.

Statements

All logical reasoning is based on statements. A **statement** is a sentence that is either true or false.

EXAMPLE 1

IDENTIFYING STATEMENTS Which of the following are statements? Why or why not?

a. Apple manufactures computers.
b. Apple manufactures the world's best computers.
c. Did you buy a Dell?
d. A $2,000 computer that is discounted 25% will cost $1,000.
e. I am telling a lie.

SOLUTION

a. The sentence "Apple manufactures computers" is true; therefore, it is a statement.
b. The sentence "Apple manufactures the world's best computers" is an opinion, and as such, it is neither true nor false. It is true for some people and false for others. Therefore, it is not a statement.
c. The sentence "Did you buy a Dell?" is a question. As such, it is neither true nor false; it is not a statement.
d. The sentence "A $2,000 computer that is discounted 25% will cost $1,000" is false; therefore, it is a statement. (A $2,000 computer that is discounted 25% would cost $1,500.)
e. The sentence "I am telling a lie" is a self-contradiction, or paradox. If it were true, the speaker would be telling a lie, but in telling the truth, the speaker would be contradicting the statement that he or she was lying; if it were false, the speaker would not be telling a lie, but in not telling a lie, the speaker would be contradicting the statement that he or she was lying. The sentence is not a statement.

By tradition, symbolic logic uses lowercase letters as labels for statements. The most frequently used letters are p, q, r, s, and t. We can label the statement "It is snowing" as statement p in the following manner:

p: It is snowing.

If it *is* snowing, p is labeled true, whereas if it is *not* snowing, p is labeled false.

Compound Statements and Logical Connectives

It is easy to determine whether a statement such as "Charles donated blood" is true or false; either he did or he didn't. However, not all statements are so simple; some are more involved. For example, the truth of "Charles donated blood and did not wash his car, or he went to the library," depends on the truth of the individual pieces that make up the larger, compound statement. A **compound statement** is a statement that contains one or more simpler statements. A compound statement can be formed by inserting the word *not* into a simpler statement or by joining two or more statements with connective words such as *and, or, if . . . then . . . , only if,* and *if and only if.* The compound statement "Charles did *not* wash his car" is formed from the simpler statement "Charles did wash his car." The compound statement "Charles donated blood *and* did *not* wash his car, *or* he went to the library" consists of three statements, each of which may be true or false.

Figure 1.28 diagrams two equivalent compound statements.

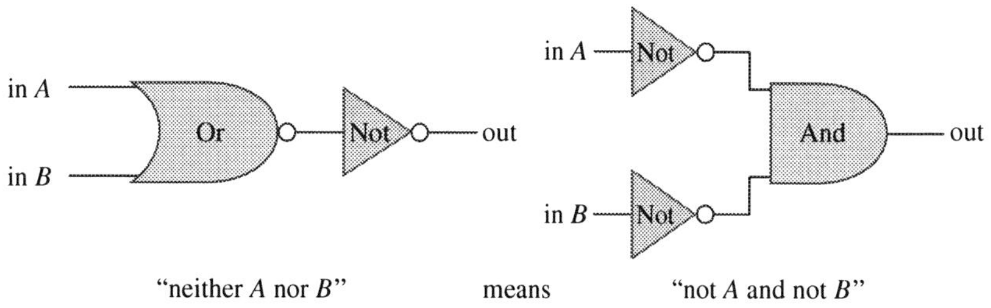

FIGURE 1.28 Technicians and engineers use compound statements and logical connectives to study the flow of electricity through switching circuits.

When is a compound statement true? Before we can answer this question, we must first examine the various ways in which statements can be connected. Depending on how the statements are connected, the resulting compound statement can be a *negation,* a *conjunction,* a *disjunction,* a *conditional,* or any combination thereof.

The Negation $\sim p$

The **negation** of a statement is the denial of the statement and is represented by the symbol $\sim$. The negation is frequently formed by inserting the word *not.* For example, given the statement "*p:* It is snowing," the negation would be "$\sim p$: It is not snowing." If it *is* snowing, p is true and $\sim p$ is false. Similarly, if it is *not* snowing, p is false and $\sim p$ is true. A statement and its negation always have opposite truth values; when one is true, the other is false. Because the truth of the negation depends on the truth of the original statement, a negation is classified as a compound statement.

EXAMPLE 2

WRITING A NEGATION Write a sentence that represents the negation of each statement:

a. The senator is a Democrat.
b. The senator is not a Democrat.
c. Some senators are Republicans.
d. All senators are Republicans.
e. No senator is a Republican.

SOLUTION

a. The negation of "The senator is a Democrat" is "The senator is not a Democrat."
b. The negation of "The senator is not a Democrat" is "The senator is a Democrat."
c. A common error would be to say that the negation of "Some senators are Republicans" is "Some senators are not Republicans." However, "Some senators are Republicans" is not denied by "Some senators are not Republicans." The statement "Some senators *are* Republicans" implies that at least one senator is a Republican. The negation of this statement is "It is not the case that at least one senator is a Republican," or (more commonly phrased) the negation is "No senator is a Republican."
d. The negation of "All senators are Republicans" is "It is not the case that all senators are Republicans," or "There exists a senator who is not a Republican," or (more commonly phrased) "Some senators are not Republicans."
e. The negation of "No senator is a Republican" is "It is not the case that no senator is a Republican" or, in other words, "There exists at least one senator who *is* a Republican." If "some" is interpreted as meaning "at least one," the negation can be expressed as "Some senators are Republicans."

The words *some, all,* and *no* (or *none*) are referred to as **quantifiers.** Parts (c) through (e) of Example 2 contain quantifiers. The linked pairs of quantified statements shown in Figure 1.29 are negations of each other.

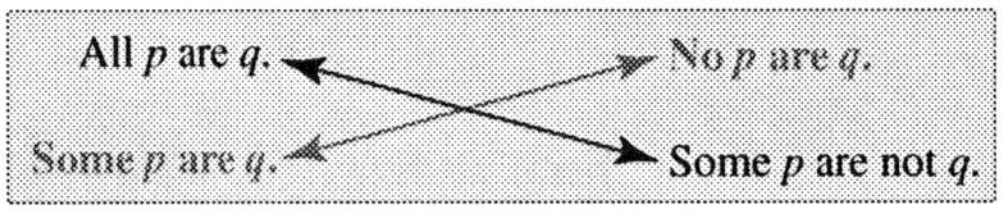

FIGURE 1.29 Negations of statements containing quantifiers.

The Conjunction $p \wedge q$

Consider the statement "Norma Rae is a union member and she is a Democrat." This is a compound statement, because it consists of two statements—"Norma Rae is a union member" and "she (Norma Rae) is a Democrat"—and the connective word *and.* Such a compound statement is referred to as a conjunction. A **conjunction** consists of two or more statements connected by the word *and.* We use the symbol $\wedge$ to represent the word *and;* thus, the conjunction "$p \wedge q$" represents the compound statement "*p* and *q*."

EXAMPLE 3

TRANSLATING WORDS INTO SYMBOLS Using the symbolic representations

p: Norma Rae is a union member.

q: Norma Rae is a Democrat.

express the following compound statements in symbolic form:

a. Norma Rae is a union member and she is a Democrat.
b. Norma Rae is a union member and she is not a Democrat.

SOLUTION

a. The compound statement "Norma Rae is a union member and she is a Democrat" can be represented as $p \wedge q$.
b. The compound statement "Norma Rae is a union member and she is not a Democrat" can be represented as $p \wedge \sim q$.

The Disjunction $p \vee q$

When statements are connected by the word *or,* a **disjunction** is formed. We use the symbol $\vee$ to represent the word *or.* Thus, the disjunction "$p \vee q$" represents the compound statement "p or q." We can interpret the word *or* in two ways. Consider the statements

p: Kaitlin is a registered Republican.

q: Paki is a registered Republican.

The statement "Kaitlin is a registered Republican or Paki is a registered Republican" can be symbolized as $p \vee q$. Notice that it is possible that *both* Kaitlin and Paki are registered Republicans. In this example, *or* includes the possibility that both things may happen. In this case, we are working with the **inclusive *or.***

Now consider the statements

p: Kaitlin is a registered Republican.

q: Kaitlin is a registered Democrat.

The statement "Kaitlin is a registered Republican or Kaitlin is a registered Democrat" does *not* include the possibility that both may happen; one statement *excludes* the other. When this happens, we are working with the **exclusive *or.*** In our study of symbolic logic (as in most mathematics), we will always use the *inclusive or.* Therefore, "p or q" means "p or q or both."

EXAMPLE 4

TRANSLATING SYMBOLS INTO WORDS Using the symbolic representations

p: Juanita is a college graduate.

q: Juanita is employed.

express the following compound statements in words:

a. $p \vee q$ **b.** $p \wedge q$

c. $p \vee \sim q$ **d.** $\sim p \wedge q$

SOLUTION

a. $p \vee q$ represents the statement "Juanita is a college graduate or Juanita is employed (or both)."

b. $p \wedge q$ represents the statement "Juanita is a college graduate and Juanita is employed."

c. $p \vee \sim q$ represents the statement "Juanita is a college graduate or Juanita is not employed."

d. $\sim p \wedge q$ represents the statement "Juanita is not a college graduate and Juanita is employed."

The Conditional $p \rightarrow q$

Consider the statement "If it is raining, then the streets are wet." This is a compound statement because it connects two statements, namely, "it is raining" and "the streets are wet." Notice that the statements are connected with "if . . . then . . ." phrasing. Any statement of the form "if p then q" is called a **conditional** (or an **implication**); p is called the **hypothesis** (or **premise**) of the conditional, and q is called the **conclusion** of the conditional. The conditional "if p then q" is represented by the symbols "$p \rightarrow q$" (p implies q). When people use conditionals in

HISTORICAL NOTE

GOTTFRIED WILHELM LEIBNIZ 1646–1716

In addition to cofounding calculus (see Chapter 13), the German-born Gottfried Wilhelm Leibniz contributed much to the development of symbolic logic. A precocious child, Leibniz was self-taught in many areas. He taught himself Latin at the age of eight and began the study of Greek when he was twelve. In the process, he was exposed to the writings of Aristotle and became intrigued by formalized logic.

At the age of fifteen, Leibniz entered the University of Leipzig to study law. He received his bachelor's degree two years later, earned his master's degree the following year, and then transferred to the University of Nuremberg.

Leibniz received his doctorate in law within a year and was immediately offered a professorship but refused it, saying that he had "other things in mind." Besides law, these "other things" included politics, religion, history, literature, metaphysics, philosophy, logic, and mathematics. Thereafter, Leibniz worked under the sponsorship of the courts of various nobles, serving as lawyer, historian, and librarian to the elite. At one point, Leibniz was offered the position of librarian at the Vatican but declined the offer.

Leibniz's affinity for logic was characterized by his search for a *characteristica universalis,* or "universal character." Leibniz believed that by combining logic and mathematics, a general symbolic language could be created in which all scientific problems could be solved with a minimum of effort. In this universal language, statements and the logical relationships between them would be represented by letters and symbols. In Leibniz's words, "All truths of reason would be reduced to a kind of calculus, and the errors would only be errors of computation." In essence, Leibniz believed that once a problem had been translated into this universal language of symbolic logic, it would be solved automatically by simply applying the mathematical rules that governed the manipulation of the symbols.

Leibniz's work in the field of symbolic logic did not arouse much academic curiosity; many say that it was too far ahead of its time. The study of symbolic logic was not systematically investigated again until the nineteenth century.

In the early 1670s, Leibniz invented one of the world's first mechanical calculating machines. Leibniz's machine could multiply and divide, whereas an earlier machine invented by Blaise Pascal (see Chapter 3) could only add and subtract.

everyday speech, they often omit the word *then,* as in "If it is raining, the streets are wet." Alternatively, the conditional "if p then q" may be phrased as "q if p" ("The streets are wet if it is raining").

EXAMPLE 5 **TRANSLATING WORDS INTO SYMBOLS** Using the symbolic representations

p: I am healthy.
q: I eat junk food.
r: I exercise regularly.

express the following compound statements in symbolic form:

a. I am healthy if I exercise regularly.
b. If I eat junk food and do not exercise, then I am not healthy.

SOLUTION

a. "I am healthy if I exercise regularly" is a conditional (*if . . . then . . .*) and can be rephrased as follows:

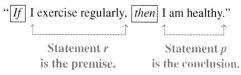

"*If* I exercise regularly, *then* I am healthy."

Statement r is the premise. Statement p is the conclusion.

The given compound statement can be expressed as $r \rightarrow p$.

b. "If I eat junk food and do not exercise, then I am not healthy" is a conditional (*if . . . then . . .*) that contains a conjunction (*and*) and two negations (*not*):

"If I eat junk food *and* do *not* exercise, then I am *not* healthy."

The premise contains a conjunction and a negation. The conclusion contains a negation.

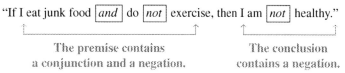

The premise of the conditional can be represented by $q \wedge \sim r$, while the conclusion can be represented by $\sim p$. Thus, the given compound statement has the symbolic form $(q \wedge \sim r) \rightarrow \sim p$.

EXAMPLE 6

TRANSLATING WORDS INTO SYMBOLS Express the following statements in symbolic form:

a. All mammals are warm-blooded.
b. No snake is warm-blooded.

SOLUTION

a. The statement "All mammals are warm-blooded" can be rephrased as "If it is a mammal, then it is warm-blooded." Therefore, we define two simple statements p and q as

p: It is a mammal.

q: It is warm-blooded.

The statement now has the form

"*If* it is a mammal, *then* it is warm-blooded."

Statement p is the premise. Statement q is the conclusion.

and can be expressed as $p \rightarrow q$. In general, any statement of the form "All p are q" can be symbolized as $p \rightarrow q$.

b. The statement "No snake is warm-blooded" can be rephrased as "If it is a snake, then it is not warm-blooded." Therefore, we define two simple statements p and q as

p: It is a snake.

q: It is warm-blooded.

The statement now has the form

"*If* it is a snake, *then* it is *not* warm-blooded."

Statement p is the premise. The negation of statement q is the conclusion.

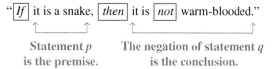

and can be expressed as $p \rightarrow \sim q$. In general, any statement of the form "No p is q" can be symbolized as $p \rightarrow \sim q$.

Necessary and Sufficient Conditions

As Example 6 shows, conditionals are not always expressed in the form "if p then q." In addition to "all p are q," other standard forms of a conditional include statements that contain the word *sufficient* or *necessary*.

Consider the statement "Being a mammal is sufficient for being warm-blooded." One definition of the word *sufficient* is "adequate." Therefore, "being a mammal" is an adequate condition for "being warm-blooded"; hence, "being a mammal" implies "being warm-blooded." Logically, the statement "Being a mammal is sufficient for being warm-blooded" is equivalent to saying "If it is a mammal, then it is warm-blooded." Consequently, the general statement "p is sufficient for q" is an alternative form of the conditional "if p then q" and can be symbolized as $p \to q$.

"Being a mammal" is a *sufficient* (adequate) condition for "being warm-blooded," but is it a *necessary* condition? Of course not: some animals are warm-blooded but are not mammals (chickens, for example). One definition of the word *necessary* is "required." Therefore, "being a mammal" is not required for "being warm-blooded." However, is "being warm-blooded" a necessary (required) condition for "being a mammal"? Of course it is: all mammals *are* warm-blooded (that is, there are no cold-blooded mammals). Logically, the statement "being warm-blooded is necessary for being a mammal" is equivalent to saying "If it is a mammal, then it is warm-blooded." Consequently, the general statement "q is necessary for p" is an alternative form of the conditional "if p then q" and can be symbolized as $p \to q$.

In summary, a sufficient condition is the hypothesis or premise of a conditional statement, whereas a necessary condition is the conclusion of a conditional statement.

Sufficient condition	→	Necessary condition
p: hypothesis	p implies q	q: conclusion

EXAMPLE 7

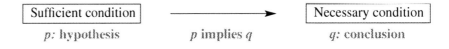

TRANSLATING WORDS INTO SYMBOLS Using the symbolic representations

p: A person obeys the law.

q: A person is arrested.

express the following compound statements in symbolic form:

a. Being arrested is necessary for not obeying the law.
b. Obeying the law is sufficient for not being arrested.

SOLUTION

a. "Being arrested" is a necessary condition; hence, "a person is arrested" is the conclusion of a conditional statement.

A person does not obey the law.	→	A person is arrested.
		A necessary condition.

Therefore, the statement "Being arrested is necessary for not obeying the law" can be rephrased as follows:

"[*If*] a person does [*not*] obey the law, [*then*] the person is arrested."

The negation of statement p is the premise. Statement q is the conclusion.

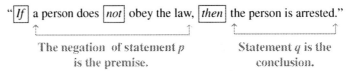

The given compound statement can be expressed as $\sim p \to q$.

b. "Obeying the law" is a sufficient condition; hence, "A person obeys the law" is the premise of a conditional statement.

A person obeys the law. → A person is not arrested.

A sufficient condition.

Therefore, the statement "Obeying the law is sufficient for not being arrested" can be rephrased as follows:

"*If* a person obeys the law, *then* the person is *not* arrested."

Statement p is the premise. — The negation of statement q is the conclusion.

The given compound statement can be expressed as $p \to \sim q$.

We have seen that a statement is a sentence that is either true or false and that connecting two or more statements forms a compound statement. Figure 1.30 summarizes the logical connectives and symbols that were introduced in this section. The various connectives have been defined; we can now proceed in our analysis of the conditions under which a compound statement is true. This analysis is carried out in the next section.

Statement	Symbol	Read as . . .
negation	$\sim p$	not p
conjunction	$p \wedge q$	p and q
disjunction	$p \vee q$	p or q
conditional (implication)	$p \to q$	if p, then q p is sufficient for q q is necessary for p

FIGURE 1.30 Logical connectives.

1.2 Exercises

1. Which of the following are statements? Why or why not?
   a. George Washington was the first president of the United States.
   b. Abraham Lincoln was the second president of the United States.
   c. Who was the first vice president of the United States?
   d. Abraham Lincoln was the best president.
2. Which of the following are statements? Why or why not?
   a. $3 + 5 = 6$
   b. Solve the equation $2x + 5 = 3$.
   c. $x^2 + 1 = 0$ has no solution.
   d. $x^2 - 1 = (x + 1)(x - 1)$
   e. Is $\sqrt{2}$ a rational number?
3. Determine which pairs of statements are negations of each other.
   a. All of the fruits are red.
   b. None of the fruits is red.
   c. Some of the fruits are red.
   d. Some of the fruits are not red.
4. Determine which pairs of statements are negations of each other.
   a. Some of the beverages contain caffeine.
   b. Some of the beverages do not contain caffeine.
   c. None of the beverages contain caffeine.
   d. All of the beverages contain caffeine.
5. Write a sentence that represents the negation of each statement.
   a. Her dress is not red.
   b. Some computers are priced under \$100.

c. All dogs are four-legged animals.
d. No sleeping bag is waterproof.

6. Write a sentence that represents the negation of each statement.
a. She is not a vegetarian.
b. Some elephants are pink.
c. All candy promotes tooth decay.
d. No lunch is free.

7. Using the symbolic representations
p: The lyrics are controversial.
q: The performance is banned.
express the following compound statements in symbolic form.
a. The lyrics are controversial, and the performance is banned.
b. If the lyrics are not controversial, the performance is not banned.
c. It is not the case that the lyrics are controversial or the performance is banned.
d. The lyrics are controversial, and the performance is not banned.
e. Having controversial lyrics is sufficient for banning a performance.
f. Noncontroversial lyrics are necessary for not banning a performance.

8. Using the symbolic representations
p: The food is spicy.
q: The food is aromatic.
express the following compound statements in symbolic form.
a. The food is aromatic and spicy.
b. If the food isn't spicy, it isn't aromatic.
c. The food is spicy, and it isn't aromatic.
d. The food isn't spicy or aromatic.
e. Being nonaromatic is sufficient for food to be nonspicy.
f. Being spicy is necessary for food to be aromatic.

9. Using the symbolic representations
p: A person plays the guitar.
q: A person rides a motorcycle.
r: A person wears a leather jacket.
express the following compound statements in symbolic form.
a. If a person plays the guitar or rides a motorcycle, then the person wears a leather jacket.
b. A person plays the guitar, rides a motorcycle, and wears a leather jacket.
c. A person wears a leather jacket and doesn't play the guitar or ride a motorcycle.
d. All motorcycle riders wear leather jackets.
e. Not wearing a leather jacket is sufficient for not playing the guitar or riding a motorcycle.
f. Riding a motorcycle or playing the guitar is necessary for wearing a leather jacket.

10. Using the symbolic representations
p: The car costs $70,000.
q: The car goes 140 mph.
r: The car is red.
express the following compound statements in symbolic form.
a. All red cars go 140 mph.
b. The car is red, goes 140 mph, and does not cost $70,000.
c. If the car does not cost $70,000, it does not go 140 mph.
d. The car is red and it does not go 140 mph or cost $70,000.
e. Being able to go 140 mph is sufficient for a car to cost $70,000 or be red.
f. Not being red is necessary for a car to cost $70,000 and not go 140 mph.

In Exercises 11–34, translate the sentence into symbolic form. Be sure to define each letter you use. (More than one answer is possible.)

11. All squares are rectangles.
12. All people born in the United States are American citizens.
13. No square is a triangle.
14. No convicted felon is eligible to vote.
15. All whole numbers are even or odd.
16. All muscle cars from the Sixties are polluters.
17. No whole number is greater than 3 and less than 4.
18. No electric-powered car is a polluter.
19. Being an orthodontist is sufficient for being a dentist.
20. Being an author is sufficient for being literate.
21. Knowing Morse code is necessary for operating a telegraph.
22. Knowing CPR is necessary for being a paramedic.
23. Being a monkey is sufficient for not being an ape.
24. Being a chimpanzee is sufficient for not being a monkey.
25. Not being a monkey is necessary for being an ape.
26. Not being a chimpanzee is necessary for being a monkey.
27. I do not sleep soundly if I drink coffee or eat chocolate.
28. I sleep soundly if I do not drink coffee or eat chocolate.
29. Your check is not accepted if you do not have a driver's license or a credit card.
30. Your check is accepted if you have a driver's license or a credit card.
31. If you drink and drive, you are fined or you go to jail.

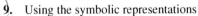

32. If you are rich and famous, you have many friends and enemies.

33. You get a refund or a store credit if the product is defective.

34. The streets are slippery if it is raining or snowing.

35. Using the symbolic representations
p: I am an environmentalist.
q: I recycle my aluminum cans.
express the following in words.
a. $p \wedge q$ **b.** $p \rightarrow q$
c. $\sim q \rightarrow \sim p$ **d.** $q \vee \sim p$

36. Using the symbolic representations
p: I am innocent.
q: I have an alibi.
express the following in words.
a. $p \wedge q$ **b.** $p \rightarrow q$
c. $\sim q \rightarrow \sim p$ **d.** $q \vee \sim p$

37. Using the symbolic representations
p: I am an environmentalist.
q: I recycle my aluminum cans.
r: I recycle my newspapers.
express the following in words.
a. $(q \vee r) \rightarrow p$ **b.** $\sim p \rightarrow \sim(q \vee r)$
c. $(q \wedge r) \vee \sim p$ **d.** $(r \wedge \sim q) \rightarrow \sim p$

38. Using the symbolic representations
p: I am innocent.
q: I have an alibi.
r: I go to jail.
express the following in words.
a. $(p \vee q) \rightarrow \sim r$ **b.** $(p \wedge \sim q) \rightarrow r$
c. $(\sim p \wedge q) \vee r$ **d.** $(p \wedge r) \rightarrow \sim q$

39. Which statement, #1 or #2, is more appropriate? Explain why.
Statement #1: "Cold weather is necessary for it to snow."
Statement #2: "Cold weather is sufficient for it to snow."

40. Which statement, #1 or #2, is more appropriate? Explain why.
Statement #1: "Being cloudy is necessary for it to rain."
Statement #2: "Being cloudy is sufficient for it to rain."

41. Which statement, #1 or #2, is more appropriate? Explain why.
Statement #1: "Having 31 days in a month is necessary for it not to be February."
Statement #2: "Having 31 days in a month is sufficient for it not to be February."

42. Which statement, #1 or #2, is more appropriate? Explain why.
Statement #1: "Being the Fourth of July is necessary for the U.S. Post Office to be closed."
Statement #2: "Being the Fourth of July is sufficient for the U.S. Post Office to be closed."

Answer the following questions using complete sentences and your own words.

• CONCEPT QUESTIONS

43. What is a negation?

44. What is a conjunction?

45. What is a disjunction?

46. What is a conditional?

47. What is a sufficient condition?

48. What is a necessary condition?

49. What is the difference between the inclusive *or* and the exclusive *or?*

50. Create a sentence that is a self-contradiction, or paradox, as in part (e) of Example 1.

• HISTORY QUESTIONS

51. In what academic field did Gottfried Leibniz receive his degrees? Why is the study of logic important in this field?

52. Who developed a formal system of logic based on syllogistic arguments?

53. What is meant by *characteristica universalis?* Who proposed this theory?

THE NEXT LEVEL

If a person wants to pursue an advanced degree (something beyond a bachelor's or four-year degree), chances are the person must take a standardized exam to gain admission to a graduate school or to be admitted into a specific program. These exams are intended to measure verbal, quantitative, and analytical skills that have developed throughout a person's life. Many classes and study guides are available to help people prepare for the exams. The following questions are typical of those found in the study guides.

Exercises 54–58 refer to the following: A culinary institute has a small restaurant in which the students prepare various dishes. The menu changes daily, and during a specific week, the following dishes are to be prepared: moussaka, pilaf, quiche, ratatouille, stroganoff, and teriyaki. During the week, the restaurant does not prepare any other kind of dish. The selection of dishes the restaurant offers is consistent with the following conditions:

- *If the restaurant offers pilaf, then it does not offer ratatouille.*
- *If the restaurant does not offer stroganoff, then it offers pilaf.*
- *If the restaurant offers quiche, then it offers both ratatouille and teriyaki.*
- *If the restaurant offers teriyaki, then it offers moussaka or stroganoff or both.*

54. Which one of the following could be a complete and accurate list of the dishes the restaurant offers on a specific day?
a. pilaf, quiche, ratatouille, teriyaki
b. quiche, stroganoff, teriyaki

c. quiche, ratatouille, teriyaki
d. ratatouille, stroganoff
e. quiche, ratatouille

55. Which one of the following cannot be a complete and accurate list of the dishes the restaurant offers on a specific day?
a. moussaka, pilaf, quiche, ratatouille, teriyaki
b. quiche, ratatouille, stroganoff, teriyaki
c. moussaka, pilaf, teriyaki
d. stroganoff, teriyaki
e. pilaf, stroganoff

56. Which one of the following could be the only kind of dish the restaurant offers on a specific day?
a. teriyaki
b. stroganoff
c. ratatouille
d. quiche
e. moussaka

57. If the restaurant does not offer teriyaki, then which one of the following must be true?
a. The restaurant offers pilaf.
b. The restaurant offers at most three different dishes.
c. The restaurant offers at least two different dishes.
d. The restaurant offers neither quiche nor ratatouille.
e. The restaurant offers neither quiche nor pilaf.

58. If the restaurant offers teriyaki, then which one of the following must be false?
a. The restaurant does not offer moussaka.
b. The restaurant does not offer ratatouille.
c. The restaurant does not offer stroganoff.
d. The restaurant offers ratatouille but not quiche.
e. The restaurant offers ratatouille but not stroganoff.

1.3 Truth Tables

Objectives

- Construct a truth table for a compound statement
- Determine whether two statements are equivalent
- Apply De Morgan's Laws

Suppose your friend Maria is a doctor, and you know that she is a Democrat. If someone told you, "Maria is a doctor and a Republican," you would say that the statement was false. On the other hand, if you were told, "Maria is a doctor or a Republican," you would say that the statement was true. Each of these statements is a compound statement—the result of joining individual statements with connective words. When is a compound statement true, and when is it false? To answer these questions, we must examine whether the individual statements are true or false and the manner in which the statements are connected.

The **truth value** of a statement is the classification of the statement as true or false and is denoted by T or F. For example, the truth value of the statement "Santa Fe is the capital of New Mexico" is T. (The statement is true.) In contrast, the truth value of "Memphis is the capital of Tennessee" is F. (The statement is false.)

A convenient way of determining whether a compound statement is true or false is to construct a truth table. A **truth table** is a listing of all possible combinations of the individual statements as true or false, along with the resulting truth value of the compound statement. As we will see, truth tables also allow us to distinguish valid arguments from invalid arguments.

	p
1.	T
2.	F

FIGURE 1.31
Truth values for a statement p.

	p	$\sim p$
1.	T	F
2.	F	T

FIGURE 1.32
Truth table for a negation $\sim p$.

	p	q
1.	T	T
2.	T	F
3.	F	T
4.	F	F

FIGURE 1.33
Truth values for two statements.

	p	q	$p \wedge q$
1.	T	T	T
2.	T	F	F
3.	F	T	F
4.	F	F	F

FIGURE 1.34
Truth table for a conjunction $p \wedge q$.

	p	q	$p \vee q$
1.	T	T	T
2.	T	F	T
3.	F	T	T
4.	F	F	F

FIGURE 1.35
Truth table for a disjunction $p \vee q$.

The Negation $\sim p$

The **negation** of a statement is the denial, or opposite, of the statement. (As was stated in the previous section, because the truth value of the negation depends on the truth value of the original statement, a negation can be classified as a compound statement.) To construct the truth table for the negation of a statement, we must first examine the original statement. A statement p may be true or false, as shown in Figure 1.31. If the statement p is true, the negation $\sim p$ is false; if p is false, $\sim p$ is true. The truth table for the compound statement $\sim p$ is given in Figure 1.32. Row 1 of the table is read "$\sim p$ is false when p is true." Row 2 is read "$\sim p$ is true when p is false."

The Conjunction $p \wedge q$

A **conjunction** is the joining of two statements with the word *and.* The compound statement "Maria is a doctor and a Republican" is a conjunction with the following symbolic representation:

p: Maria is a doctor.
q: Maria is a Republican.
$p \wedge q$: Maria is a doctor and a Republican.

The truth value of a compound statement depends on the truth values of the individual statements that make it up. How many rows will the truth table for the conjunction $p \wedge q$ contain? Because p has two possible truth values (T or F) and q has two possible truth values (T or F), we need four $(2 \cdot 2)$ rows in order to list all possible combinations of Ts and Fs, as shown in Figure 1.33.

For the conjunction $p \wedge q$ to be true, the components p and q must *both* be true; the conjunction is false otherwise. The completed truth table for the conjunction $p \wedge q$ is given in Figure 1.34. The symbols p and q can be replaced by any statements. The table gives the truth value of the statement "p and q," dependent upon the truth values of the individual statements "p" and "q." For instance, row 3 is read "The conjunction $p \wedge q$ is false when p is false and q is true." The other rows are read in a similar manner.

The Disjunction $p \vee q$

A **disjunction** is the joining of two statements with the word *or.* The compound statement "Maria is a doctor or a Republican" is a disjunction (the *inclusive or*) with the following symbolic representation:

p: Maria is a doctor.
q: Maria is a Republican.
$p \vee q$: Maria is a doctor or a Republican.

Even though your friend Maria the doctor is not a Republican, the disjunction "Maria is a doctor or a Republican" is true. For a disjunction to be true, *at least one* of the components must be true. A disjunction is false only when *both* components are false. The truth table for the disjunction $p \vee q$ is given in Figure 1.35.

EXAMPLE 1

CONSTRUCTING A TRUTH TABLE Under what specific conditions is the following compound statement true? "I have a high school diploma, or I have a full-time job and no high school diploma."

SOLUTION

First, we translate the statement into symbolic form, and then we construct the truth table for the symbolic expression. Define p and q as

p: I have a high school diploma.

q: I have a full-time job.

The given statement has the symbolic representation $p \vee (q \wedge \sim p)$.

Because there are two letters, we need $2 \cdot 2 = 4$ rows. We need to insert a column for each connective in the symbolic expression $p \vee (q \wedge \sim p)$. As in algebra, we start inside any grouping symbols and work our way out. Therefore, we need a column for $\sim p$, a column for $q \wedge \sim p$, and a column for the entire expression $p \vee (q \wedge \sim p)$, as shown in Figure 1.36.

	p	q	$\sim p$	$q \wedge \sim p$	$p \vee (q \wedge \sim p)$
1.	T	T			
2.	T	F			
3.	F	T			
4.	F	F			

FIGURE 1.36 Required columns in the truth table.

In the $\sim p$ column, fill in truth values that are opposite those for p. Next, the conjunction $q \wedge \sim p$ is true only when both components are true; enter a T in row 3 and Fs elsewhere. Finally, the disjunction $p \vee (q \wedge \sim p)$ is false only when both components p and $(q \wedge \sim p)$ are false; enter an F in row 4 and Ts elsewhere. The completed truth table is shown in Figure 1.37.

	p	q	$\sim p$	$q \wedge \sim p$	$p \vee (q \wedge \sim p)$
1.	T	T	F	F	T
2.	T	F	F	F	T
3.	F	T	T	T	T
4.	F	F	T	F	F

FIGURE 1.37 Truth table for $p \vee (q \wedge \sim p)$.

As is indicated in the truth table, the symbolic expression $p \vee (q \wedge \sim p)$ is true under all conditions except one: row 4; the expression is false when both p and q are false. Therefore, the statement "I have a high school diploma, or I have a full-time job and no high school diploma" is true in every case except when the speaker has no high school diploma and no full-time job.

If the symbolic representation of a compound statement consists of two different letters, its truth table will have $2 \cdot 2 = 4$ rows. How many rows are required if a compound statement consists of three letters—say, p, q, and r? Because each statement has two possible truth values (T and F), the truth table must contain $2 \cdot 2 \cdot 2 = 8$ rows. In general, each time a new statement is added, the number of rows doubles.

NUMBER OF ROWS

If a compound statement consists of n individual statements, each represented by a different letter, the number of rows required in its truth table is 2^n.

EXAMPLE 2

CONSTRUCTING A TRUTH TABLE Under what specific conditions is the following compound statement true? "I own a handgun, and it is not the case that I am a criminal or police officer."

SOLUTION

First, we translate the statement into symbolic form, and then we construct the truth table for the symbolic expression. Define the three simple statements as follows:

p: I own a handgun.

q: I am a criminal.

r: I am a police officer.

The given statement has the symbolic representation $p \wedge \sim(q \vee r)$. Since there are three letters, we need $2^3 = 8$ rows. We start with three columns, one for each letter. To account for all possible combinations of p, q, and r as true or false, proceed as follows:

1. Fill the first half (four rows) of column 1 with Ts and the rest with Fs, as shown in Figure 1.38(a).
2. In the next column, split each half into halves, the first half receiving Ts and the second Fs. In other words, alternate two Ts and two Fs in column 2, as shown in Figure 1.38(b).
3. Again, split each half into halves; the first half receives Ts, and the second half receives Fs. Because we are dealing with the third (last) column, the Ts and Fs will alternate, as shown in Figure 1.38(c).

	p	q	r
1.	T		
2.	T		
3.	T		
4.	T		
5.	F		
6.	F		
7.	F		
8.	F		

(a)

	p	q	r
1.	T	T	
2.	T	T	
3.	T	F	
4.	T	F	
5.	F	T	
6.	F	T	
7.	F	F	
8.	F	F	

(b)

	p	q	r
1.	T	T	T
2.	T	T	F
3.	T	F	T
4.	T	F	F
5.	F	T	T
6.	F	T	F
7.	F	F	T
8.	F	F	F

(c)

FIGURE 1.38 Truth values for three statements.

(This process of filling the first half of the first column with Ts and the second half with Fs and then splitting each half into halves with blocks of Ts and Fs applies to all truth tables.)

We need to insert a column for each connective in the symbolic expression $p \wedge \sim(q \vee r)$, as shown in Figure 1.39.

	p	q	r	$q \vee r$	$\sim(q \vee r)$	$p \wedge \sim(q \vee r)$
1.	T	T	T			
2.	T	T	F			
3.	T	F	T			
4.	T	F	F			
5.	F	T	T			
6.	F	T	F			
7.	F	F	T			
8.	F	F	F			

FIGURE 1.39 Required columns in the truth table.

Now fill in the appropriate symbol in the column under $q \vee r$. Enter F if *both* q and r are false; enter T otherwise (that is, if at least one is true). In the $\sim(q \vee r)$ column, fill in truth values that are opposite those for $q \vee r$, as in Figure 1.40.

	p	q	r	$q \vee r$	$\sim(q \vee r)$	$p \wedge \sim(q \vee r)$
1.	T	T	T	T	F	
2.	T	T	F	T	F	
3.	T	F	T	T	F	
4.	T	F	F	F	T	
5.	F	T	T	T	F	
6.	F	T	F	T	F	
7.	F	F	T	T	F	
8.	F	F	F	F	T	

FIGURE 1.40 Truth values of the expressions $q \vee r$ and $\sim(q \vee r)$.

The conjunction $p \wedge \sim(q \vee r)$ is true only when *both* p and $\sim(q \vee r)$ are true; enter a T in row 4 and Fs elsewhere. The truth table is shown in Figure 1.41.

	p	q	r	$q \vee r$	$\sim(q \vee r)$	$p \wedge \sim(q \vee r)$
1.	T	T	T	T	F	F
2.	T	T	F	T	F	F
3.	T	F	T	T	F	F
4.	T	F	F	F	T	T
5.	F	T	T	T	F	F
6.	F	T	F	T	F	F
7.	F	F	T	T	F	F
8.	F	F	F	F	T	F

FIGURE 1.41 Truth table for $p \wedge \sim(q \vee r)$.

As indicated in the truth table, the expression $p \wedge \sim(q \vee r)$ is true only when p is true and both q and r are false. Therefore, the statement "I own a handgun, and it is not the case that I am a criminal or police officer" is true only when the speaker owns a handgun, is not a criminal, and is not a police officer—in other words, the speaker is a law-abiding citizen who owns a handgun.

The Conditional $p \rightarrow q$

A **conditional** is a compound statement of the form "If p, then q" and is symbolized $p \rightarrow q$. Under what circumstances is a conditional true, and when is it false? Consider the following (compound) statement: "If you give me $50, then I will give you a ticket to the ballet." This statement is a conditional and has the following representation:

	p	q	$p \rightarrow q$
1.	T	T	T
2.	T	F	F
3.	F	T	?
4.	F	F	?

FIGURE 1.42
What if p is false?

p: You give me $50.

q: I give you a ticket to the ballet.

$p \rightarrow q$: If you give me $50, then I will give you a ticket to the ballet.

The conditional can be viewed as a promise: *If* you give me $50, *then* I will give you a ticket to the ballet. Suppose you give me $50; that is, suppose p is true. I have two options: Either I give you a ticket to the ballet (q is true), or I do not (q is false). If I do give you the ticket, the conditional $p \rightarrow q$ is true (I have kept my promise); if I do not give you the ticket, the conditional $p \rightarrow q$ is false (I have not kept my promise). These situations are shown in rows 1 and 2 of the truth table in Figure 1.42. Rows 3 and 4 require further analysis.

Suppose you do not give me $50; that is, suppose p is false. Whether or not I give you a ticket, you cannot say that I broke my promise; that is, you cannot say that the conditional $p \rightarrow q$ is false. Consequently, since a statement is either true or false, the conditional is labeled true (by default). In other words, when the premise p of a conditional is false, it does not matter whether the conclusion q is true or false. In both cases, the conditional $p \rightarrow q$ is automatically labeled true, because it is not false.

	p	q	$p \rightarrow q$
1.	T	T	T
2.	T	F	F
3.	F	T	T
4.	F	F	T

FIGURE 1.43
Truth table for $p \rightarrow q$.

The completed truth table for a conditional is given in Figure 1.43. Notice that the only circumstance under which a conditional is false is when the premise p is true and the conclusion q is false, as shown in row 2.

EXAMPLE 3

CONSTRUCTING A TRUTH TABLE Under what conditions is the symbolic expression $q \rightarrow \sim p$ true?

SOLUTION

Our truth table has $2^2 = 4$ rows and contains a column for p, q, $\sim p$, and $q \rightarrow \sim p$, as shown in Figure 1.44.

	p	q	$\sim p$	$q \rightarrow \sim p$
1.	T	T		
2.	T	F		
3.	F	T		
4.	F	F		

FIGURE 1.44 Required columns in the truth table.

	p	q	$\sim p$	$q \rightarrow \sim p$
1.	T	T	F	F
2.	T	F	F	T
3.	F	T	T	T
4.	F	F	T	T

FIGURE 1.45 Truth table for $q \rightarrow \sim p$.

In the $\sim p$ column, fill in truth values that are opposite those for p. Now, a conditional is false only when its premise (in this case, q) is true and its conclusion (in this case, $\sim p$) is false. Therefore, $q \rightarrow \sim p$ is false only in row 1; the conditional $q \rightarrow \sim p$ is true under all conditions except the condition that both p and q are true. The completed truth table is shown in Figure 1.45.

EXAMPLE 4

CONSTRUCTING A TRUTH TABLE Construct a truth table for the following compound statement: "I walk up the stairs if I want to exercise or if the elevator isn't working."

SOLUTION

Rewriting the statement so the word *if* is first, we have "If I want to exercise or (if) the elevator isn't working, then I walk up the stairs."

Now we must translate the statement into symbols and construct a truth table. Define the following:

p: I want to exercise.

q: The elevator is working.

r: I walk up the stairs.

The statement now has the symbolic representation $(p \vee \sim q) \rightarrow r$. Because we have three letters, our table must have $2^3 = 8$ rows. Inserting a column for each letter and a column for each connective, we have the initial setup shown in Figure 1.46.

	p	q	r	$\sim q$	$p \vee \sim q$	$(p \vee \sim q) \rightarrow r$
1.	T	T	T			
2.	T	T	F			
3.	T	F	T			
4.	T	F	F			
5.	F	T	T			
6.	F	T	F			
7.	F	F	T			
8.	F	F	F			

FIGURE 1.46 Required columns in the truth table.

In the column labeled $\sim q$, enter truth values that are the opposite of those of q. Next, enter the truth values of the disjunction $p \vee \sim q$ in column 5. Recall that a disjunction is false only when both components are false and is true otherwise. Consequently, enter Fs in rows 5 and 6 (since both p and $\sim q$ are false) and Ts in the remaining rows, as shown in Figure 1.47.

The last column involves a conditional; it is false only when its premise is true and its conclusion is false. Therefore, enter Fs in rows 2, 4, and 8 (since $p \vee \sim q$ is true and r is false) and Ts in the remaining rows. The truth table is shown in Figure 1.48.

As Figure 1.48 shows, the statement "I walk up the stairs if I want to exercise or if the elevator isn't working" is true in all situations except those listed in rows 2, 4, and 8. For instance, the statement is false (row 8) when the speaker does not want to exercise, the elevator is not working, and the speaker does not walk up the stairs—in other words, the speaker stays on the ground floor of the building when the elevator is broken.

	p	q	r	$\sim q$	$p \vee \sim q$	$(p \vee \sim q) \to r$
1.	T	T	T	F	T	
2.	T	T	F	F	T	
3.	T	F	T	T	T	
4.	T	F	F	T	T	
5.	F	T	T	F	F	
6.	F	T	F	F	F	
7.	F	F	T	T	T	
8.	F	F	F	T	T	

FIGURE 1.47 Truth values of the expressions.

	p	q	r	$\sim q$	$p \vee \sim q$	$(p \vee \sim q) \to r$
1.	T	T	T	F	T	T
2.	T	T	F	F	T	F
3.	T	F	T	T	T	T
4.	T	F	F	T	T	F
5.	F	T	T	F	F	T
6.	F	T	F	F	F	T
7.	F	F	T	T	T	T
8.	F	F	F	T	T	F

FIGURE 1.48 Truth table for $(p \vee \sim q) \to r$.

Equivalent Expressions

When you purchase a car, the car is either new or used. If a salesperson told you, "It is not the case that the car is not new," what condition would the car be in? This compound statement consists of one individual statement ("*p:* The car is new") and two negations:

"It is not the case that the car is not new."

$\sim$ $\qquad\qquad$ $\sim p$

	p
1.	T
2.	F

FIGURE 1.49
Truth values of p.

Does this mean that the car is new? To answer this question, we will construct a truth table for the symbolic expression $\sim(\sim p)$ and compare its truth values with those of the original p. Because there is only one letter, we need $2^1 = 2$ rows, as shown in Figure 1.49.

We must insert a column for $\sim p$ and a column for $\sim(\sim p)$. Now, $\sim p$ has truth values that are opposite those of p, and $\sim(\sim p)$ has truth values that are opposite those of $\sim p$, as shown in Figure 1.50.

	p	$\sim p$	$\sim(\sim p)$
1.	T	F	T
2.	F	T	F

FIGURE 1.50
Truth table for $\sim(\sim p)$.

Notice that the values in the column labeled $\sim(\sim p)$ are identical to those in the column labeled p. Whenever this happens, the expressions are said to be equivalent and may be used interchangeably. Therefore, the statement "It is not the case that the car is not new" is equivalent in meaning to the statement "The car is new."

Equivalent expressions are symbolic expressions that have identical truth values in each corresponding entry. The expression $p \equiv q$ is read "p is equivalent to q" or "p and q are equivalent." As we can see in Figure 1.50, an expression and its double negation are logically equivalent. This relationship can be expressed as $p \equiv \sim(\sim p)$.

EXAMPLE 5

DETERMINING WHETHER STATEMENTS ARE EQUIVALENT Are the statements "If I am a homeowner, then I pay property taxes" and "I am a homeowner, and I do not pay property taxes" equivalent?

SOLUTION

We begin by defining the statements:

p: I am a homeowner.

q: I pay property taxes.

$p \rightarrow q$: If I am a homeowner, then I pay property taxes.

$p \wedge \sim q$: I am a homeowner, and I do not pay property taxes.

The truth table contains $2^2 = 4$ rows, and the initial setup is shown in Figure 1.51.

Now enter the appropriate truth values under $\sim q$ (the opposite of q). Because the conjunction $p \wedge \sim q$ is true only when both p and $\sim q$ are true, enter a T in row 2 and Fs elsewhere. The conditional $p \rightarrow q$ is false only when p is true and q is false; therefore, enter an F in row 2 and Ts elsewhere. The completed truth table is shown in Figure 1.52.

Because the entries in the columns labeled $p \wedge \sim q$ and $p \rightarrow q$ are not the same, the statements are not equivalent. "If I am a homeowner, then I pay property taxes" is *not* equivalent to "I am a homeowner and I do not pay property taxes."

	p	q	$\sim q$	$p \wedge \sim q$	$p \rightarrow q$
1.	T	T			
2.	T	F			
3.	F	T			
4.	F	F			

FIGURE 1.51 Required columns in the truth table.

	p	q	$\sim q$	$p \wedge \sim q$	$p \rightarrow q$
1.	T	T	F	F	T
2.	T	F	T	T	F
3.	F	T	F	F	T
4.	F	F	T	F	T

FIGURE 1.52 Truth table for $p \rightarrow q$.

Notice that the truth values in the columns under $p \wedge \sim q$ and $p \rightarrow q$ in Figure 1.52 are exact opposites; when one is T, the other is F. Whenever this happens, one statement is the negation of the other. Consequently, $p \wedge \sim q$ is the negation of $p \rightarrow q$ (and vice versa). This can be expressed as $p \wedge \sim q \equiv \sim(p \rightarrow q)$. The negation of a conditional is logically equivalent to the conjunction of the premise and the negation of the conclusion.

Statements that look or sound different may in fact have the same meaning. For example, "It is not the case that the car is not new" really means the same as "The car is new," and "It is not the case that if I am a homeowner, then I pay property taxes"

HISTORICAL NOTE

GEORGE BOOLE, 1815–1864

George Boole is called "the father of symbolic logic." Computer science owes much to this self-educated mathematician. Born the son of a poor shopkeeper in Lincoln, England, Boole had very little formal education, and his prospects for rising above his family's lower-class status were dim. Like Leibniz, he taught himself Latin; at the age of twelve, he translated an ode of Horace into English, winning the attention of the local schoolmasters. (In his day, knowledge of Latin was a prerequisite to scholarly endeavors and to becoming a socially accepted gentleman.) After that, his academic desires were encouraged, and at the age of fifteen, he began his long teaching career. While teaching arithmetic, he studied advanced mathematics and physics.

In 1849, after nineteen years of teaching at elementary schools, Boole received his big break: He was appointed professor of mathematics at Queen's College in the city of Cork, Ireland. At last, he was able to research advanced mathematics, and he became recognized as a first-class mathematician. This was a remarkable feat, considering Boole's lack of formal training and degrees.

Boole's most influential work, *An Investigation of the Laws of Thought, on Which Are Founded the Mathematical Theories of Logic and Probabilities*, was published in 1854. In it, he wrote, "There exist certain general principles founded in the very nature of language and logic that exhibit laws as identical in form as with the laws of the general symbols of algebra." With this insight, Boole had taken a big step into the world of logical reasoning and abstract mathematical analysis.

Perhaps because of his lack of formal training, Boole challenged the status quo, including the Aristotelian assumption that *all* logical arguments could be reduced to syllogistic arguments. In doing so, he employed symbols to represent concepts, as did Leibniz, but he also developed systems of algebraic manipulation to accompany these symbols. Thus, Boole's creation is a marriage of logic and mathematics. However, as is the case with almost all new theories, Boole's symbolic logic was not met with total approbation. In particular, one staunch opponent of his work was Georg Cantor, whose work on the origins of set theory and the magnitude of infinity will be investigated in Chapter 2.

In the many years since Boole's original work was unveiled, various scholars have modified, improved, generalized, and extended its central concepts. Today, Boolean algebras are the essence of computer software and circuit design. After all, a computer merely manipulates predefined symbols and conforms to a set of preassigned algebraic commands.

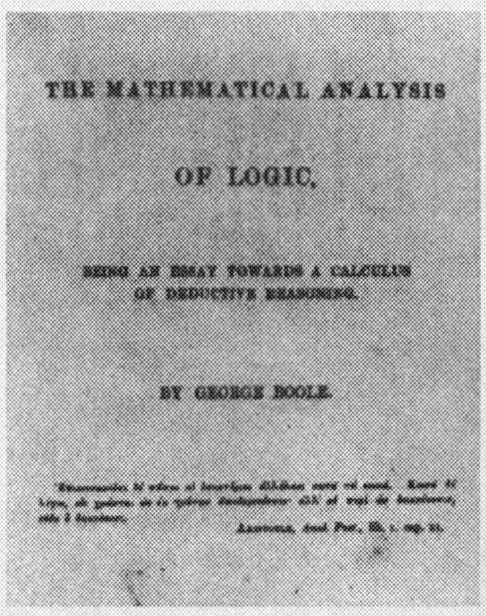

THE MATHEMATICAL ANALYSIS

OF LOGIC,

BEING AN ESSAY TOWARDS A CALCULUS OF DEDUCTIVE REASONING.

BY GEORGE BOOLE.

Through an algebraic manipulation of logical symbols, Boole revolutionized the age-old study of logic. His essay The Mathematical Analysis of Logic *laid the foundation for his later book* An Investigation of the Laws of Thought.

actually means the same as "I am a homeowner, and I do not pay property taxes." When we are working with equivalent statements, we can substitute either statement for the other without changing the truth value.

De Morgan's Laws

Earlier in this section, we saw that the negation of a negation is equivalent to the original statement; that is, $\sim(\sim p) \equiv p$. Another negation "formula" that we discovered was $\sim(p \rightarrow q) \equiv p \wedge \sim q$, that is, the negation of a conditional. Can we find similar "formulas" for the negations of the other basic connectives, namely, the conjunction and the disjunction? The answer is yes, and the results are credited to the English mathematician and logician Augustus De Morgan.

DE MORGAN'S LAWS

The negation of the conjunction $p \wedge q$ is given by $\sim(p \wedge q) \equiv \sim p \vee \sim q$.

"Not p and q" is equivalent to "not p or not q."

The negation of the disjunction $p \vee q$ is given by $\sim(p \vee q) \equiv \sim p \wedge \sim q$.

"Not p or q" is equivalent to "not p and not q."

De Morgan's Laws are easily verified through the use of truth tables and will be addressed in the exercises (see Exercises 55 and 56).

EXAMPLE 6

APPLYING DE MORGAN'S LAWS Using De Morgan's Laws, find the negation of each of the following:

a. It is Friday and I receive a paycheck.
b. You are correct or I am crazy.

SOLUTION

a. The symbolic representation of "It is Friday and I receive a paycheck" is

p: It is Friday.

q: I receive a paycheck.

$p \wedge q$: It is Friday and I receive a paycheck.

Therefore, the negation is $\sim(p \wedge q) \equiv \sim p \vee \sim q$, that is, "It is not Friday or I do not receive a paycheck."

b. The symbolic representation of "You are correct or I am crazy" is

p: You are correct.

q: I am crazy.

$p \vee q$: You are correct or I am crazy.

Therefore, the negation is $\sim(p \vee q) \equiv \sim p \wedge \sim q$, that is, "You are not correct and I am not crazy."

As we have seen, the truth value of a compound statement depends on the truth values of the individual statements that make it up. The truth tables of the basic connectives are summarized in Figure 1.53.

Equivalent statements are statements that have the same meaning. Equivalent statements for the negations of the basic connectives are given in Figure 1.54.

	p	$\sim p$
1.	T	F
2.	F	T

Negation

	p	q	$p \wedge q$
1.	T	T	T
2.	T	F	F
3.	F	T	F
4.	F	F	F

Conjunction

	p	q	$p \vee q$
1.	T	T	T
2.	T	F	T
3.	F	T	T
4.	F	F	F

Disjunction

	p	q	$p \rightarrow q$
1.	T	T	T
2.	T	F	F
3.	F	T	T
4.	F	F	T

Conditional

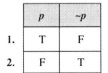

FIGURE 1.53 Truth tables for the basic connectives.

1. $\sim(\sim p) \equiv p$	the negation of a negation
2. $\sim(p \wedge q) \equiv \sim p \vee \sim q$	the negation of a conjunction
3. $\sim(p \vee q) \equiv \sim p \wedge \sim q$	the negation of a disjunction
4. $\sim(p \rightarrow q) \equiv p \wedge \sim q$	the negation of a conditional

FIGURE 1.54 Negations of the basic connectives.

1.3 Exercises

In Exercises 1–20, construct a truth table for the symbolic expressions.

1. $p \vee \sim q$
2. $p \wedge \sim q$
3. $p \vee \sim p$
4. $p \wedge \sim p$
5. $p \rightarrow \sim q$
6. $\sim p \rightarrow q$
7. $\sim q \rightarrow \sim p$
8. $\sim p \rightarrow \sim q$
9. $(p \vee q) \rightarrow \sim p$
10. $(p \wedge q) \rightarrow \sim q$
11. $(p \vee q) \rightarrow (p \wedge q)$
12. $(p \wedge q) \rightarrow (p \vee q)$
13. $p \wedge \sim(q \vee r)$
14. $p \vee \sim(q \vee r)$
15. $p \vee (\sim q \wedge r)$
16. $\sim p \vee \sim(q \wedge r)$
17. $(\sim r \vee p) \rightarrow (q \wedge p)$
18. $(q \wedge p) \rightarrow (\sim r \vee p)$
19. $(p \vee r) \rightarrow (q \wedge \sim r)$
20. $(p \wedge r) \rightarrow (q \vee \sim r)$

In Exercises 21–40, translate the compound statement into symbolic form and then construct the truth table for the expression.

21. If it is raining, then the streets are wet.
22. If the lyrics are not controversial, the performance is not banned.
23. The water supply is rationed if it does not rain.
24. The country is in trouble if he is elected.
25. All squares are rectangles.
26. All muscle cars from the Sixties are polluters.
27. No square is a triangle.
28. No electric-powered car is a polluter.
29. Being a monkey is sufficient for not being an ape.
30. Being a chimpanzee is sufficient for not being a monkey.
31. Not being a monkey is necessary for being an ape.
32. Not being a chimpanzee is necessary for being a monkey.
33. Your check is accepted if you have a driver's license or a credit card.
34. You get a refund or a store credit if the product is defective.
35. If leaded gasoline is used, the catalytic converter is damaged and the air is polluted.
36. If he does not go to jail, he is innocent or has an alibi.
37. I have a college degree and I do not have a job or own a house.
38. I surf the Internet and I make purchases and do not pay sales tax.
39. If Proposition A passes and Proposition B does not, jobs are lost or new taxes are imposed.
40. If Proposition A does not pass and the legislature raises taxes, the quality of education is lowered and unemployment rises.

In Exercises 41–50, construct a truth table to determine whether the statements in each pair are equivalent.

41. The streets are wet or it is not raining.
If it is raining, then the streets are wet.
42. The streets are wet or it is not raining.
If the streets are not wet, then it is not raining.
43. He has a high school diploma or he is unemployed.
If he does not have a high school diploma, then he is unemployed.
44. She is unemployed or she does not have a high school diploma.
If she is employed, then she does not have a high school diploma.
45. If handguns are outlawed, then outlaws have handguns.
If outlaws have handguns, then handguns are outlawed.
46. If interest rates continue to fall, then I can afford to buy a house.
If interest rates do not continue to fall, then I cannot afford to buy a house.
47. If the spotted owl is on the endangered species list, then lumber jobs are lost.

If lumber jobs are not lost, then the spotted owl is not on the endangered species list.

48. If I drink decaffeinated coffee, then I do not stay awake.
If I do stay awake, then I do not drink decaffeinated coffee.

49. The plaintiff is innocent or the insurance company does not settle out of court.
The insurance company settles out of court and the plaintiff is not innocent.

50. The plaintiff is not innocent and the insurance company settles out of court.
It is not the case that the plaintiff is innocent or the insurance company does not settle out of court.

In Exercises 51–54, construct truth tables to determine which pairs of statements are equivalent.

51. **i.** Knowing Morse code is sufficient for operating a telegraph.
ii. Knowing Morse code is necessary for operating a telegraph.
iii. Not knowing Morse code is sufficient for not operating a telegraph.
iv. Not knowing Morse code is necessary for not operating a telegraph.

52. **i.** Knowing CPR is necessary for being a paramedic.
ii. Knowing CPR is sufficient for being a paramedic.
iii. Not knowing CPR is necessary for not being a paramedic.
iv. Not knowing CPR is sufficient for not being a paramedic.

53. **i.** The water being cold is necessary for not going swimming.
ii. The water not being cold is necessary for going swimming.
iii. The water being cold is sufficient for not going swimming.
iv. The water not being cold is sufficient for going swimming.

54. **i.** The sky not being clear is sufficient for it to be raining.
ii. The sky being clear is sufficient for it not to be raining.
iii. The sky not being clear is necessary for it to be raining.
iv. The sky being clear is necessary for it not to be raining.

55. Using truth tables, verify De Morgan's Law

$$\sim(p \wedge q) \equiv \sim p \vee \sim q.$$

56. Using truth tables, verify De Morgan's Law

$$\sim(p \vee q) \equiv \sim p \wedge \sim q.$$

In Exercises 57–68, write the statement in symbolic form, construct the negation of the expression (in simplified symbolic form), and express the negation in words.

57. I have a college degree and I am not employed.
58. It is snowing and classes are canceled.
59. The television set is broken or there is a power outage.
60. The freeway is under construction or I do not ride the bus.
61. If the building contains asbestos, the original contractor is responsible.
62. If the legislation is approved, the public is uninformed.
63. The First Amendment has been violated if the lyrics are censored.
64. Your driver's license is taken away if you do not obey the laws.
65. Rainy weather is sufficient for not washing my car.
66. Drinking caffeinated coffee is sufficient for not sleeping.
67. Not talking is necessary for listening.
68. Not eating dessert is necessary for being on a diet.

Answer the following questions using complete sentences and your own words.

• CONCEPT QUESTIONS

69. **a.** Under what conditions is a disjunction true?
b. Under what conditions is a disjunction false?
70. **a.** Under what conditions is a conjunction true?
b. Under what conditions is a conjunction false?
71. **a.** Under what conditions is a conditional true?
b. Under what conditions is a conditional false?
72. **a.** Under what conditions is a negation true?
b. Under what conditions is a negation false?
73. What are equivalent expressions?
74. What is a truth table?
75. When constructing a truth table, how do you determine how many rows to create?

• HISTORY QUESTIONS

76. Who is considered "the father of symbolic logic"?
77. Boolean algebra is a combination of logic and mathematics. What is it used for?

1.4 More on Conditionals

OBJECTIVES

- Create the converse, inverse, and contrapositive of a conditional statement
- Determine equivalent variations of a conditional statement
- Interpret "only if" statements
- Interpret a biconditional statement

Conditionals differ from conjunctions and disjunctions with regard to the possibility of changing the order of the statements. In algebra, the sum $x + y$ is equal to the sum $y + x$; that is, addition is commutative. In everyday language, one realtor might say, "The house is perfect and the lot is priceless," while another says, "The lot is priceless and the house is perfect." Logically, their meanings are the same, since $(p \wedge q) \equiv (q \wedge p)$. The order of the components in a conjunction or disjunction makes no difference in regard to the truth value of the statement. This is not so with conditionals.

Variations of a Conditional

Given two statements p and q, various "if . . . then . . ." statements can be formed.

EXAMPLE 1

TRANSLATING SYMBOLS INTO WORDS Using the statements

p: You are compassionate.

q: You contribute to charities.

write an "if . . . then . . ." sentence represented by each of the following:

a. $p \rightarrow q$ **b.** $q \rightarrow p$ **c.** $\sim p \rightarrow \sim q$ **d.** $\sim q \rightarrow \sim p$

SOLUTION

a. $p \rightarrow q$: If you are compassionate, then you contribute to charities.
b. $q \rightarrow p$: If you contribute to charities, then you are compassionate.
c. $\sim p \rightarrow \sim q$: If you are not compassionate, then you do not contribute to charities.
d. $\sim q \rightarrow \sim p$: If you do not contribute to charities, then you are not compassionate.

Each part of Example 1 contains an "if . . . then . . ." statement and is called a conditional. Any given conditional has three variations: a converse, an inverse, and a contrapositive. The **converse** of the conditional "if p then q" is the compound statement "if q then p" That is, we form the converse of the conditional by interchanging the premise and the conclusion; $q \rightarrow p$ is the converse of $p \rightarrow q$. The statement in part (b) of Example 1 is the converse of the statement in part (a).

The **inverse** of the conditional "if p then q" is the compound statement "if not p then not q." We form the inverse of the conditional by negating both the premise and the conclusion; $\sim p \rightarrow \sim q$ is the inverse of $p \rightarrow q$. The statement in part (c) of Example 1 is the inverse of the statement in part (a).

The **contrapositive** of the conditional "if p then q" is the compound statement "if not q then not p." We form the contrapositive of the conditional by negating *and* interchanging both the premise and the conclusion; $\sim q \rightarrow \sim p$ is the contrapositive of $p \rightarrow q$. The statement in part (d) of Example 1 is the contrapositive of the statement in part (a). The variations of a given conditional are summarized in Figure 1.55. As we will see, some of these variations are equivalent, and some are not. Unfortunately, many people incorrectly treat them all as equivalent.

Name	Symbolic Form	Read As . . .
a (given) conditional	$p \rightarrow q$	If p, then q.
the converse (of $p \rightarrow q$)	$q \rightarrow p$	If q, then p.
the inverse (of $p \rightarrow q$)	$\sim p \rightarrow \sim q$	If not p, then not q.
the contrapositive (of $p \rightarrow q$)	$\sim q \rightarrow \sim p$	If not q, then not p.

FIGURE 1.55 Variations of a conditional.

EXAMPLE 2

CREATING VARIATIONS OF A CONDITIONAL STATEMENT Given the conditional "You did not receive the proper refund if you prepared your own income tax form," write the sentence that represents each of the following.

a. the converse of the conditional
b. the inverse of the conditional
c. the contrapositive of the conditional

SOLUTION

a. Rewriting the statement in the standard "if . . . then . . ." form, we have the conditional "If you prepared your own income tax form, then you did not receive the proper refund." The converse is formed by interchanging the premise and the conclusion. Thus, the converse is written as "If you did not receive the proper refund, then you prepared your own income tax form."
b. The inverse is formed by negating both the premise and the conclusion. Thus, the inverse is written as "If you did not prepare your own income tax form, then you received the proper refund."
c. The contrapositive is formed by negating *and* interchanging the premise and the conclusion. Thus, the contrapositive is written as "If you received the proper refund, then you did not prepare your own income tax form."

Equivalent Conditionals

We have seen that the conditional $p \rightarrow q$ has three variations: the converse $q \rightarrow p$, the inverse $\sim p \rightarrow \sim q$, and the contrapositive $\sim q \rightarrow \sim p$. Do any of these "if. . . then . . ." statements convey the same meaning? In other words, are any of these compound statements equivalent?

EXAMPLE 3

DETERMINING EQUIVALENT STATEMENTS Determine which (if any) of the following are equivalent: a conditional $p \rightarrow q$, the converse $q \rightarrow p$, the inverse $\sim p \rightarrow \sim q$, and the contrapositive $\sim q \rightarrow \sim p$.

SOLUTION

To investigate the possible equivalencies, we must construct a truth table that contains all the statements. Because there are two letters, we need $2^2 = 4$ rows. The table must have a column for $\sim p$, one for $\sim q$, one for the conditional $p \rightarrow q$, and one for each variation of the conditional. The truth values of the negations $\sim p$ and $\sim q$ are readily entered, as shown in Figure 1.56.

	p	q	$\sim p$	$\sim q$	$p \rightarrow q$	$q \rightarrow p$	$\sim p \rightarrow \sim q$	$\sim q \rightarrow \sim p$
1.	T	T	F	F				
2.	T	F	F	T				
3.	F	T	T	F				
4.	F	F	T	T				

FIGURE 1.56 Required columns in the truth table.

An "if . . . then . . ." statement is false only when the premise is true and the conclusion is false. Consequently, $p \rightarrow q$ is false only when p is T and q is F; enter an F in row 2 and Ts elsewhere in the column under $p \rightarrow q$.

Likewise, the converse $q \rightarrow p$ is false only when q is T and p is F; enter an F in row 3 and Ts elsewhere.

In a similar manner, the inverse $\sim p \rightarrow \sim q$ is false only when $\sim p$ is T and $\sim q$ is F; enter an F in row 3 and Ts elsewhere.

Finally, the contrapositive $\sim q \rightarrow \sim p$ is false only when $\sim q$ is T and $\sim p$ is F; enter an F in row 2 and Ts elsewhere.

The completed truth table is shown in Figure 1.57. Examining the entries in Figure 1.57, we can see that the columns under $p \rightarrow q$ and $\sim q \rightarrow \sim p$ are identical; each has an F in row 2 and Ts elsewhere. Consequently, a conditional and its contrapositive are equivalent: $p \rightarrow q \equiv \sim q \rightarrow \sim p$.

Likewise, we notice that $q \rightarrow p$ and $\sim p \rightarrow \sim q$ have identical truth values; each has an F in row 3 and Ts elsewhere. Thus, the converse and the inverse of a conditional are equivalent: $q \rightarrow p \equiv \sim p \rightarrow \sim q$.

	p	q	$\sim p$	$\sim q$	$p \rightarrow q$	$q \rightarrow p$	$\sim p \rightarrow \sim q$	$\sim q \rightarrow \sim p$
1.	T	T	F	F	T	T	T	T
2.	T	F	F	T	F	T	T	F
3.	F	T	T	F	T	F	F	T
4.	F	F	T	T	T	T	T	T

FIGURE 1.57 Truth table for a conditional and its variations.

We have seen that different "if . . . then . . ." statements can convey the same meaning—that is, that certain variations of a conditional are equivalent (see Figure 1.58). For example, the compound statements "If you are compassionate, then you contribute to charities" and "If you do not contribute to charities, then you are not compassionate" convey the same meaning. (The second conditional is the contrapositive of the first.) Regardless of its specific contents (p, q, $\sim p$, or $\sim q$), every "if . . . then . . ." statement has an equivalent variation formed by negating *and* interchanging the premise and the conclusion of the given conditional statement.

Equivalent Statements	Symbolic Representations
a conditional and its contrapositive	$(p \rightarrow q) \equiv (\sim q \rightarrow \sim p)$
the converse and the inverse (of the conditional $p \rightarrow q$)	$(q \rightarrow p) \equiv (\sim p \rightarrow \sim q)$

FIGURE 1.58 Equivalent "if . . . then . . ." statements.

EXAMPLE 4

CREATING A CONTRAPOSITIVE Given the statement "Being a doctor is necessary for being a surgeon," express the contrapositive in terms of the following:

a. a sufficient condition
b. a necessary condition

SOLUTION

a. Recalling that a *necessary condition* is the *conclusion* of a conditional, we can rephrase the statement "Being a doctor is necessary for being a surgeon" as follows:

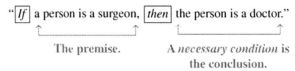

"[*If*] a person is a surgeon, [*then*] the person is a doctor."

The premise. — A *necessary condition* is the conclusion.

Therefore, by negating and interchanging the premise and conclusion, the contrapositive is

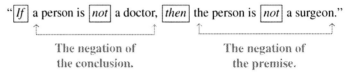

"[*If*] a person is [*not*] a doctor, [*then*] the person is [*not*] a surgeon."

The negation of the conclusion. — The negation of the premise.

Recalling that a *sufficient condition* is the *premise* of a conditional, we can phrase the contrapositive of the original statement as "*Not being a doctor is sufficient for not being a surgeon.*"

b. From part (a), the contrapositive of the original statement is the conditional statement

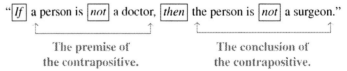

"[*If*] a person is [*not*] a doctor, [*then*] the person is [*not*] a surgeon."

The premise of the contrapositive. — The conclusion of the contrapositive.

Because a necessary condition is the conclusion of a conditional, the contrapositive of the (original) statement "Being a doctor is necessary for being a surgeon" can be expressed as "*Not being a surgeon is necessary for not being a doctor.*"

The "Only If" Connective

Consider the statement "A prisoner is paroled only if the prisoner obeys the rules." What is the premise, and what is the conclusion? Rather than using p and q (which might bias our investigation), we define

r: A prisoner is paroled.
s: A prisoner obeys the rules.

The given statement is represented by "r only if s." Now, "r only if s" means that r can happen *only* if s happens. In other words, if s does not happen, then r does not happen, or $\sim s \rightarrow \sim r$. We have seen that $\sim s \rightarrow \sim r$ is equivalent to $r \rightarrow s$. Consequently, "r only if s" is equivalent to the conditional $r \rightarrow s$. The premise of the statement "A prisoner is paroled only if the prisoner obeys the rules" is "A prisoner is paroled," and the conclusion is "The prisoner obeys the rules."

The conditional $p \rightarrow q$ can be phrased "p only if q." Even though the word *if* precedes q, q is not the premise. *Whatever follows the connective "only if" is the conclusion of the conditional.*

EXAMPLE 5

ANALYZING AN "ONLY IF" STATEMENT For the compound statement "You receive a federal grant only if your artwork is not obscene," do the following:

a. Determine the premise and the conclusion.
b. Rewrite the compound statement in the standard "if . . . then . . ." form.
c. Interpret the conditions that make the statement false.

SOLUTION

a. Because the compound statement contains an "only if" connective, the statement that follows "only if" is the conclusion of the conditional. The premise is "You receive a federal grant." The conclusion is "Your artwork is not obscene."
b. The given compound statement can be rewritten as "If you receive a grant, then your artwork is not obscene."
c. First we define the symbols.

p: You receive a federal grant.

q: Your artwork is obscene.

	p	q	$\sim q$	$p \to \sim q$
1.	T	T	F	F
2.	T	F	T	T
3.	F	T	F	T
4.	F	F	T	T

FIGURE 1.59
Truth table for the conditional $p \to \sim q$.

Then the statement has the symbolic representation $p \to \sim q$. The truth table for $p \to \sim q$ is given in Figure 1.59.

The expression $p \to q$ is false under the conditions listed in row 1 (when p and q are both true). Therefore, the statement "You receive a federal grant only if your artwork is not obscene" is false when an artist *does* receive a federal grant *and* the artist's artwork *is* obscene.

The Biconditional $p \leftrightarrow q$

What do the words *bicycle, binomial,* and *bilingual* have in common? Each word begins with the prefix *bi,* meaning "two." Just as the word *bilingual* means "two languages," the word *biconditional* means "two conditionals."

In everyday speech, conditionals often get "hooked together" in a circular fashion. For instance, someone might say, "If I am rich, then I am happy, and if I am happy, then I am rich." Notice that this compound statement is actually the conjunction (*and*) of a conditional (if rich, then happy) and its converse (if happy, then rich). Such a statement is referred to as a biconditional. A **biconditional** is a statement of the form $(p \to q) \wedge (q \to p)$ and is symbolized as $p \leftrightarrow q$. The symbol $p \leftrightarrow q$ is read "p if and only if q" and is frequently abbreviated "p iff q." A biconditional is equivalent to the conjunction of two conversely related conditionals: $p \leftrightarrow q \equiv [(p \to q) \wedge (q \to p)]$.

In addition to the phrase "if and only if," a biconditional can also be expressed by using "necessary" and "sufficient" terminology. The statement "p is sufficient for q" can be rephrased as "if p then q" (and symbolized as $p \to q$), whereas the statement "p is necessary for q" can be rephrased as "if q then p" (and symbolized as $q \to p$). Therefore, the biconditional "p if and only if q" can also be phrased as "p is necessary and sufficient for q."

EXAMPLE 6

ANALYZING A BICONDITIONAL STATEMENT Express the biconditional "A citizen is eligible to vote if and only if the citizen is at least eighteen years old" as the conjunction of two conditionals.

SOLUTION

The given biconditional is equivalent to "If a citizen is eligible to vote, then the citizen is at least eighteen years old, *and* if a citizen is at least eighteen years old, then the citizen is eligible to vote."

Under what circumstances is the biconditional $p \leftrightarrow q$ true, and when is it false? To find the answer, we must construct a truth table. Utilizing the equivalence $p \leftrightarrow q \equiv [(p \rightarrow q) \wedge (q \rightarrow p)]$, we get the completed table shown in Figure 1.60. (Recall that a conditional is false only when its premise is true and its conclusion is false and that a conjunction is true only when both components are true.) We can see that a biconditional is true only when the two components p and q have the same truth value—that is, when p and q are both true or when p and q are both false. On the other hand, a biconditional is false when the two components p and q have opposite truth value—that is, when p is true and q is false or vice versa.

	p	q	$p \rightarrow q$	$q \rightarrow p$	$(p \rightarrow q) \wedge (q \rightarrow p)$
1.	T	T	T	T	T
2.	T	F	F	T	F
3.	F	T	T	F	F
4.	F	F	T	T	T

FIGURE 1.60 Truth table for a biconditional $p \leftrightarrow q$.

Many theorems in mathematics can be expressed as biconditionals. For example, when solving a quadratic equation, we have the following: "The equation $ax^2 + bx + c = 0$ has exactly one solution if and only if the discriminant $b^2 - 4ac = 0$." Recall that the solutions of a quadratic equation are

$$x = \frac{-b \pm \sqrt{b^2 - 4ac}}{2a}$$

This biconditional is equivalent to "If the equation $ax^2 + bx + c = 0$ has exactly one solution, then the discriminant $b^2 - 4ac = 0$, and if the discriminant $b^2 - 4ac = 0$, then the equation $ax^2 + bx + c = 0$ has exactly one solution"—that is, one condition implies the other.

1.4 Exercises

In Exercises 1–2, using the given statements, write the sentence represented by each of the following.

a. $p \rightarrow q$ **b.** $q \rightarrow p$

c. $\sim p \rightarrow \sim q$ **d.** $\sim q \rightarrow \sim p$

e. Which of parts (a)–(d) are equivalent? Why?

1. *p:* She is a police officer.
q: She carries a gun.

2. *p:* I am a multimillion-dollar lottery winner.
q: I am a world traveler.

In Exercises 3–4, using the given statements, write the sentence represented by each of the following.

a. $p \rightarrow \sim q$ **b.** $\sim q \rightarrow p$

c. $\sim p \rightarrow q$ **d.** $q \rightarrow \sim p$

e. Which of parts (a)–(d) are equivalent? Why?

3. *p:* I watch television.
q: I do my homework.

4. *p:* He is an artist.
q: He is a conformist.

In Exercises 5–10, form (a) the inverse, (b) the converse, and (c) the contrapositive of the given conditional.

5. If you pass this mathematics course, then you fulfill a graduation requirement.

6. If you have the necessary tools, assembly time is less than thirty minutes.

7. The television set does not work if the electricity is turned off.

8. You do not win if you do not buy a lottery ticket.

9. You are a vegetarian if you do not eat meat.
10. If chemicals are properly disposed of, the environment is not damaged.

In Exercises 11–14, express the contrapositive of the given conditional in terms of (a) a sufficient condition and (b) a necessary condition.

11. Being an orthodontist is sufficient for being a dentist.
12. Being an author is sufficient for being literate.
13. Knowing Morse code is necessary for operating a telegraph.
14. Knowing CPR is necessary for being a paramedic.

In Exercises 15–20, (a) determine the premise and conclusion, (b) rewrite the compound statement in the standard "if . . . then . . ." form, and (c) interpret the conditions that make the statement false.

15. I take public transportation only if it is convenient.
16. I eat raw fish only if I am in a Japanese restaurant.
17. I buy foreign products only if domestic products are not available.
18. I ride my bicycle only if it is not raining.
19. You may become a U.S. senator only if you are at least thirty years old and have been a citizen for nine years.
20. You may become the president of the United States only if you are at least thirty-five years old and were born a citizen of the United States.

In Exercises 21–28, express the given biconditional as the conjunction of two conditionals.

21. You obtain a refund if and only if you have a receipt.
22. We eat at Burger World if and only if Ju Ju's Kitsch-Inn is closed.
23. The quadratic equation $ax^2 + bx + c = 0$ has two distinct real solutions if and only if $b^2 - 4ac > 0$.
24. The quadratic equation $ax^2 + bx + c = 0$ has complex solutions iff $b^2 - 4ac < 0$.
25. A polygon is a triangle iff the polygon has three sides.
26. A triangle is isosceles iff the triangle has two equal sides.
27. A triangle having a 90° angle is necessary and sufficient for $a^2 + b^2 = c^2$.
28. A triangle having three equal sides is necessary and sufficient for a triangle having three equal angles.

In Exercises 29–36, translate the two statements into symbolic form and use truth tables to determine whether the statements are equivalent.

29. I cannot have surgery if I do not have health insurance.
    If I can have surgery, then I do have health insurance.
30. If I am illiterate, I cannot fill out an application form.
    I can fill out an application form if I am not illiterate.
31. If you earn less than $12,000 per year, you are eligible for assistance.
    If you are not eligible for assistance, then you earn at least $12,000 per year.
32. If you earn less than $12,000 per year, you are eligible for assistance.
    If you earn at least $12,000 per year, you are not eligible for assistance.
33. I watch television only if the program is educational.
    I do not watch television if the program is not educational.
34. I buy seafood only if the seafood is fresh.
    If I do not buy seafood, the seafood is not fresh.
35. Being an automobile that is American-made is sufficient for an automobile having hardware that is not metric.
    Being an automobile that is not American-made is necessary for an automobile having hardware that is metric.
36. Being an automobile having metric hardware is sufficient for being an automobile that is not American-made.
    Being an automobile not having metric hardware is necessary for being an automobile that is American-made.

In Exercises 37–46, write an equivalent variation of the given conditional.

37. If it is not raining, I walk to work.
38. If it makes a buzzing noise, it is not working properly.
39. It is snowing only if it is cold.
40. You are a criminal only if you do not obey the law.
41. You are not a vegetarian if you eat meat.
42. You are not an artist if you are not creative.
43. All policemen own guns.
44. All college students are sleep deprived.
45. No convicted felon is eligible to vote.
46. No man asks for directions.

In Exercises 47–52, determine which pairs of statements are equivalent.

47. i. If Proposition 111 passes, freeways are improved.
    ii. If Proposition 111 is defeated, freeways are not improved.
    iii. If the freeways are improved, Proposition 111 passes.
    iv. If the freeways are not improved, Proposition 111 does not pass.
48. i. If the Giants win, then I am happy.
    ii. If I am happy, then the Giants win.

- **iii.** If the Giants lose, then I am unhappy.
- **iv.** If I am unhappy, then the Giants lose.

49.
- **i.** I go to church if it is Sunday.
- **ii.** I go to church only if it is Sunday.
- **iii.** If I do not go to church, it is not Sunday.
- **iv.** If it is not Sunday, I do not go to church.

50.
- **i.** I am a rebel if I do not have a cause.
- **ii.** I am a rebel only if I do not have a cause.
- **iii.** I am not a rebel if I have a cause.
- **iv.** If I am not a rebel, I have a cause.

51.
- **i.** If line 34 is greater than line 29, I use Schedule X.
- **ii.** If I use Schedule X, then line 34 is greater than line 29.
- **iii.** If I do not use Schedule X, then line 34 is not greater than line 29.
- **iv.** If line 34 is not greater than line 29, then I do not use Schedule X.

52.
- **i.** If you answer yes to all of the above, then you complete Part II.
- **ii.** If you answer no to any of the above, then you do not complete Part II.
- **iii.** If you completed Part II, then you answered yes to all of the above.
- **iv.** If you did not complete Part II, then you answered no to at least one of the above.

Answer the following questions using complete sentences and your own words.

• CONCEPT QUESTIONS

53. What is a contrapositive?

54. What is a converse?

55. What is an inverse?

56. What is a biconditional?

57. How is an "if . . . then . . ." statement related to an "only if" statement?

THE NEXT LEVEL

If a person wants to pursue an advanced degree (something beyond a bachelor's or four-year degree), chances are the person must take a standardized exam to gain admission to a graduate school or to be admitted into a specific program. These exams are intended to measure verbal, quantitative, and analytical skills that have developed throughout a person's life. Many classes and study guides are available to help people prepare for the exams. The following questions are typical of those found in the study guides.

Exercises 58–62 refer to the following: Assuming that a movie's popularity is measured by its gross box office receipts, six recently released movies—M, N, O, P, Q, and R—are ranked from most popular (first) to least popular (sixth). There are no ties. The ranking is consistent with the following conditions:

- *O is more popular than R.*
- *If N is more popular than O, then neither Q nor R is more popular than P.*
- *If O is more popular than N, then neither P nor R is more popular than Q.*
- *M is more popular than N, or else M is more popular than O, but not both.*

58. Which one of the following could be the ranking of the movies, from most popular to least popular?
- **a.** N, M, O, R, P, Q
- **b.** P, O, M, Q, N, R
- **c.** Q, P, R, O, M, N
- **d.** O, Q, M, P, N, R
- **e.** P, Q, N, O, R, M

59. If N is the second most popular movie, then which one of the following could be true?
- **a.** O is more popular than M.
- **b.** Q is more popular than M.
- **c.** R is more popular than M.
- **d.** Q is more popular than P.
- **e.** O is more popular than N.

60. Which one of the following cannot be the most popular movie?
- **a.** M
- **b.** N
- **c.** O
- **d.** P
- **e.** Q

61. If R is more popular than M, then which one of the following could be true?
- **a.** M is more popular than O.
- **b.** M is more popular than Q.
- **c.** N is more popular than P.
- **d.** N is more popular than O.
- **e.** N is more popular than R.

62. If O is more popular than P and less popular than Q, then which one of the following could be true?
- **a.** M is more popular than O.
- **b.** N is more popular than M.
- **c.** N is more popular than O.
- **d.** R is more popular than Q.
- **e.** P is more popular than R.

1.5 Analyzing Arguments

OBJECTIVES

- Identify a tautology
- Use a truth table to analyze an argument

Lewis Carroll's Cheshire Cat told Alice that he was mad (crazy). Alice then asked, "'And how do you know that you're mad?' 'To begin with,' said the cat, 'a dog's not mad. You grant that?' 'I suppose so,' said Alice. 'Well, then,' the cat went on, 'you see a dog growls when it's angry, and wags its tail when it's pleased. Now *I* growl when I'm pleased, and wag my tail when I'm angry. Therefore I'm mad!'"

Does the Cheshire Cat have a valid deductive argument? Does the conclusion follow logically from the hypotheses? To answer this question, and others like it, we will utilize symbolic logic and truth tables to account for all possible combinations of the individual statements as true or false.

Valid Arguments

When someone makes a sequence of statements and draws some conclusion from them, he or she is presenting an argument. An **argument** consists of two components: the initial statements, or hypotheses, and the final statement, or conclusion. When presented with an argument, a listener or reader may ask, "Does this person have a logical argument? Does his or her conclusion necessarily follow from the given statements?"

An argument is **valid** if the conclusion of the argument is guaranteed under its given set of hypotheses. (That is, the conclusion is inescapable in all instances.) For example, we used Venn diagrams in Section 1.1 to show the argument

"All men are mortal. Socrates is a man.	} the hypotheses
Therefore, Socrates is mortal."	} the conclusion

is a valid argument. Given the hypotheses, the conclusion is guaranteed. The term *valid* does not mean that all the statements are true but merely that the conclusion was reached via a proper deductive process. As shown in Example 3 of Section 1.1, the argument

"All doctors are men. My mother is a doctor.	} the hypotheses
Therefore, my mother is a man."	} the conclusion

is also a valid argument. Even though the conclusion is obviously false, the conclusion is guaranteed, *given the hypotheses.*

The hypotheses in a given logical argument may consist of several interrelated statements, each containing negations, conjunctions, disjunctions, and conditionals. By joining all the hypotheses in the form of a conjunction, we can form a single conditional that represents the entire argument. That is, if an argument has n hypotheses ($h_1, h_2, \ldots, h_n$) and conclusion c, the argument will have the form "if (h_1 and $h_2 \ldots$ and h_n), then c."

Using a logical argument, Lewis Carroll's Cheshire Cat tried to convince Alice that he was crazy. Was his argument valid?

CONDITIONAL REPRESENTATION OF AN ARGUMENT

An argument having n hypotheses $h_1, h_2, \cdots, h_n$ and conclusion c can be represented by the conditional $[h_1 \wedge h_2 \wedge \cdots \wedge h_n] \rightarrow c$.

If the conditional representation of an argument is always true (regardless of the actual truthfulness of the individual statements), the argument is valid. If there is at least one instance in which the conditional is false, the argument is invalid.

EXAMPLE 1

USING A TRUTH TABLE TO ANALYZE AN ARGUMENT Determine whether the following argument is valid:

"If he is illiterate, he cannot fill out the application.
He can fill out the application.
Therefore, he is not illiterate."

SOLUTION

First, number the hypotheses and separate them from the conclusion with a line:

1. If he is illiterate, he cannot fill out the application.
2. He can fill out the application.

Therefore, he is not illiterate.

Now use symbols to represent each different component in the statements:

p: He is illiterate.
q: He can fill out the application.

FEATURED IN THE NEWS

CHURCH CARVING MAY BE ORIGINAL 'CHESHIRE CAT'

London—Devotees of writer Lewis Carroll believe they have found what inspired his grinning Cheshire Cat, made famous in his book "Alice's Adventures in Wonderland."

Members of the Lewis Carroll Society made the discovery over the weekend in a church at which the author's father was once rector in the Yorkshire village of Croft in northern England.

It is a rough-hewn carving of a cat's head smiling near an altar, probably dating to the 10th century. Seen from below and from the perspective of a small boy, all that can be seen is the grinning mouth.

Carroll's Alice watched the Cheshire Cat disappear "ending with the grin, which remained for some time after the rest of the head had gone."

Alice mused: "I have often seen a cat without a grin, but not a grin without a cat. It is the most curious thing I have seen in all my life."

We could have defined q as "He *cannot* fill out the application" (as stated in premise 1), but it is customary to define the symbols with a positive sense. Symbolically, the argument has the form

1. $p \to \sim q$ } the hypotheses
2. q

$\therefore \sim p$ } conclusion

and is represented by the conditional $[(p \to \sim q) \wedge q] \to \sim p$. The symbol $\therefore$ is read "therefore."

To construct a truth table for this conditional, we need $2^2 = 4$ rows. A column is required for the following: each negation, each hypothesis, the conjunction of the hypotheses, the conclusion, and the conditional representation of the argument. The initial setup is shown in Figure 1.61.

Fill in the truth table as follows:

$\sim q$: A negation has the opposite truth values; enter a T in rows 2 and 4 and an F in rows 1 and 3.

Hypothesis 1: A conditional is false only when its premise is true and its conclusion is false; enter an F in row 1 and Ts elsewhere.

				Hypothesis 1	Hypothesis 2	Column Representing All the Hypotheses	Conclusion c	Conditional Representation of the Argument
	p	q	$\sim q$	$p \to \sim q$	q	$1 \wedge 2$	$\sim p$	$(1 \wedge 2) \to c$
1.	T	T						
2.	T	F						
3.	F	T						
4.	F	F						

FIGURE 1.61 Required columns in the truth table.

Hypothesis 2: Recopy the q column.

$1 \wedge 2$: A conjunction is true only when both components are true; enter a T in row 3 and Fs elsewhere.

Conclusion c: A negation has the opposite truth values; enter an F in rows 1 and 2 and a T in rows 3 and 4.

At this point, all that remains is the final column (see Figure 1.62).

				1	2		c	
	p	q	$\sim q$	$p \rightarrow \sim q$	q	$1 \wedge 2$	$\sim p$	$(1 \wedge 2) \rightarrow c$
1.	T	T	F	F	T	F	F	
2.	T	F	T	T	F	F	F	
3.	F	T	F	T	T	T	T	
4.	F	F	T	T	F	F	T	

FIGURE 1.62 Truth values of the expressions.

The last column in the truth table is the conditional that represents the entire argument. A conditional is false only when its premise is true and its conclusion is false. The only instance in which the premise $(1 \wedge 2)$ is true is row 3. Corresponding to this entry, the conclusion $\sim p$ is also true. Consequently, the conditional $(1 \wedge 2) \rightarrow c$ is true in row 3. Because the premise $(1 \wedge 2)$ is false in rows 1, 2, and 4, the conditional $(1 \wedge 2) \rightarrow c$ is automatically true in those rows as well. The completed truth table is shown in Figure 1.63.

				1	2		c	
	p	q	$\sim q$	$p \rightarrow \sim q$	q	$1 \wedge 2$	$\sim p$	$(1 \wedge 2) \rightarrow c$
1.	T	T	F	F	T	F	F	T
2.	T	F	T	T	F	F	F	T
3.	F	T	F	T	T	T	T	T
4.	F	F	T	T	F	F	T	T

FIGURE 1.63 Truth table for the argument $[(p \rightarrow \sim q) \wedge q] \rightarrow \sim p$.

The completed truth table shows that the conditional $[(p \rightarrow \sim q) \wedge q] \rightarrow \sim p$ is always true. The conditional represents the argument "If he is illiterate, he cannot fill out the application. He can fill out the application. Therefore, he is not illiterate." Thus, the argument is valid.

Tautologies

A **tautology** is a statement that is always true. For example, the statement

"$(a + b)^2 = a^2 + 2ab + b^2$"

is a tautology.

EXAMPLE 2

DETERMINING WHETHER A STATEMENT IS A TAUTOLOGY Determine whether the statement $(p \wedge q) \rightarrow (p \vee q)$ is a tautology.

SOLUTION

We need to construct a truth table for the statement. Because there are two letters, the table must have $2^2 = 4$ rows. We need a column for $(p \wedge q)$, one for $(p \vee q)$, and one for $(p \wedge q) \rightarrow (p \vee q)$. The completed truth table is shown in Figure 1.64.

	p	q	$p \wedge q$	$p \vee q$	$(p \wedge q) \rightarrow (p \vee q)$
1.	T	T	T	T	T
2.	T	F	F	T	T
3.	F	T	F	T	T
4.	F	F	F	F	T

FIGURE 1.64 Truth table for the statement $(p \wedge q) \rightarrow (p \vee q)$.

Because $(p \wedge q) \rightarrow (p \vee q)$ is always true, it is a tautology.

As we have seen, an argument can be represented by a single conditional. If this conditional is always true, the argument is valid (and vice versa).

VALIDITY OF AN ARGUMENT

An argument having n hypotheses $h_1, h_2, \ldots, h_n$ and conclusion c is valid if and only if the conditional $[h_1 \wedge h_2 \wedge \ldots \wedge h_n] \rightarrow c$ is a tautology.

EXAMPLE 3

USING A TRUTH TABLE TO ANALYZE AN ARGUMENT Determine whether the following argument is valid:
"If the defendant is innocent, the defendant does not go to jail. The defendant does not go to jail. Therefore, the defendant is innocent."

SOLUTION

Separating the hypotheses from the conclusion, we have

1. If the defendant is innocent, the defendant does not go to jail.
2. The defendant does not go to jail.

Therefore, the defendant is innocent.

Now we define symbols to represent the various components of the statements:

p: The defendant is innocent.
q: The defendant goes to jail.

Symbolically, the argument has the form

1. $p \rightarrow \sim q$
2. $\sim q$

$\therefore p$

and is represented by the conditional $[(p \rightarrow \sim q) \wedge \sim q] \rightarrow p$.

Now we construct a truth table with four rows, along with the necessary columns. The completed table is shown in Figure 1.65.

			2	1		c	
	p	q	$\sim q$	$p \to \sim q$	$1 \wedge 2$	p	$(1 \wedge 2) \to c$
1.	T	T	F	F	F	T	T
2.	T	F	T	T	T	T	T
3.	F	T	F	T	F	F	T
4.	F	F	T	T	T	F	F

FIGURE 1.65 Truth table for the argument $[(p \to \sim q) \wedge \sim q] \to p$.

The column representing the argument has an F in row 4; therefore, the conditional representation of the argument is *not* a tautology. In particular, the conclusion does not logically follow the hypotheses when both p and q are false (row 4). The argument is not valid. Let us interpret the circumstances expressed in row 4, the row in which the argument breaks down. Both p and q are false—that is, the defendant is guilty and the defendant does *not* go to jail. Unfortunately, this situation can occur in the real world; guilty people do not *always* go to jail! As long as it is possible for a guilty person to avoid jail, the argument is invalid.

The following argument was presented as Example 6 in Section 1.1. In that section, we constructed a Venn diagram to show that the argument was in fact valid. We now show an alternative method; that is, we construct a truth table to determine whether the argument is valid.

EXAMPLE 4

USING A TRUTH TABLE TO ANALYZE AN ARGUMENT Determine whether the following argument is valid: "No snake is warm-blooded. All mammals are warm-blooded. Therefore, snakes are not mammals."

SOLUTION

Separating the hypotheses from the conclusion, we have

1. No snake is warm-blooded.
2. All mammals are warm-blooded.

Therefore, snakes are not mammals.

These statements can be rephrased as follows:

1. If it is a snake, then it is not warm-blooded.
2. If it is a mammal, then it is warm-blooded.

Therefore, if it is a snake, then it is not a mammal.

Now we define symbols to represent the various components of the statements:

p: It is a snake.
q: It is warm-blooded.
r: It is a mammal.

Symbolically, the argument has the form

1. $p \to \sim q$
2. $r \to q$

$\therefore p \to \sim r$

and is represented by the conditional $[(p \to \sim q) \wedge (r \to q)] \to (p \to \sim r)$.

Now we construct a truth table with eight rows ($2^3 = 8$), along with the necessary columns. The completed table is shown in Figure 1.66.

						1	2		c	
	p	q	r	$\sim q$	$\sim r$	$p \to \sim q$	$r \to q$	$1 \wedge 2$	$p \to \sim r$	$(1 \wedge 2) \to c$
1.	T	T	T	F	F	F	T	F	F	T
2.	T	T	F	F	T	F	T	F	T	T
3.	T	F	T	T	F	T	F	F	F	T
4.	T	F	F	T	T	T	T	T	T	T
5.	F	T	T	F	F	T	T	T	T	T
6.	F	T	F	F	T	T	T	T	T	T
7.	F	F	T	T	F	T	F	F	T	T
8.	F	F	F	T	T	T	T	T	T	T

FIGURE 1.66 Truth table for the argument $[(p \to \sim q) \wedge (r \to q)] \to (p \to \sim r)$.

The last column of the truth table represents the argument and contains all T's. Consequently, the conditional $[(p \to \sim q) \wedge (r \to q)] \to (p \to \sim r)$ is a tautology; the argument is valid.

The preceding examples contained relatively simple arguments, each consisting of only two hypotheses and two simple statements (letters). In such cases, many people try to employ "common sense" to confirm the validity of the argument. For instance, the argument "If it is raining, the streets are wet. It is raining. Therefore, the streets are wet" is obviously valid. However, it might not be so simple to determine the validity of an argument that contains several hypotheses and many simple statements. Indeed, in such cases, the argument's truth table might become quite lengthy, as in the next example.

EXAMPLE 5

USING A TRUTH TABLE TO ANALYZE AN ARGUMENT The following whimsical argument was written by Lewis Carroll and appeared in his 1896 book *Symbolic Logic:*

> "No ducks waltz. No officers ever decline to waltz. All my poultry are ducks. Therefore, my poultry are not officers."

Construct a truth table to determine whether the argument is valid.

HISTORICAL NOTE

CHARLES LUTWIDGE DODGSON, 1832–1898

© Bettmann/CORBIS

To those who assume that it is impossible for a person to excel both in the creative worlds of art and literature and in the disciplined worlds of mathematics and logic, the life of Charles Lutwidge Dodgson is a wondrous counterexample. Known the world over as Lewis Carroll, Dodgson penned the nonsensical classics *Alice's Adventures in Wonderland* and *Through the Looking Glass*. However, many people are surprised to learn that Dodgson (from age eighteen to his death) was a permanent resident at the University at Oxford, teaching mathematics and logic. And as if that were not enough, Dodgson is now recognized as one of the leading portrait photographers of the Victorian era.

The eldest son in a family of eleven children, Charles amused his younger siblings with elaborate games, poems, stories, and humorous drawings. This attraction to entertaining children with fantastic stories manifested itself in much of his later work as Lewis Carroll. Besides his obvious interest in telling stories, the young Dodgson was also intrigued by mathematics. At the age of eight, Charles asked his father to explain a book on logarithms. When told that he was too young to understand, Charles persisted, "But please, explain!"

The Dodgson family had a strong ecclesiastical tradition; Charles's father, great-grandfather, and great-great-grandfather were all clergymen. Following in his father's footsteps, Charles attended Christ Church, the largest and most celebrated of all the Oxford colleges. After graduating in 1854, Charles remained at Oxford, accepting the position of mathematical lecturer in 1855. However, appointment to this position was conditional upon his taking Holy Orders in the Anglican church and upon his remaining celibate. Dodgson complied and was named a deacon in 1861.

The year 1856 was filled with events that had lasting effects on Dodgson. Charles Lutwidge created his pseudonym by translating his first and middle names into Latin (Carolus Ludovic), reversing their order (Ludovic Carolus), and translating them back into English (Lewis Carroll). In this same year, Dodgson began his "hobby" of photography. He is considered by many to have been an artistic pioneer in this new field (photography was invented in 1839). Most of Dodgson's work consists of portraits that chronicle the Victorian era, and over 700 photographs taken by Dodgson have been preserved. His favorite subjects were children, especially young girls.

Dodgson's affinity for children brought about a meeting in 1856 that would eventually establish his place in the history of literature. Early in the year, Dodgson met the four children of the dean of Christ Church: Harry, Lorina, Edith, and Alice Liddell. He began seeing the children on a regular basis, amusing them with stories and photographing them. Although he had a wonderful relationship with all four, Alice received his special attention.

On July 4, 1862, while rowing and picnicking with Alice and her sisters, Dodgson entertained the Liddell girls with a fantastic story of a little girl named Alice who fell into a rabbit hole. Captivated by

SOLUTION Separating the hypotheses from the conclusion, we have

1. No ducks waltz.
2. No officers ever decline to waltz.
3. All my poultry are ducks.

Therefore, my poultry are not officers.

These statements can be rephrased as

1. If it is a duck, then it does not waltz.
2. If it is an officer, then it does not decline to waltz. (Equivalently, "If it is an officer, then it will waltz.")
3. If it is my poultry, then it is a duck.

Therefore, if it is my poultry, then it is not an officer.

© Bettmann/CORBIS

Young Alice Liddell inspired Lewis Carroll to write Alice's Adventures in Wonderland. *This photo is one of the many Carroll took of Alice.*

the story, Alice Liddell insisted that Dodgson write it down for her. He complied, initially titling it *Alice's Adventure Underground.*

Dodgson's friends subsequently encouraged him to publish the manuscript, and in 1865, after editing and inserting new episodes, Lewis Carroll gave the world *Alice's Adventures in Wonderland.* Although the book appeared to be a whimsical excursion into chaotic nonsense, Dodgson's masterpiece contained many exercises in logic and metaphor. The book was a success, and in 1871, a sequel, *Through the Looking Glass,* was printed. When asked to comment on the meaning of his writings, Dodgson replied, "I'm very much afraid I didn't mean anything but nonsense! Still, you know, words mean more than we mean to express when we use them; so a whole book ought to mean a great deal more than the writer means. So, whatever good meanings are in the book, I'm glad to accept as the meaning of the book."

In addition to writing "children's stories," Dodgson wrote numerous mathematics essays and texts, including *The Fifth Book of Euclid Proved Algebraically, Formulae of Plane Trigonometry, A Guide to the Mathematical Student,* and *Euclid and His Modern Rivals.* In the field of formal logic, Dodgson's books *The Game of Logic* (1887) and *Symbolic Logic* (1896) are still used as sources of inspiration in numerous schools worldwide.

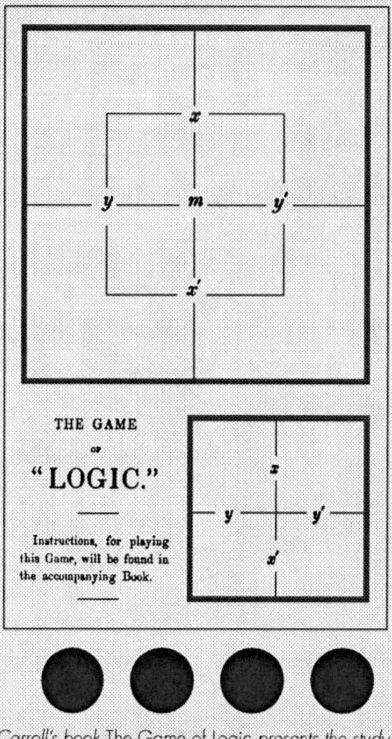

Brown University Library

Carroll's book The Game of Logic *presents the study of formalized logic in a gamelike fashion. After listing the "rules of the game" (complete with gameboard and markers), Carroll captures the reader's interest with nonsensical syllogisms.*

Now we define symbols to represent the various components of the statements:

p: It is a duck.
q: It will waltz.
r: It is an officer.
s: It is my poultry.

Symbolically, the argument has the form

1. $p \rightarrow \sim q$
2. $r \rightarrow q$
3. $s \rightarrow p$

$\therefore s \rightarrow \sim r$

							1	2	3		c	
	p	q	r	s	$\sim q$	$\sim r$	$p \to \sim q$	$r \to q$	$s \to p$	$1 \wedge 2 \wedge 3$	$s \to \sim r$	$(1 \wedge 2 \wedge 3) \to c$
1.	T	T	T	T	F	F	F	T	T	F	F	T
2.	T	T	T	F	F	F	F	T	T	F	T	T
3.	T	T	F	T	F	T	F	T	T	F	T	T
4.	T	T	F	F	F	T	F	T	T	F	T	T
5.	T	F	T	T	T	F	T	F	T	F	F	T
6.	T	F	T	F	T	F	T	F	T	F	T	T
7.	T	F	F	T	T	T	T	T	T	T	T	T
8.	T	F	F	F	T	T	T	T	T	T	T	T
9.	F	T	T	T	F	F	T	T	F	F	F	T
10.	F	T	T	F	F	F	T	T	T	T	T	T
11.	F	T	F	T	F	T	T	T	F	F	T	T
12.	F	T	F	F	F	T	T	T	T	T	T	T
13.	F	F	T	T	T	F	T	F	F	F	F	T
14.	F	F	T	F	T	F	T	F	T	F	T	T
15.	F	F	F	T	T	T	T	T	F	F	T	T
16.	F	F	F	F	T	T	T	T	T	T	T	T

FIGURE 1.67 Truth table for the argument $[(p \to \sim q) \wedge (r \to q) \wedge (s \to p)] \to (s \to \sim r)$.

Now we construct a truth table with sixteen rows ($2^4 = 16$), along with the necessary columns. The completed table is shown in Figure 1.67.

The last column of the truth table represents the argument and contains all T's. Consequently, the conditional $[(p \to \sim q) \wedge (r \to q) \wedge (s \to p)] \to (s \to \sim r)$ is a tautology; the argument is valid.

1.5 Exercises

In Exercises 1–10, use the given symbols to rewrite the argument in symbolic form.

1. p: It is raining.
q: The streets are wet. } Use these symbols.

1. If it is raining, then the streets are wet.
2. It is raining.

Therefore, the streets are wet.

2. p: I have a college degree.
q: I am lazy. } Use these symbols.

1. If I have a college degree, I am not lazy.
2. I do not have a college degree.

Therefore, I am lazy.

3. p: It is Tuesday.
q: The tour group is in Belgium. } Use these symbols.

1. If it is Tuesday, then the tour group is in Belgium.
2. The tour group is not in Belgium.

Therefore, it is not Tuesday.

4. p: You are a gambler.
q: You have financial security. } Use these symbols.

1. You do not have financial security if you are a gambler.
2. You do not have financial security.

Therefore, you are a gambler.

5. p: You exercise regularly.
q: You are healthy.
} Use these symbols.

1. You exercise regularly only if you are healthy.
2. You do not exercise regularly.

Therefore, you are not healthy.

6. p: The senator supports new taxes.
q: The senator is reelected.
} Use these symbols.

1. The senator is not reelected if she supports new taxes.
2. The senator does not support new taxes.

Therefore, the senator is reelected.

7. p: A person knows Morse code.
q: A person operates a telegraph.
r: A person is Nikola Tesla.
} Use these symbols.

1. Knowing Morse code is necessary for operating a telegraph.
2. Nikola Tesla knows Morse code.

Therefore, Nikola Tesla operates a telegraph.

HINT: Hypothesis 2 can be symbolized as $r \wedge p$.

8. p: A person knows CPR.
q: A person is a paramedic.
r: A person is David Lee Roth.
} Use these symbols.

1. Knowing CPR is necessary for being a paramedic.
2. David Lee Roth is a paramedic.

Therefore, David Lee Roth knows CPR.

HINT: Hypothesis 2 can be symbolized as $r \wedge q$.

9. p: It is a monkey.
q: It is an ape.
r: It is King Kong.
} Use these symbols.

1. Being a monkey is sufficient for not being an ape.
2. King Kong is an ape.

Therefore, King Kong is not a monkey.

10. p: It is warm-blooded.
q: It is a reptile.
r: It is Godzilla.
} Use these symbols.

1. Being warm-blooded is sufficient for not being a reptile.
2. Godzilla is not warm-blooded.

Therefore, Godzilla is a reptile.

In Exercises 11–20, use a truth table to determine the validity of the argument specified. If the argument is invalid, interpret the specific circumstances that cause it to be invalid.

11. the argument in Exercise 1
12. the argument in Exercise 2
13. the argument in Exercise 3
14. the argument in Exercise 4
15. the argument in Exercise 5
16. the argument in Exercise 6
17. the argument in Exercise 7
18. the argument in Exercise 8
19. the argument in Exercise 9
20. the argument in Exercise 10

In Exercises 21–42, define the necessary symbols, rewrite the argument in symbolic form, and use a truth table to determine whether the argument is valid. If the argument is invalid, interpret the specific circumstances that cause the argument to be invalid.

21. 1. If the Democrats have a majority, Smith is appointed and student loans are funded.
2. Smith is appointed or student loans are not funded.

Therefore, the Democrats do not have a majority.

22. 1. If you watch television, you do not read books.
2. If you read books, you are wise.

Therefore, you are not wise if you watch television.

23. 1. If you argue with a police officer, you get a ticket.
2. If you do not break the speed limit, you do not get a ticket.

Therefore, if you break the speed limit, you argue with a police officer.

24. 1. If you do not recycle newspapers, you are not an environmentalist.
2. If you recycle newspapers, you save trees.

Therefore, you are an environmentalist only if you save trees.

25. 1. All pesticides are harmful to the environment.
2. No fertilizer is a pesticide.

Therefore, no fertilizer is harmful to the environment.

26. 1. No one who can afford health insurance is unemployed.
2. All politicians can afford health insurance.

Therefore, no politician is unemployed.

27. 1. All poets are loners.
2. All loners are taxi drivers.

Therefore, all poets are taxi drivers.

28. 1. All forest rangers are environmentalists.
2. All forest rangers are storytellers.

Therefore, all environmentalists are storytellers.

29. 1. No professor is a millionaire.
2. No millionaire is illiterate.

Therefore, no professor is illiterate.

30. 1. No artist is a lawyer.
2. No lawyer is a musician.

Therefore, no artist is a musician.

31. 1. All lawyers study logic.
2. You study logic only if you are a scholar.
3. You are not a scholar.

Therefore, you are not a lawyer.

32. 1. All licensed drivers have insurance.
2. You obey the law if you have insurance.
3. You obey the law.

Therefore, you are a licensed driver.

33. 1. Drinking espresso is sufficient for not sleeping.
2. Not eating dessert is necessary for being on a diet.
3. Not eating dessert is sufficient for drinking espresso.

Therefore, not being on a diet is necessary for sleeping.

34. 1. Not being eligible to vote is sufficient for ignoring politics.
2. Not being a convicted felon is necessary for being eligible to vote.
3. Ignoring politics is sufficient for being naive.

Therefore, being naive is necessary being a convicted felon.

35. If the defendant is innocent, he does not go to jail. The defendant goes to jail. Therefore, the defendant is guilty.

36. If the defendant is innocent, he does not go to jail. The defendant is guilty. Therefore, the defendant goes to jail.

37. If you are not in a hurry, you eat at Lulu's Diner. If you are in a hurry, you do not eat good food. You eat at Lulu's. Therefore, you eat good food.

38. If you give me a hamburger today, I pay you tomorrow. If you are a sensitive person, you give me a hamburger today. You are not a sensitive person. Therefore, I do not pay you tomorrow.

39. If you listen to rock and roll, you do not go to heaven. If you are a moral person, you go to heaven. Therefore, you are not a moral person if you listen to rock and roll.

40. If you follow the rules, you have no trouble. If you are not clever, you have trouble. You are clever. Therefore, you do not follow the rules.

41. The water not being cold is sufficient for going swimming. Having goggles is necessary for going swimming. I have no goggles. Therefore, the water is cold.

42. I wash my car only if the sky is clear. The sky not being clear is necessary for it to rain. I do not wash my car. Therefore, it is raining.

The arguments given in Exercises 43–50 were written by Lewis Carroll and appeared in his 1896 book Symbolic Logic. *For each argument, define the necessary symbols, rewrite the argument in symbolic form, and use a truth table to determine whether the argument is valid.*

43. 1. All medicine is nasty.
2. Senna is a medicine.

Therefore, senna is nasty.

NOTE: Senna is a laxative extracted from the dried leaves of cassia plants.

44. 1. All pigs are fat.
2. Nothing that is fed on barley-water is fat.

Therefore, pigs are not fed on barley-water.

45. 1. Nothing intelligible ever puzzles me.
2. Logic puzzles me.

Therefore, logic is unintelligible.

46. 1. No misers are unselfish.
2. None but misers save eggshells.

Therefore, no unselfish people save eggshells.

47. 1. No Frenchmen like plum pudding.
2. All Englishmen like plum pudding.

Therefore, Englishmen are not Frenchmen.

48. 1. A prudent man shuns hyenas.
2. No banker is imprudent.

Therefore, no banker fails to shun hyenas.

49. 1. All wasps are unfriendly.
2. No puppies are unfriendly.

Therefore, puppies are not wasps.

50. 1. Improbable stories are not easily believed.
2. None of his stories are probable.

Therefore, none of his stories are easily believed.

Answer the following questions using complete sentences and your own words.

• CONCEPT QUESTIONS

51. What is a tautology?

52. What is the conditional representation of an argument?

53. Find a "logical" argument in a newspaper article, an advertisement, or elsewhere in the media. Analyze that argument and discuss the implications.

• HISTORY QUESTIONS

54. What was Charles Dodgson's pseudonym? How did he get it? What classic "children's stories" did he write?

55. What did Charles Dodgson contribute to the study of formal logic?

56. Charles Dodgson was a pioneer in what artistic field?

57. Who was Alice Liddell?

WEB PROJECT

58. Write a research paper on any historical topic referred to in this chapter or a related topic. Below is a partial list of topics.
- Aristotle
- George Boole
- Augustus De Morgan
- Charles Dodgson/Lewis Carroll
- Gottfried Wilhelm Leibniz

Some useful links for this web project are listed on the text web site: **www.cengage.com/math/johnson**

1 CHAPTER REVIEW

TERMS

argument
biconditional
compound statement
conclusion
conditional
conjunction
contrapositive
converse
deductive reasoning
disjunction
equivalent expressions
exclusive *or*
hypothesis
implication
inclusive *or*
inductive reasoning
invalid argument
inverse
logic
necessary
negation
premise
quantifier
reasoning
statement
sudoku
sufficient
syllogism
tautology
truth table
truth value
valid argument
Venn diagram

REVIEW EXERCISES

1. Classify each argument as deductive or inductive.

a. 1. Hitchcock's "Psycho" is a suspenseful movie.
2. Hitchcock's "The Birds" is a suspenseful movie.
Therefore, all Hitchcock movies are suspenseful.

b. 1. All Hitchcock movies are suspenseful.
2. "Psycho" is a Hitchcock movie.
Therefore, "Psycho" is suspenseful.

2. Explain the general rule or pattern used to assign the given letter to the given word. Fill in the blank with the letter that fits the pattern.

day	morning	afternoon	dusk	night
y	r	f	s	____

3. Fill in the blank with what is most likely to be the next number. Explain the pattern generated by your answer.
1, 6, 11, 4, ____

In Exercises 4–9, construct a Venn diagram to determine the validity of the given argument.

4. 1. All truck drivers are union members.
2. Rocky is a truck driver.
Therefore, Rocky is a union member.

5. 1. All truck drivers are union members.
2. Rocky is not a truck driver.
Therefore, Rocky is not a union member.

6. 1. All mechanics are engineers.
2. Casey Jones is an engineer.
Therefore, Casey Jones is a mechanic.

7. 1. All mechanics are engineers.
2. Casey Jones is not an engineer.
Therefore, Casey Jones is not a mechanic.

8. 1. Some animals are dangerous.
2. A gun is not an animal.
Therefore, a gun is not dangerous.

9. 1. Some contractors are electricians.
2. All contractors are carpenters.
Therefore, some electricians are carpenters.

10. Solve the following sudoku puzzle.

	2					3	4	
			3		6			7
			7			9		8
		4			1	7		
5			4	7	8			3
		1	6			2		
2		6			4			
3			5		7			
	7	8					3	

11. Explain why each of the following is or is not a statement.

a. The Golden Gate Bridge spans Chesapeake Bay.
b. The capital of Delaware is Dover.
c. Where are you spending your vacation?
d. Hawaii is the best place to spend a vacation.

12. Determine which pairs of statements are negations of each other.

a. All of the lawyers are ethical.
b. Some of the lawyers are ethical.
c. None of the lawyers is ethical.
d. Some of the lawyers are not ethical.

13. Write a sentence that represents the negation of each statement.
 a. His car is not new.
 b. Some buildings are earthquake proof.
 c. All children eat candy.
 d. I never cry in a movie theater.

14. Using the symbolic representations

 p: The television program is educational.

 q: The television program is controversial.

 express the following compound statements in symbolic form.
 a. The television program is educational and controversial.
 b. If the television program isn't controversial, it isn't educational.
 c. The television program is educational and it isn't controversial.
 d. The television program isn't educational or controversial.
 e. Not being controversial is necessary for a television program to be educational.
 f. Being controversial is sufficient for a television program not to be educational.

15. Using the symbolic representations

 p: The advertisement is effective.

 q: The advertisement is misleading.

 r: The advertisement is outdated.

 express the following compound statements in symbolic form.
 a. All misleading advertisements are effective.
 b. It is a current, honest, effective advertisement.
 c. If an advertisement is outdated, it isn't effective.
 d. The advertisement is effective and it isn't misleading or outdated.
 e. Not being outdated or misleading is necessary for an advertisement to be effective.
 f. Being outdated and misleading is sufficient for an advertisement not to be effective.

16. Using the symbolic representations

 p: It is expensive.

 q: It is undesirable.

 express the following in words.
 a. $p \to \sim q$
 b. $q \leftrightarrow \sim p$
 c. $\sim(p \vee q)$
 d. $(p \wedge \sim q) \vee (\sim p \wedge q)$

17. Using the symbolic representations

 p: The movie is critically acclaimed.

 q: The movie is a box office hit.

 r: The movie is available on DVD.

 express the following in words.
 a. $(p \vee q) \to r$
 b. $(p \wedge \sim q) \to \sim r$
 c. $\sim(p \vee q) \wedge r$
 d. $\sim r \to (\sim p \wedge \sim q)$

In Exercises 18–25, construct a truth table for the compound statement.

18. $p \vee \sim q$
19. $p \wedge \sim q$
20. $\sim p \to q$
21. $(p \wedge q) \to \sim q$
22. $q \vee \sim(p \vee r)$
23. $\sim p \to (q \vee r)$
24. $(q \wedge p) \to (\sim r \vee p)$
25. $(p \vee r) \to (q \wedge \sim r)$

In Exercises 26–30, construct a truth table to determine whether the statements in each pair are equivalent.

26. The car is unreliable or expensive.
 If the car is reliable, then it is expensive.
27. If I get a raise, I will buy a new car.
 If I do not get a raise, I will not buy a new car.
28. She is a Democrat or she did not vote.
 She is not a Democrat and she did vote.
29. The raise is not unjustified and the management opposes it.
 It is not the case that the raise is unjustified or the management does not oppose it.
30. Walking on the beach is sufficient for not wearing shoes. Wearing shoes is necessary for not walking on the beach.

In Exercises 31–38, write a sentence that represents the negation of each statement.

31. Jesse had a party and nobody came.
32. You do not go to jail if you pay the fine.
33. I am the winner or you are blind.
34. He is unemployed and he did not apply for financial assistance.
35. The selection procedure has been violated if his application is ignored.
36. The jackpot is at least $1 million.
37. Drinking espresso is sufficient for not sleeping.
38. Not eating dessert is necessary for being on a diet.
39. Given the statements

 p: You are an avid jogger.

 q: You are healthy.

 write the sentence represented by each of the following.
 a. $p \to q$
 b. $q \to p$
 c. $\sim p \to \sim q$
 d. $\sim q \to \sim p$
 e. $p \leftrightarrow q$

40. Form (a) the inverse, (b) the converse, and (c) the contrapositive of the conditional "If he is elected, the country is in big trouble."

In Exercises 41 and 42, express the contrapositive of the given conditional in terms of (a) a sufficient condition, and (b) a necessary condition.

41. Having a map is sufficient for not being lost.
42. Having syrup is necessary for eating pancakes.

In Exercises 43–48, (a) determine the premise and conclusion and (b) rewrite the compound statement in the standard "if . . . then . . ." form.

43. The economy improves only if unemployment goes down.

44. The economy improves if unemployment goes down.

45. No computer is unrepairable.

46. All gemstones are valuable.

47. Being the fourth Thursday in November is sufficient for the U.S. Post Office to be closed.

48. Having diesel fuel is necessary for the vehicle to operate.

In Exercises 49 and 50, translate the two statements into symbolic form and use truth tables to determine whether the statements are equivalent.

49. If you are allergic to dairy products, you cannot eat cheese.

If you cannot eat cheese, then you are allergic to dairy products.

50. You are a fool if you listen to me.

You are not a fool only if you do not listen to me.

51. Which pairs of statements are equivalent?

i. If it is not raining, I ride my bicycle to work.

ii. If I ride my bicycle to work, it is not raining.

iii. If I do not ride my bicycle to work, it is raining.

iv. If it is raining, I do not ride my bicycle to work.

In Exercises 52–57, define the necessary symbols, rewrite the argument in symbolic form, and use a truth table to determine whether the argument is valid.

52. 1. If you do not make your loan payment, your car is repossessed.
2. Your car is repossessed.

Therefore, you did not make your loan payment.

53. 1. If you do not pay attention, you do not learn the new method.
2. You do learn the new method.

Therefore, you do pay attention.

54. 1. If you rent DVD, you will not go to the movie theater.
2. If you go to the movie theater, you pay attention to the movie.

Therefore, you do not pay attention to the movie if you rent DVDs.

55. 1. If the Republicans have a majority, Farnsworth is appointed and no new taxes are imposed.
2. New taxes are imposed.

Therefore, the Republicans do not have a majority or Farnsworth is not appointed.

56. 1. Practicing is sufficient for making no mistakes.
2. Making a mistake is necessary for not receiving an award.
3. You receive an award.

Therefore, you practice.

57. 1. Practicing is sufficient for making no mistakes.
2. Making a mistake is necessary for not receiving an award.
3. You do not receive an award.

Therefore, you do not practice.

In Exercises 58–66, define the necessary symbols, rewrite the argument in symbolic form, and use a truth table to determine whether the argument is valid.

58. If the defendant is guilty, he goes to jail. The defendant does not go to jail. Therefore, the defendant is not guilty.

59. I will go to the concert only if you buy me a ticket. You bought me a ticket. Therefore, I will go to the concert.

60. If tuition is raised, students take out loans or drop out. If students do not take out loans, they drop out. Students do drop out. Therefore, tuition is raised.

61. If our oil supply is cut off, our economy collapses. If we go to war, our economy doesn't collapse. Therefore, if our oil supply isn't cut off, we do not go to war.

62. No professor is uneducated. No monkey is educated. Therefore, no professor is a monkey.

63. No professor is uneducated. No monkey is a professor. Therefore, no monkey is educated.

64. Vehicles stop if the traffic light is red. There is no accident if vehicles stop. There is an accident. Therefore, the traffic light is not red.

65. Not investing money in the stock market is necessary for invested money to be guaranteed. Invested money not being guaranteed is sufficient for not retiring at an early age. Therefore, you retire at an early age only if your money is not invested in the stock market.

66. Not investing money in the stock market is necessary for invested money to be guaranteed. Invested money not being guaranteed is sufficient for not retiring at an early age. You do not invest in the stock market. Therefore, you retire at an early age.

Determine the validity of the arguments in Exercises 67 and 68 by constructing a

a. *Venn diagram and a*

b. *truth table.*

c. *How do the answers to parts (a) and (b) compare? Why?*

67. 1. If you own a hybrid vehicle, then you are an environmentalist.
2. You are not an environmentalist.

Therefore, you do not own a hybrid vehicle.

68. 1. If you own a hybrid vehicle, then you are an environmentalist.
2. You are an environmentalist.

Therefore, you own a hybrid vehicle.

Answer the following questions using complete sentences and your own words.

• CONCEPT QUESTIONS

69. What is a statement?

70. **a.** What is a disjunction? Under what conditions is a disjunction true?
b. What is a conjunction? Under what conditions is a conjunction true?
c. What is a conditional? Under what conditions is a conditional true?
d. What is a negation? Under what conditions is a negation true?

71. **a.** What is a sufficient condition?
b. What is a necessary condition?

72. What is a tautology?

73. When constructing a truth table, how do you determine how many rows to create?

• HISTORY QUESTIONS

74. What role did the following people play in the development of formalized logic?
- Aristotle
- George Boole
- Augustus De Morgan
- Charles Dodgson
- Gottfried Wilhelm Leibniz

SETS AND COUNTING

2

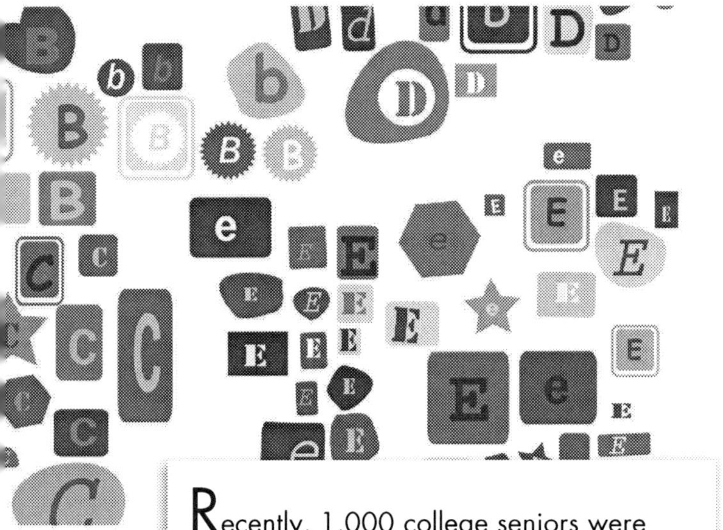

Recently, 1,000 college seniors were asked whether they favored increasing their state's gasoline tax to generate funds to improve highways and whether they favored increasing their state's alcohol tax to generate funds to improve the public education system. The responses were tallied, and the following results were printed in the campus newspaper: 746 students favored an increase in the gasoline tax, 602 favored an increase in the alcohol tax, and 449 favored increase in both taxes. How many of these 1,000 students favored an increase in at least one of the taxes? How many favored increasing only the gasoline tax? How many favored increasing only the alcohol tax? How many favored increasing neither tax?

The mathematical tool that was designed to answer questions like these is

continued

WHAT WE WILL DO IN THIS CHAPTER

WE'LL USE VENN DIAGRAMS TO DEPICT THE RELATIONSHIPS BETWEEN SETS:

- One set might be contained within another set.
- Two or more sets might, or might not, share elements in common.

WE'LL EXPLORE APPLICATIONS OF VENN DIAGRAMS:

- The results of consumer surveys, marketing analyses, and political polls can be analyzed by using Venn diagrams.
- Venn diagrams can be used to prove general formulas related to set theory.

WE'LL EXPLORE VARIOUS METHODS OF COUNTING:

- A fundamental principle of counting is used to determine the total number of possible ways of selecting specified items. For example, how many different student ID numbers are possible at your school?

continued

What We Will Do in This Chapter — *continued*

- In selecting items from a specified group, sometimes the order in which the items are selected matters (the awarding of prizes: first, second, and third), and sometimes it does not (selecting numbers in a lottery or people for a committee). How does this affect your method of counting?

WE'LL USE SETS IN VARIOUS CONTEXTS:

- In this text, we will use set theory extensively in Chapter 3 on probability.
- Many standardized admissions tests, such as the Graduate Record Exam (GRE) and the Law School Admissions Test (LSAT), ask questions that can be answered with set theory.

WE'LL EXPLORE SETS THAT HAVE AN INFINITE NUMBER OF ELEMENTS:

- One-to-one correspondences are used to "count" and compare the number of elements in infinite sets.
- Not all infinite sets have the same number of elements; some infinite sets are countable, and some are not.

the *set. Webster's New World College Dictionary* defines a **set** as "a prescribed collection of points, numbers, or other objects that satisfy a given condition." Although you might be able to answer the questions about taxes without any formal knowledge of sets, the mental reasoning involved in obtaining your answers uses some of the basic principles of sets. (Incidentally, the answers to the above questions are 899, 297, 153, and 101, respectively.)

The branch of mathematics that deals with sets is called **set theory.** Set theory can be helpful in solving both mathematical and nonmathematical problems. We will explore set theory in the first half of this chapter. As the above example shows, set theory often involves the analysis of the relationships between sets and counting the number of elements in a specific category. Consequently, various methods of counting, collectively known as **combinatorics,** will be developed and discussed in the second half of this chapter. Finally, what if a set has too many elements to count by using finite numbers? For example, how many integers are there? How many real numbers? The chapter concludes with an exploration of infinite sets and various "levels of infinity."

2.1 Sets and Set Operations

OBJECTIVES

- Learn the basic vocabulary and notation of set theory
- Learn and apply the union, intersection, and complement operations
- Draw Venn diagrams

A **set** is a collection of objects or things. The objects or things in the set are called **elements** (or *members*) of the set. In our example above, we could talk about the *set* of students who favor increasing only the gasoline tax or the *set* of students who do not favor increasing either tax. In geography, we can talk about the *set* of all state capitals or the *set* of all states west of the Mississippi. It is easy to determine whether something is in these sets; for example, Des Moines is an element of the set of state capitals, whereas Dallas is not. Such sets are called **well-defined** because there is a way of determining for sure whether a particular item is an element of the set.

EXAMPLE 1

DETERMINING WELL-DEFINED SETS Which of the following sets are well-defined?

a. the set of all movies directed by Alfred Hitchcock
b. the set of all great rock-and-roll bands
c. the set of all possible two-person committees selected from a group of five people

SOLUTION

a. This set is well-defined; either a movie was directed by Hitchcock, or it was not.
b. This set is *not* well-defined; membership is a matter of opinion. Some people would say that the Ramones (one of the pioneer punk bands of the late 1970s) are a member, while others might say they are not. (Note: The Ramones were inducted into the Rock and Roll Hall of Fame in 2002.)
c. This set is well-defined; either the two people are from the group of five, or they are not.

Notation

By tradition, a set is denoted by a capital letter, frequently one that will serve as a reminder of the contents of the set. **Roster notation** (also called *listing notation*) is a method of describing a set by listing each element of the set inside the symbols { and }, which are called *set braces.* In a listing of the elements of a set, each distinct element is listed only once, and the order of the elements doesn't matter.

The symbol $\in$ stands for the phrase *is an element of,* and $\notin$ stands for *is not an element of.* The **cardinal number** of a set A is the number of elements in the set and is denoted by $n(A)$. Thus, if R is the set of all letters in the name "Ramones," then $R = \{r, a, m, o, n, e, s\}$. Notice that m is an element of the set R, x is not an element of R, and R has 7 elements. In symbols, $m \in R$, $x \notin R$, and $n(R) = 7$.

Roberta Bayley/Evening Standard/Hulton Archive/Getty Images

The "Ramones" or The "Moaners"? The set R of all letters in the name "Ramones" is the same as the set M of all letters in the name "Moaners." Consequently, the sets are equal; $M = R = \{a, e, m, n, o, r, s\}$. (R.I.P. Joey Ramone 1951–2001, Dee Dee Ramone 1952–2002, Johnny Ramone 1948–2004.)

Two sets are **equal** if they contain exactly the same elements. *The order in which the elements are listed does not matter.* If M is the set of all letters in the name "Moaners," then $M = \{m, o, a, n, e, r, s\}$. This set contains exactly the same elements as the set R of letters in the name "Ramones." Therefore, $M = R = \{a, e, m, n, o, r, s\}$.

Often, it is not appropriate or not possible to describe a set in roster notation. For extremely large sets, such as the set V of all registered voters in Detroit, or for sets that contain an infinite number of elements, such as the set G of all negative real numbers, the roster method would be either too cumbersome or impossible to use. Although V could be expressed via the roster method (since each county compiles a list of all registered voters in its jurisdiction), it would take hundreds or even thousands of pages to list everyone who is registered to vote in Detroit! In the case of the set G of all negative real numbers, no list, no matter how long, is capable of listing all members of the set; there is an infinite number of negative numbers.

In such cases, it is often necessary, or at least more convenient, to use **set-builder notation,** which lists the rules that determine whether an object is an element of the set rather than the actual elements. A set-builder description of set G above is

$$G = \{x \mid x < 0 \quad \text{and} \quad x \in \Re\}$$

which is read as "the set of all x such that x is less than zero and x is a real number." A set-builder description of set V above is

$$V = \{\text{persons} \mid \text{the person is a registered voter in Detroit}\}$$

which is read as "the set of all persons such that the person is a registered voter in Detroit." In set-builder notation, the vertical line stands for the phrase "such that."

Whatever is on the left side of the line is the general type of thing in the set, while the rules about set membership are listed on the right.

EXAMPLE 2

READING SET-BUILDER NOTATION Describe each of the following in words.

a. $\{x \mid x > 0 \text{ and } x \in \Re\}$
b. {persons | the person is a living former U.S. president}
c. {women | the woman is a former U.S. president}

SOLUTION

a. the set of all x such that x is a positive real number
b. the set of all people such that the person is a living former U.S. president
c. the set of all women such that the woman is a former U.S. president

The set listed in part (c) of Example 2 has no elements; there are no women who are former U.S. presidents. If we let W equal "the set of all women such that the woman is a former U.S. president," then $n(W) = 0$. A set that has no elements is called an **empty set** and is denoted by $\varnothing$ or by $\{\}$. Notice that since the empty set has no elements, $n(\varnothing) = 0$. In contrast, the set $\{0\}$ is not empty; it has one element, the number zero, so $n(\{0\}) = 1$.

Universal Set and Subsets

When we work with sets, we must define a universal set. For any given problem, the **universal set,** denoted by U, is the set of all possible elements of any set used in the problem. For example, when we spell words, U is the set of all letters in the alphabet. When every element of one set is also a member of another set, we say that the first set is a *subset* of the second; for instance, {p, i, n} is a subset of {p, i, n, e}. In general, we say that A is a **subset** of B, denoted by $A \subseteq B$, if for every $x \in A$ it follows that $x \in B$. Alternatively, $A \subseteq B$ if A contains no elements that are not in B. If A contains an element that is not in B, then A is not a subset of B (symbolized as $A \not\subseteq B$).

EXAMPLE 3

DETERMINING SUBSETS Let B = {countries | the country has a permanent seat on the U.N. Security Council}. Determine whether A is a subset of B.

a. A = {Russian Federation, United States}
b. A = {China, Japan}
c. A = {United States, France, China, United Kingdom, Russian Federation}
d. $A = \{\ \}$

SOLUTION

We use the roster method to list the elements of set B.

B = {China, France, Russian Federation, United Kingdom, United States}

a. Since every element of A is also an element of B, A is a subset of B; $A \subseteq B$.
b. Since A contains an element (Japan) that is not in B, A is not a subset of B; $A \not\subseteq B$.
c. Since every element of A is also an element of B (note that $A = B$), A is a subset of B (and B is a subset of A); $A \subseteq B$ (and $B \subseteq A$). In general, every set is a subset of itself; $A \subseteq A$ for any set A.
d. Does A contain an element that is not in B? No! Therefore, A (an empty set) is a subset of B; $A \subseteq B$. In general, the empty set is a subset of all sets; $\varnothing \subseteq A$ for any set A.

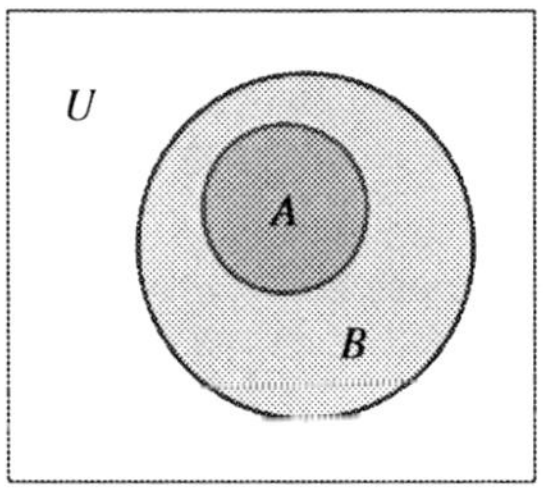

FIGURE 2.1
A is a subset of B. $A \subseteq B$.

We can express the relationship $A \subseteq B$ visually by drawing a Venn diagram, as shown in Figure 2.1. A **Venn diagram** consists of a rectangle, representing the universal set, and various closed figures within the rectangle, each representing a set. Recall that Venn diagrams were used in Section 1.1 to determine whether an argument was valid.

If two sets are equal, they contain exactly the same elements. It then follows that each is a subset of the other. For example, if $A = B$, then every element of A is an element of B (and vice versa). In this case, A is called an **improper subset** of B. (Likewise, B is an improper subset of A.) Every set is an improper subset of itself; for example, $A \subseteq A$. On the other hand, if A is a subset of B and B contains an element not in A (that is, $A \neq B$), then A is called a **proper subset** of B. To indicate a proper subset, the symbol $\subset$ is used. While it is acceptable to write $\{1, 2\} \subseteq \{1, 2, 3\}$, the relationship of a proper subset is stressed when it is written $\{1, 2\} \subset \{1, 2, 3\}$. Notice the similarities between the subset symbols, $\subset$ and $\subseteq$, and the inequality symbols, $<$ and $\leq$, used in algebra; it is acceptable to write $1 \leq 3$, but writing $1 < 3$ is more informative.

Intersection of Sets

Sometimes an element of one set is also an element of another set; that is, the sets may overlap. This overlap is called the **intersection** of the sets. If an element is in two sets *at the same time,* it is in the intersection of the sets.

INTERSECTION OF SETS

The **intersection** of set A and set B, denoted by $A \cap B$, is

$$A \cap B = \{x \mid x \in A \quad \text{and} \quad x \in B\}$$

The intersection of two sets consists of those elements that are common to both sets.

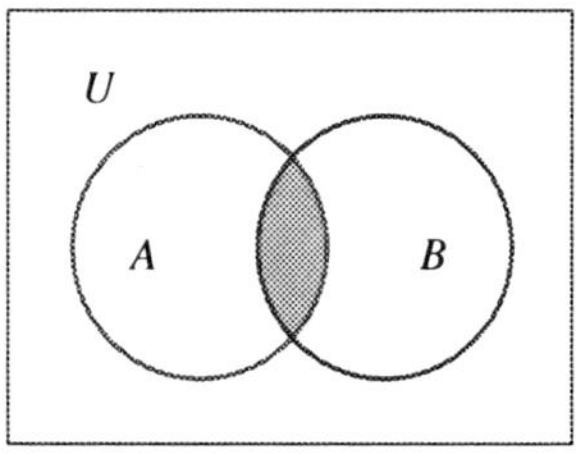

FIGURE 2.2
The intersection $A \cap B$ is represented by the (overlapping) shaded region.

For example, given the sets $A = \{\text{Buffy, Spike, Willow, Xander}\}$ and $B = \{\text{Angel, Anya, Buffy, Giles, Spike}\}$, their intersection is $A \cap B = \{\text{Buffy, Spike}\}$.

Venn diagrams are useful in depicting the relationship between sets. The Venn diagram in Figure 2.2 illustrates the intersection of two sets; the shaded region represents $A \cap B$.

Mutually Exclusive Sets

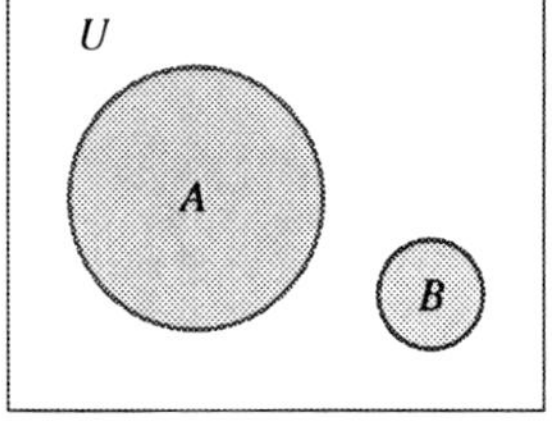

FIGURE 2.3
Mutually exclusive sets have no elements in common ($A \cap B = \varnothing$).

Sometimes a pair of sets has no overlap. Consider an ordinary deck of playing cards. Let $D = \{\text{cards} \mid \text{the card is a diamond}\}$ and $S = \{\text{cards} \mid \text{the card is a spade}\}$. Certainly, *no* cards are both diamonds and spades *at the same time;* that is, $S \cap D = \varnothing$.

Two sets A and B are **mutually exclusive** (or *disjoint*) if they have no elements in common, that is, if $A \cap B = \varnothing$. The Venn diagram in Figure 2.3 illustrates mutually exclusive sets.

Union of Sets

What does it mean when we ask, "How many of the 500 college students in a transportation survey own an automobile or a motorcycle?" Does it mean "How many students own either an automobile or a motorcycle *or both*?" or does it mean "How many students own either an automobile or a motorcycle, *but not both*?" The former is called the *inclusive or,* because it includes the possibility of owning both; the latter is called the *exclusive or.* In logic and in mathematics, the word *or* refers to the *inclusive or,* unless you are told otherwise.

The meaning of the word *or* is important to the concept of union. The **union** of two sets is a new set formed by joining those two sets together, just as the union of the states is the joining together of fifty states to form one nation.

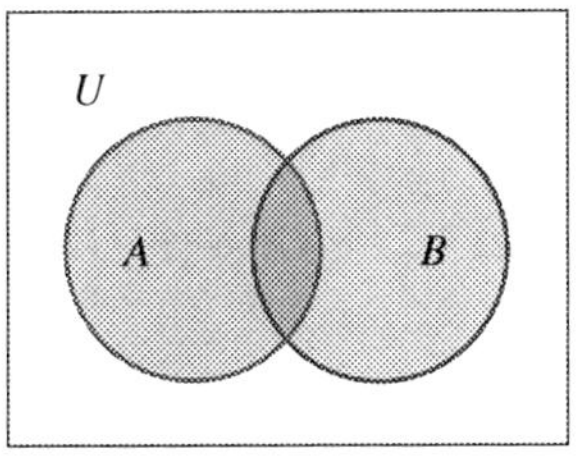

FIGURE 2.4

The union $A \cup B$ is represented by the (entire) shaded region.

UNION OF SETS

The **union** of set A and set B, denoted by $A \cup B$, is

$$A \cup B = \{x \mid x \in A \quad \text{or} \quad x \in B\}$$

The union of A and B consists of all elements that are in either A or B or both, that is, all elements that are in at least one of the sets.

For example, given the sets A = {Conan, David} and B = {Ellen, Katie, Oprah}, their union is $A \cup B$ = {Conan, David, Ellen, Katie, Oprah}, and their intersection is $A \cap B = \emptyset$. Note that because they have no elements in common, A and B are mutually exclusive sets. The Venn diagram in Figure 2.4 illustrates the union of two sets; the entire shaded region represents $A \cup B$.

EXAMPLE 4

FINDING THE INTERSECTION AND UNION OF SETS Given the sets $A = \{1, 2, 3\}$ and $B = \{2, 4, 6\}$, find the following.

a. $A \cap B$ (the intersection of A and B)
b. $A \cup B$ (the union of A and B)

SOLUTION

a. The intersection of two sets consists of those elements that are common to both sets; therefore, we have

$$\begin{aligned} A \cap B &= \{1, 2, 3\} \cap \{2, 4, 6\} \\ &= \{2\} \end{aligned}$$

b. The union of two sets consists of all elements that are in at least one of the sets; therefore, we have

$$\begin{aligned} A \cup B &= \{1, 2, 3\} \cup \{2, 4, 6\} \\ &= \{1, 2, 3, 4, 6\} \end{aligned}$$

The Venn diagram in Figure 2.5 shows the composition of each set and illustrates the intersection and union of the two sets.

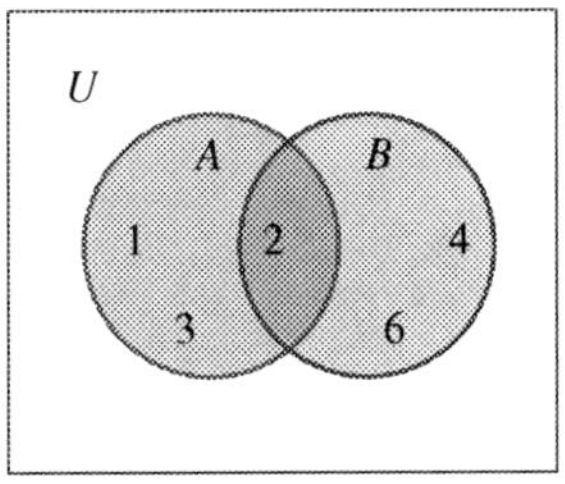

FIGURE 2.5

The composition of sets A and B in Example 4.

Because $A \cup B$ consists of all elements that are in A or B (or both), to find $n(A \cup B)$, we add $n(A)$ plus $n(B)$. However, doing so results in an answer that might be too big; that is, if A and B have elements in common, these elements will be counted twice (once as a part of A and once as a part of B). Therefore, to find the cardinal number of $A \cup B$, we add the cardinal number of A to the cardinal number of B and then *subtract* the cardinal number of $A \cap B$ (so that the overlap is not counted twice).

CARDINAL NUMBER FORMULA FOR THE UNION/INTERSECTION OF SETS

For any two sets A and B, the number of elements in their union is $n(A \cup B)$, where

$$n(A \cup B) = n(A) + n(B) - n(A \cap B)$$

and $n(A \cap B)$ is the number of elements in their intersection.

As long as any three of the four quantities in the general formula are known, the missing quantity can be found by algebraic manipulation.

EXAMPLE 5

ANALYZING THE COMPOSITION OF A UNIVERSAL SET Given $n(U) = 169$, $n(A) = 81$, and $n(B) = 66$, find the following.

a. If $n(A \cap B) = 47$, find $n(A \cup B)$ and draw a Venn diagram depicting the composition of the universal set.

b. If $n(A \cup B) = 147$, find $n(A \cap B)$ and draw a Venn diagram depicting the composition of the universal set.

SOLUTION

a. We must use the Union/Intersection Formula. Substituting the three given quantities, we have

$$\begin{aligned} n(A \cup B) &= n(A) + n(B) - n(A \cap B) \\ &= 81 + 66 - 47 \\ &= 100 \end{aligned}$$

The Venn diagram in Figure 2.6 illustrates the composition of U.

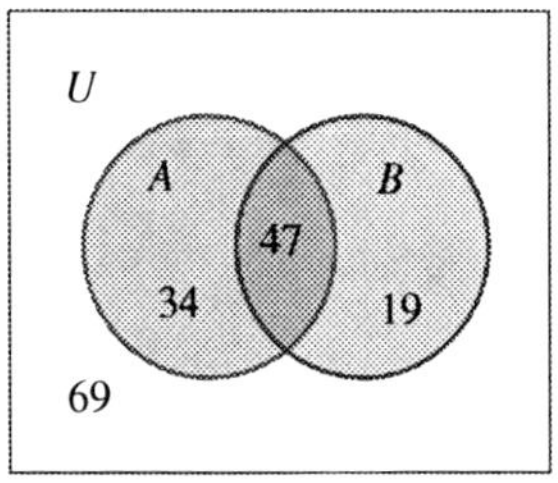

FIGURE 2.6 $n(A \cap B) = 47$.

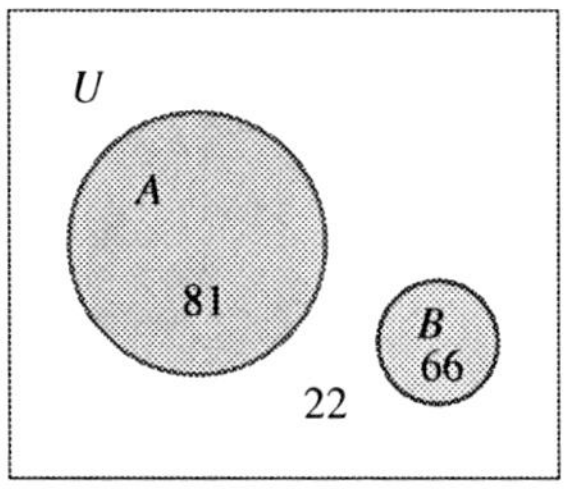

FIGURE 2.7 $n(A \cup B) = 147$.

b. We must use the Union/Intersection Formula. Substituting the three given quantities, we have

$$\begin{aligned} n(A \cup B) &= n(A) + n(B) - n(A \cap B) \\ 147 &= 81 + 66 - n(A \cap B) \\ 147 &= 147 - n(A \cap B) \\ n(A \cap B) &= 147 - 147 \\ n(A \cap B) &= 0 \end{aligned}$$

Therefore, A and B have no elements in common; they are mutually exclusive. The Venn diagram in Figure 2.7 illustrates the composition of U.

EXAMPLE **6**

ANALYZING THE RESULTS OF A SURVEY A recent transportation survey of 500 college students (the universal set U) yielded the following information: 291 students own an automobile (A), 179 own a motorcycle (M), and 85 own both an automobile and a motorcycle ($A \cap M$). What percent of these students own an automobile or a motorcycle?

SOLUTION

Recall that "automobile or motorcycle" means "automobile or motorcycle or both" (the inclusive *or*) and that *or* implies union. Hence, we must find $n(A \cup M)$, the cardinal number of the union of sets A and M. We are given that $n(A) = 291$, $n(M) = 179$, and $n(A \cap M) = 85$. Substituting the given values into the Union/Intersection Formula, we have

$$\begin{aligned} n(A \cup M) &= n(A) + n(M) - n(A \cap M) \\ &= 291 + 179 - 85 \\ &= 385 \end{aligned}$$

Therefore, 385 of the 500 students surveyed own an automobile or a motorcycle. Expressed as a percent, $385/500 = 0.77$; therefore, 77% of the students own an automobile or a motorcycle (or both).

Complement of a Set

In certain situations, it might be important to know how many things are *not* in a given set. For instance, when playing cards, you might want to know how many cards are not ranked lower than a five; or when taking a survey, you might want to know how many people did not vote for a specific proposition. The set of all elements in the universal set that are *not* in a specific set is called the *complement* of the set.

COMPLEMENT OF A SET

The **complement** of set A, denoted by A' (read "A prime" or "the complement of A"), is

$$A' = \{x \mid x \in U \quad \text{and} \quad x \notin A\}$$

The complement of a set consists of all elements that are in the universal set but not in the given set.

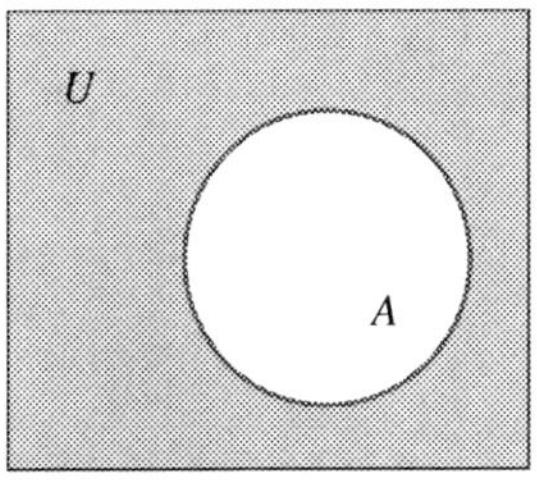

FIGURE 2.8

The complement A' is represented by the shaded region.

For example, given that $U = \{1, 2, 3, 4, 5, 6, 7, 8, 9\}$ and $A = \{1, 3, 5, 7, 9\}$, the complement of A is $A' = \{2, 4, 6, 8\}$. What is the complement of A'? Just as $-(-x) = x$ in algebra, $(A')' = A$ in set theory. The Venn diagram in Figure 2.8 illustrates the complement of set A; the shaded region represents A'.

Suppose A is a set of elements, drawn from a universal set U. If x is an element of the universal set ($x \in U$), then exactly one of the following must be true: (1) x is an element of A ($x \in A$), or (2) x is not an element of A ($x \notin A$). Since no element of the universal set can be in both A and A' at the same time, it follows that A and A' are mutually exclusive sets whose union equals the entire universal set. Therefore, the sum of the cardinal numbers of A and A' equals the cardinal number of U.

HISTORICAL NOTE

JOHN VENN, 1834–1923

Courtesy The Masters and Fellows of Gonville and Caius College, Cambridge

John Venn is considered by many to be one of the originators of modern symbolic logic. Venn received his degree in mathematics from the University at Cambridge at the age of twenty-three. He was then elected a fellow of the college and held this fellowship until his death, some 66 years later. Two years after receiving his degree, Venn accepted a teaching position at Cambridge: college lecturer in moral sciences.

During the latter half of the nineteenth century, the study of logic experienced a rebirth in England. Mathematicians were attempting to symbolize and quantify the central concepts of logical thought. Consequently, Venn chose to focus on the study of logic during his tenure at Cambridge. In addition, he investigated the field of probability and published *The Logic of Chance,* his first major work, in 1866.

Venn was well read in the works of his predecessors, including the noted logicians Augustus De Morgan, George Boole, and Charles Dodgson (a.k.a. Lewis Carroll). Boole's pioneering work on the marriage of logic and algebra proved to be a strong influence on Venn; in fact, Venn used the type of diagram that now bears his name in an 1876 paper in which he examined Boole's system of symbolic logic.

Venn was not the first scholar to use the diagrams that now bear his name. Gottfried Leibniz, Leonhard Euler, and others utilized similar diagrams years before Venn did. Examining each author's diagrams, Venn was critical of their lack of uniformity. He developed a consistent, systematic explanation of the general use of geometrical figures in the analysis of logical arguments. Today, these geometrical figures are known by his name and are used extensively in elementary set theory and logic.

Venn's writings were held in high esteem. His textbooks, *Symbolic Logic* (1881) and *The Principles of Empirical Logic* (1889), were used during the late nineteenth and early twentieth centuries. In addition to his works on logic and probability, Venn conducted much research into historical records, especially those of his college and those of his family.

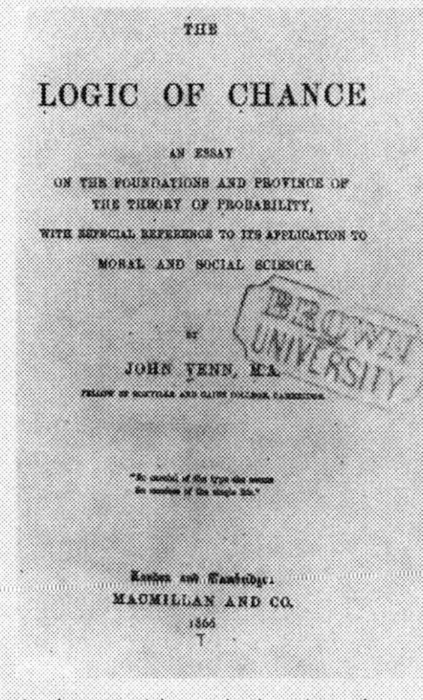

THE

LOGIC OF CHANCE

AN ESSAY

ON THE FOUNDATIONS AND PROVINCE OF

THE THEORY OF PROBABILITY,

WITH ESPECIAL REFERENCE TO ITS APPLICATION TO

MORAL AND SOCIAL SCIENCE.

BY

JOHN VENN, M.A.

London and Cambridge:

MACMILLAN AND CO.

1866

Brown University Library

Set theory and the cardinal numbers of sets are used extensively in the study of probability. Although he was a professor of logic, Venn investigated the foundations and applications of theoretical probability. Venn's first major work, The Logic of Chance, *exhibited the diversity of his academic interests.*

It is often quicker to count the elements that are *not* in a set than to count those that are. Consequently, to find the cardinal number of a set, we can subtract the cardinal number of its complement from the cardinal number of the universal set; that is, $n(A) = n(U) - n(A')$.

CARDINAL NUMBER FORMULA FOR THE COMPLEMENT OF A SET

For any set A and its complement A',

$$n(A) + n(A') = n(U)$$

where U is the universal set.

Alternatively,

$$n(A) = n(U) - n(A') \quad \text{and} \quad n(A') = n(U) - n(A)$$

EXAMPLE 7

USING THE COMPLEMENT FORMULA How many letters in the alphabet precede the letter w?

SOLUTION

Rather than counting all the letters that precede w, we will take a shortcut by counting all the letters that do *not* precede w. Let $L = \{\text{letters} \mid \text{the letter precedes w}\}$. Therefore, $L' = \{\text{letter} \mid \text{the letter does not precede w}\}$. Now $L' = \{w, x, y, z\}$, and $n(L') = 4$; therefore, we have

$$\begin{aligned} n(L) &= n(U) - n(L') \quad \text{Complement Formula} \\ &= 26 - 4 \\ &= 22 \end{aligned}$$

There are twenty-two letters preceding the letter w.

Shading Venn Diagrams

In an effort to visualize the results of operations on sets, it may be necessary to shade specific regions of a Venn diagram. The following example shows a systematic method for shading the intersection or union of any two sets.

EXAMPLE 8

SHADING VENN DIAGRAMS On a Venn diagram, shade in the region corresponding to the indicated set.

a. $A \cap B'$ **b.** $A \cup B'$

SOLUTION

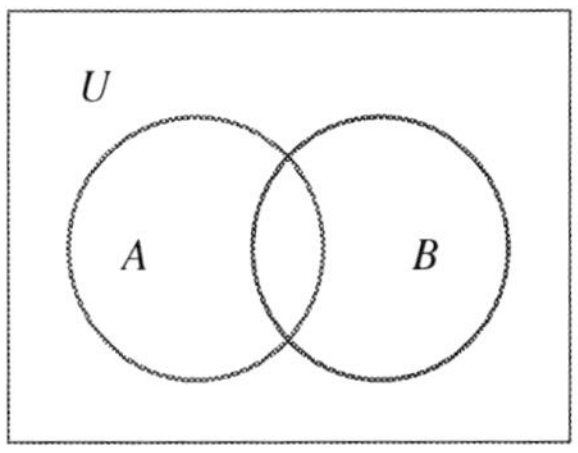

FIGURE 2.9
Two overlapping circles.

a. First, draw and label two overlapping circles as shown in Figure 2.9. The two "components" of the operation $A \cap B'$ are "A" and "B'." Shade each of these components in contrasting ways; shade one of them, say A, with horizontal lines, and the other with vertical lines as in Figure 2.10. Be sure to include a legend, or key, identifying each type of shading.

To be in the intersection of two sets, an element must be in *both* sets at the same time. Therefore, the intersection of A and B' is the region that is shaded in *both* directions (horizontal and vertical) at the same time. A final diagram depicting $A \cap B'$ is shown in Figure 2.11.

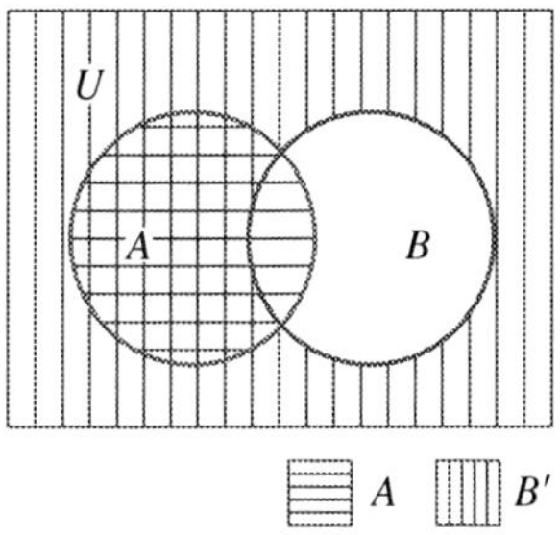

FIGURE 2.10

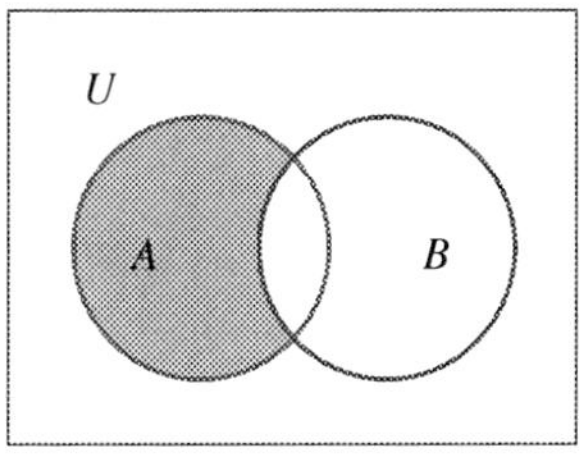

$A \cap B'$

FIGURE 2.11

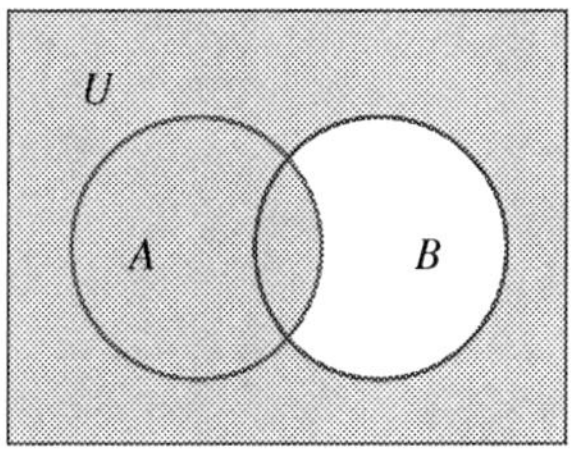

$A \cup B'$

FIGURE 2.12

b. Refer to Figure 2.10. To be in the union of two sets, an element must be in *at least one* of the sets. Therefore, the union of A and B' consists of all regions that are shaded in *any* direction whatsoever (horizontal or vertical or both). A final diagram depicting $A \cup B'$ is shown in Figure 2.12.

Set Theory and Logic

If you have read Chapter 1, you have probably noticed that set theory and logic have many similarities. For instance, the union symbol ∪ and the disjunction symbol ∨ have the same meaning, but they are used in different circumstances; ∪ goes between sets, while ∨ goes between logical expressions. The ∪ and ∨ symbols are similar in appearance because their usages are similar. A comparison of the terms and symbols used in set theory and logic is given in Figure 2.13.

Set Theory		Logic		Common Wording
Term	Symbol	Term	Symbol	
union	$\cup$	disjunction	$\vee$	or
intersection	$\cap$	conjunction	$\wedge$	and
complement	$'$	negation	$\sim$	not
subset	$\subseteq$	conditional	$\rightarrow$	if . . . then . . .

FIGURE 2.13 Comparison of terms and symbols used in set theory and logic.

Applying the concepts and symbols of Chapter 1, we can define the basic operations of set theory in terms of logical biconditionals. The biconditionals in Figure 2.14 are tautologies (expressions that are always true); the first biconditional is read as "x is an element of the union of sets A and B if and only if x is an element of set A or x is an element of set B."

Basic Operations in Set Theory	Logical Biconditional
union	$[x \in (A \cup B)] \leftrightarrow [x \in A \vee x \in B]$
intersection	$[x \in (A \cap B)] \leftrightarrow [x \in A \wedge x \in B]$
complement	$(x \in A') \leftrightarrow \sim (x \in A)$
subset	$(A \subseteq B) \leftrightarrow (x \in A \rightarrow x \in B)$

FIGURE 2.14 Set theory operations as logical biconditionals.

2.1 Exercises

1. State whether the given set is well defined.
   a. the set of all black automobiles
   b. the set of all inexpensive automobiles
   c. the set of all prime numbers
   d. the set of all large numbers

2. Suppose $A = \{2, 5, 7, 9, 13, 25, 26\}$.
   a. Find $n(A)$
   b. True or false: $7 \in A$
   c. True or false: $9 \notin A$
   d. True or false: $20 \notin A$

In Exercises 3–6, list all subsets of the given set. Identify which subsets are proper and which are improper.

3. $B = \{\text{Lennon, McCartney}\}$

4. $N = \{0\}$

5. $S = \{\text{yes, no, undecided}\}$

6. $M = \{\text{classical, country, jazz, rock}\}$

In Exercises 7–10, the universal set is $U = \{0, 1, 2, 3, 4, 5, 6, 7, 8, 9\}$.

7. If $A = \{1, 2, 3, 4, 5\}$ and $B = \{4, 5, 6, 7, 8\}$, find the following.

a. $A \cap B$ **b.** $A \cup B$
c. A' **d.** B'

8. If $A = \{2, 3, 5, 7\}$ and $B = \{2, 4, 6, 7\}$, find the following.

a. $A \cap B$ **b.** $A \cup B$
c. A' **d.** B'

9. If $A = \{1, 3, 5, 7, 9\}$ and $B = \{0, 2, 4, 6, 8\}$, find the following.

a. $A \cap B$ **b.** $A \cup B$
c. A' **d.** B'

10. If $A = \{3, 6, 9\}$ and $B = \{4, 8\}$, find the following.

a. $A \cap B$ **b.** $A \cup B$
c. A' **d.** B'

In Exercises 11–16, the universal set is U = {Monday, Tuesday, Wednesday, Thursday, Friday, Saturday, Sunday}. If A = {Monday, Tuesday, Wednesday, Thursday, Friday} and B = {Friday, Saturday, Sunday}, find the indicated set.

11. $A \cap B$ **12.** $A \cup B$
13. B' **14.** A'
15. $A' \cup B$ **16.** $A \cap B'$

In Exercises 17–26, use a Venn diagram like the one in Figure 2.15 to shade in the region corresponding to the indicated set.

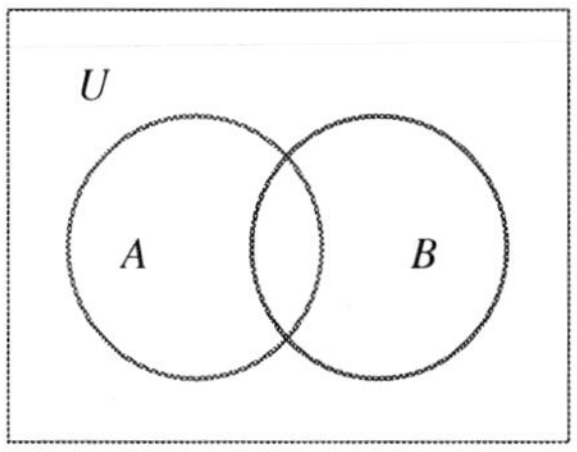

FIGURE 2.15 Two overlapping circles.

17. $A \cap B$ **18.** $A \cup B$
19. A' **20.** B'
21. $A \cup B'$ **22.** $A' \cup B$
23. $A' \cap B$ **24.** $A \cap B'$
25. $A' \cup B'$ **26.** $A' \cap B'$

27. Suppose $n(U) = 150$, $n(A) = 37$, and $n(B) = 84$.

a. If $n(A \cup B) = 100$, find $n(A \cap B)$ and draw a Venn diagram illustrating the composition of U.

b. If $n(A \cup B) = 121$, find $n(A \cap B)$ and draw a Venn diagram illustrating the composition of U.

28. Suppose $n(U) = w$, $n(A) - x$, $n(B) = y$, and $n(A \cup B) = z$.

a. Why must x be less than or equal to z?

b. If $A \neq U$ and $B \neq U$, fill in the blank with the most appropriate symbol: $<$, $>$, $\leq$, or $\geq$.

w_____z, w_____y, y_____z, x_____w

c. Find $n(A \cap B)$ and draw a Venn diagram illustrating the composition of U.

29. In a recent transportation survey, 500 high school seniors were asked to check the appropriate box or boxes on the following form:

☐ I own an automobile.
☐ I own a motorcycle.

The results were tabulated as follows: 102 students checked the automobile box, 147 checked the motorcycle box, and 21 checked both boxes.

a. Draw a Venn diagram illustrating the results of the survey.

b. What percent of these students own an automobile or a motorcycle?

30. In a recent market research survey, 500 married couples were asked to check the appropriate box or boxes on the following form:

☐ We own a DVD player.
☐ We own a microwave oven.

The results were tabulated as follows: 301 couples checked the DVD player box, 394 checked the microwave oven box, and 217 checked both boxes.

a. Draw a Venn diagram illustrating the results of the survey.

b. What percent of these couples own a DVD player or a microwave oven?

31. In a recent socioeconomic survey, 700 married women were asked to check the appropriate box or boxes on the following form:

☐ I have a career.
☐ I have a child.

The results were tabulated as follows: 285 women checked the child box, 316 checked the career box, and 196 were blank (no boxes were checked).

a. Draw a Venn diagram illustrating the results of the survey.

b. What percent of these women had both a child and a career?

32. In a recent health survey, 700 single men in their twenties were asked to check the appropriate box or boxes on the following form:

> ☐ I am a member of a private gym.
> ☐ I am a vegetarian.

The results were tabulated as follows: 349 men checked the gym box, 101 checked the vegetarian box, and 312 were blank (no boxes were checked).

a. Draw a Venn diagram illustrating the results of the survey.

b. What percent of these men were both members of a private gym and vegetarians?

For Exercises 33–36, let

$U = \{x \mid x$ is the name of one of the states in the United States$\}$

$A = \{x \mid x \in U$ and x begins with the letter A$\}$

$I = \{x \mid x \in U$ and x begins with the letter I$\}$

$M = \{x \mid x \in U$ and x begins with the letter M$\}$

$N = \{x \mid x \in U$ and x begins with the letter N$\}$

$O = \{x \mid x \in U$ and x begins with the letter O$\}$

33. Find $n(M')$.

34. Find $n(A \cup N)$.

35. Find $n(I' \cap O')$.

36. Find $n(M \cap I)$.

For Exercises 37–40, let

$U = \{x \mid x$ is the name of one of the months in a year$\}$

$J = \{x \mid x \in U$ and x begins with the letter J$\}$

$Y = \{x \mid x \in U$ and x ends with the letter Y$\}$

$V = \{x \mid x \in U$ and x begins with a vowel$\}$

$R = \{x \mid x \in U$ and x ends with the letter R$\}$

37. Find $n(R')$.

38. Find $n(J \cap V)$.

39. Find $n(J \cup Y)$.

40. Find $n(V \cap R)$.

In Exercises 41–50, determine how many cards, in an ordinary deck of fifty-two, fit the description. (If you are unfamiliar with playing cards, see the end of Section 3.1 for a description of a standard deck.)

41. spades or aces

42. clubs or 2's

43. face cards or black

44. face cards or diamonds

45. face cards and black

46. face cards and diamonds

47. aces or 8's

48. 3's or 6's

49. aces and 8's

50. 3's and 6's

51. Suppose $A = \{1, 2, 3\}$ and $B = \{1, 2, 3, 4, 5, 6\}$.

a. Find $A \cap B$.

b. Find $A \cup B$.

c. In general, if $E \cap F = E$, what must be true concerning sets E and F?

d. In general, if $E \cup F = F$, what must be true concerning sets E and F?

52. Fill in the blank, and give an example to support your answer.

a. If $A \subset B$, then $A \cap B =$ _____.

b. If $A \subset B$, then $A \cup B =$ _____.

53. **a.** List all subsets of $A = \{a\}$. How many subsets does A have?

b. List all subsets of $A = \{a, b\}$. How many subsets does A have?

c. List all subsets of $A = \{a, b, c\}$. How many subsets does A have?

d. List all subsets of $A = \{a, b, c, d\}$. How many subsets does A have?

e. Is there a relationship between the cardinal number of set A and the number of subsets of set A?

f. How many subsets does $A = \{a, b, c, d, e, f\}$ have?
HINT: Use your answer to part (e).

54. Prove the Cardinal Number Formula for the Complement of a Set.
HINT: Apply the Union/Intersection Formula to A and A'.

Answer the following questions using complete sentences and your own words.

• CONCEPT QUESTIONS

55. If $A \cap B = \emptyset$, what is the relationship between sets A and B?

56. If $A \cup B = \emptyset$, what is the relationship between sets A and B?

57. Explain the difference between $\{0\}$ and $\emptyset$.

58. Explain the difference between 0 and $\{0\}$.

59. Is it possible to have $A \cap A = \emptyset$?

60. What is the difference between proper and improper subsets?

61. A set can be described by two methods: the roster method and set-builder notation. When is it advantageous to use the roster method? When is it advantageous to use set-builder notation?

62. Translate the following symbolic expressions into English sentences.

a. $x \in (A \cap B) \leftrightarrow (x \in A \wedge x \in B)$

b. $(x \in A') \leftrightarrow \sim (x \in A)$

c. $(A \subseteq B) \leftrightarrow (x \in A \rightarrow x \in B)$

• HISTORY QUESTIONS

63. In what academic field was John Venn a professor? Where did he teach?

64. What was one of John Venn's main contributions to the field of logic? What new benefits did it offer?

THE NEXT LEVEL

If a person wants to pursue an advanced degree (something beyond a bachelor's or four-year degree), chances are the person must take a standardized exam to gain admission to a school or to be admitted into a specific program. These exams are intended to measure verbal, Quantitative, and analytical skills that have developed throughout a person's life. Many classes and study guides are available to help people prepare for the exams. The following questions are typical of those found in the study guides.

Exercises 65–69 refer to the following: Two collectors, John and Juneko, are each selecting a group of three posters from a group of seven movie posters: J, K, L, M, N, O, *and* P. *No poster can be in both groups. The selections made by John and Juneko are subject to the following restrictions:*

- If K is in John's group, M must be in Juneko's group.
- If N is in John's group, P must be in Juneko's group.
- J and P cannot be in the same group.
- M and O cannot be in the same group.

65. Which of the following pairs of groups selected by John and Juneko conform to the restrictions?

	John	Juneko
a.	J, K, L	M, N, O
b.	J, K, P	L, M, N
c.	K, N, P	J, M, O
d.	L, M, N	K, O, P
e.	M, O, P	J, K, N

66. If N is in John's group, which of the following could not be in Juneko's group?

a. J **b.** K **c.** L **d.** M **e.** P

67. If K and N are in John's group, Juneko's group must consist of which of the following?

a. J, M, and O

b. J, O, and P

c. L, M, and P

d. L, O, and P

e. M, O, and P

68. If J is in Juneko's group, which of the following is true?

a. K cannot be in John's group.

b. N cannot be in John's group.

c. O cannot be in Juneko's group.

d. P must be in John's group.

e. P must be in Juneko's group.

69. If K is in John's group, which of the following is true?

a. J must be in John's group.

b. O must be in John's group.

c. L must be in Juneko's group.

d. N cannot be in John's group.

e. O cannot be in Juneko's group.

2.2 Applications of Venn Diagrams

OBJECTIVES

- Use Venn diagrams to analyze the results of surveys
- Develop and apply De Morgan's Laws of complements

As we have seen, Venn diagrams are very useful tools for visualizing the relationships between sets. They can be used to establish general formulas involving set operations and to determine the cardinal numbers of sets. Venn diagrams are particularly useful in survey analysis.

Surveys

Surveys are often used to divide people or objects into categories. Because the categories sometimes overlap, people can fall into more than one category. Venn diagrams and the formulas for cardinal numbers can help researchers organize the data.

EXAMPLE 1

ANALYZING THE RESULTS OF A SURVEY: TWO SETS Has the advent of the DVD affected attendance at movie theaters? To study this question, Professor Redrum's film class conducted a survey of people's movie-watching habits. He had his students ask hundreds of people between the ages of sixteen and forty-five to check the appropriate box or boxes on the following form:

☐ I watched a movie in a theater during the past month.
☐ I watched a movie on a DVD during the past month.

After the professor had collected the forms and tabulated the results, he told the class that 388 people had checked the theater box, 495 had checked the DVD box, 281 had checked both boxes, and 98 of the forms were blank. Giving the class only this information, Professor Redrum posed the following three questions.

a. What percent of the people surveyed watched a movie in a theater or on a DVD during the past month?
b. What percent of the people surveyed watched a movie in a theater only?
c. What percent of the people surveyed watched a movie on a DVD only?

SOLUTION

a. To calculate the desired percentages, we must determine $n(U)$, the total number of people surveyed. This can be accomplished by drawing a Venn diagram. Because the survey divides people into two categories (those who watched a movie in a theater and those who watched a movie on a DVD), we need to define two sets. Let

$$T = \{\text{people} \mid \text{the person watched a movie in a theater}\}$$
$$D = \{\text{people} \mid \text{the person watched a movie on a DVD}\}$$

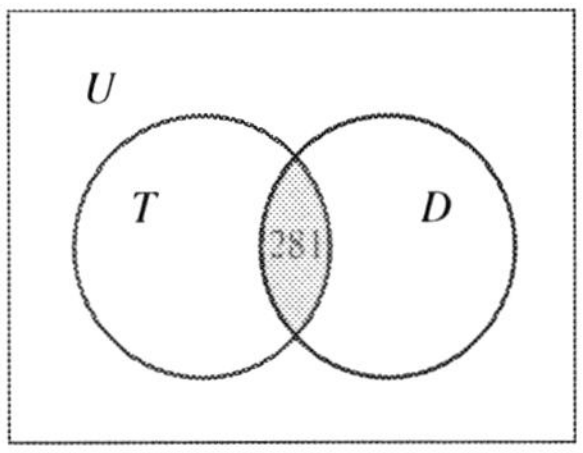

FIGURE 2.16
$n(T \cap D) = 281$.

Now translate the given survey information into the symbols for the sets and attach their given cardinal numbers: $n(T) = 388$, $n(D) = 495$, and $n(T \cap D) = 281$.

Our first goal is to find $n(U)$. To do so, we will fill in the cardinal numbers of all regions of a Venn diagram consisting of two overlapping circles (because we are dealing with two sets). The intersection of T and D consists of 281 people, so we draw two overlapping circles and fill in 281 as the number of elements in common (see Figure 2.16).

Because we were given $n(T) = 388$ and know that $n(T \cap D) = 281$, the difference $388 - 281 = 107$ tells us that 107 people watched a movie in a theater but did not watch a movie on a DVD. We fill in 107 as the number of people who watched a movie only in a theater (see Figure 2.17).

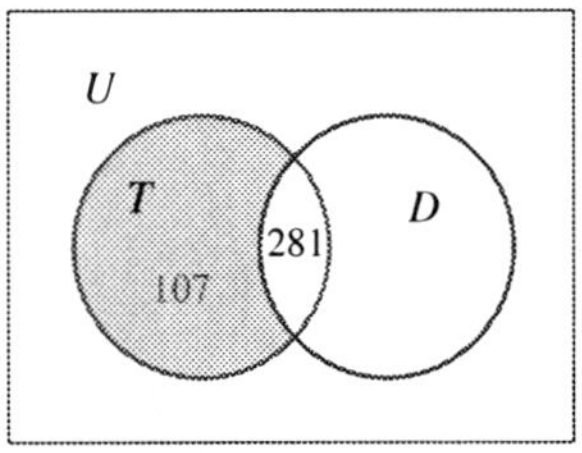

FIGURE 2.17
$n(T \cup D') = 107$.

Because $n(D) = 495$, the difference $495 - 281 = 214$ tells us that 214 people watched a movie on a DVD but not in a theater. We fill in 214 as the number of people who watched a movie only on a DVD (see Figure 2.18).

The only region remaining to be filled in is the region outside both circles. This region represents people who didn't watch a movie in a theater or on a DVD and is symbolized by $(T \cup D)'$. Because 98 people didn't check either box on the form, $n[(T \cup D)'] = 98$ (see Figure 2.19).

After we have filled in the Venn diagram with all the cardinal numbers, we readily see that $n(U) = 98 + 107 + 281 + 214 = 700$. Therefore, 700 people were in the survey.

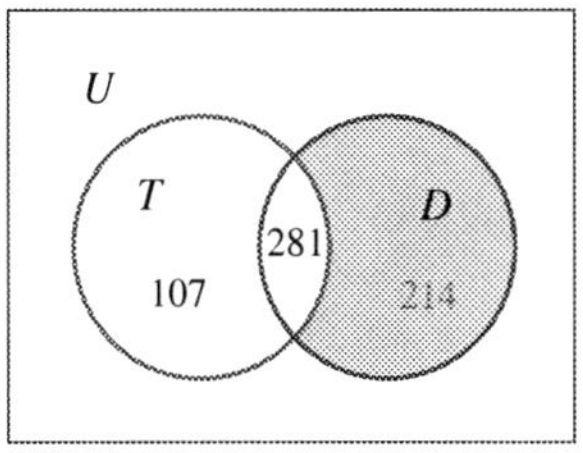

FIGURE 2.18
$n(D \cap T') = 214$.

To determine what *percent* of the people surveyed watched a movie in a theater *or* on a DVD during the past month, simply divide $n(T \cup D)$ by $n(U)$:

$$\frac{n(T \cup D)}{n(U)} = \frac{107 + 281 + 214}{700}$$
$$= \frac{602}{700}$$
$$= 0.86$$

Therefore, exactly 86% of the people surveyed watched a movie in a theater or on a DVD during the past month.

b. To find what *percent* of the people surveyed watched a movie in a theater only, divide 107 (the number of people who watched a movie in a theater only) by $n(U)$:

$$\frac{107}{700} = 0.152857142\ldots$$
$$\approx 0.153 \text{ (rounding off to three decimal places)}$$

Approximately 15.3% of the people surveyed watched a movie in a theater only.

c. Because 214 people watched a movie on DVD only, $214/700 = 0.305714285\ldots$, or approximately 30.6%, of the people surveyed watched a movie on DVD only.

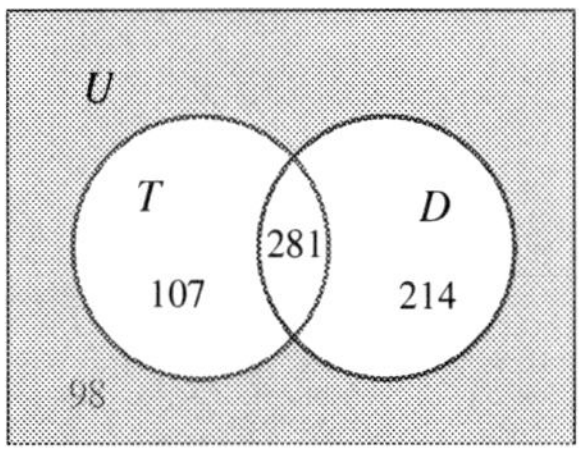

FIGURE 2.19
Completed Venn diagram.

When you solve a cardinal number problem (a problem that asks, "How many?" or "What percent?") involving a universal set that is divided into various categories (for instance, a survey), use the following general steps.

SOLVING A CARDINAL NUMBER PROBLEM

A cardinal number problem is a problem in which you are asked, "How many?" or "What percent?"

1. Define a set for each category in the universal set. If a category and its negation are both mentioned, define one set A and utilize its complement A'.
2. Draw a Venn diagram with as many overlapping circles as the number of sets you have defined.
3. Write down all the given cardinal numbers corresponding to the various given sets.
4. Starting with the innermost overlap, fill in each region of the Venn diagram with its cardinal number.
5. In answering a "what percent" problem, round off your answer to the nearest tenth of a percent.

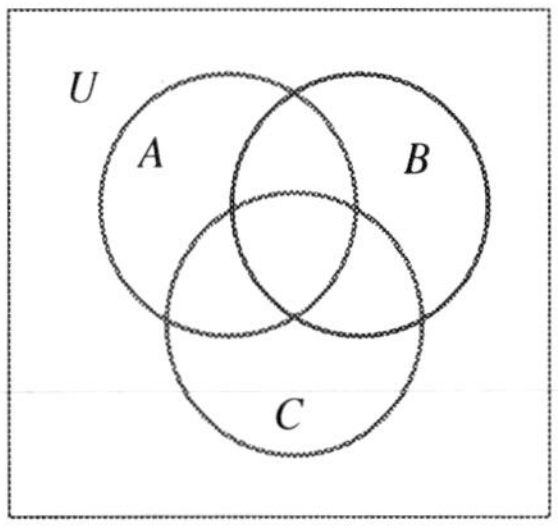

FIGURE 2.20
Three overlapping circles.

When we are working with three sets, we must account for all possible intersections of the sets. Hence, in such cases, we will use the Venn diagram shown in Figure 2.20

EXAMPLE 2

ANALYZING THE RESULTS OF A SURVEY: THREE SETS A consumer survey was conducted to examine patterns in ownership of notebook computers, cellular telephones, and DVD players. The following data were obtained: 213 people had notebook computers, 294 had cell phones, 337 had DVD players, 109 had all three, 64 had none, 198 had cell phones and DVD players, 382 had cell phones or notebook computers, and 61 had notebook computers and DVD players but no cell phones.

a. What percent of the people surveyed owned a notebook computer but no DVD player or cell phone?

b. What percent of the people surveyed owned a DVD player but no notebook computer or cell phone?

SOLUTION

a. To calculate the desired percentages, we must determine $n(U)$, the total number of people surveyed. This can be accomplished by drawing a Venn diagram. Because the survey divides people into three categories (those who own a notebook computer, those who own a cell phone, and those who own a DVD player), we need to define three sets. Let

$C = \{\text{people} \mid \text{the person owns a notebook computer}\}$

$T = \{\text{people} \mid \text{the person owns a cellular telephone}\}$

$D = \{\text{people} \mid \text{the person owns a DVD player}\}$

Now translate the given survey information into the symbols for the sets and attach their given cardinal numbers:

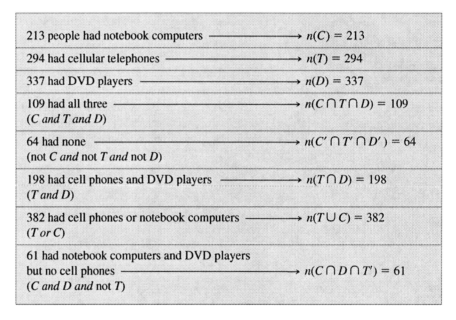

213 people had notebook computers →	$n(C) = 213$
294 had cellular telephones →	$n(T) = 294$
337 had DVD players →	$n(D) = 337$
109 had all three (*C and T and D*) →	$n(C \cap T \cap D) = 109$
64 had none (not *C and* not *T and* not *D*) →	$n(C' \cap T' \cap D') = 64$
198 had cell phones and DVD players (*T and D*) →	$n(T \cap D) = 198$
382 had cell phones or notebook computers (*T or C*) →	$n(T \cup C) = 382$
61 had notebook computers and DVD players but no cell phones (*C and D and* not *T*) →	$n(C \cap D \cap T') = 61$

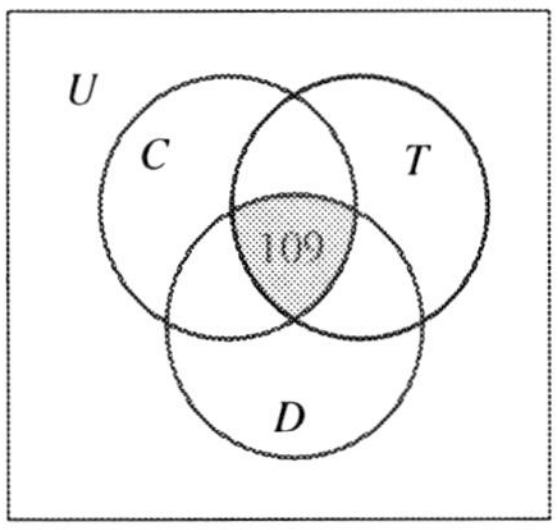

FIGURE 2.21
$n(C \cap T \cap D) = 109$.

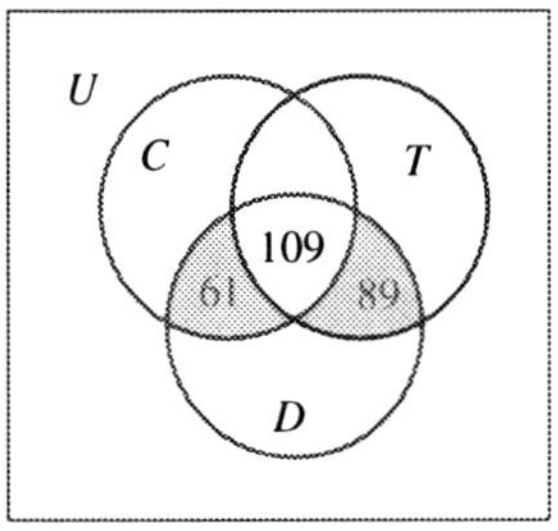

FIGURE 2.22
Determining cardinal numbers in a Venn diagram.

Our first goal is to find $n(U)$. To do so, we will fill in the cardinal numbers of all regions of a Venn diagram like that in Figure 2.20. We start by using information concerning membership in all three sets. Because the intersection of all three sets consists of 109 people, we fill in 109 in the region common to C and T and D (see Figure 2.21).

Next, we utilize any information concerning membership in two of the three sets. Because $n(T \cap D) = 198$, a total of 198 people are common to both T and D; some are in C, and some are not in C. Of these 198 people, 109 are in C (see Figure 2.21). Therefore, the difference $198 - 109 = 89$ gives the number not in C. Eighty-nine people are in T and D and *not* in C; that is, $n(T \cap D \cap C') = 89$. Concerning membership in the two sets C and D, we are given $n(C \cap D \cap T') = 61$. Therefore, we know that 61 people are in C and D and not in T (see Figure 2.22).

We are given $n(T \cup C) = 382$. From this number, we can calculate $n(T \cap C)$ by using the Union/Intersection Formula:

$$n(T \cup C) = n(T) + n(C) - n(T \cap C)$$
$$382 = 294 + 213 - n(T \cap C)$$
$$n(T \cap C) = 125$$

Therefore, a total of 125 people are in T and C; some are in D, and some are not in D. Of these 125 people, 109 are in D (see Figure 2.21). Therefore, the difference

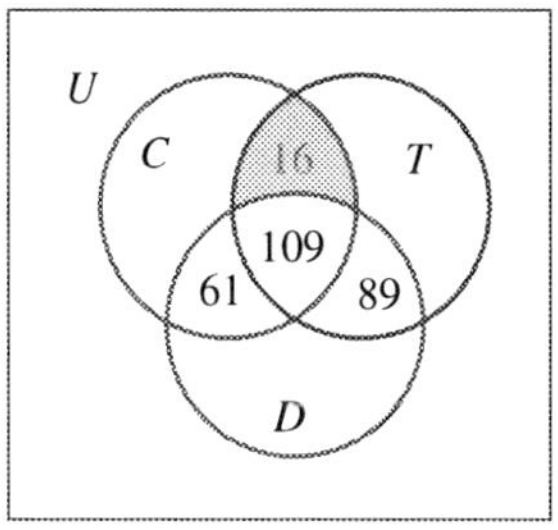

FIGURE 2.23 Determining cardinal numbers in a Venn diagram.

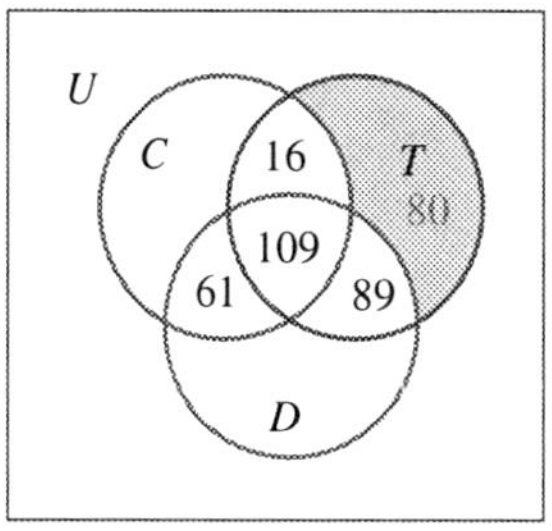

FIGURE 2.24 Determining cardinal numbers in a Venn diagram.

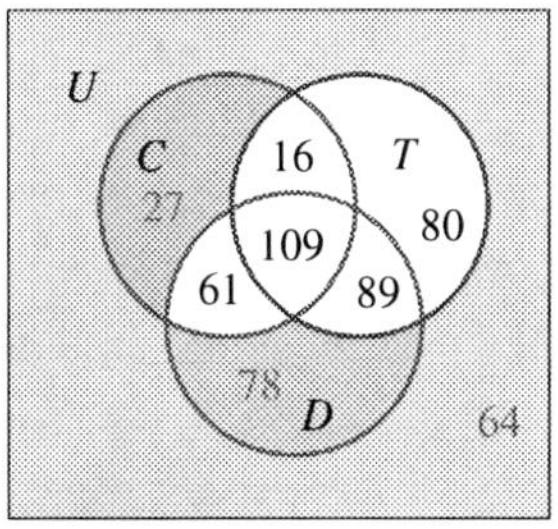

FIGURE 2.25

A completed Venn diagram.

$125 - 109 = 16$ gives the number not in D. Sixteen people are in T and C and *not* in D; that is, $n(C \cap T \cap D') = 16$ (see Figure 2.23).

Knowing that a total of 294 people are in T (given $n(T) = 294$), we are now able to fill in the last region of T. The missing region (people in T only) has $294 - 109 - 89 - 16 = 80$ members; $n(T \cap C' \cap D') = 80$ (see Figure 2.24).

In a similar manner, we subtract the known pieces of C from $n(C) = 213$, which is given, and obtain $213 - 61 - 109 - 16 = 27$; therefore, 27 people are in C only. Likewise, to find the last region of D, we use $n(D) = 337$ (given) and obtain $337 - 89 - 109 - 61 = 78$; therefore, 78 people are in D only. Finally, the 64 people who own none of the items are placed "outside" the three circles (see Figure 2.25).

By adding up the cardinal numbers of all the regions in Figure 2.25, we find that the total number of people in the survey is 524; that is, $n(U) = 524$.

Now, to determine what *percent* of the people surveyed owned only a notebook computer (no DVD player and no cell phone), we simply divide $n(C \cap D' \cap T')$ by $n(U)$:

$$\frac{n(C \cap D' \cap T')}{n(U)} = \frac{27}{524}$$
$$= 0.051526717\ldots$$

Approximately 5.2% of the people surveyed owned a notebook computer and did not own a DVD player or a cellular telephone.

b. To determine what *percent* of the people surveyed owned only a DVD player (no notebook computer and no cell phone), we divide $n(D \cap C' \cap T')$ by $n(U)$:

$$\frac{n(D \cap C' \cap T')}{n(U)} = \frac{78}{524}$$
$$= 0.148854961\ldots$$

Approximately 14.9% of the people surveyed owned a DVD player and did not own a notebook computer or a cell phone.

De Morgan's Laws

One of the basic properties of algebra is the distributive property:

$$a(b + c) = ab + ac$$

Given $a(b + c)$, the operation outside the parentheses can be distributed over the operation inside the parentheses. It makes no difference whether you add b and c first and then multiply the sum by a or first multiply each pair, a and b, a and c, and then add their products; the same result is obtained. Is there a similar property for the complement, union, and intersection of sets?

EXAMPLE 3

INVESTIGATING THE COMPLEMENT OF A UNION Suppose $U = \{1, 2, 3, 4, 5\}$, $A = \{1, 2, 3\}$, and $B = \{2, 3, 4\}$.

a. For the given sets, does $(A \cup B)' = A' \cup B'$?
b. For the given sets, does $(A \cup B)' = A' \cap B'$?

SOLUTION

a. To find $(A \cup B)'$, we must first find $A \cup B$:

$$\begin{aligned} A \cup B &= \{1, 2, 3\} \cup \{2, 3, 4\} \\ &= \{1, 2, 3, 4\} \end{aligned}$$

The complement of $A \cup B$ (relative to the given universal set U) is

$$(A \cup B)' = \{5\}$$

To find $A' \cup B'$, we must first find A' and B':

$$A' = \{4, 5\} \quad \text{and} \quad B' = \{1, 5\}$$

The union of A' and B' is

$$\begin{aligned} A' \cup B' &= \{4, 5\} \cup \{1, 5\} \\ &= \{1, 4, 5\} \end{aligned}$$

Now, $\{5\} \neq \{1, 4, 5\}$; therefore, $(A \cup B)' \neq A' \cup B'$.

b. We find $(A \cup B)'$ as in part (a): $(A \cup B)' = \{5\}$. Now,

$$\begin{aligned} A' \cap B' &= \{4, 5\} \cap \{1, 5\} \\ &= \{5\} \end{aligned}$$

For the given sets, $(A \cup B)' = A' \cap B'$.

Part (a) of Example 3 shows that the operation of complementation *cannot* be explicitly distributed over the operation of union; that is, $(A \cup B)' \neq A' \cup B'$. However, part (b) of the example implies that there *may* be some relationship between the complement, union, and intersection of sets. The fact that $(A \cup B)' = A' \cap B'$ *for the given sets* A and B does not mean that it is true *for all sets* A and B. We will use a general Venn diagram to examine the validity of the statement $(A \cup B)' = A' \cap B'$.

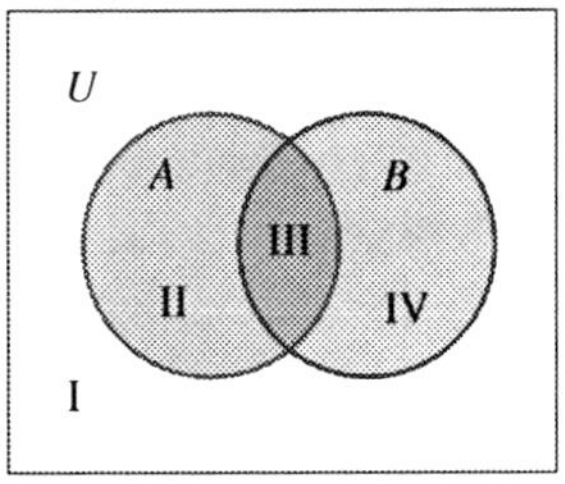

FIGURE 2.26
Four regions in a universal set U.

When we draw two overlapping circles within a universal set, four regions are formed. Every element of the universal set U is in exactly one of the following regions, as shown in Figure 2.26:

I in neither A nor B
II in A and not in B
III in both A and B
IV in B and not in A

The set $A \cup B$ consists of all elements in regions II, III, and IV. Therefore, the complement $(A \cup B)'$ consists of all elements in region I. A' consists of all elements in regions I and IV, and B' consists of the elements in regions I and II. Therefore, the elements common to both A' and B' are those in region I; that is, the set $A' \cap B'$ consists of all elements in region I. Since $(A \cup B)'$ and $A' \cap B'$ contain exactly the same elements (those in region I), the sets are equal; that is, $(A \cup B)' = A' \cap B'$ is true for all sets A and B.

The relationship $(A \cup B)' = A' \cap B'$ is known as one of **De Morgan's Laws.** Simply stated, "the complement of a union is the intersection of the complements." In a similar manner, it can be shown that $(A \cap B)' = A' \cup B'$ (see Exercise 33).

HISTORICAL NOTE

AUGUSTUS DE MORGAN, 1806–1871

© Bettmann/CORBIS

Being born blind in one eye did not stop Augustus De Morgan from becoming a well-read philosopher, historian, logician, and mathematician. De Morgan was born in Madras, India, where his father was working for the East India Company. On moving to England, De Morgan was educated at Cambridge, and at the age of twenty-two, he became the first professor of mathematics at the newly opened University of London (later renamed University College).

De Morgan viewed all of mathematics as an abstract study of symbols and of systems of operations applied to these symbols. While studying the ramifications of symbolic logic, De Morgan formulated the general properties of complementation that now bear his name. Not limited to symbolic logic, De Morgan's many works include books and papers on the foundations of algebra, differential calculus, and probability. He was known to be a jovial person who was fond of puzzles, and his witty and amusing book *A Budget of Paradoxes* still entertains readers today. Besides his accomplishments in the academic arena, De Morgan was an expert flutist, spoke five languages, and thoroughly enjoyed big-city life.

Knowing of his interest in probability, an actuary (someone who studies life expectancies and determines payments of premiums for insurance companies) once asked De Morgan a question concerning the probability that a certain group of people would be alive at a certain time. In his response, De Morgan employed a formula containing the number π. In amazement, the actuary responded, "That must surely be a delusion! What can a circle have to do with the number of people alive at a certain time?" De Morgan replied that π has numerous applications and occurrences in many diverse areas of mathematics. Because it was first defined and used in geometry, people are conditioned to accept the mysterious number only in reference to a circle. However, in the history of mathematics, if probability had been systematically studied before geometry and circles, our present-day interpretation of the number π would be entirely different. In addition to his accomplishments in logic and higher-level mathematics, De Morgan introduced a convention with which we are all familiar: In a paper written in 1845, he suggested the use of a slanted line to represent a fraction, such as 1/2 or 3/4.

De Morgan was a staunch defender of academic freedom and religious tolerance. While he was a student at Cambridge, his application for a fellowship was refused because he would not take and sign a theological oath. Later in life, he resigned his professorship as a protest against religious bias. (University College gave preferential treatment to members of the Church of England when textbooks were selected and did not have an open policy on religious philosophy.) Augustus De Morgan was a man who was unafraid to take a stand and make personal sacrifices when it came to principles he believed in.

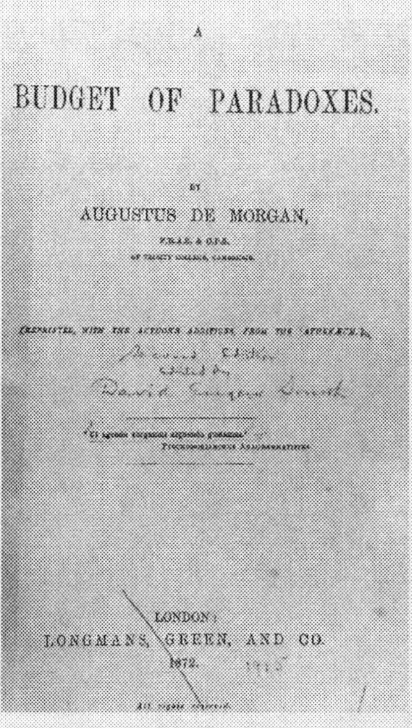

A

BUDGET OF PARADOXES.

BY

AUGUSTUS DE MORGAN,

LONDON:
LONGMANS, GREEN, AND CO.
1872.

Gematria is a mystic pseudoscience in which numbers are substituted for the letters in a name. De Morgan's book A Budget of Paradoxes contains several gematria puzzles, such as, "Mr. Davis Thom found a young gentleman of the name of St. Claire busy at the Beast number: he forthwith added the letters in στκλαιρε (the Greek spelling of St. Claire) and found 666." (Verify this by using the Greek numeral system.)

DE MORGAN'S LAWS

For any sets A and B,

$$(A \cup B)' = A' \cap B'$$

That is, the complement of a union is the intersection of the complements. Also,

$$(A \cap B)' = A' \cup B'$$

That is, the complement of an intersection is the union of the complements.

TOPIC X BLOOD TYPES: *SET THEORY IN THE REAL WORLD*

Human blood types are a classic example of set theory. As you may know, there are four categories (or sets) of blood types: A, B, AB, and O. Knowing someone's blood type is extremely important in case a blood transfusion is required; if blood of two different types is combined, the blood cells may begin to clump together, with potentially fatal consequences! (Do you know your blood type?)

What exactly are "blood types"? In the early 1900s, the Austrian scientist Karl Landsteiner observed the presence (or absence) of two distinct chemical molecules on the surface of all red blood cells in numerous samples of human blood. Consequently, he labeled one molecule "A" and the other "B." The presence or absence of these specific molecules is the basis of the universal classification of blood types. Specifically, blood samples containing only the A molecule are labeled type A, whereas those containing only the B molecule are labeled type B. If a blood sample contains both molecules (A and B) it is labeled type AB; and if neither is present, the blood is typed as O. The presence (or absence) of these molecules can be depicted in a standard Venn diagram as shown in Figure 2.27. In the notation of set operations, type A blood is denoted $A \cap B'$, type B is $B \cap A'$, type AB is $A \cap B$, and type O is $A' \cap B'$.

If a specific blood sample is mixed with blood containing a blood molecule (A or B) that it does not already have, the presence of the foreign molecule may cause the mixture of blood to clump. For example, type A blood cannot be mixed with any blood containing the B molecule (type B or type AB). Therefore, a person with type A blood can receive a transfusion only of type A or type O blood. Consequently, a person with type AB blood may receive a transfusion of any blood type; type AB is referred to as the "universal receiver." Because type O blood contains neither the A nor the B molecule, all blood types are compatible with type O blood; type O is referred to as the "universal donor."

It is not uncommon for scientists to study rhesus monkeys in an effort to learn more about human physiology. In so doing, a certain blood protein was discovered in rhesus monkeys. Subsequently, scientists found that the blood of some people contained this protein, whereas the blood of others did not. The presence, or absence, of this protein in human blood is referred to as the *Rh factor;* blood containing the protein is labeled "Rh+", whereas "Rh−" indicates the absence of the protein. The Rh factor of human blood is especially important for expectant mothers; a fetus can develop problems if its parents have opposite Rh factors.

When a person's blood is typed, the designation includes both the regular blood type and the Rh factor. For instance, type AB− indicates the presence of both the A and B molecules (type AB), along with the absence of the rhesus protein; type O+ indicates the absence of both the A and B molecules (type O), along with the presence of the rhesus protein. Utilizing the Rh factor, there are eight possible blood types as shown in Figure 2.28.

We will investigate the occurrence and compatibility of the various blood types in Example 5 and in Exercises 35–43.

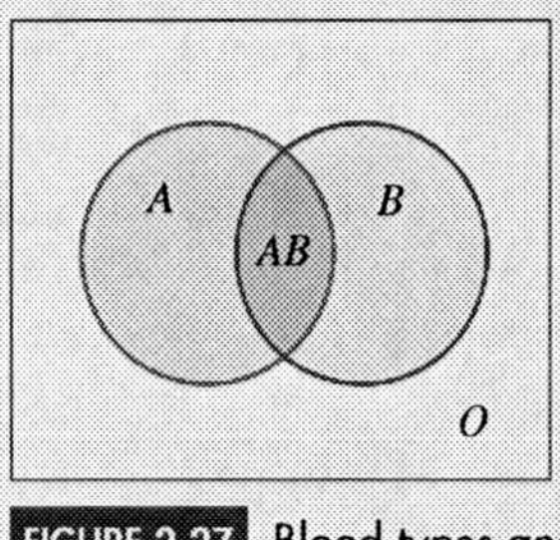

FIGURE 2.27 Blood types and the presence of the A and B molecules.

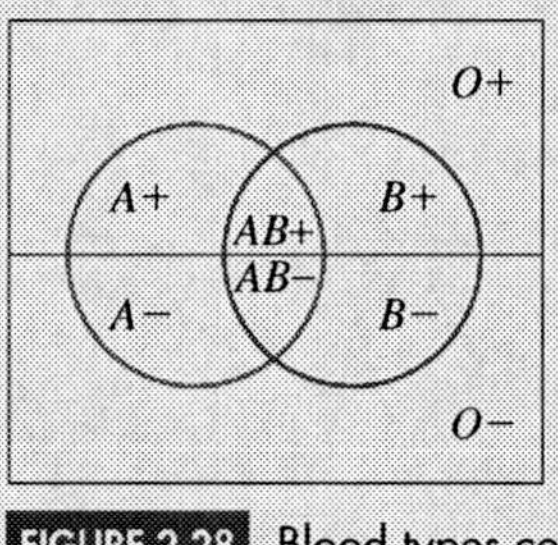

FIGURE 2.28 Blood types combined with the Rh factor.

EXAMPLE 4

APPLYING DE MORGAN'S LAW Suppose $U = \{0, 1, 2, 3, 4, 5, 6, 7, 8, 9\}$, $A = \{2, 3, 7, 8\}$, and $B = \{0, 4, 5, 7, 8, 9\}$. Use De Morgan's Law to find $(A' \cup B)'$.

SOLUTION

The complement of a union is equal to the intersection of the complements; therefore, we have

$$
\begin{aligned}
(A' \cup B)' &= (A')' \cap B' && \text{De Morgan's Law} \\
&= A \cap B' && (A')' = A \\
&= \{2, 3, 7, 8\} \cap \{1, 2, 3, 6\} \\
&= \{2, 3\}
\end{aligned}
$$

Notice that this problem could be done without using De Morgan's Law, but solving it would then involve finding first A', then $A' \cup B$, and finally $(A' \cup B)'$. This method would involve more work. (Try it!)

EXAMPLE 5

INVESTIGATING BLOOD TYPES IN THE UNITED STATES The American Red Cross has compiled a massive database of the occurrence of blood types in the United States. Their data indicate that on average, out of every 100 people in the United States, 44 have the A molecule, 15 have the B molecule, and 45 have neither the A nor the B molecule. What percent of the U.S. population have the following blood types?

a. Type O? **b.** Type AB? **c.** Type A? **d.** Type B?

SOLUTION

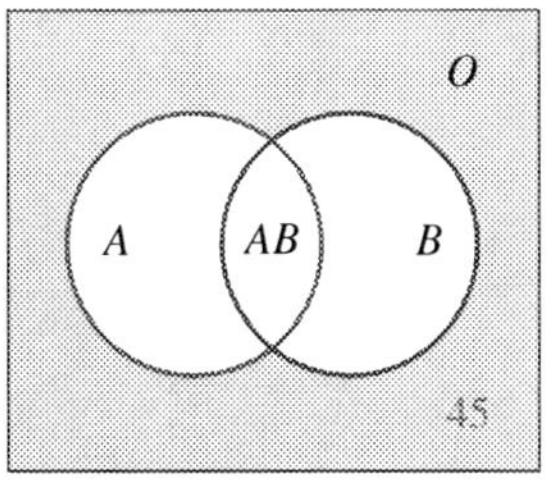

FIGURE 2.29 Forty-five of 100 (45%) have type O blood.

a. First, we define the appropriate sets. Let

$$A = \{\text{Americans} \mid \text{the person has the A molecule}\}$$
$$B = \{\text{Americans} \mid \text{the person has the B molecule}\}$$

We are given the following cardinal numbers: $n(U) = 100$, $n(A) = 44$, $n(B) = 15$, and $n(A' \cap B') = 45$. Referring to Figure 2.27, and given that 45 people (out of 100) have neither the A molecule nor the B molecule, we conclude that 45 of 100 people, or 45%, have type O blood as shown in Figure 2.29.

b. Applying De Morgan's Law to the Complement Formula, we have the following.

$$n(A \cup B) + n[(A \cup B)'] = n(U) \quad \text{Complement Formula}$$
$$n(A \cup B) + n(A' \cap B') = n(U) \quad \text{applying De Morgan's Law}$$
$$n(A \cup B) + 45 = 100 \quad \text{substituting known values}$$

Therefore, $n(A \cup B) = 55$.

Now, use the Union/Intersection Formula.

$$n(A \cup B) = n(A) + n(B) - n(A \cap B) \quad \text{Union/Intersection Formula}$$
$$55 = 44 + 15 - n(A \cap B) \quad \text{substituting known values}$$
$$n(A \cap B) = 44 + 15 - 55 \quad \text{adding } n(A \cap B) \text{ and subtracting 55}$$

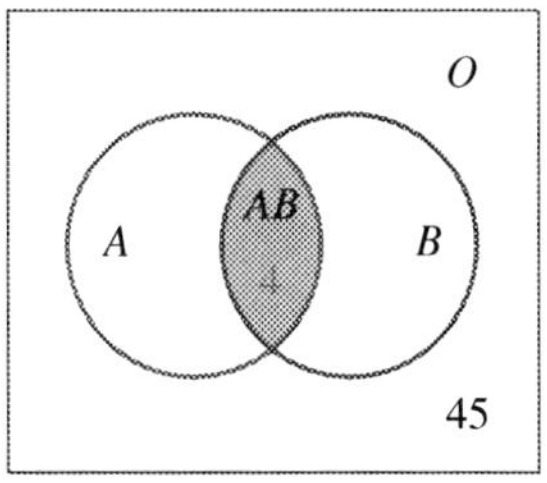

FIGURE 2.30 Four in 100 (4%) have type AB blood.

Therefore, $n(A \cap B) = 4$. This means that 4 people (of 100) have both the A and the B molecules; that is, 4 of 100 people, or 4%, have type AB blood. See Figure 2.30.

c. Knowing that a total of 44 people have the A molecule, that is, $n(A) = 44$, we subtract $n(A \cap B) = 4$ and conclude that 40 have *only* the A molecule.

Therefore, $n(A \cap B') = 40$. This means that 40 people (of 100) have only the A molecule; that is, 40 of 100 people, or 40%, have type A blood. See Figure 2.31.

d. Knowing that a total of 15 people have the B molecule, that is, $n(B) = 15$, we subtract $n(A \cap B) = 4$ and conclude that 11 have *only* the B molecule.

Therefore, $n(B \cap A') = 11$. This means that 11 people (of 100) have only the B molecule; that is, 11 of 100 people, or 11%, have type B blood (see Figure 2.32).

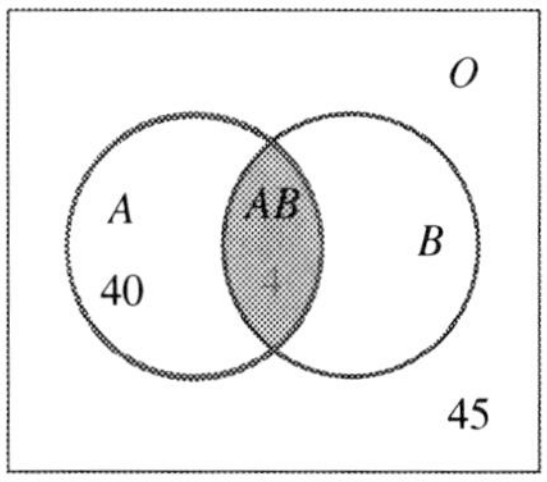

FIGURE 2.31 Forty in 100 (40%) have type A blood.

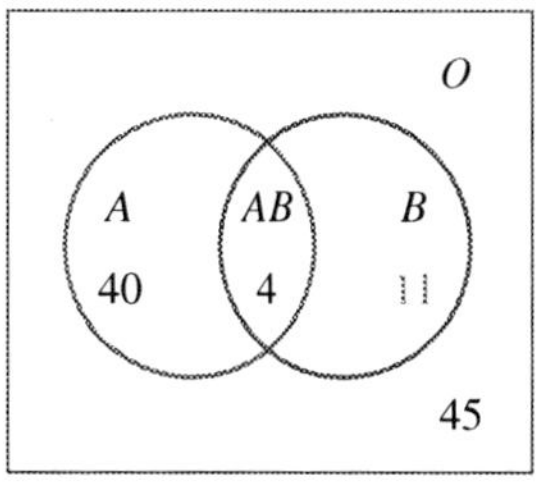

FIGURE 2.32 Eleven in 100 (11%) have type B blood.

The occurrence of blood types in the United States is summarized in Figure 2.33.

Blood Type	O	A	B	AB
Occurrence	45%	40%	11%	4%

FIGURE 2.33 Occurrence of blood types in the United States.

The occurrence of blood types given in Figure 2.33 can be further categorized by including the Rh factor. According to the American Red Cross, out of every 100 people in the United States, blood types and Rh factors occur at the rates shown in Figure 2.34.

Blood Type	O+	O–	A+	A–	B+	B–	AB+	AB–
Occurrence	38	7	34	6	9	2	3	1

FIGURE 2.34 Blood types per 100 people in the United States. (*Source:* American Red Cross.)

2.2 Exercises

1. A survey of 200 people yielded the following information: 94 people owned a DVD player, 127 owned a microwave oven, and 78 owned both. How many people owned the following?
   a. a DVD player or a microwave oven
   b. a DVD player but not a microwave oven
   c. a microwave oven but not a DVD player
   d. neither a DVD player nor a microwave oven
2. A survey of 300 workers yielded the following information: 231 workers belonged to a union, and 195 were Democrats. If 172 of the union members were Democrats, how many workers were in the following situations?
   a. belonged to a union or were Democrats
   b. belonged to a union but were not Democrats
   c. were Democrats but did not belong to a union
   d. neither belonged to a union nor were Democrats
3. The records of 1,492 high school graduates were examined, and the following information was obtained: 1,072 graduates took biology, and 679 took geometry. If 271 of those who took geometry did not take biology, how many graduates took the following?
   a. both classes
   b. at least one of the classes
   c. biology but not geometry
   d. neither class
4. A department store surveyed 428 shoppers, and the following information was obtained: 214 shoppers made a purchase, and 299 were satisfied with the service they received. If 52 of those who made a purchase were not satisfied with the service, how many shoppers did the following?
   a. made a purchase and were satisfied with the service
   b. made a purchase or were satisfied with the service
   c. were satisfied with the service but did not make a purchase
   d. were not satisfied and did not make a purchase
5. In a survey, 674 adults were asked what television programs they had recently watched. The following information was obtained: 226 adults watched neither the Big Game nor the New Movie, and 289 watched the New Movie. If 183 of those who watched the New Movie did not watch the Big Game, how many of the surveyed adults watched the following?
   a. both programs
   b. at least one program
   c. the Big Game
   d. the Big Game but not the New Movie

6. A survey asked 816 college freshmen whether they had been to a movie or eaten in a restaurant during the past week. The following information was obtained: 387 freshmen had been to neither a movie nor a restaurant, and 266 had been to a movie. If 92 of those who had been to a movie had not been to a restaurant, how many of the surveyed freshmen had been to the following?
   a. both a movie and a restaurant
   b. a movie or a restaurant
   c. a restaurant
   d. a restaurant but not a movie
7. A local 4-H club surveyed its members, and the following information was obtained: 13 members had rabbits, 10 had goats, 4 had both rabbits and goats, and 18 had neither rabbits nor goats.
   a. What percent of the club members had rabbits or goats?
   b. What percent of the club members had only rabbits?
   c. What percent of the club members had only goats?
8. A local anime fan club surveyed its members regarding their viewing habits last weekend, and the following information was obtained: 30 members had watched an episode of *Naruto,* 44 had watched an episode of *Death Note,* 21 had watched both an episode of *Naruto* and an episode of *Death Note,* and 14 had watched neither *Naruto* nor *Death Note.*
   a. What percent of the club members had watched *Naruto* or *Death Note?*
   b. What percent of the club members had watched only *Naruto?*
   c. What percent of the club members had watched only *Death Note?*
9. A recent survey of w shoppers (that is, $n(U) = w$) yielded the following information: x shoppers shopped at Sears, y shopped at JCPenney's, and z shopped at both. How many people shopped at the following?
   a. Sears or JCPenney's
   b. only Sears
   c. only JCPenney's
   d. neither Sears nor JCPenney's
10. A recent transportation survey of w urban commuters (that is, $n(U) = w$) yielded the following information: x commuters rode neither trains nor buses, y rode trains, and z rode only trains. How many people rode the following?
   a. trains and buses
   b. only buses
   c. buses
   d. trains or buses
11. A consumer survey was conducted to examine patterns in ownership of laptop computers, cellular telephones, and DVD players. The following data were obtained: 313 people had laptop computers, 232 had cell phones, 269 had DVD players, 69 had all three, 64 had none, 98 had cell phones and DVD players, 57 had cell phones but no computers or DVD players, and 104 had computers and DVD players but no cell phones.
   a. What percent of the people surveyed owned a cell phone?
   b. What percent of the people surveyed owned only a cell phone?
12. In a recent survey of monetary donations made by college graduates, the following information was obtained: 95 graduates had donated to a political campaign, 76 had donated to assist medical research, 133 had donated to help preserve the environment, 25 had donated to all three, 22 had donated to none of the three, 38 had donated to a political campaign and to medical research, 46 had donated to medical research and to preserve the environment, and 54 had donated to a political campaign and to preserve the environment.
   a. What percent of the college graduates donated to none of the three listed causes?
   b. What percent of the college graduates donated to exactly one of the three listed causes?
13. Recently, Green Day, the Kings of Leon, and the Black Eyed Peas had concert tours in the United States. A large group of college students was surveyed, and the following information was obtained: 381 students saw Black Eyed Peas, 624 saw the Kings of Leon, 712 saw Green Day, 111 saw all three, 513 saw none, 240 saw only Green Day, 377 saw Green Day and the Kings of Leon, and 117 saw the Kings of Leon and Black Eyed Peas but not Green Day.
   a. What percent of the college students saw at least one of the bands?
   b. What percent of the college students saw exactly one of the bands?
14. Dr. Hawk works in an allergy clinic, and his patients have the following allergies: 68 patients are allergic to dairy products, 93 are allergic to pollen, 91 are allergic to animal dander, 31 are allergic to all three, 29 are allergic only to pollen, 12 are allergic only to dairy products, and 40 are allergic to dairy products and pollen.
   a. What percent of Dr. Hawk's patients are allergic to animal dander?
   b. What percent of Dr. Hawk's patients are allergic only to animal dander?
15. When the members of the Eye and I Photo Club discussed what type of film they had used during the past month, the following information was obtained: 77 members used black and white, 24 used only black and white, 65 used color, 18 used only

color, 101 used black and white or color, 27 used infrared, 9 used all three types, and 8 didn't use any film during the past month.

a. What percent of the members used only infrared film?

b. What percent of the members used at least two of the types of film?

16. After leaving the polls, many people are asked how they voted. (This is called an *exit poll.*) Concerning Propositions A, B, and C, the following information was obtained: 294 people voted yes on A, 90 voted yes only on A, 346 voted yes on B, 166 voted yes only on B, 517 voted yes on A or B, 339 voted yes on C, no one voted yes on all three, and 72 voted no on all three.

a. What percent of the voters in the exit poll voted no on A?

b. What percent of the voters voted yes on more than one proposition?

17. In a recent survey, consumers were asked where they did their gift shopping. The following results were obtained: 621 consumers shopped at Macy's, 513 shopped at Emporium, 367 shopped at Nordstrom, 723 shopped at Emporium or Nordstrom, 749 shopped at Macy's or Nordstrom, 776 shopped at Macy's or Emporium, 157 shopped at all three, 96 shopped at neither Macy's nor Emporium nor Nordstrom.

a. What percent of the consumers shopped at more than one store?

b. What percent of the consumers shopped exclusively at Nordstrom?

18. A company that specializes in language tutoring lists the following information concerning its English-speaking employees: 23 employees speak German; 25 speak French; 31 speak Spanish; 43 speak Spanish or French; 38 speak French or German; 46 speak German or Spanish; 8 speak Spanish, French, and German; and 7 speak English only.

a. What percent of the employees speak at least one language other than English?

b. What percent of the employees speak at least two languages other than English?

19. In a recent survey, people were asked which radio station they listened to on a regular basis. The following results were obtained: 140 people listened to WOLD (oldies), 95 listened to WJZZ (jazz), 134 listened to WTLK (talk show news), 235 listened to WOLD or WJZZ, 48 listened to WOLD and WTLK, 208 listened to WTLK or WJZZ, and 25 listened to none.

a. What percent of people in the survey listened only to WTLK on a regular basis?

b. What percent of people in the survey did not listen to WTLK on a regular basis?

20. In a recent health insurance survey, employees at a large corporation were asked, "Have you been a patient in a hospital during the past year, and if so, for what reason?" The following results were obtained: 494 employees had an injury, 774 had an illness, 1,254 had tests, 238 had an injury and an illness and tests, 700 had an illness and tests, 501 had tests and no injury or illness, 956 had an injury or illness, and 1,543 had not been a patient.

a. What percent of the employees had been patients in a hospital?

b. What percent of the employees had tests in a hospital?

In Exercises 21 and 22, use a Venn diagram like the one in Figure 2.35.

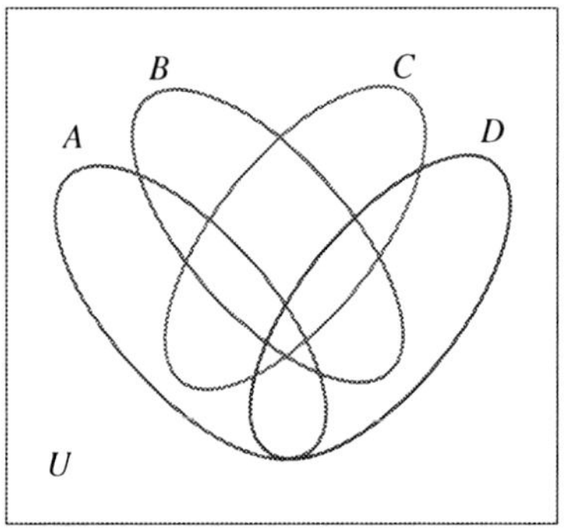

FIGURE 2.35 Four overlapping regions.

21. A survey of 136 pet owners yielded the following information: 49 pet owners own fish; 55 own a bird; 50 own a cat; 68 own a dog; 2 own all four; 11 own only fish; 14 own only a bird; 10 own fish and a bird; 21 own fish and a cat; 26 own a bird and a dog; 27 own a cat and a dog; 3 own fish, a bird, a cat, and no dog; 1 owns fish, a bird, a dog, and no cat; 9 own fish, a cat, a dog, and no bird; and 10 own a bird, a cat, a dog, and no fish. How many of the surveyed pet owners have no fish, no birds, no cats, and no dogs? (They own other types of pets.)

22. An exit poll of 300 voters yielded the following information regarding voting patterns on Propositions A, B, C, and D: 119 voters voted yes on A; 163 voted yes on B; 129 voted yes on C; 142 voted yes on D; 37 voted yes on all four; 15 voted yes on A only; 50 voted yes on B only; 59 voted yes on A and B; 70 voted yes on A and C; 82 voted yes on B and D; 93 voted yes on C and D; 10 voted yes on A, B, and C and no on D; 2 voted yes on A, B, and D and no on C; 16 voted yes on A, C, and D and no on B; and 30 voted yes on B, C, and D and no on A. How many of the surveyed voters voted no on all four propositions?

In Exercises 23–26, given the sets $U = \{0, 1, 2, 3, 4, 5, 6, 7, 8, 9\}$, $A = \{0, 2, 4, 5, 9\}$, and $B = \{1, 2, 7, 8, 9\}$, use De Morgan's Laws to find the indicated sets.

23. $(A' \cup B)'$

24. $(A' \cap B)'$

25. $(A \cap B')'$

26. $(A \cup B')'$

In Exercises 27–32, use a Venn diagram like the one in Figure 2.36 to shade in the region corresponding to the indicated set.

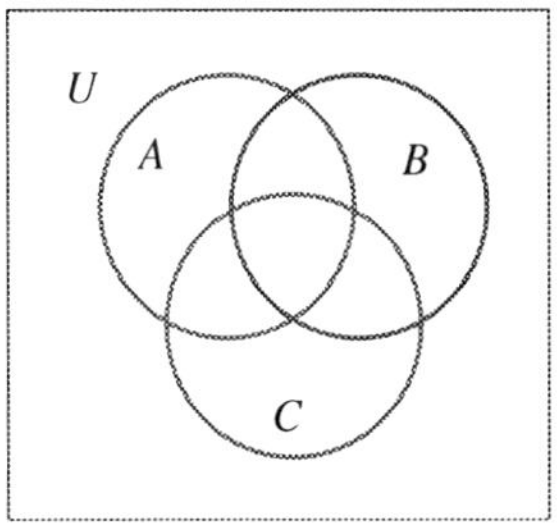

FIGURE 2.36 Three overlapping circles.

27. $A \cap B \cap C$

28. $A \cup B \cup C$

29. $(A \cup B)' \cap C$

30. $A \cap (B \cup C)'$

31. $B \cap (A \cup C')$

32. $(A' \cup B) \cap C'$

33. Using Venn diagrams, prove De Morgan's Law $(A \cap B)' = A' \cup B'$.

34. Using Venn diagrams, prove $A \cup (B \cap C) = (A \cup B) \cap (A \cup C)$.

Use the data in Figure 2.34 to complete Exercises 35–39. Round off your answers to a tenth of a percent.

35. What percent of all people in the United States have blood that is

a. Rh positive? **b.** Rh negative?

36. Of all people in the United States who have type O blood, what percent are

a. Rh positive? **b.** Rh negative?

37. Of all people in the United States who have type A blood, what percent are

a. Rh positive? **b.** Rh negative?

38. Of all people in the United States who have type B blood, what percent are

a. Rh positive? **b.** Rh negative?

39. Of all people in the United States who have type AB blood, what percent are

a. Rh positive? **b.** Rh negative?

40. If a person has type A blood, what blood types may the person receive in a transfusion?

41. If a person has type B blood, what blood types may the person receive in a transfusion?

42. If a person has type AB blood, what blood types may the person receive in a transfusion?

43. If a person has type O blood, what blood types may the person receive in a transfusion?

Answer the following questions using complete sentences and your own words.

• HISTORY QUESTIONS

44. What notation did De Morgan introduce in regard to fractions?

45. Why did De Morgan resign his professorship at University College?

THE NEXT LEVEL

If a person wants to pursue an advanced degree (something beyond a bachelor's or four-year degree), chances are the person must take a standardized exam to gain admission to a school or to be admitted into a specific program. These exams are intended to measure verbal, quantitative, and analytical skills that have developed throughout a person's life. Many classes and study guides are available to help people prepare for the exams. The following questions are typical of those found in the study guides.

Exercises 46–52 refer to the following: A nonprofit organization's board of directors, composed of four women (Angela, Betty, Carmen, and Delores) and three men (Ed, Frank, and Grant), holds frequent meetings. A meeting can be held at Betty's house, at Delores's house, or at Frank's house.

- Delores cannot attend any meetings at Betty's house.
- Carmen cannot attend any meetings on Tuesday or on Friday.
- Angela cannot attend any meetings at Delores's house.
- Ed can attend only those meetings that Grant also attends.
- Frank can attend only those meetings that both Angela and Carmen attend.

46. If all members of the board are to attend a particular meeting, under which of the following circumstances can it be held?

a. Monday at Betty's

b. Tuesday at Frank's

c. Wednesday at Delores's

d. Thursday at Frank's

e. Friday at Betty's

47. Which of the following can be the group that attends a meeting on Wednesday at Betty's?

a. Angela, Betty, Carmen, Ed, and Frank
b. Angela, Betty, Ed, Frank, and Grant
c. Angela, Betty, Carmen, Delores, and Ed
d. Angela, Betty, Delores, Frank, and Grant
e. Angela, Betty, Carmen, Frank, and Grant

48. If Carmen and Angela attend a meeting but Grant is unable to attend, which of the following could be true?

a. The meeting is held on Tuesday.
b. The meeting is held on Friday.
c. The meeting is held at Delores's.
d. The meeting is held at Frank's.
e. The meeting is attended by six of the board members.

49. If the meeting is held on Tuesday at Betty's, which of the following pairs can be among the board members who attend?

a. Angela and Frank
b. Ed and Betty
c. Carmen and Ed
d. Frank and Delores
e. Carmen and Angela

50. If Frank attends a meeting on Thursday that is not held at his house, which of the following must be true?

a. The group can include, at most, two women.
b. The meeting is at Betty's house.
c. Ed is not at the meeting.
d. Grant is not at the meeting.
e. Delores is at the meeting.

51. If Grant is unable to attend a meeting on Tuesday at Delores's, what is the largest possible number of board members who can attend?

a. 1 **b.** 2 **c.** 3
d. 4 **e.** 5

52. If a meeting is held on Friday, which of the following board members *cannot* attend?

a. Grant **b.** Delores **c.** Ed
d. Betty **e.** Frank

WEB PROJECT

53. A person's Rh factor will limit the person's options regarding the blood types he or she may receive during a transfusion. Fill in the following chart. How does a person's Rh factor limit that person's options regarding compatible blood?

If Your Blood Type Is:	You Can Receive:
O+	
O–	
A+	
A–	
B+	
B–	
AB+	
AB–	

Some useful links for this web project are listed on the text web site:
www.cengage.com/math/johnson

2.3 Introduction to Combinatorics

OBJECTIVES

- Develop and apply the Fundamental Principle of Counting
- Develop and evaluate factorials

If you went on a shopping spree and bought two pairs of jeans, three shirts, and two pairs of shoes, how many new outfits (consisting of a new pair of jeans, a new shirt, and a new pair of shoes) would you have? A compact disc buyers' club sends you a brochure saying that you can pick any five CDs from a group of 50 of today's hottest sounds for only $1.99. How many different combinations can you choose? Six local bands have volunteered to perform at a benefit concert, and there is some concern over the order in which the bands will perform. How many different lineups are possible? The answers to questions like these can be obtained by listing all the possibilities or by using three shortcut counting methods: the **Fundamental Principle of Counting, combinations,** and **permutations**. Collectively, these methods are known as **combinatorics**. (Incidentally, the answers to the questions above are 12 outfits, 2,118,760 CD combinations, and 720 lineups.) In this section, we consider the first shortcut method.

The Fundamental Principle of Counting

Daily life requires that we make many decisions. For example, we must decide what food items to order from a menu, what items of clothing to put on in the morning, and what options to order when purchasing a new car. Often, we are asked to make a series of decisions: "Do you want soup or salad? What type of dressing? What type of vegetable? What entrée? What beverage? What dessert?" These individual components of a complete meal lead to the question "Given all the choices of soups, salads, dressings, vegetables, entrées, beverages, and desserts, what is the total number of possible dinner combinations?"

When making a series of decisions, how can you determine the total number of possible selections? One way is to list all the choices for each category and then match them up in all possible ways. To ensure that the choices are matched up in all possible ways, you can construct a **tree diagram.** A tree diagram consists of clusters of line segments, or *branches,* constructed as follows: A cluster of branches is drawn for each decision to be made such that the number of branches in each cluster equals the number of choices for the decision. For instance, if you must make two decisions and there are two choices for decision 1 and three choices for decision 2, the tree diagram would be similar to the one shown in Figure 2.37.

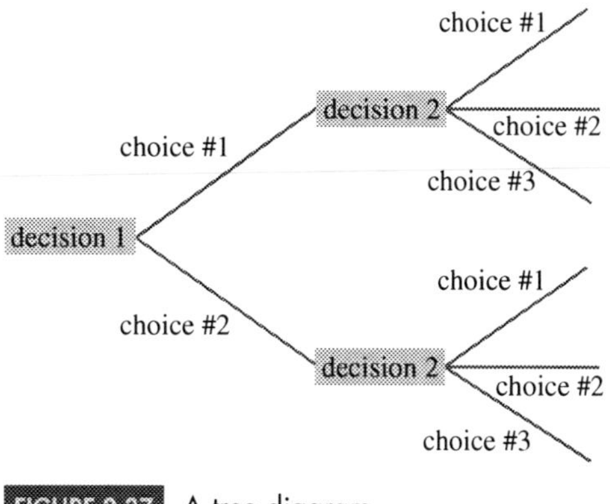

FIGURE 2.37 A tree diagram.

Although this method can be applied to all problems, it is very time consuming and impractical when you are dealing with a series of many decisions, each of which contains numerous choices. Instead of actually listing all possibilities via a tree diagram, using a shortcut method might be desirable. The following example gives a clue to finding such a shortcut.

EXAMPLE 1

DETERMINING THE TOTAL NUMBER OF POSSIBLE CHOICES IN A SERIES OF DECISIONS If you buy two pairs of jeans, three shirts, and two pairs of shoes, how many new outfits (consisting of a new pair of jeans, a new shirt, and a new pair of shoes) would you have?

SOLUTION

Because there are three categories, selecting an outfit requires a series of three decisions: You must select one pair of jeans, one shirt, and one pair of shoes. We will make our three decisions in the following order: jeans, shirt, and shoes. (The order in which the decisions are made does not affect the overall outfit.)

Our first decision (jeans) has two choices (jeans 1 or jeans 2); our tree starts with two branches, as in Figure 2.38.

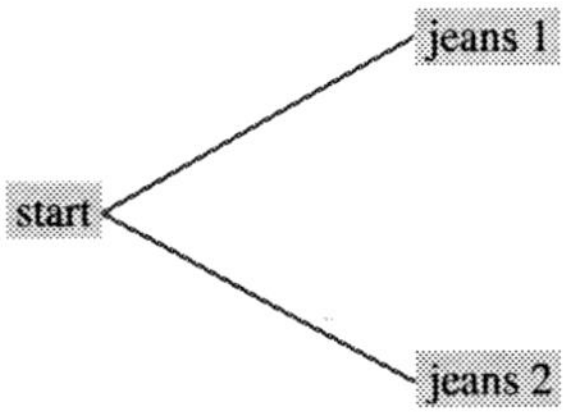

FIGURE 2.38 The first decision.

Our second decision is to select a shirt, for which there are three choices. At each pair of jeans on the tree, we draw a cluster of three branches, one for each shirt, as in Figure 2.39.

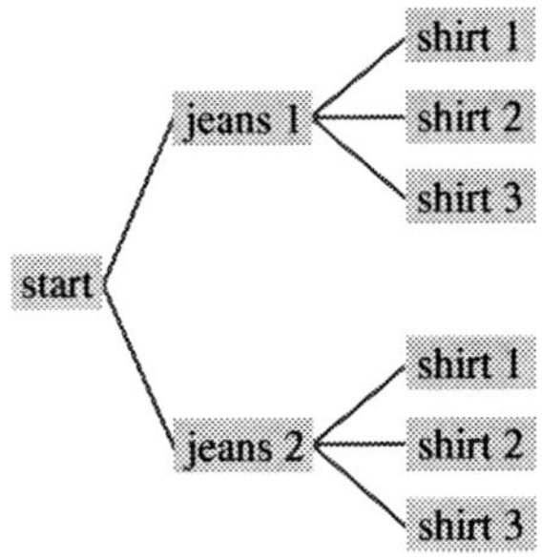

FIGURE 2.39 The second decision.

Our third decision is to select a pair of shoes, for which there are two choices. At each shirt on the tree, we draw a cluster of two branches, one for each pair of shoes, as in Figure 2.40.

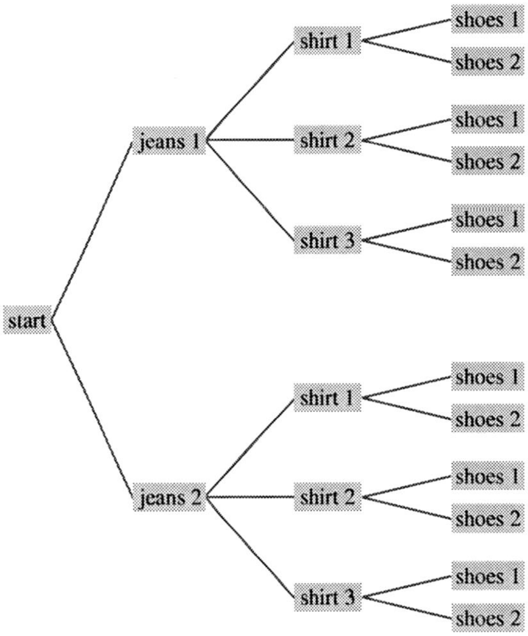

Possible Outfits

jeans 1, shirt 1, shoes 1
jeans 1, shirt 1, shoes 2
jeans 1, shirt 2, shoes 1
jeans 1, shirt 2, shoes 2
jeans 1, shirt 3, shoes 1
jeans 1, shirt 3, shoes 2
jeans 2, shirt 1, shoes 1
jeans 2, shirt 1, shoes 2
jeans 2, shirt 2, shoes 1
jeans 2, shirt 2, shoes 2
jeans 2, shirt 3, shoes 1
jeans 2, shirt 3, shoes 2

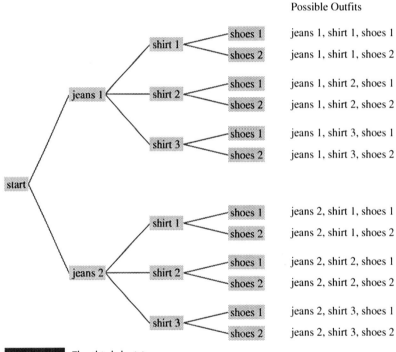

FIGURE 2.40 The third decision.

We have now listed all possible ways of putting together a new outfit; twelve outfits can be formed from two pairs of jeans, three shirts, and two pairs of shoes.

Referring to Example 1, note that each time a decision had to be made, the number of branches on the tree diagram was *multiplied* by a factor equal to the number of choices for the decision. Therefore, the total number of outfits could have been obtained by *multiplying* the number of choices for each decision:

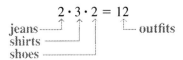

$$2 \cdot 3 \cdot 2 = 12$$

jeans, shirts, shoes (the factors 2, 3, 2); outfits (12)

The generalization of this process of multiplication is called the Fundamental Principle of Counting.

THE FUNDAMENTAL PRINCIPLE OF COUNTING

The total number of possible outcomes of a series of decisions (making selections from various categories) is found by multiplying the number of choices for each decision (or category) as follows:

1. Draw a box for each decision.
2. Enter the number of choices for each decision in the appropriate box and multiply.

EXAMPLE 2

APPLYING THE FUNDAMENTAL PRINCIPLE OF COUNTING A serial number consists of two consonants followed by three nonzero digits followed by a vowel (A, E, I, O, U): for example, "ST423E" and "DD666E." Determine how many serial numbers are possible given the following conditions.

a. Letters and digits cannot be repeated in the same serial number.
b. Letters and digits can be repeated in the same serial number.

SOLUTION

a. Because the serial number has six symbols, we must make six decisions. Consequently, we must draw six boxes:

$\square \; \square \; \square \; \square \; \square \; \square$

There are twenty-one different choices for the first consonant. Because the letters cannot be repeated, there are only twenty choices for the second consonant. Similarly, there are nine different choices for the first nonzero digit, eight choices for the second, and seven choices for the third. There are five different vowels, so the total number of possible serial numbers is

$$\underbrace{\boxed{21} \times \boxed{20}}_{\text{consonants}} \times \underbrace{\boxed{9} \times \boxed{8} \times \boxed{7}}_{\text{nonzero digits}} \times \underbrace{\boxed{5}}_{\text{vowel}} = 1{,}058{,}400$$

There are 1,058,400 possible serial numbers when the letters and digits cannot be repeated within a serial number.

b. Because letters and digits can be repeated, the number of choices does not decrease by one each time as in part (a). Therefore, the total number of possibilities is

$$\underbrace{\boxed{21} \times \boxed{21}}_{\text{consonants}} \times \underbrace{\boxed{9} \times \boxed{9} \times \boxed{9}}_{\text{nonzero digits}} \times \underbrace{\boxed{5}}_{\text{vowel}} = 1{,}607{,}445$$

There are 1,607,445 possible serial numbers when the letters and digits can be repeated within a serial number.

Factorials

EXAMPLE 3

APPLYING THE FUNDAMENTAL PRINCIPLE OF COUNTING Three students rent a three-bedroom house near campus. One of the bedrooms is very desirable (it has its own bath), one has a balcony, and one is undesirable (it is very small). In how many ways can the housemates choose the bedrooms?

SOLUTION

Three decisions must be made: who gets the room with the bath, who gets the room with the balcony, and who gets the small room. Using the Fundamental Principle of Counting, we draw three boxes and enter the number of choices for each decision. There are three choices for who gets the room with the bath. Once that decision has been made, there are two choices for who gets the room with the balcony, and finally, there is only one choice for the small room.

$$\boxed{3} \times \boxed{2} \times \boxed{1} = 6$$

There are six different ways in which the three housemates can choose the three bedrooms.

Combinatorics often involve products of the type $3 \cdot 2 \cdot 1 = 6$, as seen in Example 3. This type of product is called a **factorial,** and the product $3 \cdot 2 \cdot 1$ is written as $3!$. In this manner, $4! = 4 \cdot 3 \cdot 2 \cdot 1$ ($= 24$), and $5! = 5 \cdot 4 \cdot 3 \cdot 2 \cdot 1$ ($= 120$).

FACTORIALS

If n is a positive integer, then *n factorial,* denoted by $n!$, is the product of all positive integers less than or equal to n.

$$n! = n \cdot (n-1) \cdot (n-2) \cdot \cdots \cdot 2 \cdot 1$$

As a special case, we define $0! = 1$.

Many scientific calculators have a button that will calculate a factorial. Depending on your calculator, the button will look like $\boxed{x!}$ or $\boxed{n!}$, and you might have to press a [shift] or [2nd] button first. For example, to calculate 6!, type the number 6, press the factorial button, and obtain 720. To calculate a factorial on most graphing calculators, do the following:

- Type the value of n. (For example, type the number 6.)
- Press the [MATH] button.
- Press the right arrow button [→] as many times as necessary to highlight [PRB].
- Press the down arrow [↓] as many times as necessary to highlight the "!" symbol, and press [ENTER].
- Press [ENTER] to execute the calculation.

To calculate a factorial on a Casio graphing calculator, do the following:

- Press the [MENU] button; this gives you access to the main menu.
- Press 1 to select the RUN mode; this mode is used to perform arithmetic operations.
- Type the value of n. (For example, type the number 6.)
- Press the [OPTN] button; this gives you access to various options displayed at the bottom of the screen.
- Press the [F6] button to see more options (i.e., [→]).
- Press the [F3] button to select probability options (i.e., [PROB]).
- Press the [F1] button to select factorial (i.e., [x!]).
- Press the [EXE] button to execute the calculation.

The factorial symbol "$n!$" was first introduced by Christian Kramp (1760–1826) of Strasbourg in his *Élements d'Arithmétique Universelle* (1808). Before the introduction of this "modern" symbol, factorials were commonly denoted by $\underline{|n}$. However, printing presses of the day had difficulty printing this symbol; consequently, the symbol $n!$ came into prominence because it was relatively easy for a typesetter to use.

EXAMPLE 4

EVALUATING FACTORIALS Find the following values.

a. $6!$ **b.** $\frac{8!}{5!}$ **c.** $\frac{8!}{3! \cdot 5!}$

SOLUTION

a. $6! = 6 \cdot 5 \cdot 4 \cdot 3 \cdot 2 \cdot 1$
$= 720$

Therefore, $6! = 720$.

6 [x!]

6 [MATH] [PRB] [!] [ENTER]

Casio 6 [OPTN] [→] (i.e., [F6]) [PROB] (i.e., [F3]) [x!] (i.e., [F1]) [EXE]

b.
$$\frac{8!}{5!} = \frac{8 \cdot 7 \cdot 6 \cdot 5 \cdot 4 \cdot 3 \cdot 2 \cdot 1}{5 \cdot 4 \cdot 3 \cdot 2 \cdot 1}$$
$$= \frac{8 \cdot 7 \cdot 6 \cdot \cancel{5} \cdot \cancel{4} \cdot \cancel{3} \cdot \cancel{2} \cdot \cancel{1}}{\cancel{5} \cdot \cancel{4} \cdot \cancel{3} \cdot \cancel{2} \cdot \cancel{1}}$$
$$= 8 \cdot 7 \cdot 6$$
$$= 336$$

Therefore, $\frac{8!}{5!} = 336$.

Using a calculator, we obtain the same result.

8 [x!] [÷] 5 [x!] [=]

8 [MATH] [PRB] [!] [÷] 5 [MATH] [PRB] [!] [ENTER]

c.
$$\frac{8!}{3! \cdot 5!} = \frac{8 \cdot 7 \cdot 6 \cdot 5 \cdot 4 \cdot 3 \cdot 2 \cdot 1}{(3 \cdot 2 \cdot 1)(5 \cdot 4 \cdot 3 \cdot 2 \cdot 1)}$$
$$= \frac{8 \cdot 7 \cdot 6 \cdot \cancel{5} \cdot \cancel{4} \cdot \cancel{3} \cdot \cancel{2} \cdot \cancel{1}}{(3 \cdot 2 \cdot 1)\,(\cancel{5} \cdot \cancel{4} \cdot \cancel{3} \cdot \cancel{2} \cdot \cancel{1})}$$
$$= \frac{8 \cdot 7 \cdot 6}{3 \cdot 2 \cdot 1}$$
$$= 56$$

Therefore, $\frac{8!}{3! \cdot 5!} = 56$.

Using a calculator, we obtain the same result.

8 [x!] [÷] [(] 3 [x!] [×] 5 [x!] [)] [=]

8 [MATH] [PRB] [!] [÷] [(] 3 [MATH] [PRB] [!] [×] 5 [MATH] [PRB] [!] [)] [ENTER]

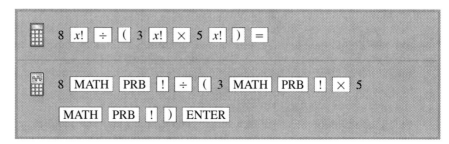

2.3 EXERCISES

1. A nickel, a dime, and a quarter are tossed.
   a. Use the Fundamental Principle of Counting to determine how many different outcomes are possible.
   b. Construct a tree diagram to list all possible outcomes.
2. A die is rolled, and a coin is tossed.
   a. Use the Fundamental Principle of Counting to determine how many different outcomes are possible.
   b. Construct a tree diagram to list all possible outcomes.
3. Jamie has decided to buy either a Mega or a Better Byte desktop computer. She also wants to purchase either Big Word, Word World, or Great Word word-processing software and either Big Number or Number World spreadsheet software.
   a. Use the Fundamental Principle of Counting to determine how many different packages of a computer and software Jamie has to choose from.
   b. Construct a tree diagram to list all possible packages of a computer and software.
4. Sammy's Sandwich Shop offers a soup, sandwich, and beverage combination at a special price. There are three sandwiches (turkey, tuna, and tofu), two soups (minestrone and split pea), and three beverages (coffee, milk, and mineral water) to choose from.
   a. Use the Fundamental Principle of Counting to determine how many different meal combinations are possible.
   b. Construct a tree diagram to list all possible soup, sandwich, and beverage combinations.
5. If you buy three pairs of jeans, four sweaters, and two pairs of boots, how many new outfits (consisting of a new pair of jeans, a new sweater, and a new pair of boots) will you have?
6. A certain model of automobile is available in six exterior colors, three interior colors, and three interior styles. In addition, the transmission can be either manual or automatic, and the engine can have either four or six cylinders. How many different versions of the automobile can be ordered?
7. To fulfill certain requirements for a degree, a student must take one course each from the following groups: health, civics, critical thinking, and elective. If there are four health, three civics, six critical thinking, and ten elective courses, how many different options for fulfilling the requirements does a student have?
8. To fulfill a requirement for a literature class, a student must read one short story by each of the following authors: Stephen King, Clive Barker, Edgar Allan Poe, and H. P. Lovecraft. If there are twelve King, six Barker, eight Poe, and eight Lovecraft stories to choose from, how many different combinations of reading assignments can a student choose from to fulfill the reading requirement?
9. A sporting goods store has fourteen lines of snow skis, seven types of bindings, nine types of boots, and three types of poles. Assuming that all items are compatible with each other, how many different complete ski equipment packages are available?
10. An audio equipment store has ten different amplifiers, four tuners, six turntables, eight tape decks, six compact disc players, and thirteen speakers. Assuming that all components are compatible with each other, how many different complete stereo systems are available?
11. A cafeteria offers a complete dinner that includes one serving each of appetizer, soup, entrée, and dessert for \$6.99. If the menu has three appetizers, four soups, six entrées, and three desserts, how many different meals are possible?
12. A sandwich shop offers a "U-Chooz" special consisting of your choice of bread, meat, cheese, and special sauce (one each). If there are six different breads, eight meats, five cheeses, and four special sauces, how many different sandwiches are possible?
13. How many different Social Security numbers are possible? (A Social Security number consists of nine digits that can be repeated.)
14. To use an automated teller machine (ATM), a customer must enter his or her four-digit Personal Identification Number (PIN). How many different PINs are possible?
15. Every book published has an International Standard Book Number (ISBN). The number is a code used to identify the specific book and is of the form X-XXX-XXXXX-X, where X is one of digits 0, 1, 2, . . . , 9. How many different ISBNs are possible?
16. How many different Zip Codes are possible using (a) the old style (five digits) and (b) the new style (nine digits)? Why do you think the U.S. Postal Service introduced the new system?
17. Telephone area codes are three-digit numbers of the form XXX.
   a. Originally, the first and third digits were neither 0 nor 1 and the second digit was always a 0 or a 1. How many three-digit numbers of this type are possible?
   b. Over time, the restrictions listed in part (a) have been altered; currently, the only requirement is that the first digit is neither 0 nor 1. How many three-digit numbers of this type are possible?

c. Why were the original restrictions listed in part (a) altered?

18. Major credit cards such as VISA and MasterCard have a sixteen-digit account number of the form XXXX-XXXX-XXXX-XXXX. How many different numbers of this type are possible?

19. The serial number on a dollar bill consists of a letter followed by eight digits and then a letter. How many different serial numbers are possible, given the following conditions?

a. Letters and digits cannot be repeated.
b. Letters and digits can be repeated.
c. The letters are nonrepeated consonants and the digits can be repeated.

20. The serial number on a new twenty-dollar bill consists of two letters followed by eight digits and then a letter. How many different serial numbers are possible, given the following conditions?

a. Letters and digits cannot be repeated.
b. Letters and digits can be repeated.
c. The first and last letters are repeatable vowels, the second letter is a consonant, and the digits can be repeated.

21. Each student at State University has a student I.D. number consisting of four digits (the first digit is nonzero, and digits may be repeated) followed by three of the letters A, B, C, D, and E (letters may not be repeated). How many different student numbers are possible?

22. Each student at State College has a student I.D. number consisting of five digits (the first digit is nonzero, and digits may be repeated) followed by two of the letters A, B, C, D, and E (letters may not be repeated). How many different student numbers are possible?

In Exercises 23–38, find the indicated value.

23. $4!$ **24.** $5!$

25. $10!$ **26.** $8!$

27. $20!$ **28.** $25!$

29. $6! \cdot 4!$ **30.** $8! \cdot 6!$

31. **a.** $\frac{6!}{4!}$ **b.** $\frac{6!}{2!}$ **32.** **a.** $\frac{8!}{6!}$ **b.** $\frac{8!}{2!}$

33. $\frac{8!}{5! \cdot 3!}$ **34.** $\frac{9!}{5! \cdot 4!}$

35. $\frac{8!}{4! \cdot 4!}$ **36.** $\frac{6!}{3! \cdot 3!}$

37. $\frac{82!}{80! \cdot 2!}$ **38.** $\frac{77!}{74! \cdot 3!}$

39. Find the value of $\frac{n!}{(n-r)!}$ when $n = 16$ and $r = 14$.

40. Find the value of $\frac{n!}{(n-r)!}$ when $n = 19$ and $r = 16$.

41. Find the value of $\frac{n!}{(n-r)!}$ when $n = 5$ and $r = 5$.

42. Find the value of $\frac{n!}{(n-r)!}$ when $n = r$.

43. Find the value of $\frac{n!}{(n-r)!r!}$ when $n = 7$ and $r = 3$.

44. Find the value of $\frac{n!}{(n-r)!r!}$ when $n = 7$ and $r = 4$.

45. Find the value of $\frac{n!}{(n-r)!r!}$ when $n = 5$ and $r = 5$.

46. Find the value of $\frac{n!}{(n-r)!r!}$ when $n = r$.

Answer the following questions using complete sentences and your own words.

• CONCEPT QUESTIONS

47. What is the Fundamental Principle of Counting? When is it used?

48. What is a factorial?

• HISTORY QUESTIONS

49. Who invented the modern symbol denoting a factorial? What symbol did it replace? Why?

THE NEXT LEVEL

If a person wants to pursue an advanced degree (something beyond a bachelor's or four-year degree), chances are the person must take a standardized exam to gain admission to a school or to be admitted into a specific program. These exams are intended to measure verbal, quantitative, and analytical skills that have developed throughout a person's life. Many classes and study guides are available to help people prepare for the exams. The following questions are typical of those found in the study guides.

Exercises 50–54 refer to the following: In an executive parking lot, there are six parking spaces in a row, labeled 1 through 6. Exactly five cars of five different colors—black, gray, pink, white, and yellow—are to be parked in the spaces. The cars can park in any of the spaces as long as the following conditions are met:

- The pink car must be parked in space 3.
- The black car must be parked in a space next to the space in which the yellow car is parked.
- The gray car cannot be parked in a space next to the space in which the white car is parked.

50. If the yellow car is parked in space 1, how many acceptable parking arrangements are there for the five cars?

a. 1 **b.** 2 **c.** 3 **d.** 4 **e.** 5

51. Which of the following must be true of any acceptable parking arrangement?

a. One of the cars is parked in space 2.
b. One of the cars is parked in space 6.

c. There is an empty space next to the space in which the gray car is parked.
d. There is an empty space next to the space in which the yellow car is parked.
e. Either the black car or the yellow car is parked in a space next to space 3.

52. If the gray car is parked in space 2, none of the cars can be parked in which space?
a. 1 **b.** 3 **c.** 4 **d.** 5 **e.** 6

53. The white car could be parked in any of the spaces except which of the following?
a. 1 **b.** 2 **c.** 4 **d.** 5 **e.** 6

54. If the yellow car is parked in space 2, which of the following must be true?
a. None of the cars is parked in space 5.
b. The gray car is parked in space 6.
c. The black car is parked in a space next to the space in which the white car is parked.
d. The white car is parked in a space next to the space in which the pink car is parked.
e. The gray car is parked in a space next to the space in which the black car is parked.

2.4 Permutations and Combinations

OBJECTIVES

- Develop and apply the Permutation Formula
- Develop and apply the Combination Formula
- Determine the number of distinguishable permutations

The Fundamental Principle of Counting allows us to determine the total number of possible outcomes when a series of decisions (making selections from various categories) must be made. In Section 2.3, the examples and exercises involved selecting *one item each* from various categories; if you buy two pairs of jeans, three shirts, and two pairs of shoes, you will have twelve ($2 \cdot 3 \cdot 2 = 12$) new outfits (consisting of a new pair of jeans, a new shirt, and a new pair of shoes). In this section, we examine the situation when *more than one* item is selected from a category. If more than one item is selected, the selections can be made either *with* or *without* replacement.

With versus Without Replacement

Selecting items *with replacement* means that the same item *can* be selected more than once; after a specific item has been chosen, it is put back into the pool of future choices. Selecting items *without replacement* means that the same item *cannot* be selected more than once; after a specific item has been chosen, it is not replaced.

Suppose you must select a four-digit Personal Identification Number (PIN) for a bank account. In this case, the digits are selected with replacement; each time a specific digit is selected, the digit is put back into the pool of choices for the next selection. (Your PIN can be 3666; the same digit can be selected more than once.) When items are selected with replacement, we use the Fundamental Principle of Counting to determine the total number of possible outcomes; there are $10 \cdot 10 \cdot 10 \cdot 10 = 10{,}000$ possible four-digit PINs.

In many situations, items cannot be selected more than once. For instance, when selecting a committee of three people from a group of twenty, you cannot select the same person more than once. Once you have selected a specific person (say, Lauren), you do not put her back into the pool of choices. When selecting items without replacement, depending on whether the order of selection is important, *permutations* or *combinations* are used to determine the total number of possible outcomes.

Permutations

When more than one item is selected (without replacement) from a single category, and the order of selection *is* important, the various possible outcomes are called **permutations.** For example, when the rankings (first, second, and third place) in a talent contest are announced, the order of selection is important; Monte in first, Lynn in second, and Ginny in third place is different from Ginny in first, Monte in second, and Lynn in third. "Monte, Lynn, Ginny" and "Ginny, Monte, Lynn" are different permutations of the contestants. Naturally, these selections are made without replacement; we cannot select Monte for first place and reselect him for second place.

EXAMPLE 1

FINDING THE NUMBER OF PERMUTATIONS Six local bands have volunteered to perform at a benefit concert, but there is enough time for only four bands to play. There is also some concern over the order in which the chosen bands will perform. How many different lineups are possible?

SOLUTION

We must select four of the six bands and put them in a specific order. The bands are selected without replacement; a band cannot be selected to play and then be reselected to play again. Because we must make four decisions, we draw four boxes and put the number of choices for each decision in each appropriate box. There are six choices for the opening band. Naturally, the opening band could not be the follow-up act, so there are only five choices for the next group. Similarly, there are four candidates for the third group and three choices for the closing band. The total number of different lineups possible is found by multiplying the number of choices for each decision:

$$\boxed{6} \times \boxed{5} \times \boxed{4} \times \boxed{3} = 360$$

↑ opening band ↑ closing band

With four out of six bands playing in the performance, 360 lineups are possible. *Because the order of selecting the bands is important, the various possible outcomes, or lineups, are called permutations;* there are 360 permutations of six items when the items are selected four at a time.

The computation in Example 1 is similar to a factorial, but the factors do not go all the way down to 1; the product $6 \cdot 5 \cdot 4 \cdot 3$ is a "truncated" (cut-off) factorial. We can change this truncated factorial into a complete factorial in the following manner:

$$6 \cdot 5 \cdot 4 \cdot 3 = \frac{6 \cdot 5 \cdot 4 \cdot 3 \cdot (2 \cdot 1)}{(2 \cdot 1)} \quad \text{multiplying by } \frac{2}{2} \text{ and } \frac{1}{1}$$

$$= \frac{6!}{2!}$$

Notice that this last expression can be written as $\frac{6!}{2!} = \frac{6!}{(6-4)!}$. (Recall that we were selecting four out of six bands.) This result is generalized as follows.

PERMUTATION FORMULA

The number of **permutations,** or arrangements, of r items selected without replacement from a pool of n items ($r \leq n$), denoted by ${}_nP_r$, is

$$ {}_nP_r = \frac{n!}{(n-r)!} $$

Permutations are used whenever more than one item is selected (without replacement) from a category and the order of selection is important.

Using the notation above and referring to Example 1, we note that 360 possible lineups of four bands selected from a pool of six can be denoted by ${}_6P_4 = \frac{6!}{(6-4)!} = 360$. Other notations can be used to represent the number of permutations of a group of items. In particular, the notations ${}_nP_r$, $P(n, r)$, P_r^n, and $P_{n,r}$ all represent the number of possible permutations (or arrangements) of r items selected (without replacement) from a pool of n items.

EXAMPLE 2

FINDING THE NUMBER OF PERMUTATIONS Three door prizes (first, second, and third) are to be awarded at a ten-year high school reunion. Each of the 112 attendees puts his or her name in a hat. The first name drawn wins a two-night stay at the Chat 'n' Rest Motel, the second name wins dinner for two at Juju's Kitsch-Inn, and the third wins a pair of engraved mugs. In how many different ways can the prizes be awarded?

SOLUTION

We must select 3 out of 112 people (without replacement), and the order in which they are selected *is* important. (Winning dinner is different from winning the mugs.) Hence, we must find the number of permutations of 3 items selected from a pool of 112:

$$ \begin{aligned} {}_{112}P_3 &= \frac{112!}{(112-3)!} \\ &= \frac{112!}{109!} \\ &= \frac{112 \cdot 111 \cdot 110 \cdot 109 \cdot 108 \cdot \cdots \cdot 2 \cdot 1}{109 \cdot 108 \cdot \cdots \cdot 2 \cdot 1} \\ &= 112 \cdot 111 \cdot 110 \\ &= 1{,}367{,}520 \end{aligned} $$

There are 1,367,520 different ways in which the three prizes can be awarded to the 112 people.

In Example 2, if you try to use a calculator to find $\frac{112!}{109!}$ directly, you will not obtain an answer. Entering 112 and pressing $\boxed{x!}$ results in a calculator error. (Try it.) Because factorials get very large very quickly, most calculators are not able to find any factorial over 69!. ($69! = 1.711224524 \times 10^{98}$.)

EXAMPLE **3**

FINDING THE NUMBER OF PERMUTATIONS A bowling league has ten teams. In how many different ways can the teams be ranked in the standings at the end of a tournament? (Ties are not allowed.)

SOLUTION

Because order is important, we find the number of permutations of ten items selected from a pool of ten items:

$$\begin{aligned} {}_{10}P_{10} &= \frac{10!}{(10-10)!} \\ &= \frac{10!}{0!} \qquad \text{Recall that } 0! = 1. \\ &= \frac{10!}{1} \\ &= 3{,}628{,}800 \end{aligned}$$

In a league containing ten teams, 3,628,800 different standings are possible at the end of a tournament.

Combinations

When items are selected from a group, the order of selection may or may not be important. If the order is important (as in Examples 1, 2, and 3), permutations are used to determine the total number of selections possible. What if the order of selection is *not* important? When more than one item is selected (without replacement) from a single category and the order of selection is not important, the various possible outcomes are called **combinations.**

EXAMPLE **4**

LISTING ALL POSSIBLE COMBINATIONS Two adults are needed to chaperone a daycare center's field trip. Marcus, Vivian, Frank, and Keiko are the four managers of the center. How many different groups of chaperones are possible?

SOLUTION

In selecting the chaperones, the order of selection is *not* important; "Marcus and Vivian" is the same as "Vivian and Marcus." Hence, the permutation formula cannot be used. Because we do not yet have a shortcut for finding the total number of possibilities when the order of selection is not important, we must list all the possibilities:

Marcus and Vivian	Marcus and Frank	Marcus and Keiko
Vivian and Frank	Vivian and Keiko	Frank and Keiko

Therefore, six different groups of two chaperones are possible from the group of four managers. Because the order in which the people are selected is not important, the various possible outcomes, or groups of chaperones, are called *combinations;* there are six combinations when two items are selected from a pool of four.

Just as ${}_nP_r$ denotes the number of *permutations* of r elements selected from a pool of n elements, ${}_nC_r$ denotes the number of *combinations* of r elements selected from a pool of n elements. In Example 4, we found that there are six combinations of two people selected from a pool of four by listing all six of the combinations; that is, ${}_4C_2 = 6$. If we had a larger pool, listing each combination to find out how many there are would be extremely time consuming and tedious! Instead of listing, we take a different approach. We first find the number of permutations (with the permutation formula) and then alter that number to account for the distinction between permutations and combinations.

To find the number of combinations of two people selected from a pool of four, we first find the number of permutations:

$${}_4P_2 = \frac{4!}{(4-2)!} = \frac{4!}{2!} = 12$$

This figure of 12 must be altered to account for the distinction between permutations and combinations.

In Example 4, we listed combinations; one such combination was "Marcus and Vivian." If we had listed permutations, we would have had to list both "Marcus and Vivian" and "Vivian and Marcus," because the *order* of selection matters with permutations. In fact, each combination of two chaperones listed in Example 4 generates two permutations; each pair of chaperones can be given in two different orders. Thus, there are twice as many permutations of two people selected from a pool of four as there are combinations. Alternatively, there are half as many combinations of two people selected from a pool of four as there are permutations. We used the permutation formula to find that ${}_4P_2 = 12$; thus,

$${}_4C_2 = \frac{1}{2} \cdot {}_4P_2 = \frac{1}{2}(12) = 6$$

This answer certainly fits with Example 4; we listed exactly six combinations.

What if three of the four managers were needed to chaperone the daycare center's field trip? Rather than finding the number of combinations by listing each possibility, we first find the number of permutations and then alter that number to account for the distinction between permutations and combinations.

The number of permutations of three people selected from a pool of four is

$${}_4P_3 = \frac{4!}{(4-3)!} = \frac{4!}{1!} = 24$$

We know that some of these permutations represent the same combination. For example, the combination "Marcus and Vivian and Keiko" generates $3! = 6$ different permutations (using initials, they are: MVK, MKV, KMV, KVM, VMK, VKM). Because each combination of three people generates six different permutations, there are one-sixth as many combinations as permutations. Thus,

$${}_4C_3 = \frac{1}{6} \cdot {}_4P_3 = \frac{1}{6}(24) = 4$$

This means that if three of the four managers were needed to chaperone the daycare center's field trip, there would be ${}_4C_3 = 4$ possible combinations.

We just saw that when two items are selected from a pool of n items, each combination of two generates $2! = 2$ permutations, so

$${}_nC_2 = \frac{1}{2!} \cdot {}_nP_2$$

We also saw that when three items are selected from a pool of n items, each combination of three generates $3! = 6$ permutations, so

$${}_nC_3 = \frac{1}{3!} \cdot {}_nP_3$$

More generally, when r items are selected from a pool of n items, each combination of r items generates $r!$ permutations, so

$$\begin{aligned} {}_nC_r &= \frac{1}{r!} \cdot {}_nP_r \\ &= \frac{1}{r!} \cdot \frac{n!}{(n-r)!} && \text{using the Permutation Formula} \\ &= \frac{n!}{r! \cdot (n-r)!} && \text{multiplying the fractions together} \end{aligned}$$

COMBINATION FORMULA

The number of distinct **combinations** of r items selected without replacement from a pool of n items ($r \leq n$), denoted by ${}_nC_r$, is

$${}_nC_r = \frac{n!}{(n-r)!\,r!}$$

Combinations are used whenever one or more items are selected (without replacement) from a category and the order of selection is not important.

EXAMPLE 5

FINDING THE NUMBER OF COMBINATIONS A DVD club sends you a brochure that offers any five DVDs from a group of fifty of today's hottest releases. How many different selections can you make?

SOLUTION

Because the order of selection is *not* important, we find the number of combinations when five items are selected from a pool of fifty:

$$\begin{aligned} {}_{50}C_5 &= \frac{50!}{(50-5)!\,5!} \\ &= \frac{50!}{45!\,5!} \\ &= \frac{50 \cdot 49 \cdot 48 \cdot 47 \cdot 46}{5 \cdot 4 \cdot 3 \cdot 2 \cdot 1} \\ &= 2{,}118{,}760 \end{aligned}$$

50 [x!] [÷] [(] 45 [x!] [×] 5 [x!] [)] [=]

Graphing calculators have buttons that will calculate ${}_nP_r$ and ${}_nC_r$. To use them, do the following:

- Type the value of n. (For example, type the number 50.)
- Press the [MATH] button.
- Press the right arrow button [→] as many times as necessary to highlight [PRB].
- Press the down arrow button [↓] as many times as necessary to highlight the appropriate symbol— [${}_nP_r$] for permutations, [${}_nC_r$] for combinations—and press [ENTER].
- Type the value of r. (For example, type the number 5.)
- Press [ENTER] to execute the calculation.

On a Casio graphing calculator, do the following:

- Press the MENU button; this gives you access to the main menu.
- Press 1 to select the RUN mode; this mode is used to perform arithmetic operations.
- Type the value of n. (For example, type the number 50.)
- Press the OPTN button; this gives you access to various options displayed at the bottom of the screen.
- Press the F6 button to see more options (i.e., →).
- Press the F3 button to select probability options (i.e., PROB).
- Press the F3 button to select combinations (i.e., $_nC_r$) or the F2 button to select permutations (i.e., $_nP_r$).
- Type the value of r. (For example, type the number 5.)
- Press the EXE button to execute the calculation.

TI	50 MATH PRB $_nC_r$ 5 ENTER
Casio	50 OPTN → (i.e., F6) PROB (i.e., F3) nCr (i.e., F3) 5 EXE

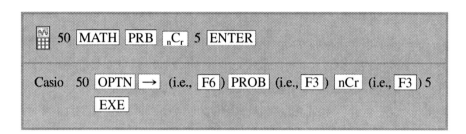

In choosing five out of fifty DVDs, 2,118,760 combinations are possible.

EXAMPLE 6

FINDING THE NUMBER OF COMBINATIONS A group consisting of twelve women and nine men must select a five-person committee. How many different committees are possible if it must consist of the following?

a. three women and two men **b.** any mixture of men and women

SOLUTION

a. Our problem involves two categories: women and men. The Fundamental Principle of Counting tells us to draw two boxes (one for each category), enter the number of choices for each, and multiply:

the number of ways in which we can select three out of twelve women	×	the number of ways in which we can select two out of nine men	= ?

Because the order of selecting the members of a committee is not important, we will use combinations:

$$
\begin{aligned}
({}_{12}C_3)\cdot({}_9C_2) &= \frac{12!}{(12-3)!\cdot 3!}\cdot\frac{9!}{(9-2)!\cdot 2!}\\
&= \frac{12!}{9!\cdot 3!}\cdot\frac{9!}{7!\cdot 2!}\\
&= \frac{12\cdot 11\cdot 10}{3\cdot 2\cdot 1}\cdot\frac{9\cdot 8}{2\cdot 1}\\
&= 220\cdot 36\\
&= 7{,}920
\end{aligned}
$$

(12 [x!] [÷] (9 [x!] [×] 3 [x!])) [×] (9 [x!] [÷] (7 [x!] [×] 2 [x!])) [=]

12 [MATH] [PRB] [nCr] 3 [×] 9 [MATH] [PRB] [nCr] 2 [ENTER]

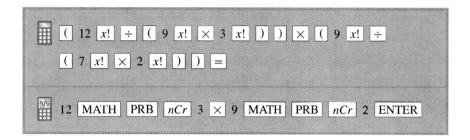

There are 7,920 different committees consisting of three women and two men.

b. Because the gender of the committee members doesn't matter, our problem involves only one category: people. We must choose five out of the twenty-one people, and the order of selection is not important:

$$\begin{aligned} {}_{21}C_5 &= \frac{21!}{(21-5)! \cdot 5!} \\ &= \frac{21!}{16! \cdot 5!} \\ &= \frac{21 \cdot 20 \cdot 19 \cdot 18 \cdot 17}{5 \cdot 4 \cdot 3 \cdot 2 \cdot 1} \\ &= 20{,}349 \end{aligned}$$

21 [x!] [÷] (16 [x!] [×] 5 [x!]) [=]

21 [MATH] [PRB] [nCr] 5 [ENTER]

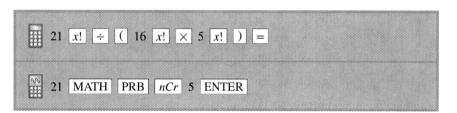

There are 20,349 different committees consisting of five people.

EXAMPLE 7

EVALUATING THE COMBINATION FORMULA Find the value of ${}_5C_r$ for the following values of r:

a. $r = 0$ **b.** $r = 1$ **c.** $r = 2$ **d.** $r = 3$ **e.** $r = 4$ **f.** $r = 5$

SOLUTION

a. $${}_5C_0 = \frac{5!}{(5-0)! \cdot 0!} = \frac{5!}{5! \cdot 0!} = 1$$

b. $${}_5C_1 = \frac{5!}{(5-1)! \cdot 1!} = \frac{5!}{4! \cdot 1!} = 5$$

c. $${}_5C_2 = \frac{5!}{(5-2)! \cdot 2!} = \frac{5!}{3! \cdot 2!} = 10$$

d. $${}_5C_3 = \frac{5!}{(5-3)! \cdot 3!} = \frac{5!}{2! \cdot 3!} = 10$$

e. $${}_5C_4 = \frac{5!}{(5-4)! \cdot 4!} = \frac{5!}{1! \cdot 4!} = 5$$

f. $${}_5C_5 = \frac{5!}{(5-5)! \cdot 5!} = \frac{5!}{0! \cdot 5!} = 1$$

The combinations generated in Example 7 exhibit a curious pattern. Notice that the values of $_5C_r$ are symmetric: $_5C_0 = {}_5C_5$, $_5C_1 = {}_5C_4$, and $_5C_2 = {}_5C_3$. Now examine the diagram in Figure 2.41. Each number in this "triangle" of numbers is the sum of two numbers in the row immediately above it. For example, $2 = 1 + 1$ and $10 = 4 + 6$, as shown by the inserted arrows. It is no coincidence that the values of $_5C_r$ found in Example 7 also appear as a row of numbers in this "magic" triangle. In fact, the sixth row contains all the values of $_5C_r$ for $r = 0, 1, 2, 3, 4$, and 5. In general, the $(n + 1)^{\text{th}}$ row of the triangle contains all the values of $_nC_r$ for $r = 0, 1, 2, \ldots, n$; alternatively, the nth row of the triangle contains all the values of $_{n-1}C_r$ for $r = 0, 1, 2, \ldots, n-1$. For example, the values of $_9C_r$, for $r = 0, 1, 2, \ldots, 9$, are in the tenth row, and vice versa, the entries in the tenth row are the values of $_9C_r$, for $r = 0, 1, 2, \ldots, 9$.

1st row 1

2nd row 1 1

3rd row 1 2 1

4th row 1 3 3 1

5th row 1 4 6 4 1

6th row 1 5 10 10 5 1

and so on

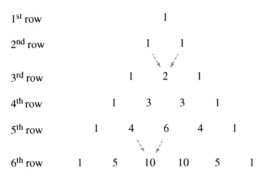

FIGURE 2.41 Pascal's triangle.

Historically, this triangular pattern of numbers is referred to as *Pascal's Triangle,* in honor of the French mathematician, scientist, and philosopher Blaise Pascal (1623–1662). Pascal is a cofounder of probability theory (see the Historical Note in Section 3.1). Although the triangle has Pascal's name attached to it, this "magic" arrangement of numbers was known to other cultures hundreds of years before Pascal's time.

The most important part of any problem involving combinatorics is deciding which counting technique (or techniques) to use. The following list of general steps and the flowchart in Figure 2.42 can help you to decide which method or methods to use in a specific problem.

WHICH COUNTING TECHNIQUE?

1. What is being selected?
2. If the selected items can be repeated, use the **Fundamental Principle of Counting** and multiply the number of choices for each category.
3. If there is only one category, use:
   combinations if the order of selection does not matter—that is, r items can be selected from a pool of n items in $_nC_r = \frac{n!}{(n-r)! \cdot r!}$ ways.
   permutations if the order of selection does matter—that is, r items can be selected from a pool of n items in $_nP_r = \frac{n!}{(n-r)!}$ ways.
4. If there is more than one category, use the **Fundamental Principle of Counting** with one box per category.
   a. If you are selecting one item per category, the number in the box for that category is the number of choices for that category.
   b. If you are selecting more than one item per category, the number in the box for that category is found by using step 3.

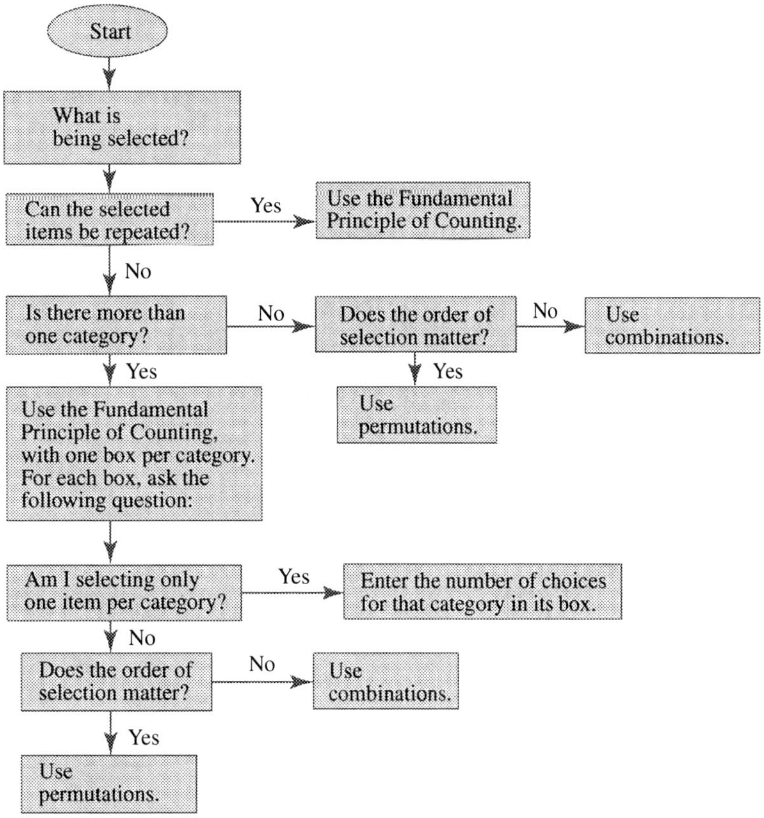

FIGURE 2.42 Which counting technique?

EXAMPLE 8

USING THE "WHICH COUNTING TECHNIQUE?" FLOWCHART A standard deck of playing cards contains fifty-two cards.

a. How many different five-card hands containing four kings are possible?
b. How many different five-card hands containing four queens are possible?
c. How many different five-card hands containing four kings or four queens are possible?
d. How many different five-card hands containing four of a kind are possible?

SOLUTION

a. We use the flowchart in Figure 2.42 and answer the following questions.

Q. What is being selected?
A. Playing cards.

Q. Can the selected items be repeated?
A. No.

Q. Is there more than one category?
A. Yes: Because we must have five cards, we need four kings and one non-king. Therefore, we need two boxes:

kings × non-kings

Q. Am I selecting only one item per category?
A. *Kings:* no. Does the order of selection matter? No: Use combinations. Because there are $n = 4$ kings in the deck and we want to select $r = 4$, we must compute ${}_4C_4$. *Non-kings:* yes. Enter the number of choices for that category: There are 48 non-kings.

HISTORICAL NOTE

CHU SHIH-CHIEH, CIRCA 1280–1303

Chu Shih-chieh was the last and most acclaimed mathematician of the Sung Dynasty in China. Little is known of his personal life; the actual dates of his birth and death are unknown. His work appears to have flourished during the close of the thirteenth century. It is believed that Chu Shih-chieh spent many years as a wandering scholar, earning a living by teaching mathematics to those who wanted to learn.

Two of Chu Shih-chieh's works have survived the centuries. The first, *Suan-hsüeh ch'i-meng (Introduction to Mathematical Studies)*, was written in 1299 and contains elementary mathematics. This work was very influential in Japan and Korea, although it was lost in China until the nineteenth century. Written in 1303, Chu's second work *Ssu-yüan yü-chien (Precious Mirror of the Four Elements)* contains more advanced mathematics. The topics of *Precious Mirror* include the solving of simultaneous equations and the solving of equations up to the fourteenth degree.

Of the many diagrams in *Precious Mirror,* one has special interest: the arithmetic triangle. Chu Shih-chieh's triangle contains the first eight rows of what is known in the West as Pascal's Triangle. However, Chu does not claim credit for the triangle; he refers to it as "a diagram of the *old* method for finding eighth and lower powers." "Pascal's" Triangle was known to the Chinese well over 300 years before Pascal was born!

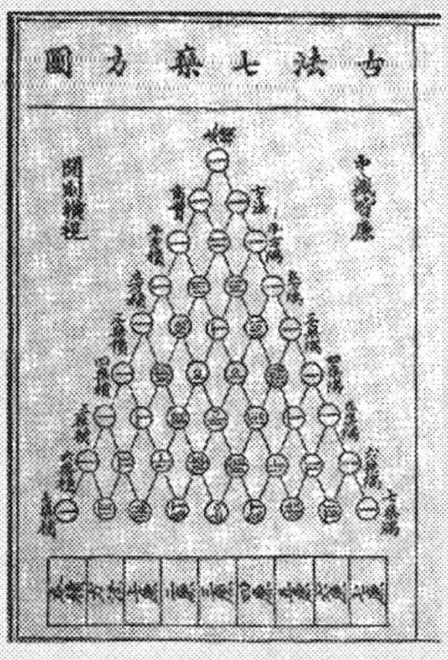

Brown University Library

The "Pascal" Triangle as depicted in 1303 at the front of Chu Shih-chieh's Ssu-yüan yü-chien. *It is entitled "The Old Method Chart of the Seven Multiplying Squares" and tabulates the binomial coefficients up to the eighth power.*

$$
\begin{aligned}
\boxed{\text{kings}} \times \boxed{\text{non-kings}} &= \boxed{{}_4C_4} \times \boxed{48} \\
&= \frac{4!}{(4-4)! \cdot 4!} \cdot 48 \\
&= \frac{4!}{0! \cdot 4!} \cdot 48 \\
&= 1 \cdot 48 \\
&= 48
\end{aligned}
$$

There are forty-eight different five-card hands containing four kings.

b. Using the same method as in part (a), we would find that there are forty-eight different five-card hands containing four queens; the number of five-card hands containing four queens is the same as the number of five-card hands containing four kings.

c. To find the number of five-card hands containing four kings or four queens, we define the following sets:

$$A = \{\text{five-card hands} \mid \text{the hand contains four kings}\}$$
$$B = \{\text{five-card hands} \mid \text{the hand contains four queens}\}$$

Consequently,

$$A \cup B = \{\text{five-card hands} \mid \text{the hand contains four kings } \textit{or} \text{ four queens}\}$$
$$A \cap B = \{\text{five-card hands} \mid \text{the hand contains four kings } \textit{and} \text{ four queens}\}$$

(Recall that the union symbol, $\cup$, may be interpreted as the word "or," while the intersection symbol, $\cap$, may be interpreted as the word "and." See Figure 2.13 for a comparison of set theory and logic.)

Because there are no five-card hands that contain four kings *and* four queens, we note that $n(A \cap B) = 0$.

Using the Union/Intersection Formula for the Union of Sets, we obtain

$$\begin{aligned} n(A \cup B) &= n(A) + n(B) - n(A \cap B) \\ &= 48 + 48 - 0 \\ &= 96 \end{aligned}$$

There are ninety-six different five-card hands containing four kings or four queens.

d. *Four of a kind* means four cards of the same "denomination," that is, four 2's, or four 3's, or four 4's, or . . . , or four kings, or four aces. Now, regardless of the denomination of the card, there are forty-eight different five-card hands that contain four of any specific denomination; there are forty-eight different five-card hands that contain four 2's, there are forty-eight different five-card hands that contain four 3's, there are forty-eight different five-card hands that contain four 4's, and so on. As is shown in part (c), the word "or" implies that we *add* cardinal numbers. Consequently,

$$\begin{aligned} n(\text{four of a kind}) &= n(\text{four 2's or four 3's or} \ldots \text{or four kings or four aces}) \\ &= n(\text{four 2's}) + n(\text{four 3's} + \cdots + n(\text{four kings}) \\ &\quad + n(\text{four aces}) \\ &= 48 + 48 + \cdots + 48 + 48 \quad (\text{thirteen times}) \\ &= 13 \times 48 \\ &= 624 \end{aligned}$$

There are 624 different five-card hands containing four of a kind.

As is shown in Example 8, there are 624 possible five-card hands that contain four of a kind. When you are dealt five cards, what is the likelihood (or probability) that you will receive one of these hands? This question, and its answer, will be explored in Section 3.4, "Combinatorics and Probability."

Permutations of Identical Items

In how many different ways can the three letters in the word "SAW" be arranged? As we know, arrangements are referred to as *permutations,* so we can apply the Permutation Formula, ${}_nP_r = \frac{n!}{(n-r)!}$.

Therefore,

$${}_3P_3 = \frac{3!}{(3-3)!} = \frac{3!}{0!} = \frac{3 \cdot 2 \cdot 1}{1} = 6$$

The six permutations of the letters in SAW are

SAW SWA AWS ASW WAS WSA

In general, if we have three *different* items (the letters in SAW), we can arrange them in $3! = 6$ ways. However, this method applies only if the items are all different (distinct).

What happens if some of the items are the same (identical)? For example, in how many different ways can the three letters in the word "SEE" be arranged? Because two of the letters are identical (E), we cannot use the Permutation Formula directly; we take a slightly different approach. Temporarily, let us assume that the E's are written in different colored inks, say, red and blue. Therefore, SEE could be expressed as SEE. These three symbols could be arranged in $3! = 6$ ways as follows:

SEE SEE ESE ESE EES EES

If we now remove the color, the arrangements are

SEE SEE ESE ESE EES EES

Some of these arrangements are duplicates of others; as we can see, there are only three different or **distinguishable permutations,** namely, SEE, ESE, and EES. Notice that when $n = 3$ (the total number of letters in SEE) and $x = 2$ (the number of identical letters), we can divide $n!$ by $x!$ to obtain the number of distinguishable permutations; that is,

$$\frac{n!}{x!} = \frac{3!}{2!} = \frac{3\cdot 2\cdot 1}{2\cdot 1} = \frac{3\cdot \cancel{2}\cdot \cancel{1}}{\cancel{2}\cdot \cancel{1}} = 3$$

This method is applicable because dividing by the factorial of the repeated letter eliminates the duplicate arrangements; the method may by generalized as follows.

DISTINGUISHABLE PERMUTATIONS OF IDENTICAL ITEMS

The number of **distinguishable permutations** (or arrangements) of n items in which x items are identical, y items are identical, z items are identical, and so on, is $\frac{n!}{x!y!z!\cdots}$. That is, to find the number of distinguishable permutations, divide the total factorial by the factorial of each repeated item.

EXAMPLE 9

FINDING THE NUMBER OF DISTINGUISHABLE PERMUTATIONS
Find the number of distinguishable permutations of the letters in the word "MISSISSIPPI."

SOLUTION

The word "MISSISSIPPI" has $n = 11$ letters; I is repeated $x = 4$ times, S is repeated $y - 4$ times, and P is repeated $z = 2$ times. Therefore, we divide the total factorial by the factorial of each repeated letter and obtain

$$\frac{n!}{x!y!z!} = \frac{11!}{4!4!2!} = 34{,}650$$

The letters in the word MISSISSIPPI can be arranged in 34,650 ways. (Note that if the 11 letters were all different, there would be $11! = 39{,}96{,}800$ permutations.)

2.4 EXERCISES

In Exercises 1–12, find the indicated value:

1. a. ${}_7P_3$ b. ${}_7C_3$
2. a. ${}_8P_4$ b. ${}_8C_4$
3. a. ${}_5P_5$ b. ${}_5C_5$
4. a. ${}_9P_0$ b. ${}_9C_0$
5. a. ${}_{14}P_1$ b. ${}_{14}C_1$
6. a. ${}_{13}C_3$ b. ${}_{13}C_{10}$
7. a. ${}_{100}P_3$ b. ${}_{100}C_3$
8. a. ${}_{80}P_4$ b. ${}_{80}C_4$
9. a. ${}_xP_{x-1}$ b. ${}_xC_{x-1}$
10. a. ${}_xP_1$ b. ${}_xC_1$
11. a. ${}_xP_2$ b. ${}_xC_2$
12. a. ${}_xP_{x-2}$ b. ${}_xC_{x-2}$

13. **a.** Find ${}_3P_2$.
    b. List all of the permutations of {a, b, c} when the elements are taken two at a time.
14. **a.** Find ${}_3C_2$.
    b. List all of the combinations of {a, b, c} when the elements are taken two at a time.
15. **a.** Find ${}_4C_2$.
    b. List all of the combinations of {a, b, c, d} when the elements are taken two at a time.
16. **a.** Find ${}_4P_2$.
    b. List all of the permutations of {a, b, c, d} when the elements are taken two at a time.
17. An art class consists of eleven students. All of them must present their portfolios and explain their work to the instructor and their classmates at the end of the semester.
    a. If their names are drawn from a hat to determine who goes first, second, and so on, how many presentation orders are possible?

b. If their names are put in alphabetical order to determine who goes first, second, and so on, how many presentation orders are possible?

18. An English class consists of twenty-three students, and three are to be chosen to give speeches in a school competition. In how many different ways can the teacher choose the team, given the following conditions?
    a. The order of the speakers is important.
    b. The order of the speakers is not important.

19. In how many ways can the letters in the word "school" be arranged? (See the photograph below.)

20. A committee of four is to be selected from a group of sixteen people. How many different committees are possible, given the following conditions?
    a. There is no distinction between the responsibilities of the members.
    b. One person is the chair, and the rest are general members.
    c. One person is the chair, one person is the secretary, one person is responsible for refreshments, and one person cleans up after meetings.

21. A softball league has thirteen teams. If every team must play every other team once in the first round of league play, how many games must be scheduled?

22. In a group of eighteen people, each person shakes hands once with each other person in the group. How many handshakes will occur?

23. A softball league has thirteen teams. How many different end-of-the-season rankings of first, second, and third place are possible (disregarding ties)?

24. Three hundred people buy raffle tickets. Three winning tickets will be drawn at random.
    a. If first prize is \$100, second prize is \$50, and third prize is \$20, in how many different ways can the prizes be awarded?
    b. If each prize is \$50, in how many different ways can the prizes be awarded?

25. A group of nine women and six men must select a four-person committee. How many committees are possible if it must consist of the following?
    a. two women and two men
    b. any mixture of men and women
    c. a majority of women

26. A group of ten seniors, eight juniors, six sophomores, and five freshmen must select a committee of four. How many committees are possible if the committee must contain the following:
    a. one person from each class
    b. any mixture of the classes
    c. exactly two seniors

Exercises 27–32 refer to a deck of fifty-two playing cards (jokers not allowed). If you are unfamiliar with playing cards, see the end of Section 3.1 for a description of a standard deck.

27. How many five-card poker hands are possible?

28. a. How many five-card poker hands consisting of all hearts are possible?
    b. How many five-card poker hands consisting of all cards of the same suit are possible?

AP Photo/News & Record, Joseph Rodriguez

Exercise 19: Right letters, wrong order. SHCOOL is painted along the newly paved road leading to Southern Guilford High School on Drake Road Monday, August 9, 2010, in Greensboro, N.C.

29. **a.** How many five-card poker hands containing exactly three aces are possible?
    b. How many five-card poker hands containing three of a kind are possible?
30. **a.** How many five-card poker hands consisting of three kings and two queens are possible?
    b. How many five-card poker hands consisting of three of a kind and a pair (a *full house*) are possible?
31. How many five-card poker hands containing two pair are possible?

    HINT: You must select two of the thirteen ranks, then select a pair of each, and then one of the remaining cards.
32. How many five-card poker hands containing exactly one pair are possible?

    HINT: After selecting a pair, you must select three of the remaining twelve ranks and then select one card of each.
33. A 6/53 lottery requires choosing six of the numbers 1 through 53. How many different lottery tickets can you choose? (Order is not important, and the numbers do not repeat.)
34. A 7/39 lottery requires choosing seven of the numbers 1 through 39. How many different lottery tickets can you choose? (Order is not important, and the numbers do not repeat.)
35. A 5/36 lottery requires choosing five of the numbers 1 through 36. How many different lottery tickets can you choose? (Order is not important, and the numbers do not repeat.)
36. A 6/49 lottery requires choosing six of the numbers 1 through 49. How many different lottery tickets can you choose? (Order is not important, and the numbers do not repeat.)
37. Which lottery would be easier to win, a 6/53 or a 5/36? Why?

    HINT: See Exercises 33 and 35.
38. Which lottery would be easier to win, a 6/49 or a 7/39? Why?

    HINT: See Exercises 34 and 36.
39. **a.** Find the sum of the entries in the first row of Pascal's Triangle.
    b. Find the sum of the entries in the second row of Pascal's Triangle.
    c. Find the sum of the entries in the third row of Pascal's Triangle.
    d. Find the sum of the entries in the fourth row of Pascal's Triangle.
    e. Find the sum of the entries in the fifth row of Pascal's Triangle.
    f. Is there a pattern to the answers to parts (a)–(e)? If so, describe the pattern you see.
    g. Use the pattern described in part (f) to predict the sum of the entries in the sixth row of Pascal's triangle.
    h. Find the sum of the entries in the sixth row of Pascal's Triangle. Was your prediction in part (g) correct?
    i. Find the sum of the entries in the nth row of Pascal's Triangle.
40. **a.** Add adjacent entries of the sixth row of Pascal's Triangle to obtain the seventh row.
    b. Find ${}_6C_r$ for $r = 0, 1, 2, 3, 4, 5$, and 6.
    c. How are the answers to parts (a) and (b) related?
41. Use Pascal's Triangle to answer the following.
    a. In which row would you find the value of ${}_4C_2$?
    b. In which row would you find the value of ${}_nC_r$?
    c. Is ${}_4C_2$ the second number in the fourth row?
    d. Is ${}_4C_2$ the third number in the fifth row?
    e. What is the location of ${}_nC_r$? Why?
42. Given the set $S = \{a, b, c, d\}$, answer the following.
    a. How many one-element subsets does S have?
    b. How many two-element subsets does S have?
    c. How many three-element subsets does S have?
    d. How many four-element subsets does S have?
    e. How many zero-element subsets does S have?
    f. How many subsets does S have?
    g. If $n(S) = k$, how many subsets will S have?

In Exercises 43–50, find the number of permutations of the letters in each word.

43. ALASKA
44. ALABAMA
45. ILLINOIS
46. HAWAII
47. INDIANA
48. TENNESSEE
49. TALLAHASSEE
50. PHILADELPHIA

The words in each of Exercises 51–54 are homonyms *(words that are pronounced the same but have different meanings). Find the number of permutations of the letters in each word.*

51. **a.** PIER **b.** PEER
52. **a.** HEAR **b.** HERE
53. **a.** STEAL **b.** STEEL
54. **a.** SHEAR **b.** SHEER

Answer the following questions using complete sentences and your own words.

• CONCEPT QUESTIONS

55. Suppose you want to know how many ways r items can be selected from a group of n items. What determines whether you should calculate ${}_nP_r$ or ${}_nC_r$?
56. For any given values of n and r, which is larger, ${}_nP_r$ or ${}_nC_r$? Why?

THE NEXT LEVEL

If a person wants to pursue an advanced degree (something beyond a bachelor's or four-year degree), chances are the person must take a standardized exam to gain admission to a school or to be admitted into a specific program. These exams are intended to measure verbal, quantitative, and analytical skills that have developed throughout a person's life. Many classes and study guides are available to help people prepare for the exams. The following questions are typical of those found in the study guides.

Exercises 57–61 refer to the following: A baseball league has six teams: A, B, C, D, E, and F. All games are played at 7:30 P.M. *on Fridays, and there are sufficient fields for each team to play a game every Friday night. Each team must play each other team exactly once, and the following conditions must be met:*

- Team A plays team D first and team F second.
- Team B plays team E first and team C third.
- Team C plays team F first.

57. What is the total number of games that each team must play during the season?
   a. 3 **b.** 4 **c.** 5 **d.** 6 **e.** 7

58. On the first Friday, which of the following pairs of teams play each other?
   a. A and B; C and F; D and E
   b. A and B; C and E; D and F
   c. A and C; B and E; D and F
   d. A and D; B and C; E and F
   e. A and D; B and E; C and F

59. Which of the following teams must team B play second?
   a. A **b.** C **c.** D **d.** E **e.** F

60. The last set of games could be between which teams?
   a. A and B; C and F; D and E
   b. A and C; B and F; D and E
   c. A and D; B and C; E and F
   d. A and E; B and C; D and F
   e. A and F; B and E; C and D

61. If team D wins five games, which of the following must be true?
   a. Team A loses five games.
   b. Team A wins four games.
   c. Team A wins its first game.
   d. Team B wins five games.
   e. Team B loses at least one game.

WEB PROJECT

62. Write a research paper on any historical topic referred to in this section or in a previous section. Following is a partial list of topics:
   - John Venn
   - Augustus De Morgan
   - Chu Shih-chieh

   Some useful links for this web project are listed on the text web site: **www.cengage.com/math/johnson**

2.5 Infinite Sets

OBJECTIVES

- Determine whether two sets are equivalent
- Establish a one-to-one correspondence between the elements of two sets
- Determine the cardinality of various infinite sets.

WARNING: Many leading nineteenth-century mathematicians and philosophers claim that the study of infinite sets may be dangerous to your mental health.

Consider the sets E and N, where $E = \{2, 4, 6, \ldots\}$ and $N = \{1, 2, 3, \ldots\}$. Both are examples of infinite sets (they "go on forever"). E is the set of all even counting numbers, and N is the set of all counting (or natural) numbers. Because every element of E is an element of N, E is a subset of N. In addition, N contains elements not in E; therefore, E is a *proper* subset of N. Which set is "bigger," E or N? Intuition might lead many people to think that N is twice as big as E because N contains all the even counting numbers *and* all the odd counting numbers. Not so! According to the work of Georg Cantor (considered by many to be the father of set theory), N and E have exactly the same number of elements! This seeming paradox, a *proper* subset that has the *same number* of elements as the set from which it came, caused a philosophic uproar in the late nineteenth century. (Hence, the warning at the beginning of this section.) To study Cantor's work (which is now accepted and considered a cornerstone in modern mathematics), we must first investigate the meaning of a one-to-one correspondence and equivalent sets.

One-to-One Correspondence

Is there any relationship between the sets A = {one, two, three} and B = {Pontiac, Chevrolet, Ford}? Although the sets contain different types of things (numbers versus automobiles), each contains the same number of things; they are the same size. This relationship (being the same size) forms the basis of a one-to-one correspondence. A **one-to-one correspondence** between the sets A and B is a pairing up of the elements of A and B such that each element of A is paired up with exactly one element of B, and vice versa, with no element left out. For instance, the elements of A and B might be paired up as follows:

one	two	three
$\updownarrow$	$\updownarrow$	$\updownarrow$
Pontiac	Chevrolet	Ford

(Other correspondences, or matchups, are possible.) If two sets have the same cardinal number, their elements can be put into a one-to-one correspondence. Whenever a one-to-one correspondence exists between the elements of two sets A and B, the sets are **equivalent** (denoted by $A \sim B$). Hence, equivalent sets have the same number of elements.

If two sets have different cardinal numbers, it is not possible to construct a one-to-one correspondence between their elements. The sets C = {one, two} and B = {Pontiac, Chevrolet, Ford} do *not* have a one-to-one correspondence; no matter how their elements are paired up, one element of B will always be left over (B has more elements; it is "bigger"):

one	two	
$\updownarrow$	$\updownarrow$	
Pontiac	Chevrolet	Ford

The sets C and B are *not* equivalent.

Given two sets A and B, if any one of the following statements is true, then the other statements are also true:

1. There exists a one-to-one correspondence between the elements of A and B.
2. A and B are equivalent sets.
3. A and B have the same cardinal number; that is, $n(A) = n(B)$.

EXAMPLE 1

DETERMINING WHETHER TWO SETS ARE EQUIVALENT Determine whether the sets in each of the following pairs are equivalent. If they are equivalent, list a one-to-one correspondence between their elements.

a. A = {John, Paul, George, Ringo};
B = {Lennon, McCartney, Harrison, Starr}
b. $C = \{\alpha, \beta, \chi, \delta\}$; $D = \{\text{I, O}, \Delta\}$
c. $A = \{1, 2, 3, \ldots, 48, 49, 50\}$; $B = \{1, 3, 5, \ldots, 95, 97, 99\}$

HISTORICAL NOTE

GEORG CANTOR, 1845–1918

© CORBIS

Georg Ferdinand Ludwig Philip Cantor was born in St. Petersburg, Russia. His father was a stockbroker and wanted his son to become an engineer; his mother was an artist and musician. Several of Cantor's maternal relatives were accomplished musicians; in his later years, Cantor often wondered how his life would have turned out if he had become a violinist instead of pursuing a controversial career in mathematics.

Following his father's wishes, Cantor began his engineering studies at the University of Zurich in 1862. However, after one semester, he decided to study philosophy and pure mathematics. He transferred to the prestigious University of Berlin, studied under the famed mathematicians Karl Weierstrass, Ernst Kummer, and Leopold Kronecker, and received his doctorate in 1867. Two years later, Cantor accepted a teaching position at the University of Halle and remained there until he retired in 1913.

Cantor's treatises on set theory and the nature of infinite sets were first published in 1874 in *Crelle's Journal*, which was influential in mathematical circles. On their publication, Cantor's theories generated much controversy among mathematicians and philosophers. Paradoxes concerning the cardinal numbers of infinite sets, the nature of infinity, and Cantor's form of logic were unsettling to many, including Cantor's former teacher Leopold Kronecker. In fact, some felt that Cantor's work was not just revolutionary but actually dangerous. Kronecker led the attack on Cantor's theories. He was an editor of *Crelle's Journal* and held up the publication of one of Cantor's subsequent articles for so long that Cantor refused to publish ever again in the *Journal*. In addition, Kronecker blocked Cantor's efforts to obtain a teaching position at the University of Berlin. Even though Cantor was attacked by Kronecker and his followers, others respected him. Realizing the importance of communication among scholars, Cantor founded the Association of German Mathematicians in 1890 and served as its president for many years. In addition, Cantor was instrumental in organizing the first International Congress of Mathematicians, held in Zurich in 1897.

As a result of the repeated attacks on him and his work, Cantor suffered many nervous breakdowns, the first when he was thirty-nine. He died in a mental hospital in Halle at the age of seventy-three, never having received proper recognition for the true value of his discoveries. Modern mathematicians believe that Cantor's form of logic and his concepts of infinity revolutionized all of mathematics, and his work is now considered a cornerstone in its development.

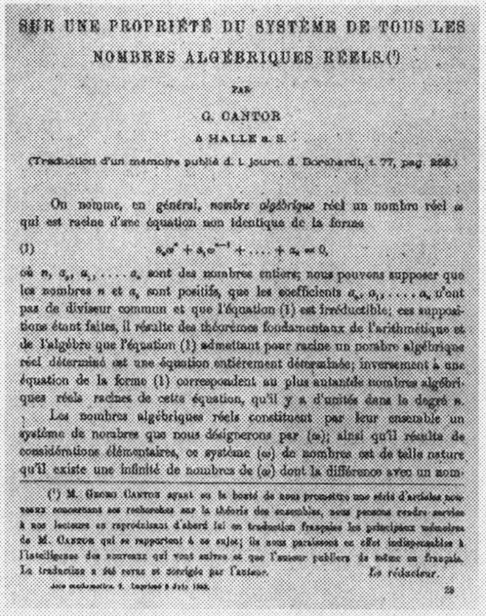

SUR UNE PROPRIÉTÉ DU SYSTÈME DE TOUS LES NOMBRES ALGÉBRIQUES RÉELS.[1]

PAR

G. CANTOR

À HALLE a. S.

(Traduction d'un mémoire publié d. l. journ. d. Borchardt, t. 77, pag. 258.)

On nomme, en général, nombre algébrique réel un nombre réel ω qui est racine d'une équation non identique de la forme

(1) $$a_0\omega^n + a_1\omega^{n-1} + \dots + a_n = 0,$$

où n, a_0, a_1, ... a_n sont des nombres entiers; nous pouvons supposer que les nombres n et a_0 sont positifs, que les coefficients a_0, a_1, ... a_n n'ont pas de diviseur commun et que l'équation (1) est irréductible; ces suppositions étant faites, il résulte des théorèmes fondamentaux de l'arithmétique et de l'algèbre que l'équation (1) admettant pour racine un nombre algébrique réel déterminé est une équation entièrement déterminée; inversement à une équation de la forme (1) correspondent au plus autant de nombres algébriques réels racines de cette équation, qu'il y a d'unités dans le degré n.

Les nombres algébriques réels constituent par leur ensemble un système de nombres que nous désignerons par (ω); ainsi qu'il résulte de considérations élémentaires, ce système (ω) de nombres est de telle nature qu'il existe une infinité de nombres de (ω) dont la différence avec un nom-

[1] M. Georg Cantor ayant eu la bonté de nous promettre une série d'articles nouveaux concernant ses recherches sur la théorie des ensembles, nous pensons rendre service à nos lecteurs en reproduisant d'abord ici en traduction française les principaux mémoires de M. Cantor qui se rapportent à ce sujet; ils nous paraissent en effet indispensables à l'intelligence des nouveaux qui vont suivre et que l'auteur publiera de même en français. La traduction a été revue et corrigée par l'auteur. *Le rédacteur.*

Brown University Library

Written in 1874, Cantor's first major paper on the theory of sets, Über eine Eigenshaft des Inbegriffes aller reellen algebraischen Zahlen *(On a Property of the System of All the Real Algebraic Numbers), sparked a major controversy concerning the nature of infinite sets. To gather international support for his theory, Cantor had his papers translated into French. This 1883 French version of Cantor's work was published in the newly formed journal* Acta Mathematica. *Cantor's works were translated into English during the early twentieth century.*

SOLUTION

a. If sets have the same cardinal number, they are equivalent. Now, $n(A) = 4$ and $n(B) = 4$; therefore, $A \sim B$.

Because A and B are equivalent, their elements can be put into a one-to-one correspondence. One such correspondence follows:

John	Paul	George	Ringo
$\updownarrow$	$\updownarrow$	$\updownarrow$	$\updownarrow$
Lennon	McCartney	Harrison	Starr

b. Because $n(C) = 4$ and $n(D) = 3$, C and D are not equivalent.

c. A consists of all natural numbers from 1 to 50, inclusive. Hence, $n(A) = 50$. B consists of all odd natural numbers from 1 to 99, inclusive. Since half of the natural numbers from 1 to 100 are odd (and half are even), there are fifty ($100 \div 2 = 50$) odd natural numbers less than 100; that is, $n(B) = 50$. Because A and B have the same cardinal number, $A \sim B$.

Many different one-to-one correspondences may be established between the elements of A and B. One such correspondence follows:

$$\begin{array}{l} A = \{1, 2, 3, \ldots, \quad n, \quad \ldots, 48, 49, 50\} \\ \quad\;\; \updownarrow\; \updownarrow\; \updownarrow \;\ldots\quad \updownarrow \quad \ldots\; \updownarrow\; \updownarrow\; \updownarrow \\ B = \{1, 3, 5, \ldots, (2n - 1), \ldots, 95, 97, 99\} \end{array}$$

That is, each natural number $n \in A$ is paired up with the odd number $(2n - 1) \in B$. The $n \leftrightarrow (2n - 1)$ part is crucial because it shows *each* individual correspondence. For example, it shows that $13 \in A$ corresponds to $25 \in B$ ($n = 13$, so $2n = 26$ and $2n - 1 = 25$). Likewise, $69 \in B$ corresponds to $35 \in A$ ($2n - 1 = 69$, so $2n = 70$ and $n = 35$).

As we have seen, if two sets have the same cardinal number, they are equivalent, and their elements can be put into a one-to-one correspondence. Conversely, if the elements of two sets can be put into a one-to-one correspondence, the sets have the same cardinal number and are equivalent. Intuitively, this result appears to be quite obvious. However, when Georg Cantor applied this relationship to infinite sets, he sparked one of the greatest philosophical debates of the nineteenth century.

Countable Sets

Consider the set of all counting numbers $N = \{1, 2, 3, \ldots\}$, which consists of an infinite number of elements. Each of these numbers is either odd or even. Defining O and E as $O = \{1, 3, 5, \ldots\}$ and $E = \{2, 4, 6, \ldots\}$, we have $O \cap E = \varnothing$ and $O \cup E = N$; the sets O and E are mutually exclusive, and their union forms the entire set of all counting numbers. Obviously, N contains elements that E does not. As we mentioned earlier, the fact that E is a *proper* subset of N might lead people to think that N is "bigger" than E. In fact, N and E are the "same size"; N and E each contain the same number of elements.

Recall that two sets are equivalent and have the same cardinal number if the elements of the sets can be matched up via a one-to-one correspondence. To show the existence of a one-to-one correspondence between the elements of two sets of numbers, we must find an explicit correspondence between the general elements of the two sets. In Example 1(c), we expressed the general correspondence as $n \leftrightarrow (2n - 1)$.

EXAMPLE 2

FINDING A ONE-TO-ONE CORRESPONDENCE BETWEEN TWO INFINITE SETS

a. Show that $E = \{2, 4, 6, 8, \ldots\}$ and $N = \{1, 2, 3, 4, \ldots\}$ are equivalent sets.
b. Find the element of N that corresponds to $1430 \in E$.
c. Find the element of N that corresponds to $x \in E$.

SOLUTION

a. To show that $E \sim N$, we must show that there exists a one-to-one correspondence between the elements of E and N. The elements of E and N can be paired up as follows:

$$\begin{array}{l} N = \{1, 2, 3, 4, \ldots, n, \ldots\} \\ \quad\;\;\; \updownarrow\; \updownarrow\; \updownarrow\; \updownarrow \;\ldots\; \updownarrow \\ E = \{2, 4, 6, 8, \ldots, 2n, \ldots\} \end{array}$$

Any natural number $n \in N$ corresponds with the even natural number $2n \in E$. Because there exists a one-to-one correspondence between the elements of E and N, the sets E and N are equivalent; that is, $E \sim N$.

b. $1430 = 2n \in E$, so $n = \frac{1430}{2} = 715 \in N$. Therefore, $715 \in N$ corresponds to $1430 \in E$.

c. $x = 2n \in E$, so $n = \frac{x}{2} \in N$. Therefore, $n = \frac{x}{2} \in N$ corresponds to $x = 2n \in E$.

We have just seen that the set of *even* natural numbers is equivalent to the set of *all* natural numbers. This equivalence implies that the two sets have the same number of elements! Although E is a proper subset of N, both sets have the same cardinal number; that is, $n(E) = n(N)$. Settling the controversy sparked by this seeming paradox, mathematicians today define a set to be an **infinite set** if it can be placed in a one-to-one correspondence with a proper subset of itself.

How many counting numbers are there? How many even counting numbers are there? We know that each set contains an infinite number of elements and that $n(N) = n(E)$, but how many is that? In the late nineteenth century, Georg Cantor defined the cardinal number of the set of counting numbers to be $\aleph_0$ (read "**aleph-null**"). Cantor utilized Hebrew letters, of which aleph, $\aleph$, is the first. Consequently, the proper response to "How many counting numbers are there?" is "There are aleph-null of them"; $n(N) = \aleph_0$. Any set that is equivalent to the set of counting numbers has cardinal number $\aleph_0$. A set is **countable** if it is finite or if it has cardinality $\aleph_0$.

Cantor was not the first to ponder the paradoxes of infinite sets. Hundreds of years before, Galileo had observed that part of an infinite set contained as many elements as the whole set. In his monumental *Dialogue Concerning the Two Chief World Systems* (1632), Galileo made a prophetic observation: "There are as many (perfect) squares as there are (natural) numbers because they are just as numerous as their roots." In other words, the elements of the sets $N = \{1, 2, 3, \ldots, n, \ldots\}$ and $S = \{1^2, 2^2, 3^2, \ldots, n^2, \ldots\}$ can be put into a one-to-one correspondence ($n \leftrightarrow n^2$). Galileo pondered which of the sets (perfect squares or natural numbers) was "larger" but abandoned the subject because he could find no practical application of this puzzle.

EXAMPLE 3

SHOWING THAT THE SET OF INTEGERS IS COUNTABLE Consider the following one-to-one correspondence between the set I of all integers and the set N of all natural numbers:

$$\begin{array}{l} N = \{1,\ 2,\ \ 3,\ 4,\ \ 5, \ldots\} \\ \quad\ \ \ \updownarrow\ \ \updownarrow\ \ \ \updownarrow\ \ \updownarrow\ \ \updownarrow \\ I = \{0, 1, -1, 2, -2, \ldots\} \end{array}$$

where an odd natural number n corresponds to a nonpositive integer $\frac{1-n}{2}$ and an even natural number n corresponds to a positive integer $\frac{n}{2}$.

a. Find the 613th integer; that is, find the element of I that corresponds to $613 \in N$.
b. Find the element of N that corresponds to $853 \in I$.
c. Find the element of N that corresponds to $-397 \in I$.
d. Find $n(I)$, the cardinal number of the set I of all integers.
e. Is the set of integers countable?

SOLUTION

a. $613 \in N$ is odd, so it corresponds to $\frac{1-613}{2} = \frac{-612}{2} = -306$. If you continued counting the integers as shown in the above correspondence, -306 would be the 613th integer in your count.

b. $853 \in I$ is positive, so

$$853 = \frac{n}{2} \quad \text{multiplying by 2}$$
$$n = 1{,}706$$

$1{,}706 \in N$ corresponds to $853 \in I$.

This means that 853 is the 1,706th integer.

c. $-397 \in I$ is negative, so

$$\begin{aligned} -397 &= \frac{1-n}{2} \\ -794 &= 1-n && \text{multiplying by 2} \\ -795 &= -n && \text{subtracting 1} \\ n &= 795 && \text{multiplying by } -1 \end{aligned}$$

$795 \in N$ corresponds to $-397 \in I$.

This means that -397 is the 795th integer.

d. The given one-to-one correspondence shows that I and N have the same (infinite) number of elements; $n(I) = n(N)$. Because $n(N) = \aleph_0$, the cardinal number of the set of all integers is $n(I) = \aleph_0$.

e. By definition, a set is called countable if it is finite or if it has cardinality $\aleph_0$. The set of integers has cardinality $\aleph_0$, so it is countable. This means that we can "count off" all of the integers, as we did in parts (a), (b), and (c).

We have seen that the sets N (all counting numbers), E (all even counting numbers), and I (all integers) contain the same number of elements, $\aleph_0$. What about a set containing fractions?

EXAMPLE 4

DETERMINING WHETHER THE SET OF POSITIVE RATIONAL NUMBERS IS COUNTABLE Determine whether the set P of all positive rational numbers is countable.

SOLUTION

The elements of P can be systematically listed in a table of rows and columns as follows: All positive rational numbers whose denominator is 1 are listed in the first row, all positive rational numbers whose denominator is 2 are listed in the second row, and so on, as shown in Figure 2.43.

Each positive rational number will appear somewhere in the table. For instance, $\frac{125}{66}$ will be in row 66 and column 125. Note that not all the entries in Figure 2.43 are in lowest terms; for instance, $\frac{2}{4}, \frac{3}{6}, \frac{4}{8}$, and so on are all equal to $\frac{1}{2}$.

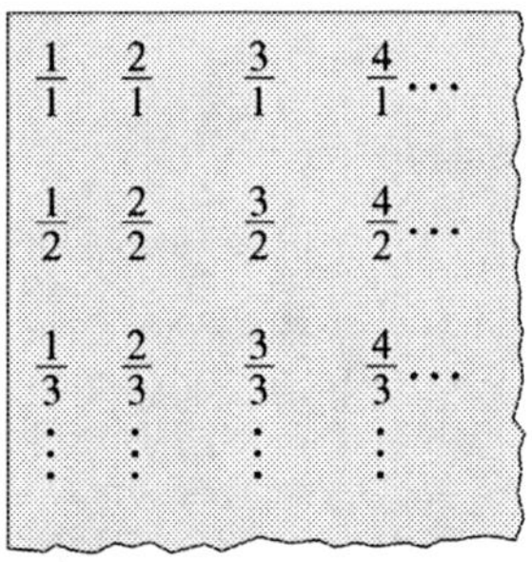

FIGURE 2.43 A list of all positive rational numbers.

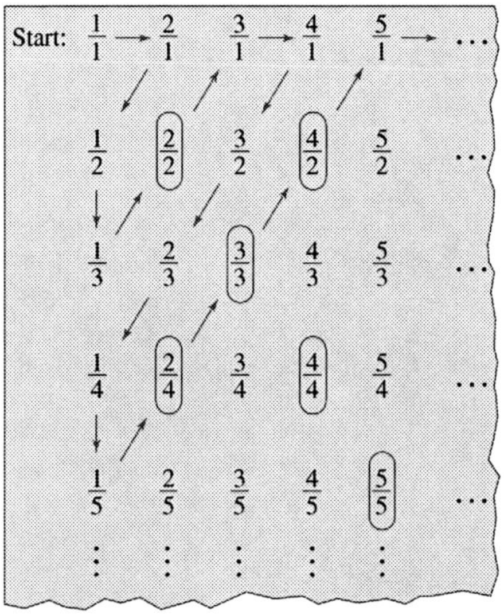

FIGURE 2.44 The circled rational numbers are not in lowest terms; they are omitted from the list.

Consequently, to avoid listing the same number more than once, an entry that is not in lowest terms must be eliminated from our list. To establish a one-to-one correspondence between P and N, we can create a zigzag diagonal pattern as shown by the arrows in Figure 2.44. Starting with $\frac{1}{1}$, we follow the arrows and omit any number that is not in lowest terms (the circled numbers in Figure 2.44). In this manner, a list of all positive rational numbers with no repetitions is created. Listing the elements of P in this order, we can put them in a one-to-one correspondence with N:

$$\begin{array}{rl} N = & \{1, 2, 3, 4, 5, 6, 7, 8, 9, 10, 11, \ldots\} \\ & \;\updownarrow\;\updownarrow\;\updownarrow\;\updownarrow\;\updownarrow\;\updownarrow\;\updownarrow\;\updownarrow\;\updownarrow\;\updownarrow \\ P = & \left\{1, 2, \frac{1}{2}, \frac{1}{3}, 3, 4, \frac{3}{2}, \frac{2}{3}, \frac{1}{4}, \frac{1}{5}, 5, \ldots\right\} \end{array}$$

Any natural number n is paired up with the positive rational number found by counting through the "list" given in Figure 2.44. Conversely, any positive rational number is located somewhere in the list and is paired up with the counting number corresponding to its place in the list.

Therefore, $P \sim N$, so the set of all positive rational numbers is countable.

Uncountable Sets

Every infinite set that we have examined so far is countable; each can be put into a one-to-one correspondence with the set of all counting numbers and consequently has cardinality $\aleph_0$. Do not be misled into thinking that all infinite sets are countable! By utilizing a "proof by contradiction," Georg Cantor showed that the infinite set $A = \{x \mid 0 \leq x < 1\}$ is *not* countable. This proof involves logic that is different from what you are used to. Do not let that intimidate you.

Assume that the set $A = \{x \mid 0 \leq x < 1\}$ is countable; that is, assume that $n(A) = \aleph_0$. This assumption implies that the elements of A and N can be put into a one-to-one correspondence; each $a \in A$ can be listed and counted. Because the elements of A are nonnegative real numbers less than 1, each $a_n = 0.\square\square\square\square\square\ldots$. Say, for instance, the numbers in our list are

$a_1 = 0.3750000\ldots$ the first element of A

$a_2 = 0.7071067\ldots$ the second element of A

$a_3 = 0.5000000\ldots$ the third element of A

$a_4 = 0.6666666\ldots$ and so on.

The *assumption* that A is countable implies that every element of A appears somewhere in the above list. However, we can create an element of A (call it b) that is *not* in the list. We build b according to the "diagonal digits" of the numbers in our list and the following rule: If the digit "on the diagonal" is not zero, put a 0 in the corresponding place in b; if the digit "on the diagonal" is zero, put a 1 in the corresponding place in b.

The "diagonal digits" of the numbers in our list are as follows:

$a_1 = 0.\boxed{3}750000\ldots$

$a_2 = 0.7\boxed{0}71067\ldots$

$a_3 = 0.50\boxed{0}0000\ldots$

$a_4 = 0.666\boxed{6}666\ldots$

Because the first digit on the diagonal is 3, the first digit of b is 0. Because the second digit on the diagonal is 0, the second digit of b is 1. Using all the "diagonal digits" of the numbers in our list, we obtain $b = 0.0110\ldots$. Because $0 \le b < 1$, it follows that $b \in A$. However, the number b is not on our list of all elements of A. This is because

$b \ne a_1$ (b and a_1 differ in the first decimal place)
$b \ne a_2$ (b and a_2 differ in the second decimal place)
$b \ne a_3$ (b and a_3 differ in the third decimal place), and so on

This contradicts the assumption that the elements of A and N can be put into a one-to-one correspondence. Since the assumption leads to a contradiction, the assumption must be false; $A = \{x \mid 0 \le x < 1\}$ is not countable. Therefore, $n(A) \ne n(N)$. That is, A is an infinite set and $n(A) \ne \aleph_0$.

An infinite set that cannot be put into a one-to-one correspondence with N is said to be **uncountable.** Consequently, an uncountable set has *more* elements than the set of all counting numbers. This implies that there are different magnitudes of infinity! To distinguish the magnitude of A from that of N, Cantor denoted the cardinality of $A = \{x \mid 0 \le x < 1\}$ as $n(A) = c$ (c for **continuum**). Thus, Cantor showed that $\aleph_0 < c$. Cantor went on to show that A was equivalent to the entire set of all real numbers, that is, $A \sim \Re$. Therefore, $n(\Re) = c$.

Although he could not prove it, Cantor hypothesized that no set could have a cardinality between $\aleph_0$ and c. This famous unsolved problem, labeled the *Continuum Hypothesis,* baffled mathematicians throughout the first half of the twentieth century. It is said that Cantor suffered a devastating nervous breakdown in 1884 when he announced that he had a proof of the Continuum Hypothesis only to declare the next day that he could show the Continuum Hypothesis to be false!

The problem was finally "solved" in 1963. Paul J. Cohen demonstrated that the Continuum Hypothesis is independent of the entire framework of set theory; that is, it can be neither proved nor disproved by using the theorems of set theory. Thus, the Continuum Hypothesis is not provable.

Although no one has produced a set with cardinality between $\aleph_0$ and c, many sets with cardinality greater than c have been constructed. In fact, modern mathematicians have shown that there are *infinitely* many magnitudes of infinity! Using subscripts, these magnitudes, or cardinalities, are represented by $\aleph_0, \aleph_1, \aleph_2, \ldots$ and have the property that $\aleph_0 < \aleph_1 < \aleph_2 < \ldots$. In this sense, the set N of all natural numbers forms the "smallest" infinite set. Using this subscripted notation, the Continuum Hypothesis implies that $c = \aleph_1$; that is, given that N forms the smallest infinite set, the set $\Re$ of all real numbers forms the next "larger" infinite set.

Points on a Line

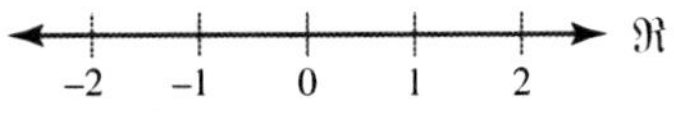

FIGURE 2.45
The real number line.

When students are first exposed to the concept of the real number system, a number line like the one in Figure 2.45 is inevitably introduced. The real number system, denoted by $\Re$, can be put into a one-to-one correspondence with all points on a line, such that every real number corresponds to exactly one point on a line and every point on a line corresponds to exactly one real number. Consequently, any (infinite) line contains c points. What about a line segment? For example, how many points does the segment [0, 1] contain? Does the segment [0, 2] contain twice as many points as the segment [0, 1]? Once again, intuition can lead to erroneous conclusions when people are dealing with infinite sets.

EXAMPLE 5

SHOWING THAT LINE SEGMENTS OF DIFFERENT LENGTHS ARE EQUIVALENT SETS OF POINTS Show that the line segments [0, 1] and [0, 2] are equivalent sets of points.

SOLUTION

Because the segment [0, 2] is twice as long as the segment [0, 1], intuition might tell us that it contains twice as many points. Not so! Recall that two sets are

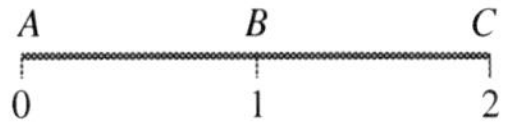

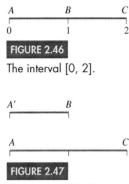

FIGURE 2.46
The interval [0, 2].

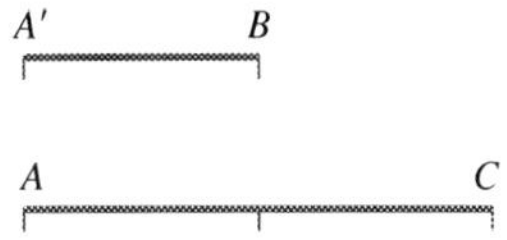

FIGURE 2.47
The intervals [*A*, *B*] and [*A*, *C*].

equivalent (and have the same cardinal number) if their elements can be put into a one-to-one correspondence.

On a number line, let *A* represent the point 0, let *B* represent 1, and let *C* represent 2, as shown in Figure 2.46. Our goal is to develop a one-to-one correspondence between the elements of the segments *AB* and *AC*. Now draw the segments separately, with *AB* above *AC*, as shown in Figure 2.47. (To distinguish the segments from each other, point *A* of segment *AB* has been relabeled as point *A*′.)

Extend segments *AA*′ and *CB* so that they meet at point *D*, as shown in Figure 2.48. Any point *E* on *A*′*B* can be paired up with the unique point *F* on *AC* formed by the intersection of lines *DE* and *AC*, as shown in Figure 2.49. Conversely, any point *F* on segment *AC* can be paired up with the unique point *E* on *A*′*B* formed by the intersection of lines *DF* and *A*′*B*. Therefore, a one-to-one correspondence exists between the two segments, so [0, 1] ~ [0, 2]. Consequently, the interval [0, 1] contains exactly the same number of points as the interval [0, 2]!

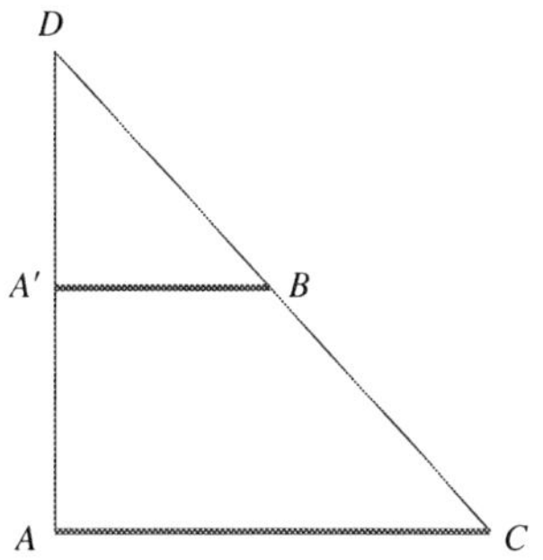

FIGURE 2.48 Extending *AA*′ and *CB* to form point *D*.

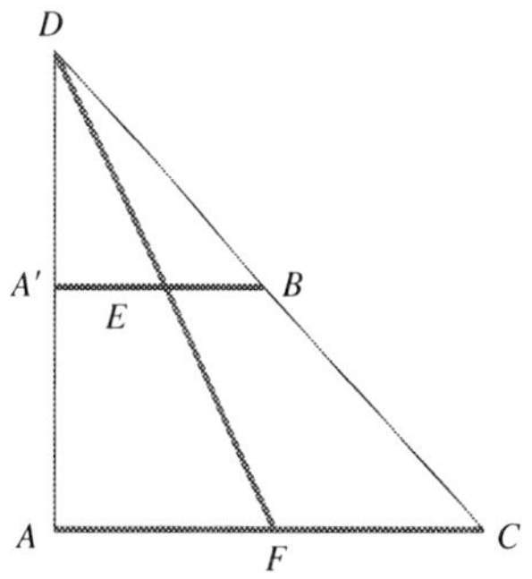

FIGURE 2.49 A one-to-one correspondence between line segments.

Even though the segment [0, 2] is twice as long as the segment [0, 1], each segment contains exactly the same number of points. The method used in Example 5 can be applied to any two line segments. Consequently, all line segments, regardless of their length, contain exactly the same number of points; a line segment 1 inch long has exactly the same number of points as a segment 1 mile long! Once again, it is easy to see why Cantor's work on the magnitude of infinity was so unsettling to many scholars.

Having concluded that all line segments contain the same number of points, we might ask how many points that is. What is the cardinal number? Given any line segment *AB*, it can be shown that $n(AB) = c$; the points of a line segment can be put into a one-to-one correspondence with the points of a line. Consequently, the interval [0, 1] contains the same number of elements as the entire real number system.

If things seem rather strange at this point, keep in mind that Cantor's pioneering work produced results that puzzled even Cantor himself. In a paper written in 1877, Cantor constructed a one-to-one correspondence between the points in a square (a two-dimensional figure) and the points on a line segment (a one-dimensional figure). Extending this concept, he concluded that a line segment and the entire two-dimensional plane contain exactly the same number of points, *c*. Communicating with his colleague Richard Dedekind, Cantor wrote, "I see it, but I do not believe it." Subsequent investigation has shown that the number of points contained in the interval [0, 1] is the same as the number of points contained in all of three-dimensional space! Needless to say, Cantor's work on the cardinality of infinity revolutionized the world of modern mathematics.

2.5 Exercises

In Exercises 1–10, find the cardinal numbers of the sets in each given pair to determine whether the sets are equivalent. If they are equivalent, list a one-to-one correspondence between their elements.

1. $S = \{$Sacramento, Lansing, Richmond, Topeka$\}$
   $C = \{$California, Michigan, Virginia, Kansas$\}$
2. $T = \{$Wyoming, Ohio, Texas, Illinois, Colorado$\}$
   $P = \{$Cheyenne, Columbus, Austin, Springfield, Denver$\}$
3. $R = \{a, b, c\}$; $G = \{\alpha, \beta, \chi, \delta\}$
4. $W = \{$I, II, III$\}$; $H = \{$one, two$\}$
5. $C = \{3, 6, 9, 12, \ldots, 63, 66\}$
   $D = \{4, 8, 12, 16, \ldots, 84, 88\}$
6. $A = \{2, 4, 6, 8, \ldots, 108, 110\}$
   $B = \{5, 10, 15, 20, \ldots, 270, 275\}$
7. $G = \{2, 4, 6, 8, \ldots, 498, 500\}$
   $H = \{1, 3, 5, 7, \ldots, 499, 501\}$
8. $E = \{2, 4, 6, 8, \ldots, 498, 500\}$
   $F = \{3, 6, 9, 12, \ldots, 750, 753\}$
9. $A = \{1, 3, 5, \ldots, 121, 123\}$
   $B = \{125, 127, 129, \ldots, 245, 247\}$
10. $S = \{4, 6, 8, \ldots, 664, 666\}$
    $T = \{5, 6, 7, \ldots, 335, 336\}$
11. **a.** Show that the set O of all odd counting numbers, $O = \{1, 3, 5, 7, \ldots\}$, and $N = \{1, 2, 3, 4, \ldots\}$ are equivalent sets.
    b. Find the element of N that corresponds to $1{,}835 \in O$.
    c. Find the element of N that corresponds to $x \in O$.
    d. Find the element of O that corresponds to $782 \in N$.
    e. Find the element of O that corresponds to $n \in N$.
12. **a.** Show that the set W of all whole numbers, $W = \{0, 1, 2, 3, \ldots\}$, and $N = \{1, 2, 3, 4, \ldots\}$ are equivalent sets.
    b. Find the element of N that corresponds to $932 \in W$.
    c. Find the element of N that corresponds to $x \in W$.
    d. Find the element of W that corresponds to $932 \in N$.
    e. Find the element of W that corresponds to $n \in N$.
13. **a.** Show that the set T of all multiples of 3, $T = \{3, 6, 9, 12, \ldots\}$, and $N = \{1, 2, 3, 4, \ldots\}$ are equivalent sets.
    b. Find the element of N that corresponds to $936 \in T$.
    c. Find the element of N that corresponds to $x \in T$.
    d. Find the element of T that corresponds to $936 \in N$.
    e. Find the element of T that corresponds to $n \in N$.
14. **a.** Show that the set F of all multiples of 5, $F = \{5, 10, 15, 20, \ldots\}$, and $N = \{1, 2, 3, 4, \ldots\}$ are equivalent sets.
    b. Find the element of N that corresponds to $605 \in F$.
    c. Find the element of N that corresponds to $x \in F$.
    d. Find the element of F that corresponds to $605 \in N$.
    e. Find the element of F that corresponds to $n \in N$.
15. Consider the following one-to-one correspondence between the set A of all even integers and the set N of all natural numbers:

$$\begin{array}{l} N = \{1, 2, 3, 4, 5, \ldots\} \\ \quad\;\; \updownarrow\; \updownarrow\; \updownarrow\; \updownarrow\; \updownarrow \\ A = \{0, 2, -2, 4, -4, \ldots\} \end{array}$$

where an odd natural number n corresponds to the nonpositive integer $1 - n$ and an even natural number n corresponds to the positive even integer n.
    a. Find the 345th even integer; that is, find the element of A that corresponds to $345 \in N$.
    b. Find the element of N that corresponds to $248 \in A$.
    c. Find the element of N that corresponds to $-754 \in A$.
    d. Find $n(A)$.
16. Consider the following one-to-one correspondence between the set B of all odd integers and the set N of all natural numbers:

$$\begin{array}{l} N = \{1, 2, 3, 4, 5, \ldots\} \\ \quad\;\; \updownarrow\; \updownarrow\; \updownarrow\; \updownarrow\; \updownarrow \\ B = \{1, -1, 3, -3, 5, \ldots\} \end{array}$$

where an even natural number n corresponds to the negative odd integer $1 - n$ and an odd natural number n corresponds to the odd integer n.
    a. Find the 345th odd integer; that is, find the element of B that corresponds to $345 \in N$.
    b. Find the element of N that corresponds to $241 \in B$.
    c. Find the element of N that corresponds to $-759 \in B$.
    d. Find $n(B)$.

In Exercises 17–22, show that the given sets of points are equivalent by establishing a one-to-one correspondence.

17. the line segments $[0, 1]$ and $[0, 3]$
18. the line segments $[1, 2]$ and $[0, 3]$
19. the circle and square shown in Figure 2.50

    HINT: Draw one figure inside the other.

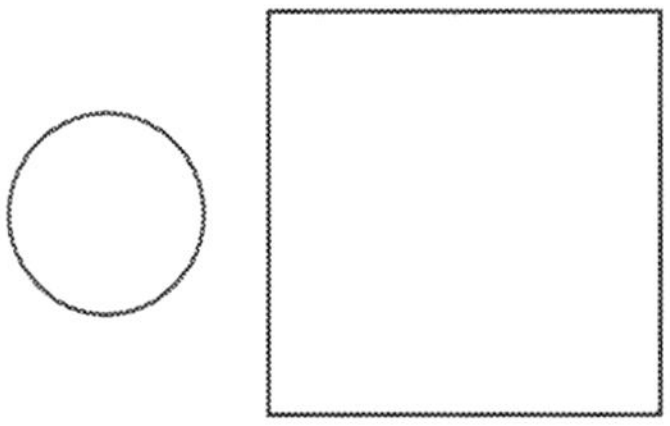

FIGURE 2.50

20. the rectangle and triangle shown in Figure 2.51

HINT: Draw one figure inside the other.

FIGURE 2.51

21. a circle of radius 1 cm and a circle of 5 cm

HINT: Draw one figure inside the other.

22. a square of side 1 cm and a square of side 5 cm

HINT: Draw one figure inside the other.

23. Show that the set of all real numbers between 0 and 1 has the same cardinality as the set of all real numbers.

HINT: Draw a semicircle to represent the set of real numbers between 0 and 1 and a line to represent the set of all real numbers, and use the method of Example 5.

Answer the following questions using complete sentences and your own words.

CONCEPT QUESTIONS

24. What is the cardinal number of the "smallest" infinite set? What set or sets have this cardinal number?

HISTORY QUESTIONS

25. What aspect of Georg Cantor's set theory caused controversy among mathematicians and philosophers?
26. What contributed to Cantor's breakdown in 1884?
27. Who demonstrated that the Continuum Hypothesis cannot be proven? When?

WEB PROJECT

28. Write a research paper on any historical topic referred to in this section. Following is a partial list of topics:
    - Georg Cantor
    - Richard Dedekind
    - Paul J. Cohen
    - Bernhard Bolzano
    - Leopold Kronecker
    - the Continuum Hypothesis

    Some useful links for this web project are listed on the text web site: **www.cengage.com/math/johnson**

2 CHAPTER REVIEW

TERMS

aleph-null
cardinal number
combination
combinatorics
complement
continuum
countable set
De Morgan's Laws
distinguishable permutations
element
empty set
equal sets
equivalent sets
factorial
Fundamental Principle of Counting
improper subset
infinite set
intersection
mutually exclusive
one-to-one correspondence
permutation
proper subset
roster notation
set
set-builder notation
set theory
subset
tree diagram
uncountable set
union
universal set
Venn diagram
well-defined set

Review Exercises

1. State whether the given set is well-defined.
   a. the set of all multiples of 5.
   b. the set of all difficult math problems
   c. the set of all great movies
   d. the set of all Oscar-winning movies
2. Given the sets
   $U = \{0, 1, 2, 3, 4, 5, 6, 7, 8, 9\}$
   $A = \{0, 2, 4, 6, 8\}$
   $B = \{1, 3, 5, 7, 9\}$
   find the following using the roster method.
   a. A' b. B'
   c. $A \cup B$ d. $A \cap B$
3. Given the sets $A = \{$Maria, Nobuko, Leroy, Mickey, Kelly$\}$ and $B = \{$Rachel, Leroy, Deanna, Mickey$\}$, find the following.
   a. $A \cup B$ b. $A \cap B$
4. List all subsets of $C = \{$Dallas, Chicago, Tampa$\}$. Identify which subsets are proper and which are improper.
5. Given $n(U) = 61$, $n(A) = 32$, $n(B) = 26$, and $n(A \cup B) = 40$, do the following.
   a. Find $n(A \cap B)$.
   b. Draw a Venn diagram illustrating the composition of U.

In Exercises 6 and 7, use a Venn diagram like the one in Figure 2.15 to shade in the region corresponding to the indicated set.

6. $(A' \cup B)'$
7. $(A \cap B')'$

In Exercises 8 and 9, use a Venn diagram like the one in Figure 2.36 to shade in the region corresponding to the indicated set.

8. $A' \cap (B \cup C')$
9. $(A' \cap B) \cup C'$
10. A survey of 1,000 college seniors yielded the following information: 396 seniors favored capital punishment, 531 favored stricter gun control, and 237 favored both.
    a. How many favored capital punishment or stricter gun control?
    b. How many favored capital punishment but not stricter gun control?
    c. How many favored stricter gun control but not capital punishment?
    d. How many favored neither capital punishment nor stricter gun control?
11. A survey of recent college graduates yielded the following information: 70 graduates earned a degree in mathematics, 115 earned a degree in education, 23 earned degrees in both mathematics and education, and 358 earned a degree in neither mathematics nor education.
    a. What percent of the college graduates earned a degree in mathematics or education?
    b. What percent of the college graduates earned a degree in mathematics only?
    c. What percent of the college graduates earned a degree in education only?
12. A local anime fan club surveyed 67 of its members regarding their viewing habits last weekend, and the following information was obtained: 30 members watched an episode of *Naruto*, 44 watched an episode of *Death Note*, 23 watched an episode of *Inuyasha*, 20 watched both *Naruto* and *Inuyasha*, 5 watched *Naruto* and *Inuyasha* but not *Death Note*, 15 watched both *Death Note* and *Inuyasha*, and 23 watched only *Death Note*.
    a. How many of the club members watched exactly one of the shows?
    b. How many of the club members watched all three shows?
    c. How many of the club members watched none of the three shows?
13. An exit poll yielded the following information concerning people's voting patterns on Propositions A, B, and C: 305 people voted yes on A, 95 voted yes only on A, 393 voted yes on B, 192 voted yes only on B, 510 voted yes on A or B, 163 voted yes on C, 87 voted yes on all three, and 249 voted no on all three. What percent of the voters voted yes on more than one proposition?
14. Given the sets $U = \{a, b, c, d, e, f, g, h, i\}$, $A = \{b, d, f, g\}$, and $B = \{a, c, d, g, i\}$, use De Morgan's Laws to find the following.
    a. $(A' \cup B)'$ b. $(A \cap B')'$
15. Refer to the Venn diagram depicted in Figure 2.32.
    a. In a group of 100 Americans, how many have type O or type A blood?
    b. In a group of 100 Americans, how many have type O and type A blood?
    c. In a group of 100 Americans, how many have neither type O nor type A blood?
16. Refer to the Venn diagram depicted in Figure 2.28.
    a. For a typical group of 100 Americans, fill in the cardinal number of each region in the diagram.
    b. In a group of 100 Americans, how many have type O blood or are Rh+?
    c. In a group of 100 Americans, how many have type O blood and are Rh+?
    d. In a group of 100 Americans, how many have neither type O blood nor are Rh+?

17. Sid and Nancy are planning their anniversary celebration, which will include viewing an art exhibit, having dinner, and going dancing. They will go either to the Museum of Modern Art or to the New Photo Gallery; dine either at Stars, at Johnny's, or at the Chelsea; and go dancing either at Le Club or at Lizards.

a. In how many different ways can Sid and Nancy celebrate their anniversary?

b. Construct a tree diagram to list all possible ways in which Sid and Nancy can celebrate their anniversary.

18. A certain model of pickup truck is available in five exterior colors, three interior colors, and three interior styles. In addition, the transmission can be either manual or automatic, and the truck can have either two-wheel or four-wheel drive. How many different versions of the pickup truck can be ordered?

19. Each student at State University has a student I.D. number consisting of five digits (the first digit is nonzero, and digits can be repeated) followed by two of the letters *A*, *B*, *C*, and *D* (letters cannot be repeated). How many different student numbers are possible?

20. Find the value of each of the following.

a. $(17 - 7)!$ **b.** $(17 - 17)!$

c. $\frac{82!}{79!}$ **d.** $\frac{27!}{20!7!}$

21. In how many ways can you select three out of eleven items under the following conditions?

a. Order of selection is not important.

b. Order of selection is important.

22. Find the value of each of the following.

a. ${}_{15}P_4$ **b.** ${}_{15}C_4$ **c.** ${}_{15}P_{11}$

23. A group of ten women and twelve men must select a three-person committee. How many committees are possible if it must consist of the following?

a. one woman and two men

b. any mixture of men and women

c. a majority of men

24. A volleyball league has ten teams. If every team must play every other team once in the first round of league play, how many games must be scheduled?

25. A volleyball league has ten teams. How many different end-of-the-season rankings of first, second, and third place are possible (disregarding ties)?

26. Using a standard deck of fifty-two cards (no jokers), how many seven-card poker hands are possible?

27. Using a standard deck of fifty-two cards and two jokers, how many seven-card poker hands are possible?

28. A 6/42 lottery requires choosing six of the numbers 1 through 42. How many different lottery tickets can you choose?

In Exercises 29 and 30, find the number of permutations of the letters in each word.

29. **a.** FLORIDA **b.** ARIZONA **c.** MONTANA

30. **a.** AFFECT **b.** EFFECT

31. What is the major difference between permutations and combinations?

32. Use Pascal's Triangle to answer the following.

a. In which entry in which row would you find the value of ${}_7C_3$?

b. In which entry in which row would you find the value of ${}_7C_4$?

c. How is the value of ${}_7C_3$ related to the value of ${}_7C_4$? Why?

d. What is the location of ${}_nC_r$? Why?

33. Given the set $S = \{a, b, c\}$, answer the following.

a. How many one-element subsets does S have?

b. How many two-element subsets does S have?

c. How many three-element subsets does S have?

d. How many zero-element subsets does S have?

e. How many subsets does S have?

f. How is the answer to part (e) related to $n(S)$?

In Exercises 34–36, find the cardinal numbers of the sets in each given pair to determine whether the sets are equivalent. If they are equivalent, list a one-to-one correspondence between their elements.

34. $A = \{\text{I, II, III, IV, V}\}$ and $B = \{\text{one, two, three, four, five}\}$

35. $C = \{3, 5, 7, \ldots, 899, 901\}$ and $D = \{2, 4, 6, \ldots, 898, 900\}$

36. $E = \{\text{Ronald}\}$ and $F = \{\text{Reagan, McDonald}\}$

37. **a.** Show that the set S of perfect squares, $S = \{1, 4, 9, 16, \ldots\}$, and $N = \{1, 2, 3, 4, \ldots\}$ are equivalent sets.

b. Find the element of N that corresponds to $841 \in S$.

c. Find the element of N that corresponds to $x \in S$.

d. Find the element of S that corresponds to $144 \in N$.

e. Find the element of S that corresponds to $n \in N$.

38. Consider the following one-to-one correspondence between the set A of all integer multiples of 3 and the set N of all natural numbers:

$$\begin{array}{l} N = \{1, 2, 3, 4, 5, \ldots\} \\ \quad\;\;\; \updownarrow\; \updownarrow\; \updownarrow\; \updownarrow\; \updownarrow \\ A = \{0, 3, -3, 6, -6, \ldots\} \end{array}$$

where an odd natural number n corresponds to the nonpositive integer $\frac{3}{2}(1 - n)$, and an even natural number n corresponds to the positive even integer $\frac{3}{2}n$.

a. Find the element of A that corresponds to $396 \in N$.

b. Find the element of N that corresponds to $396 \in A$.

c. Find the element of N that corresponds to $-153 \in A$.

d. Find $n(A)$.

39. Show that the line segments [0, 1] and $[0, \pi]$ are equivalent sets of points by establishing a one-to-one correspondence.

Answer the following questions using complete sentences and your own words.

• CONCEPT QUESTIONS

40. What is the difference between proper and improper subsets?
41. Explain the difference between {0} and $\varnothing$.
42. What is a factorial?
43. What is the difference between permutations and combinations?

• HISTORY QUESTIONS

44. What roles did the following people play in the development of set theory and combinatorics?
    - Georg Cantor
    - Augustus De Morgan
    - Christian Kramp
    - Chu Shih-chieh
    - John Venn

THE NEXT LEVEL

If a person wants to pursue an advanced degree (something beyond a bachelor's or four-year degree), chances are the person must take a standardized exam to gain admission to a school or to be admitted into a specific program. These exams are intended to measure verbal, quantitative, and analytical skills that have developed throughout a person's life. Many classes and study guides are available to help people prepare for the exams. The following questions are typical of those found in the study guides.

Exercises 45–48 refer to the following: Two doctors in a local clinic are determining which days of the week they will be on call. Each day, Monday through Sunday, is to be assigned to one of two doctors, A and B, such that the assignment is consistent with the following conditions:

- No day is assigned to both doctors.
- Neither doctor has more than four days.
- Monday and Thursday must be assigned to the same doctor.
- If Tuesday is assigned to doctor A, then so is Sunday.
- If Saturday is assigned to doctor B, then Friday is not assigned to doctor B.

45. Which one of the following could be a complete and accurate list of the days assigned to doctor A?
    a. Monday, Thursday
    b. Monday, Tuesday, Sunday
    c. Monday, Thursday, Sunday
    d. Monday, Tuesday, Thursday
    e. Monday, Thursday, Friday, Sunday
46. Which of the following cannot be true?
    a. Thursday and Sunday are assigned to doctor A.
    b. Friday and Saturday are assigned to doctor A.
    c. Monday and Tuesday are assigned to doctor B.
    d. Monday, Wednesday, and Sunday are assigned to doctor A.
    e. Tuesday, Wednesday, and Saturday are assigned to doctor B.
47. If Friday and Sunday are both assigned to doctor B, how many different ways are there to assign the other five days to the doctors?
    a. 1 **b.** 2 **c.** 3
    d. 5 **e.** 6
48. If doctor A has four days, none of which is Monday, which of the following days must be assigned to doctor A?
    a. Tuesday **b.** Wednesday **c.** Friday
    d. Saturday **e.** Sunday

3 PROBABILITY

© Wasabi/Alamy

Uncertainty is a part of our lives. When we wake up, we check the weather report to see whether it might rain. We check the traffic report to see whether we might get stuck in a jam. We check to see whether the interest rate has changed for that car, boat, or home loan. **Probability theory** is the branch of mathematics that analyzes uncertainty.

Dennis Flaherty/photodisk/Jupiter Images

WHAT WE WILL DO IN THIS CHAPTER

GAMBLING IS ALL ABOUT PROBABILITIES:

- Gambling involves uncertainty. Which number will be the winning number?
- All casinos use probability theory to set house odds so that they will make a profit.
- We will use probability theory to determine which casino games of chance are better for you to play and which to avoid.
- Probability theory's roots in gambling are deep: Probability theory got its start when a French nobleman and gambler asked his mathematician friend why he had lost so much money at dice.

THE SCIENCE OF GENETICS COULDN'T EXIST WITHOUT PROBABILITIES:

- Humans inherit traits, such as hair color or an increased risk for a certain disease. A child will inherit some traits but not others, and geneticists use probability theory to analyze this uncertainty.

continued

- We will use probability theory to analyze the inheritance of hair color—the color of the parents' hair determines a range of possibilities for the color of their child's hair.
- We will also analyze the inheritance of sickle-cell anemia, cystic fibrosis, and other inherited diseases. Your future children might be at risk for these diseases because they can be inherited from disease-free parents.
- Genetics got its start when Gregor Mendel, a nineteenth-century monk, used probability theory to analyze the effect of randomness on heredity.

BELIEVE IT OR NOT, YOUR CELL PHONE USES PROBABILITIES:

- Your cell phone's predictive text feature ("T9" and "iTap" are two examples) allows you to type text messages with a single keypress for each letter, rather than the multitap approach required by older phones. When you type "4663," you could be wanting the word "good" or "home" or "gone." The software looks "4663" up in a dictionary and selects the word that you use with the highest probability.

BUSINESSES USE PROBABILITIES:

- All insurance companies use probability theory to set fees so that the companies will make a profit.
- Investment firms use probability theory to assess financial risk.
- Medical firms use probability theory in diagnosing ailments and determining appropriate treatments.

3.1 History of Probability

OBJECTIVES

- Understand the types of problems that motivated the invention of probability theory
- Become familiar with dice, cards, and the game of roulette

Origins in Gambling

Probability theory is generally considered to have began in 1654 when the French nobleman (and successful gambler) Antoine Gombauld, the Chevalier de Méré, asked his mathematician friend Blaise Pascal why he had typically made money with one dice bet but lost money with another bet that he thought was similar. Pascal wrote a letter about this to another prominent French mathematician, Pierre de Fermat. In answering the Chevalier's question, the two launched probability theory.

More than a hundred years earlier, in 1545, the Italian physician, mathematician, and gambler Gerolamo Cardano had used probabilities to help him win more often. His *Liber de ludo aleae* (*Book on Games of Chance*) was the first book on probabilities and gambling. It also contained a section on effective cheating methods. Cardano's work was ignored until interest in Pascal's and Fermat's work prompted its publication.

For a long time, probability theory was not viewed as a serious branch of mathematics because of its early association with gambling. This started to change when Jacob Bernoulli, a prominent Swiss mathematician, wrote *Ars Conjectandi* (*The Art of Conjecture*), published in 1713. In it, Bernoulli developed the mathematical theory of probabilities. Its focus was on gambling, but it also suggested applications of probability to government, economics, law, and genetics.

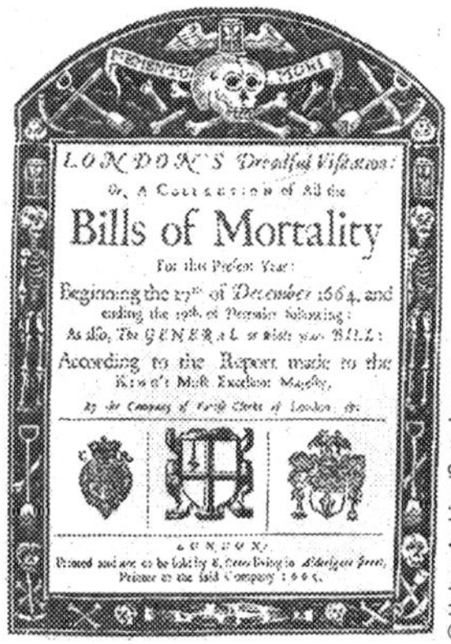

This **Bills of Mortality** (records of death) was published shortly after John Graunt's analysis.

Other Early Uses: Insurance and Genetics

In 1662, the Englishman John Graunt published his *Natural and Political Observations on the Bills of Mortality,* which used probabilities to analyze death records and was the first important incidence of the use of probabilities for a purpose other than gambling. (It preceded Bernoulli's book.) English firms were selling the first life insurance policies, and they used mortality tables to set fees that were appropriate to the risks involved but still allowed the companies to make a profit.

In the 1860s, Gregor Mendel, an Austrian monk, experimented with pea plants in the abbey garden. His experiments and theories allowed for randomness in the passing on of traits from parent to child, and he used probabilities to analyze the effect of that randomness. In combining biology and mathematics, Mendel founded the science of genetics.

Roulette

Probability theory was originally created to aid gamblers, and games of chance are the most easily and universally understood topic to which probability theory can be applied. Learning about probability and games of chance may convince you not to be a gambler. As we will see in Section 3.5, there is not a single good bet in a casino game of chance.

Roulette is the oldest casino game still being played. Its invention has variously been credited to Pascal, the ancient Chinese, a French monk, and the Italian mathematician Don Pasquale. It became popular when a French policeman introduced it to Paris in 1765 in an attempt to take the advantage away from dishonest gamblers. When the Monte Carlo casino opened in 1863, roulette was the most popular game, especially among the aristocracy. The American roulette wheel has thirty-eight numbered compartments around its circumference. Thirty-six of these compartments are numbered from 1 to 36 and are colored red or black. The remaining two are numbered 0 and 00 and are colored green. Players place their bets by putting their chips on an appropriate spot on the roulette table (see Figure 3.1) on page 134. The dealer spins the wheel and then drops a ball onto the spinning wheel. The ball eventually comes to rest in one of the compartments, and that

Roulette quickly became a favorite game of the French upper class, as shown in this 1910 photo of Monte Carlo.

compartment's number is the winning number. (See on page 183 for a photo of the roulette wheel and table.)

For example, if a player wanted to bet \$10 that the ball lands in compartment number 7, she would place \$10 worth of chips on the number 7 on the roulette table. This is a single-number bet, so house odds are 35 to 1 (see Figure 3.1 for house odds). This means that if the player wins, she wins $\$10 \cdot 35 = \$350$, and if she loses, she loses her \$10.

Similarly, if a player wanted to bet \$5 that the ball lands either on 13 or 14, he would place \$5 worth of chips on the line separating numbers 13 and 14 on the roulette table. This is a two-numbers bet, so house odds are 17 to 1. This means that if the player wins, he wins $\$5 \cdot 17 = \$85$, and if he loses, he loses his \$5.

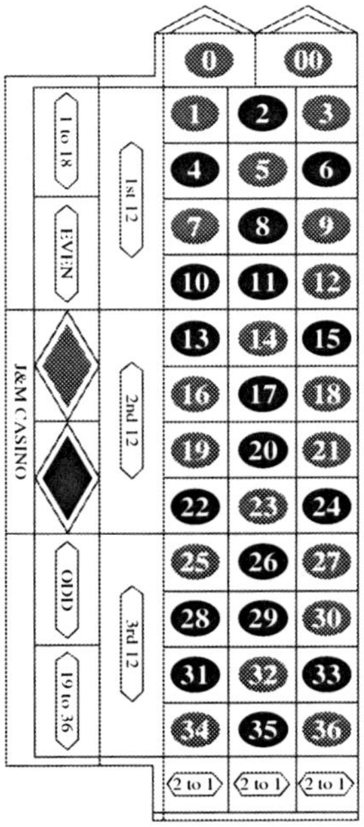

Bet	House Odds
single number	35 to 1
two numbers ("split")	17 to 1
three numbers ("street")	11 to 1
four numbers ("square")	8 to 1
five numbers ("line")	6 to 1
six numbers ("line")	5 to 1
twelve numbers (column or section)	2 to 1
low or high (1 to 18 or 19 to 36, respectively)	1 to 1
even or odd (0 and 00 are neither even nor odd)	1 to 1
red or black	1 to 1

FIGURE 3.1 The roulette table and house odds for the various roulette bets.

HISTORICAL NOTE

BLAISE PASCAL, 1623–1662

© Sheila Terry/Photo Researchers

As a child, Blaise Pascal showed an early aptitude for science and mathematics, even though he was discouraged from studying in order to protect his poor health. Acute digestive problems and chronic insomnia made his life miserable. Few of Pascal's days were without pain.

At age sixteen, Pascal wrote a paper on geometry that won the respect of the French mathematical community and the jealousy of the prominent French mathematician René Descartes. It has been suggested that the animosity between Descartes and Pascal was in part due to their religious differences. Descartes was a Jesuit, and Pascal was a Jansenist. Although Jesuits and Jansenists were both Roman Catholics, Jesuits believed in free will and supported the sciences, while Jansenists believed in predestination and mysticism and opposed the sciences and the Jesuits.

At age nineteen, to assist his father, a tax administrator, Pascal invented the first calculating machine. Besides cofounding probability theory with Pierre de Fermat, Pascal contributed to the advance of calculus, and his studies in physics culminated with a law on the effects of pressure on fluids that bears his name.

At age thirty-one, after a close escape from death in a carriage accident, Pascal turned his back on mathematics and the sciences and focused on defending Jansenism against the Jesuits. Pascal came to be the greatest Jansenist, and he aroused a storm with his anti-Jesuit *Provincial Letters*. This work is still famous for its polite irony.

Pascal turned so far from the sciences that he came to believe that reason is inadequate to solve humanity's difficulties or to satisfy our hopes, and he regarded the pursuit of science as a vanity. Even so, he still occasionally succumbed to its lure. Once, to distract himself from pain, he concentrated on a geometry problem. When the pain stopped, he decided that it was a signal from God that he had not sinned by thinking of mathematics instead of his soul. This incident resulted in the only scientific work of his last few years—and his last work. He died later that year, at age thirty-nine.

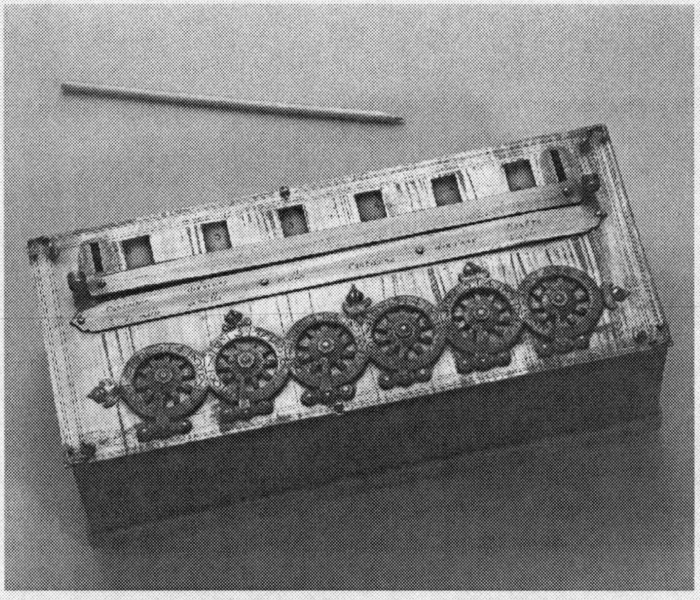
© SSPL/The Image Works

Pascal's calculator, the Pascaline.

HISTORICAL NOTE

GEROLAMO CARDANO, 1501–1576

Gerolamo Cardano is the subject of much disagreement among historians. Some see him as a man of tremendous accomplishments, while others see him as a plagiarist and a liar. All agree that he was a compulsive gambler.

Cardano was trained as a medical doctor, but he was initially denied admission to the College of Physicians of Milan. That denial was ostensibly due to his illegitimate birth, but some suggest that the denial was in fact due to his unsavory reputation as a gambler, since illegitimacy was neither a professional nor a social obstacle in sixteenth-century Italy. His lack of professional success left him with much free time, which he spent gambling and reading. It also resulted in a stay in the poorhouse.

Cardano's luck changed when he obtained a lectureship in mathematics, astronomy, and astrology at the University of Milan. He wrote a number of books on mathematics and became famous for publishing a method of solving third-degree equations. Some claim that Cardano's mathematical success was due not to his own abilities but rather to those of Ludovico Ferrari, a servant of his who went on to become a mathematics professor.

While continuing to teach mathematics, Cardano returned to the practice of medicine. (He was finally allowed to join the College of Physicians, perhaps owing to his success as a mathematician.) Cardano wrote books on medicine and the natural sciences that were well thought of. He became one of the most highly regarded physicians in Europe and counted many prominent people among his patients. He designed a tactile system, somewhat like braille, for the blind. He also designed an undercarriage suspension device that was later adapted as a universal joint for automobiles and that is still called a *cardan* in Europe.

Cardano's investment in gambling was enormous. He not only wagered (and lost) a great deal of money but also spent considerable time and effort calculating probabilities and devising strategies. His *Book on Games of Chance* contains the first correctly calculated theoretical probabilities.

Cardano's autobiography, *The Book of My Life*, reveals a unique personality. He admitted that he loved talking about himself, his accomplishments, and his illnesses and diseases. He frequently wrote of injuries done him by others and followed these complaints with gleeful accounts of his detractors' deaths. Chapter titles include "Concerning my friends and patrons," "Calumny, defamations, and treachery of my unjust accusers," "Gambling and dicing," "Religion and piety," "The disasters of my sons," "Successes in my practice," "Things absolutely supernatural," and "Things of worth which I have achieved in various studies."

A die. The other three faces have four, five, and six spots.

Dice and Craps

Dice have been cast since the beginning of time, for both divination and gambling purposes. The earliest **die** (singular of *dice*) was an animal bone, usually a knucklebone or foot bone. The Romans were avid dice players. The Roman emperor Claudius I wrote a book titled *How to Win at Dice*. During the Middle Ages, dicing schools and guilds of dicers were quite popular among the knights and ladies.

Hazard, an ancestor of the dice game craps, is an English game that was supposedly invented by the Crusaders in an attempt to ward off boredom during long, drawn-out sieges. It became quite popular in England and France in the nineteenth century. The English called a throw of 2, 3, or 12 *crabs*, and it is believed that *craps* is a French mispronunciation of that term. The game came to America with the French colonization of New Orleans and spread up the Mississippi.

Cards

The invention of playing cards has been credited to the Indians, the Arabs, the Egyptians, and the Chinese. During the Crusades, Arabs endured lengthy sieges by

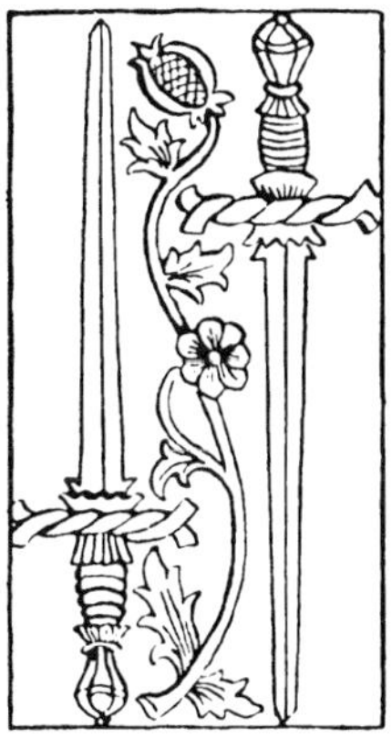

Brooks/Cole

A sixteenth-century 2 of swords.

Brooks/Cole

A sixteenth-century king of clubs.

playing card games. Their European foes acquired the cards and introduced them to their homelands. Cards, like dice, were used for divination as well as gambling. In fact, the modern deck is derived from the Tarot deck, which is composed of four suits plus twenty-two *atouts* that are not part of any suit. Each suit represents a class of medieval society: swords represent the nobility; coins, the merchants; batons or clubs, the peasants; and cups or chalices, the church. These suits are still used in regular playing cards in southern Europe. The Tarot deck also includes a joker and, in each suit, a king, a queen, a knight, a knave, and ten numbered cards.

Around 1500, the French dropped the knights and *atouts* from the deck and changed the suits from swords, coins, clubs, and chalices to *piques* (soldiers' pikes), *carreaux* (diamond-shaped building tiles), *trèfles* (clover leaf-shaped trefoils), and *coeurs* (hearts). In sixteenth-century Spain, *piques* were called *espados,* from which we get our term *spades.* Our diamonds are so named because of the shape of the carreaux. Clubs was an original Tarot suit, and hearts is a translation of *coeurs.*

The pictures on the cards were portraits of actual people. In fourteenth-century Europe, the kings were Charlemagne (hearts), the biblical David (spades), Julius Caesar (diamonds), and Alexander the Great (clubs); the queens included Helen of Troy (hearts), Pallas Athena (spades), and the biblical Rachel (diamonds). Others honored as "queen for a day" included Joan of Arc, Elizabeth I, and Elizabeth of York, wife of Henry VII. Jacks were usually famous warriors, including Sir Lancelot (clubs) and Roland, Charlemagne's nephew (diamonds).

A modern deck of cards contains fifty-two cards (thirteen in each of four suits). The four suits are hearts, diamonds, clubs, and spades (♥, ♦, ♣, ♠). Hearts and diamonds are red, and clubs and spades are black. Each suit consists of cards labeled 2 through 10, followed by jack, queen, king, and ace. **Face cards** are the jack, queen, and king; and **picture cards** are the jack, queen, king, and ace.

Two of the most popular card games are poker and blackjack. Poker's ancestor was a Persian game called *dsands,* which became popular in eighteenth-century Paris. It was transformed into a game called *poque,* which spread to America via the French colony in New Orleans. *Poker* is an American mispronunciation of the word *poque.*

The origins of blackjack (also known as *vingt-et-un* or twenty-one) are unknown. The game is called twenty-one because high odds are paid if a player's first two cards total 21 points (an ace counts as either 1 or 11, and the ten, jack, queen, and king each count as 10). A special bonus used to be paid if those two cards were a black jack and a black ace, hence the name *blackjack.*

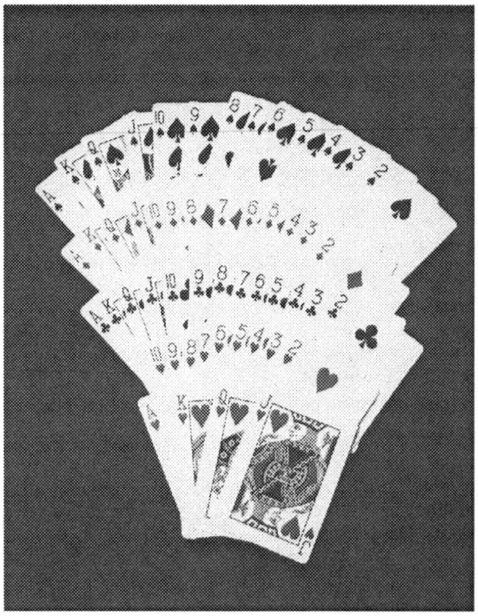

© Kathleen Olson

A modern deck of cards.

3.1 Exercises

1. The Chevalier de Méré generally made money betting that he could roll at least one 6 in four rolls of a single die. Roll a single die four times, and record the number of times a 6 comes up. Repeat this ten times. If you had made the Chevalier de Méré's bet (at $10 per game), would you have won or lost money? How much?
2. The Chevalier de Méré generally lost money betting that he could roll at least one pair of 6's in twenty-four rolls of a pair of dice. Roll a pair of dice twenty-four times, and record the number of times a 6 comes up. Repeat this five times. If you had made the Chevalier de Méré's bet (at $10 per game), would you have won or lost money? How much?
3. **a.** If you were to flip a pair of coins thirty times, approximately how many times do you think a pair of heads would come up? A pair of tails? One head and one tail?

   b. Flip a pair of coins thirty times, and record the number of times a pair of heads comes up, the number of times a pair of tails comes up, and the number of times one head and one tail come up. How closely do the results agree with your guess?
4. **a.** If you were to flip a single coin twenty times, approximately how many times do you think heads would come up? Tails?

   b. Flip a single coin twenty times, and record the number of times heads comes up and the number of times tails comes up. How closely do the results agree with your guesses?
5. **a.** If you were to roll a single die twenty times, approximately how many times do you think an even number would come up? An odd number?

   b. Roll a single die twenty times, and record the number of times an even number comes up and the number of times an odd number comes up. How closely do the results agree with your guess?
6. **a.** If you were to roll a pair of dice thirty times, approximately how many times do you think the total would be 7? Approximately how many times do you think the total would be 12?

   b. Roll a single die thirty times, and record the number of times the total is 7 and the number of times the total is 12. How closely do the results agree with your guess?
7. **a.** If you were to deal twenty-six cards from a complete deck (without jokers), approximately how many cards do you think would be red? Approximately how many do you think would be aces?

   b. Deal twenty-six cards from a complete deck (without jokers), and record the number of times a red card is dealt and the number of times an ace is dealt. How closely do the results agree with your guess?
8. **a.** If you were to deal twenty-six cards from a complete deck (without jokers), approximately how many cards do you think would be black? Approximately how many do you think would be jacks, queens, or kings?

   b. Deal twenty-six cards from a complete deck (without jokers), and record the number of times a black card is dealt and the number of times a jack, queen, or king is dealt. How closely do the results agree with your guess?

In Exercises 9–24, use Figure 3.1 to find the outcome of the bets in roulette, given the results listed.

9. You bet $10 on the 25.

   a. The ball lands on number 25.

   b. The ball lands on number 14.
10. You bet $15 on the 17.

    a. The ball lands on 18.

    b. The ball lands on 17.
11. You bet $5 on 17-20 split.

    a. The ball lands on number 17.

    b. The ball lands on number 20.

    c. The ball lands on number 32.
12. You bet $30 on the 22-23-24 street.

    a. The ball lands on number 19.

    b. The ball lands on number 22.

    c. The ball lands on number 0.
13. You bet $20 on the 8-9-11-12 square.

    a. The ball lands on number 15.

    b. The ball lands on number 9.

    c. The ball lands on number 00.
14. You bet $100 on the 0-00-1-2-3 line (the only five-number line on the table).

    a. The ball lands on number 29.

    b. The ball lands on number 2.
15. You bet $10 on the 31-32-33-34-35-36 line.

    a. The ball lands on number 5.

    b. The ball lands on number 33.
16. You bet $20 on the 13 through 24 section.

    a. The ball lands on number 00.

    b. The ball lands on number 15.

17. You bet \$25 on the first column.
   a. The ball lands on number 13.
   b. The ball lands on number 14.
18. You bet \$30 on the low numbers.
   a. The ball lands on number 8.
   b. The ball lands on number 30.
19. You bet \$50 on the odd numbers.
   a. The ball lands on number 00.
   b. The ball lands on number 5.
20. You bet \$20 on the black numbers.
   a. The ball lands on number 11.
   b. The ball lands on number 12.
21. You make a \$20 single-number bet on number 14 and also a \$25 single-number bet on number 15.
   a. The ball lands on number 16.
   b. The ball lands on number 15.
   c. The ball lands on number 14.
22. You bet \$10 on the low numbers and also bet \$20 on the 16-17-19-20 square.
   a. The ball lands on number 16.
   b. The ball lands on number 19.
   c. The ball lands on number 14.
23. You bet \$30 on the 1-2 split and also bet \$15 on the even numbers.
   a. The ball lands on number 1.
   b. The ball lands on number 2.
   c. The ball lands on number 3.
   d. The ball lands on number 4.
24. You bet \$40 on the 1-12 section and also bet \$10 on number 10.
   a. The ball lands on number 7.
   b. The ball lands on number 10.
   c. The ball lands on number 21.

25. How much must you bet on a single number to be able to win at least \$100? (Bets must be in \$1 increments.)
26. How much must you bet on a two-number split to be able to win at least \$200? (Bets must be in \$1 increments.)
27. How much must you bet on a twelve-number column to be able to win at least \$1000? (Bets must be in \$1 increments.)
28. How much must you bet on a four-number square to win at least \$600? (Bets must be in \$1 increments.)

29. a. How many hearts are there in a deck of cards?
   b. What fraction of a deck is hearts?
30. a. How many red cards are there in a deck of cards?
   b. What fraction of a deck is red?
31. a. How many face cards are there in a deck of cards?
   b. What fraction of a deck is face cards?
32. a. How many black cards are there in a deck of cards?
   b. What fraction of a deck is black?
33. a. How many kings are there in a deck of cards?
   b. What fraction of a deck is kings?

Answer the following questions using complete sentences and your own words.

• HISTORY QUESTIONS

34. Who started probability theory? How?
35. Why was probability theory not considered a serious branch of mathematics?
36. Which authors established probability theory as a serious area of interest? What are some of the areas to which these authors applied probability theory?
37. What did Gregor Mendel do with probabilities?
38. Who was Antoine Gombauld, and what was his role in probability theory?
39. Who was Gerolamo Cardano, and what was his role in probability theory?
40. Which games of chance came to America from France via New Orleans?
41. Which implements of gambling were also used for divination?
42. What is the oldest casino game still being played?
43. How were cards introduced to Europe?
44. What is the modern deck of cards derived from?
45. What game was supposed to take the advantage away from dishonest gamblers?

WEB PROJECT

46. Write a research paper on any historical topic referred to in this section or a related topic. Following is a partial list of topics:
   - Jacob Bernoulli
   - Gerolamo Cardano
   - Pierre de Fermat
   - Blaise Pascal
   - The Marquis de Laplace
   - John Graunt
   - Gregor Mendel

   Some useful links for this web project are listed on the text web site: **www.cengage.com/math/johnson**

3.2 Basic Terms of Probability

Objectives

- Learn the basic terminology of probability theoıy
- Be able to calculate simple probabilities
- Understand how probabilities are used in genetics

Much of the terminology and many of the computations of probability theory have their basis in set theory, because set theory contains the mathematical way of describing collections of objects and the size of those collections.

> **BASIC PROBABILITY TERMS**
>
> **experiment:** a process by which an observation, or **outcome,** is obtained
> **sample space:** the set S of all possible outcomes of an experiment
> **event:** any subset E of the sample space S

If a single die is rolled, the *experiment* is the rolling of the die. The possible *outcomes* are 1, 2, 3, 4, 5, and 6. The *sample space* (set of all possible outcomes) is $S = \{1, 2, 3, 4, 5, 6\}$. (The term *sample space* really means the same thing as *universal set;* the only distinction between the two ideas is that *sample space* is used only in probability theory, while *universal set* is used in any situation in which sets are used.) There are several possible *events* (subsets of the sample space), including the following:

$E_1 = \{3\}$	"a three comes up"
$E_2 = \{2, 4, 6\}$	"an even number comes up"
$E_3 = \{1, 2, 3, 4, 5, 6\}$	"a number between 1 and 6 inclusive comes up"

Notice that an event is not the same as an outcome. An *event* is a subset of the sample space; an *outcome* is an element of the sample space. "Rolling an odd number" is an event, not an outcome. It is the set $\{1, 3, 5\}$ that is composed of three separate outcomes. Some events are distinguished from outcomes only in that set brackets are used with events and not with outcomes. For example, $\{5\}$ is an event, and 5 is an outcome; either refers to "rolling a five."

The event E_3 ("a number between 1 and 6 inclusive comes up") is called a *certain event,* since $E_3 = S$. That is, E_3 is a sure thing. "Getting 17" is an *impossible event.* No outcome in the sample space $S = \{1, 2, 3, 4, 5, 6\}$ would result in a 17, so this event is actually the empty set.

> **MORE PROBABILITY TERMS**
>
> A **certain event** is an event that is equal to the sample space.
> An **impossible event** is an event that is equal to the empty set.

THE FAR SIDE® By GARY LARSON

Early shell games

An early certain event.

Finding Probabilities and Odds

The **probability** of an event is a measure of the likelihood that the event will occur. If a single die is rolled, the outcomes are equally likely; a 3 is just as likely to come up as any other number. There are six possible outcomes, so a 3 should come up about one out of every six rolls. That is, the probability of event E_1 ("a 3 comes up") is $\frac{1}{6}$. The 1 in the numerator is the number of elements in $E_1 = \{3\}$. The 6 in the denominator is the number of elements in $S = \{1, 2, 3, 4, 5, 6\}$.

If an experiment's outcomes are equally likely, then the probability of an event E is the number of outcomes in the event divided by the number of outcomes in the sample space, or $n(E)/n(S)$. (In this chapter, we discuss only experiments with equally likely outcomes.) Probability can be thought of as "success over a total."

PROBABILITY OF AN EVENT

The **probability** of an event E, denoted by $p(E)$, is

$$p(E) = \frac{n(E)}{n(S)}$$

if the experiment's outcomes are equally likely.
(*Think: Success over total.*)

Many people use the words *probability* and *odds* interchangeably. However, the words have different meanings. The **odds** in favor of an event are the number of ways the event can occur compared to the number of ways the event *can fail to occur,* or "success compared to *failure*" (if the experiment's outcomes are equally likely). The odds of event E_1 ("a 3 comes up") are 1 to 5 (or 1:5), since a three can come up in one way and can fail to come up in five ways. Similarly, the odds of event E_3 ("a number between 1 and 6 inclusive comes up") are 6 to 0 (or 6:0), since a number between 1 and 6 inclusive can come up in six ways and can fail to come up in zero ways.

ODDS OF AN EVENT

The **odds** of an event E with equally likely outcomes, denoted by $o(E)$, are given by

$$o(E) = n(E):n(E')$$

(Think: Success compared with failure.)

In addition to the above meaning, the word *odds* can also refer to "house odds," which has to do with how much you will be paid if you win a bet at a casino. The odds of an event are sometimes called the **true odds** to distinguish them from the house odds.

EXAMPLE 1

FLIPPING A COIN A coin is flipped. Find the following.

a. the sample space

b. the probability of event E_1, "getting heads"

c. the odds of event E_1, "getting heads"

d. the probability of event E_2, "getting heads or tails"

e. the odds of event E_2, "getting heads or tails"

SOLUTION

a. *Finding the sample space S:* The experiment is flipping a coin. The only possible outcomes are heads and tails. The sample space S is the set of all possible outcomes, so $S = \{h, t\}$.

b. *Finding the probability of heads:*

$$E_1 = \{h\} \text{ ("getting heads")}$$
$$p(E_1) = \frac{n(E_1)}{n(S)} = \frac{1}{2}$$

This means that one out of every two possible outcomes is a success.

c. *Finding the odds of heads:*

$$E_1' = \{t\}$$
$$o(E_1) = n(E_1) : n(E_1') = 1:1$$

This means that for every one possible success, there is one possible failure.

d. *Finding the probability of heads or tails:*

$$E_2 = \{h, t\}$$
$$p(E_2) = \frac{n(E_2)}{n(S)} = \frac{2}{2} = \frac{1}{1}$$

This means that every outcome is a success. Notice that E_2 is a certain event.

e. *Finding the odds of heads or tails:*

$$E_2' = \varnothing$$
$$o(E_2) = n(E_2) : n(E_2') = 2 : 0 = 1:0$$

This means that there are no possible failures.

EXAMPLE 2

ROLLING A DIE A die is rolled. Find the following.

a. the sample space
b. the event "rolling a 5"
c. the probability of rolling a 5
d. the odds of rolling a 5
e. the probability of rolling a number below 5
f. the odds of rolling a number below 5

SOLUTION

a. The sample space is $S = \{1, 2, 3, 4, 5, 6\}$.
b. Event E_1 "rolling a 5" is $E_1 = \{5\}$.
c. The probability of E_1 is

$$p(E_1) = \frac{n(E_1)}{n(S)} = \frac{1}{6}$$

This means that one out of every six possible outcomes is a success (that is, a 5).

d. $E_1' = \{1, 2, 3, 4, 6\}$. The odds of E_1 are

$$o(E_1) = n(E_1):n(E_1') = 1:5$$

This means that there is one possible success for every five possible failures.

e. $E_2 = \{1, 2, 3, 4\}$ ("rolling a number below 5"). The probability of E_2 is

$$p(E_2) = \frac{n(E_2)}{n(S)} = \frac{4}{6} = \frac{2}{3}$$

This means that two out of every three possible outcomes are a success.

f. $E_2' = \{5, 6\}$. The odds of E_2 are

$$o(E_2) = n(E_2):n(E_2') = 4:2 = 2:1$$

This means that there are two possible successes for every one possible failure. Notice that odds are reduced in the same manner that a fraction is reduced.

Relative Frequency versus Probability

So far, we have discussed probabilities only in a theoretical way. When we found that the probability of heads was $\frac{1}{2}$, we never actually tossed a coin. It does not always make sense to calculate probabilities theoretically; sometimes they must be found empirically, the way a batting average is calculated. For example, in 8,389 times at bat, Babe Ruth had 2,875 hits. His batting average was $\frac{2,875}{8,389} \approx 0.343$. In other words, his probability of getting a hit was 0.343.

Sometimes a probability can be found either theoretically or empirically. We have already found that the theoretical probability of heads is $\frac{1}{2}$. We could also flip a coin a number of times and calculate (number of heads)/(total number of flips); this can be called the **relative frequency** of heads, to distinguish it from the theoretical probability of heads.

The Law of Large Numbers

Usually, the relative frequency of an outcome is not equal to its probability, but if the number of trials is large, the two tend to be close. If you tossed a coin a couple of times, anything could happen, and the fact that the probability of heads is

$\frac{1}{2}$ would have no impact on the results. However, if you tossed a coin 100 times, you would probably find that the relative frequency of heads was close to $\frac{1}{2}$. If your friend tossed a coin 1,000 times, she would probably find the relative frequency of heads to be even closer to $\frac{1}{2}$ than in your experiment. This relationship between probabilities and relative frequencies is called the **Law of Large Numbers.** Cardano stated this law in his *Book on Games of Chance,* and Bernoulli proved that it must be true.

LAW OF LARGE NUMBERS

If an experiment is repeated a large number of times, the relative frequency of an outcome will tend to be close to the probability of that outcome.

The graph in Figure 3.2 shows the result of a simulated coin toss, using a computer and a random number generator rather than an actual coin. Notice that when the number of tosses is small, the relative frequency achieves values such as 0, 0.67, and 0.71. These values are not that close to the theoretical probability of 0.5. However, when the number of tosses is large, the relative frequency achieves values such as 0.48. These values are very close to the theoretical probability.

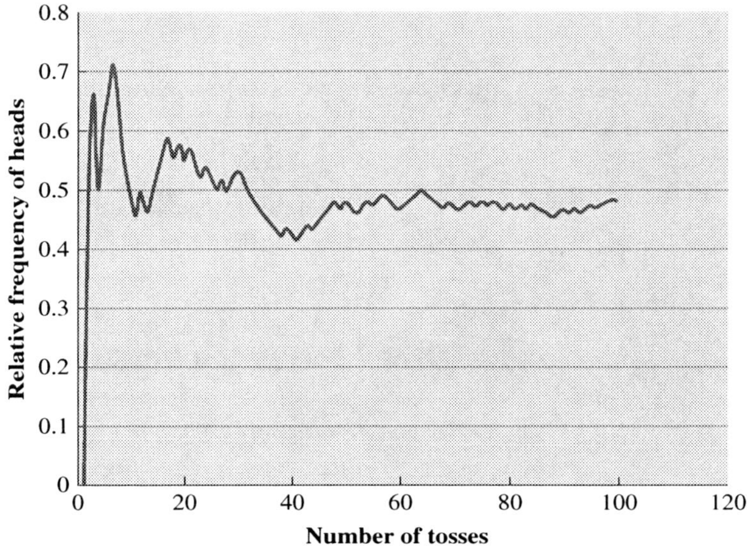

FIGURE 3.2 The relative frequency of heads after 100 simulated coin tosses.

What if we used a real coin, rather than a computer, and we tossed the coin a lot more? Three different mathematicians have performed such an experiment:

- In the eighteenth century, Count Buffon tossed a coin 4,040 times. He obtained 2,048 heads, for a relative frequency of $2,048/4,040 \approx 0.5069$.
- During World War II, the South African mathematician John Kerrich tossed a coin 10,000 times while he was imprisoned in a German concentration camp. He obtained 5,067 heads, for a relative frequency of $5,067/10,000 = 0.5067$.
- In the early twentieth century, the English mathematician Karl Pearson tossed a coin 24,000 times! He obtained 12,012 heads, for a relative frequency of $12,012/24,000 = 0.5005$.

At a casino, probabilities are much more useful to the casino (the "house") than to an individual gambler, because the house performs the experiment for a much larger number of trials (in other words, plays the game more often). In fact, the house plays the game so many times that the relative frequencies will be almost exactly the same as the probabilities, with the result that the house is not gambling at all—it knows what is going to happen. Similarly, a gambler with a "system" has to play the game for a long time for the system to be of use.

EXAMPLE 3

FLIPPING A PAIR OF COINS A pair of coins is flipped.

a. Find the sample space.
b. Find the event E "getting exactly one heads."
c. Find the probability of event E.
d. Use the Law of Large Numbers to interpret the probability of event E.

SOLUTION

The outcomes (h, h), (h, t), (t, h), and (t, t).

a. The experiment is the flipping of a pair of coins. One possible outcome is that one coin is heads and the other is tails. A second and *different* outcome is that one coin is tails and the other is heads. These two outcomes seem the same. However, if one coin were painted, it would be easy to tell them apart. Outcomes of the experiment can be described by using ordered pairs in which the first component refers to the first coin and the second component refers to the second coin. The two different ways of getting one heads and one tails are (h, t) and (t, h).

The sample space, or set of all possible outcomes, is the set $S = \{(h, h), (t, t), (h, t), (t, h)\}$. These outcomes are equally likely.

b. The event E "getting exactly one heads" is

$$E = \{(h, t), (t, h)\}$$

c. The probability of event E is

$$P(E) = \frac{n(E)}{n(S)} = \frac{2}{4} = \frac{1}{2} = 50\%$$

d. According to the Law of Large Numbers, if an experiment is repeated a large number of times, the relative frequency of that outcome will tend to be close to the probability of the outcome. Here, that means that if we were to toss a pair of coins many times, we should expect to get exactly one heads about half (or 50%) of the time. Realize that this is only a prediction, and we might never get exactly one heads.

Mendel's Use of Probabilities

In his experiments with plants, Gregor Mendel pollinated peas until he produced pure-red plants (that is, plants that would produce only red-flowered offspring) and pure-white plants. He then cross-fertilized these pure reds and pure whites and obtained offspring that had only red flowers. This amazed him, because the accepted theory of the day incorrectly predicted that these offspring would all have pink flowers.

He explained this result by postulating that there is a "determiner" that is responsible for flower color. These determiners are now called **genes.** Each plant has two flower color genes, one from each parent. Mendel reasoned that these offspring had to have inherited a red gene from one parent and a white gene from the other. These plants had one red flower gene and one white flower gene, but they had red flowers. That is, the red flower gene is **dominant,** and the white flower gene is **recessive.**

If we use R to stand for the red gene and w to stand for the white gene (with the capital letter indicating dominance), then the results can be described with a **Punnett square** in Figure 3.3 on page 146.

	R	R	← first parent's genes
w	(R, w)	(R, w)	← possible offspring
w	(R, w)	(R, w)	← possible offspring

↑
second parent's genes

FIGURE 3.3 A Punnett square for the first generation.

When the offspring of this experiment were cross-fertilized, Mendel found that approximately three-fourths of the offspring had red flowers and one-fourth had white flowers (see Figure 3.4). Mendel successfully used probability theory to analyze this result.

	R	w	← first parent's genes
R	(R, R)	(w, R)	← possible offspring
w	(R, w)	(w, w)	← possible offspring

↑
second parent's genes

FIGURE 3.4 A Punnett square for the second generation.

Only one of the four possible outcomes, (w, w), results in a white-flowered plant, so event E_1 that the plant has white flowers is $E_1 = \{(w, w)\}$; therefore,

$$p(E_1) - \frac{n(E_1)}{n(S)} = \frac{1}{4}$$

This means that we should expect the actual relative frequency of white-flowered plants to be close to $\frac{1}{4}$. That is, about one-fourth, or 25%, of the second-generation plants should have white flowers (if we have a lot of plants).

Each of the other three outcomes, (R, R), (R, w), and (w, R), results in a red-flowered plant, because red dominates white. The event E_2 that the plant has red flowers is $E_2 = \{(R, R), (R, w), (w, R)\}$; therefore,

$$p(E_2) = \frac{n(E_2)}{n(S)} = \frac{3}{4} = 75\%$$

Thus, we should expect the actual relative frequency of red-flowered plants to be close to $\frac{3}{4}$. That is, about three-fourths, or 75%, of the plants should have red flowers.

Outcomes (R, w) and (w, R) are genetically identical; it does not matter which gene is inherited from which parent. For this reason, geneticists do not use the ordered-pair notation but instead refer to each of these two outcomes as "Rw." The only difficulty with this convention is that it makes the sample space appear to be S = {RR, Rw, ww}, which consists of only three elements, when in fact it consists of four elements. This distinction is important; if the sample space consisted of three equally likely elements, then the probability of a red-flowered offspring would be $\frac{2}{3}$ rather than $\frac{3}{4}$. Mendel knew that the sample space had to have four elements, because his cross-fertilization experiments resulted in a relative frequency very close to $\frac{3}{4}$, not $\frac{2}{3}$.

Ronald Fisher, a noted British statistician, used statistics to deduce that Mendel fudged his data. Mendel's relative frequencies were unusually close to the theoretical probabilities, and Fisher found that there was only about a 0.00007 chance of such close agreement. Others have suggested that perhaps Mendel did not willfully change his results but rather continued collecting data until the numbers were in close agreement with his expectations.*

*R. A. Fisher, "Has Mendel's Work Been Rediscovered?" *Annals of Science* 1, 1936, pp. 115–137.

HISTORICAL NOTE

GREGOR JOHANN MENDEL, 1822–1884

© Bettmann/CORBIS

Johann Mendel was born to an Austrian peasant family. His interest in botany began on the family farm, where he helped his father graft fruit trees. He studied philosophy, physics, and mathematics at the University Philosophical Institute in Olmütz. He was unsuccessful in finding a job, so he quit school and returned to the farm. Depressed by the prospects of a bleak future, he became ill and stayed at home for a year.

Mendel later returned to Olmütz. After two years of study, he found the pressures of school and work to be too much, and his health again broke down. On the advice of his father and a professor, he entered the priesthood, even though he did not feel called to serve the church. His name was changed from Johann to Gregor.

Relieved of his financial difficulties, he was able to continue his studies. However, his nervous disposition interfered with his pastoral duties, and he was assigned to substitute teaching. He enjoyed this work and was popular with the staff and students, but he failed the examination for certification as a teacher. Ironically, his lowest grades were in biology. The Augustinians then sent him to the University of Vienna, where he became particularly interested in his plant physiology professor's unorthodox belief that new plant varieties can be caused by naturally arising variations. He was also fascinated by his classes in physics, where he was exposed to the physicists' experimental and mathematical approach to their subject.

After further breakdowns and failures, Mendel returned to the monastery and was assigned the low-stress job of keeping the abbey garden. There he combined the experimental and mathematical approach of a physicist with his background in biology and performed a series of experiments designed to determine whether his professor was correct in his beliefs regarding the role of naturally arising variants in plants.

Mendel studied the transmission of specific traits of the pea plant—such as flower color and stem length—from parent plant to offspring. He pollinated the plants by hand and separated them until he had isolated each trait. For example, in his studies of flower color, he pollinated the plants until he produced pure-red plants (plants that would produce only red-flowered offspring) and pure-white plants.

At the time, the accepted theory of heredity was that of blending. In this view, the characteristics of both parents blend together to form an individual. Mendel reasoned that if the blending theory was correct, the union of a pure-red pea plant and a pure-white pea plant would result in a pink-flowered offspring. However, his experiments showed that such a union consistently resulted in red-flowered offspring.

Mendel crossbred a large number of peas that had different characteristics. In many cases, an offspring would have a characteristic of one of its parents, undiluted by that of the other parent. Mendel concluded that the question of which parent's characteristics would be passed on was a matter of chance, and he successfully used probability theory to estimate the frequency with which characteristics would be passed on. In so doing, Mendel founded modern genetics. Mendel attempted similar experiments with bees, but these experiments were unsuccessful because he was unable to control the mating behavior of the queen bee.

Mendel was ignored when he published his paper "Experimentation in Plant Hybridization." Sixteen years after his death, his work was rediscovered by three European botanists who had reached similar conclusions in plant breeding, and the importance of Mendel's work was finally recognized.

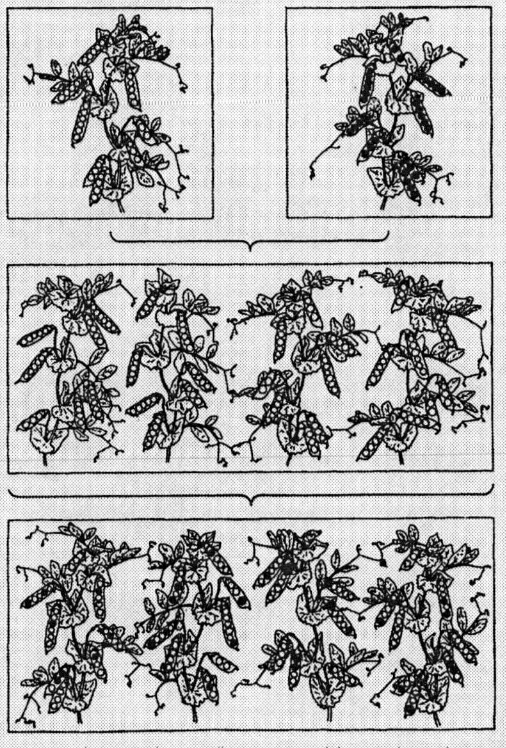
© The Granger Collection, NY

A nineteenth-century drawing illustrating Mendel's pea plants, showing the original cross, the first generation, and the second generation.

Probabilities in Genetics: Inherited Diseases

Cystic fibrosis is an inherited disease characterized by abnormally functioning exocrine glands that secrete a thick mucus, clogging the pancreatic ducts and lung passages. Most patients with cystic fibrosis die of chronic lung disease; until recently, most died in early childhood. This early death made it extremely unlikely that an afflicted person would ever parent a child. Only after the advent of Mendelian genetics did it become clear how a child could inherit the disease from two healthy parents.

In 1989, a team of Canadian and American doctors announced the discovery of the gene that is responsible for most cases of cystic fibrosis. As a result of that discovery, a new therapy for cystic fibrosis is being developed. Researchers splice a therapeutic gene into a cold virus and administer it through an affected person's nose. When the virus infects the lungs, the gene becomes active. It is hoped that this will result in normally functioning cells, without the damaging mucus.

In April 1993, a twenty-three-year-old man with advanced cystic fibrosis became the first patient to receive this therapy. In September 1996, a British team announced that eight volunteers with cystic fibrosis had received this therapy; six were temporarily cured of the disease's debilitating symptoms. In March 1999, another British team announced a new therapy that involves administering the therapeutic gene through an aerosol spray. Thanks to other advances in treatment, in 2006, infants born in the United States with cystic fibrosis had a life expectancy of twenty-seven years.

Cystic fibrosis occurs in about 1 out of every 2,000 births in the Caucasian population and only in about 1 in 250,000 births in the non-Caucasian population. It is one of the most common inherited diseases in North America. One in 25 Americans carries a single gene for cystic fibrosis. Children who inherit two such genes develop the disease; that is, cystic fibrosis is recessive.

There are tests that can be used to determine whether a person carries the gene. However, they are not accurate enough to use for the general population. They are much more accurate with people who have a family history of cystic fibrosis, so The American College of Obstetricians and Gynecologists recommends testing only for couples with a personal or close family history of cystic fibrosis.

EXAMPLE 4

PROBABILITIES AND CYSTIC FIBROSIS Each of two prospective parents carries one cystic fibrosis gene.

a. Find the probability that their child would have cystic fibrosis.
b. Find the probability that their child would be free of symptoms.
c. Find the probability that their child would be free of symptoms but could pass the cystic fibrosis gene on to his or her own child.

SOLUTION

We will denote the recessive cystic fibrosis gene with the letter c and the normal disease-free gene with an N. Each parent is Nc and therefore does not have the disease. Figure 3.5 shows the Punnett square for the child.

	N	c
N	**(N, N)**	**(c, N)**
c	**(N, c)**	**(c, c)**

FIGURE 3.5 A Punnett square for Example 4.

a. Cystic fibrosis is recessive, so only the (c, c) child will have the disease. The probability of such an event is 1/4.

b. The (N, N), (c, N), and (N, c) children will be free of symptoms. The probability of this event is

$$p(\text{healthy}) = p((\text{N, N})) + p((\text{c, N})) + p((\text{N, c}))$$
$$= \frac{1}{4} + \frac{1}{4} + \frac{1}{4} = \frac{3}{4}$$

c. The (c, N) and (N, c) children would never suffer from any symptoms but could pass the cystic fibrosis gene on to their own children. The probability of this event is

$$p((\text{c, N})) + p((\text{N, c})) = \frac{1}{4} + \frac{1}{4} = \frac{1}{2}$$

In Example 4, the Nc child is called a **carrier** because that child would never suffer from any symptoms but could pass the cystic fibrosis gene on to his or her own child. Both of the parents were carriers.

Sickle-cell anemia is an inherited disease characterized by a tendency of the red blood cells to become distorted and deprived of oxygen. Although it varies in severity, the disease can be fatal in early childhood. More often, patients have a shortened life span and chronic organ damage.

Newborns are now routinely screened for sickle-cell disease. The only true cure is a bone marrow transplant from a sibling without sickle-cell anemia; however, this can cause the patient's death, so it is done only under certain circumstances. Until recently, about 10% of the children with sickle-cell anemia had a stroke before they were twenty-one. But in 2009, it was announced that the rate of these strokes has been cut in half thanks to a new specialized ultrasound scan that identifies the individuals who have a high stroke risk. There are also medications that can decrease the episodes of pain.

Approximately 1 in every 500 black babies is born with sickle-cell anemia, but only 1 in 160,000 nonblack babies has the disease. This disease is **codominant:** A person with two sickle-cell genes will have the disease, while a person with one sickle-cell gene will have a mild, nonfatal anemia called **sickle-cell trait.** Approximately 8–10% of the black population has sickle-cell trait.

Huntington's disease, caused by a dominant gene, is characterized by nerve degeneration causing spasmodic movements and progressive mental deterioration. The symptoms do not usually appear until well after reproductive age has been reached; the disease usually hits people in their forties. Death typically follows 20 years after the onset of the symptoms. No effective treatment is available, but physicians can now assess with certainty whether someone will develop the disease, and they can estimate when the disease will strike. Many of those who are at risk choose not to undergo the test, especially if they have already had children. Folk singer Arlo Guthrie is in this situation; his father, Woody Guthrie, died of Huntington's disease. Woody's wife Marjorie formed the Committee to Combat Huntington's Disease, which has stimulated research, increased public awareness, and provided support for families in many countries.

In August 1999, researchers in Britain, Germany, and the United States discovered what causes brain cells to die in people with Huntington's disease. This discovery may eventually lead to a treatment. In 2008, a new drug that reduces the uncontrollable spasmodic movements was approved.

Tay-Sachs disease is a recessive disease characterized by an abnormal accumulation of certain fat compounds in the spinal cord and brain. Most typically,

HISTORICAL NOTE

NANCY WEXLER

Courtesy of Dr. Nancy Wexler

In 1993, scientists working together at six major research centers located most genes that cause Huntington's disease. This discovery will enable people to learn whether they carry a Huntington's gene, and it will allow pregnant women to determine whether their child carries the gene. The discovery could eventually lead to a treatment.

The collaboration of research centers was organized largely by Nancy Wexler, a Columbia University professor of neuropsychology, who is herself at risk for Huntington's disease—her mother died of it in 1978. Dr. Wexler, President of the Hereditary Disease Foundation, has made numerous trips to study and aid the people of the Venezuelan village of Lake Maracaibo, many of whom suffer from the disease or are at risk for it. All are related to one woman who died of the disease in the early 1800s. Wexler took blood and tissue samples and gave neurological and psychoneurological tests to the inhabitants of the village. The samples and test results enabled the researchers to find the single gene that causes Huntington's disease.

In October 1993, Wexler received an Albert Lasker Medical Research Award, a prestigious honor that is often a precursor to a Nobel Prize. The award was given in recognition for her contribution to the international effort that culminated in the discovery of the Huntington's disease gene. At the awards ceremony, she explained to then first lady Hillary Clinton that her genetic heritage has made her uninsurable—she would lose her health coverage if she switched jobs. She told Mrs. Clinton that more Americans will be in the same situation as more genetic discoveries are made, unless the health care system is reformed. The then first lady incorporated this information into her speech at the awards ceremony: "It is likely that in the next years, every one of us will have a pre-existing condition and will be uninsurable. . . . What will happen as we discover those genes for breast cancer, or prostate cancer, or osteoporosis, or any of the thousands of other conditions that affect us as human beings?"

AP Photo

Woody Guthrie's most famous song is "This Land is Your Land." This folksinger, guitarist, and composer was a friend of Leadbelly, Pete Seeger, and Ramblin' Jack Elliott and exerted a strong influence on Bob Dylan. Guthrie died at the age of fifty-five of Huntington's disease.

a child with Tay-Sachs disease starts to deteriorate mentally and physically at six months of age. After becoming blind, deaf, and unable to swallow, the child becomes paralyzed and dies before the age of four years. There is no effective treatment. The disease occurs once in 3,600 births among Ashkenazi Jews (Jews from central and eastern Europe), Cajuns, and French Canadians but only once in 600,000 births in other populations. Carrier-detection tests and fetal-monitoring tests are available. The successful use of these tests, combined with an aggressive counseling program, has resulted in a decrease of 90% of the incidence of this disease.

Genetic Screening

At this time, there are no conclusive tests that will tell a parent whether he or she is a cystic fibrosis carrier, nor are there conclusive tests that will tell whether a fetus has the disease. A new test resulted from the 1989 discovery of the location of most cystic fibrosis genes, but that test will detect only 85% to 95% of the cystic fibrosis genes, depending on the individual's ethnic background. The extent to which this test will be used has created quite a controversy.

Individuals who have relatives with cystic fibrosis are routinely informed about the availability of the new test. The controversial question is whether a massive genetic screening program should be instituted to identify cystic fibrosis carriers in the general population, regardless of family history. This is an important question, considering that four in five babies with cystic fibrosis are born to couples with no previous family history of the condition.

Opponents of routine screening cite a number of important concerns. The existing test is rather inaccurate; 5% to 15% of the cystic fibrosis carriers would be missed. It is not known how health insurers would use this genetic information—insurance firms could raise rates or refuse to carry people if a screening test indicated a presence of cystic fibrosis. Also, some experts question the adequacy of quality assurance for the diagnostic facilities and for the tests themselves.

Supporters of routine testing say that the individual should be allowed to decide whether to be screened. Failing to inform people denies them the opportunity to make a personal choice about their reproductive future. An individual who is found to be a carrier could choose to avoid conception, to adopt, to use artificial insemination by a donor, or to use prenatal testing to determine whether a fetus is affected—at which point the additional controversy regarding abortion could enter the picture.

The Failures of Genetic Screening

The history of genetic screening programs is not an impressive one. In the 1970s, mass screening of blacks for sickle-cell anemia was instituted. This program caused unwarranted panic; those who were told they had sickle-cell trait feared that they would develop symptoms of the disease and often did not understand the probability that their children would inherit the disease (see Exercises 73 and 74). Some people with sickle-cell trait were denied health insurance and life insurance. See the 1973 Newsweek article on "The Row over Sickle-Cell."

Featured in the News

THE ROW OVER SICKLE-CELL

. . . Two years ago, President Nixon listed sickle-cell anemia along with cancer as diseases requiring special Federal attention. . . . Federal spending for sickle-cell anemia programs has risen from a scanty $1 million a year to $15 million for 1973. At the same time, in what can only be described as a head-long rush, at least a dozen states have passed laws requiring sickle-cell screening for blacks.

While all these efforts have been undertaken with the best intentions of both whites and blacks, in recent months the campaign has begun to stir widespread and bitter controversy. Some of the educational programs have been riddled with misinformation and have unduly frightened the black community. To quite a few Negroes, the state laws are discriminatory—and to the extent that they might inhibit childbearing, even genocidal. . . .

Parents whose children have the trait often misunderstand and assume they have the disease. In some cases, airlines have allegedly refused to hire black stewardesses who have the trait, and some carriers have been turned down by life-insurance companies—or issued policies at high-risk rates.

Because of racial overtones and the stigma that attaches to persons found to have the sickle-cell trait, many experts seriously object to mandatory screening programs. They note, for example, that there are no laws requiring testing for Cooley's anemia [or other disorders that have a hereditary basis]. . . . Moreover, there is little that a person who knows he has the disease or the trait can do about it. "I don't feel," says Dr. Robert L. Murray, a black geneticist at Washington's Howard University, "that people should be required by law to be tested for something that will provide information that is more negative than positive." . . . Fortunately, some of the mandatory laws are being repealed.

3.2 Exercises

In Exercises 1–14, use this information: A jar on your desk contains twelve black, eight red, ten yellow, and five green jellybeans. You pick a jellybean without looking.

1. What is the experiment?
2. What is the sample space?

In Exercises 3–14, find the following. Write each probability as a reduced fraction and as a percent, rounded to the nearest 1%.

3. the probability that it is black
4. the probability that it is green
5. the probability that it is red or yellow
6. the probability that it is red or black
7. the probability that it is not yellow
8. the probability that it is not red
9. the probability that it is white
10. the probability that it is not white
11. the odds of picking a black jellybean
12. the odds of picking a green jellybean
13. the odds of picking a red or yellow jellybean
14. the odds of picking a red or black jellybean

In Exercises 15–28, one card is drawn from a well-shuffled deck of fifty-two cards (no jokers).

15. What is the experiment?
16. What is the sample space?

In Exercises 17–28, (a) find the probability and (b) the odds of drawing the given cards. Also, (c) use the Law of Large Numbers to interpret both the probability and the odds. (You might want to review the makeup of a deck of cards in Section 3.1.)

17. a black card
18. a heart
19. a queen
20. a 2 of clubs
21. a queen of spades
22. a club
23. a card below a 5 (count an ace as high)
24. a card below a 9 (count an ace as high)
25. a card above a 4 (count an ace as high)
26. a card above an 8 (count an ace as high)
27. a face card
28. a picture card

Age (years)	0–4	5–19	20–44	45–64	65–84	85+	total
Male	10,748	31,549	53,060	38,103	14,601	1,864	149,925
Female	10,258	30,085	51,432	39,955	18,547	3,858	154,135

FIGURE 3.6 2008 U.S. population, in thousands, by age and gender. *Source:* 2008 Census, U.S. Bureau of the Census.

Age	White	Black	American Indian, Eskimo, Aleut	Asian, Pacific Islander
0–4	15,041	2,907	222	1,042
5–19	47,556	9,445	709	3,036
20–44	79,593	14,047	1,005	5,165
45–64	60,139	8,034	497	2,911
65–84	27,426	2,826	166	987
85+	4,467	362	29	110

FIGURE 3.7 2005 U.S. population (projected), in thousands, by age and race. *Source:* Annual Population Estimates by Age Group and Sex, U.S. Bureau of the Census

In Exercises 29–38, (a) find the probability and (b) the odds of winning the given bet in roulette. Also, (c) use the Law of Large Numbers to interpret both the probability and the odds. (You might want to review the description of the game in Section 3.1.)

29. the single-number bet
30. the two-number bet
31. the three-number bet
32. the four-number bet
33. the five-number bet
34. the six-number bet
35. the twelve-number bet
36. the low-number bet
37. the even-number bet
38. the red-number bet

39. Use the information in Figure 3.6 from the U.S. Census Bureau to answer the following questions.
   a. Find the probability that in the year 2008, a U.S. resident was female.
   b. Find the probability that in the year 2008, a U.S. resident was male and between 20 and 44 years of age, inclusive.

40. Use the information in Figure 3.7 from the U.S. Census Bureau to answer the following questions.
   a. Find the probability that in the year 2005, a U.S. resident was American Indian, Eskimo, or Aleut.
   b. Find the probability that in the year 2005, a U.S. resident was Asian or Pacific Islander and between 20 and 44 years of age.

41. The dartboard in Figure 3.8 is composed of circles with radii 1 inch, 3 inches, and 5 inches.
   a. What is the probability that a dart hits the red region if the dart hits the target randomly?
   b. What is the probability that a dart hits the yellow region if the dart hits the target randomly?
   c. What is the probability that a dart hits the green region if the dart hits the target randomly?
   d. Use the Law of Large Numbers to interpret the probabilities in parts (a), (b), and (c).

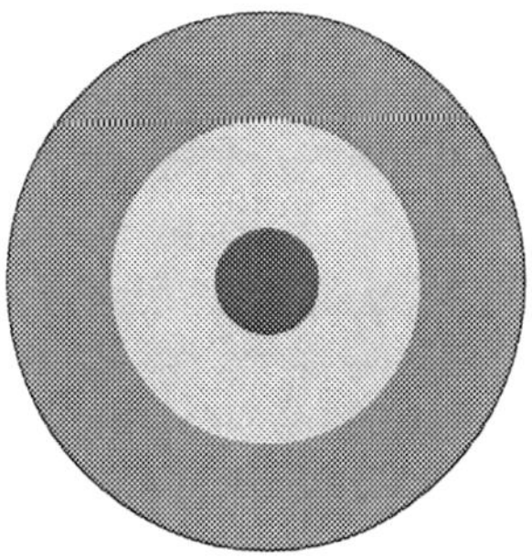

FIGURE 3.8 A dartboard for Exercise 41.

42. **a.** What is the probability of getting red on the spinner shown in Figure 3.9?
   b. What is the probability of getting blue?
   c. Use the Law of Large Numbers to interpret the probabilities in parts (a) and (b).

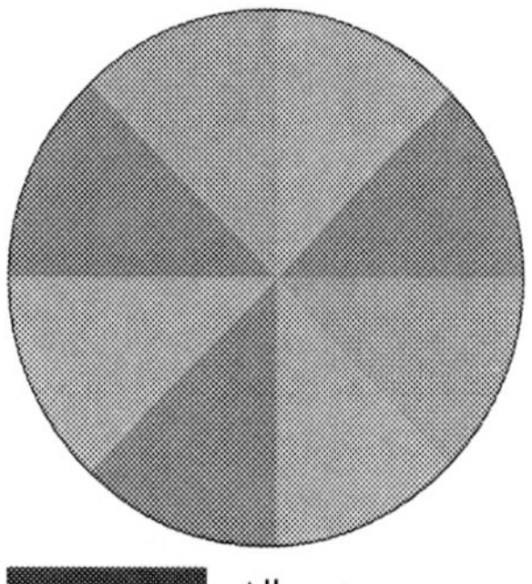

FIGURE 3.9 All sectors are equal in size and shape.

43. Amtrack's *Empire Service* train starts in Albany, New York, at 6:00 A.M., Mondays through Fridays. It arrives at New York City at 8:25 A.M. If the train breaks down, find the probability that it breaks down in the first half hour of its run. Assume that the train breaks down at random times.

44. Amtrack's *Downeaster* train leaves Biddeford, Maine, at 6:42 A.M., Mondays through Fridays. It arrives at Boston at 8:50 A.M. If the train breaks down, find the probability that it breaks down in the last 45 minutes of its run. Assume that the train breaks down at random times.

45. If $p(E) = \frac{1}{5}$, find $o(E)$.

46. If $p(E) = \frac{8}{9}$, find $o(E)$.

47. If $o(E) = 3:2$, find $p(E)$.

48. If $o(E) = 4:7$, find $p(E)$.

49. If $p(E) = \frac{a}{b}$, find $o(E)$.

In Exercises 50–52, find (a) the probability and (b) the odds of winning the following bets in roulette. In finding the odds, use the formula developed in Exercise 49. Also, (c) use the Law of Large Numbers to interpret both the probability and the odds.

50. the high-number bet

51. the odd-number bet

52. the black-number bet

53. In June 2009, sportsbook.com gave odds on who would win the 2009 World Series. They gave the New York Yankees 5:2 odds and the New York Mets 7:1 odds.

a. Are these house odds or true odds? Why?

b. Convert the odds to probabilities.

c. Who did they think was more likely to win the World Series: the Yankees or the Mets? Why?

54. In June 2009, linesmaker.com gave odds on who would win the 2009 World Series. They gave the New York Yankees 9:2 odds and the New York Mets 15:1 odds.

a. Are these house odds or true odds? Why?

b. Convert the odds to probabilities.

c. Who did they think was more likely to win the World Series: the Yankees or the Mets? Why?

d. Use the information in Exercise 53 to determine which source thought it was more probable that the Yankees would win the World Series: sportsbook .com or linesmaker.com.

In Exercises 55–60, use Figure 3.10.

55. **a.** Of the specific causes listed, which is the most likely cause of death in one year?

b. Which is the most likely cause of death in a lifetime? Justify your answers.

56. **a.** Of the causes listed, which is the least likely cause of death in one year?

Type of Accident or Injury	Odds of Dying in 1 Year	Lifetime Odds of Dying
All transportation accidents	1:6,121	1:79
Pedestrian transportation accident	1:48,816	1:627
Bicyclist transportation accident	1:319,817	1:4,111
Motorcyclist transportation accident	1:67,588	1:869
Car transportation accident	1:20,331	1:261
Airplane and space transportation accident	1:502,554	1:6,460
All nontransportation accidents	1:4,274	1:55
Falling	1:15,085	1:194
Drowning	1:82,777	1:1,064
Fire	1:92,745	1:1,192
Venomous animals and poisonous plants	1:2,823,877	1:36,297
Lightning	1:6,177,230	1:79,399
Earthquake	1:8,013,704	1:103,004
Storm	1:339,253	1:4,361
Intentional self-harm	1:9,085	1:117
Assault	1:16,360	1:210
Legal execution	1:5,490,872	1:70,577
War	1:10,981,743	1:141,154
Complications of medical care	1:111,763	1:1,437

FIGURE 3.10 The odds of dying due to an accident or injury in the United States in 2005. *Source: The odds of dying from . . .*, 2009 edition, www.nsc.org.

b. Which is the least likely cause of death in a lifetime? Justify your answers.

57. a. What is the probability that a person will die of a car transportation accident in one year?
b. What is the probability that a person will die of an airplane transportation accident in one year?
Write your answers as fractions.

58. a. What is the probability that a person will die from lightning in a lifetime?
b. What is the probability that a person will die from an earthquake in a lifetime?
Write your answers as fractions.

59. a. Of all of the specific forms of transportation accidents, which has the highest probability of causing death in one year? What is that probability?
b. Which has the lowest probability of causing death in one year? What is that probability?
Write your answers as fractions.

60. a. Of all of the specific forms of nontransportation accidents, which has the highest probability of causing death in a lifetime? What is that probability?
b. Which has the lowest probability of causing death in a lifetime? What is that probability?
Write your answers as fractions.

61. A family has two children. Using b to stand for boy and g for girl in ordered pairs, give each of the following.
a. the sample space
b. the event E that the family has exactly one daughter
c. the event F that the family has at least one daughter
d. the event G that the family has two daughters
e. $p(E)$ f. $p(F)$ g. $p(G)$
h. $o(E)$ i. $o(F)$ j. $o(G)$
(Assume that boys and girls are equally likely.)

62. Two coins are tossed. Using ordered pairs, give the following.
a. the sample space
b. the event E that exactly one is heads
c. the event F that at least one is heads
d. the event G that two are heads
e. $p(E)$ f. $p(F)$ g. $p(G)$
h. $o(E)$ i. $o(F)$ j. $o(G)$

63. A family has three children. Using b to stand for boy and g for girl and using ordered triples such as (b, b, g), give the following.
a. the sample space
b. the event E that the family has exactly two daughters
c. the event F that the family has at least two daughters
d. the event G that the family has three daughters
e. $p(E)$ f. $p(F)$ g. $p(G)$
h. $o(E)$ i. $o(F)$ j. $o(G)$
(Assume that boys and girls are equally likely.)

64. Three coins are tossed. Using ordered triples, give the following.
a. the sample space
b. the event E that exactly two are heads
c. the event F that at least two are heads
d. the event G that all three are heads
e. $p(E)$ f. $p(F)$ g. $p(G)$
h. $o(E)$ i. $o(F)$ j. $o(G)$

65. A couple plans on having two children.
a. Find the probability of having two girls.
b. Find the probability of having one girl and one boy.
c. Find the probability of having two boys.
d. Which is more likely: having two children of the same sex or two of different sexes? Why?
(Assume that boys and girls are equally likely.)

66. Two coins are tossed.
a. Find the probability that both are heads.
b. Find the probability that one is heads and one is tails.
c. Find the probability that both are tails.
d. Which is more likely: that the two coins match or that they don't match? Why?

67. A couple plans on having three children. Which is more likely: having three children of the same sex or of different sexes? Why? (Assume that boys and girls are equally likely.)

68. Three coins are tossed. Which is more likely: that the three coins match or that they don't match? Why?

69. A pair of dice is rolled. Using ordered pairs, give the following.
a. the sample space
HINT: S has 36 elements, one of which is (1, 1).
b. the event E that the sum is 7
c. the event F that the sum is 11
d. the event G that the roll produces doubles
e. $p(E)$ f. $p(F)$ g. $p(G)$
h. $o(E)$ i. $o(F)$ j. $o(G)$

70. Mendel found that snapdragons have no color dominance; a snapdragon with one red gene and one white gene will have pink flowers. If a pure-red snapdragon is crossed with a pure-white one, find the probability of the following.
a. a red offspring b. a white offspring
c. a pink offspring

71. If two pink snapdragons are crossed (see Exercise 70), find the probability of the following.
a. a red offspring b. a white offspring
c. a pink offspring

72. One parent is a cystic fibrosis carrier, and the other has no cystic fibrosis gene. Find the probability of each of the following.
a. The child would have cystic fibrosis.
b. The child would be a carrier.

c. The child would not have cystic fibrosis and not be a carrier.

d. The child would be healthy (i.e., free of symptoms).

73. If carrier-detection tests show that two prospective parents have sickle-cell trait (and are therefore carriers), find the probability of each of the following.

a. Their child would have sickle-cell anemia.

b. Their child would have sickle-cell trait.

c. Their child would be healthy (i.e., free of symptoms).

74. If carrier-detection tests show that one prospective parent is a carrier of sickle-cell anemia and the other has no sickle-cell gene, find the probability of each of the following.

a. The child would have sickle-cell anemia.

b. The child would have sickle-cell trait.

c. The child would be healthy (i.e., free of symptoms).

75. If carrier-detection tests show that one prospective parent is a carrier of Tay-Sachs and the other has no Tay-Sachs gene, find the probability of each of the following.

a. The child would have the disease.

b. The child would be a carrier.

c. The child would be healthy (i.e., free of symptoms).

76. If carrier-detection tests show that both prospective parents are carriers of Tay-Sachs, find the probability of each of the following.

a. Their child would have the disease.

b. Their child would be a carrier.

c. Their child would be healthy (i.e., free of symptoms).

77. If a parent started to exhibit the symptoms of Huntington's disease after the birth of his or her child, find the probability of each of the following. (Assume that one parent carries a single gene for Huntington's disease and the other carries no such gene.)

a. The child would have the disease.

b. The child would be a carrier.

c. The child would be healthy (i.e., free of symptoms).

Answer the following questions using complete sentences and your own words.

• CONCEPT QUESTIONS

78. Explain how you would find the theoretical probability of rolling an even number on a single die. Explain how you would find the relative frequency of rolling an even number on a single die.

79. Give five examples of events whose probabilities must be found empirically rather than theoretically.

80. Does the theoretical probability of an event remain unchanged from experiment to experiment? Why or why not? Does the relative frequency of an event remain unchanged from experiment to experiment? Why or why not?

81. Consider a "weighted die"—one that has a small weight in its interior. Such a weight would cause the face closest to the weight to come up less frequently and the face farthest from the weight to come up more frequently. Would the probabilities computed in Example 1 still be correct? Why or why not? Would the definition $p(E) = \frac{n(E)}{n(S)}$ still be appropriate? Why or why not?

82. Some dice have spots that are small indentations; other dice have spots that are filled with the same material but of a different color. Which of these two types of dice is not fair? Why? What would be the most likely outcome of rolling this type of die? Why?

HINT: 1 and 6 are on opposite faces, as are 2 and 5, and 3 and 4.

83. In the United States, 52% of the babies are boys, and 48% are girls. Do these percentages contradict an assumption that boys and girls are equally likely? Why?

84. Compare and contrast theoretical probability and relative frequency. Be sure that you discuss both the similarities and differences between the two.

85. Compare and contrast probability and odds. Be sure that you discuss both the similarities and differences between the two.

• HISTORY QUESTIONS

86. What prompted Dr. Nancy Wexler's interest in Huntington's disease?

87. What resulted from Dr. Nancy Wexler's interest in Huntington's disease?

• PROJECTS

88. a. In your opinion, what is the probability that the last digit of a phone number is odd? Justify your answer.

b. Randomly choose one page from the residential section of your local phone book. Count how many phone numbers on that page have an odd last digit and how many have an even last digit. Then compute (number of phone numbers with odd last digits)/(total number of phone numbers).

c. Is your answer to part (b) a theoretical probability or a relative frequency? Justify your answer.

d. How do your answers to parts (a) and (b) compare? Are they exactly the same, approximately the same, or dissimilar? Discuss this comparison, taking into account the ideas of probability theory.

89. a. Flip a coin ten times, and compute the relative frequency of heads.

b. Repeat the experiment described in part (a) nine more times. Each time, compute the relative frequency of heads. After finishing parts (a) and (b), you will have flipped a coin 100 times, and you will have computed ten different relative frequencies.

c. Discuss how your ten relative frequencies compare with each other and with the theoretical probability heads. In your discussion, use the ideas discussed under the heading "Relative Frequency versus Probability" in this section. Do not plagiarize; use your own words.

d. Combine the results of parts (a) and (b), and find the relative frequency of heads for all 100 coin tosses. Discuss how this relative frequency compares with those found in parts (a) and (b) and with the theoretic probability of heads. Be certain to incorporate the Law of Large Numbers into your discussion.

90. a. Roll a single die twelve times, and compute the relative frequency with which you rolled a number below a 3.

b. Repeat the experiment described in part (a) seven more times. Each time, compute the relative frequency with which you rolled a number below a 3. After finishing parts (a) and (b), you will have rolled a die ninety-six times, and you will have computed eight different relative frequencies.

c. Discuss how your eight relative frequencies compare with each other and with the theoretical probability rolling a number below a 3. In your discussion, use the ideas discussed under the heading "Relative Frequency versus Probability" in this section. Do not plagiarize; use your own words.

d. Combine the results of parts (a) and (b), and find the relative frequency of rolling a number below a 3 for all ninety-six die rolls. Discuss how this relative frequency compares with those found in parts (a) and (b) and with the theoretic probability of rolling a number less than a 3. Be certain to incorporate the Law of Large Numbers into your discussion.

91. Stand a penny upright on its edge on a smooth, hard, level surface. Then spin the penny on its edge. To do this, gently place a finger on the top of the penny. Then snap the penny with another finger (and immediately remove the holding finger) so that the penny spins rapidly before falling. Repeat this experiment fifty times, and compute the relative frequency of heads. Discuss whether or not the outcomes of this experiment are equally likely.

92. Stand a penny upright on its edge on a smooth, hard, level surface. Pound the surface with your hand so that the penny falls over. Repeat this experiment fifty times, and compute the relative frequency of heads. Discuss whether or not the outcomes of this experiment are equally likely.

WEB PROJECT

93. In the 1970s, there was a mass screening of blacks for sickle-cell anemia and a mass screening of Jews for Tay-Sachs disease. One of these was a successful program. One was not. Write a research paper on these two programs.

Some useful links for this web project are listed on the text web site:
www.cengage.com/math/johnson

3.3 Basic Rules of Probability

OBJECTIVES

- Understand what type of number a probability can be
- Learn about the relationships between probabilities, unions, and intersections

In this section, we will look at the basic rules of probability theory. The first three rules focus on why a probability can be a number like 5/7 or 32% but not a number like 9/5 or −27%. Knowing what type of number a probability can be is key to understanding what a probability means in real life. It also helps in checking your answers.

How Big or Small Can a Probability Be?

No event occurs less than 0% of the time. How could an event occur negative 15% of the time? Also, no event occurs more than 100% of the time. How could an event occur 125% of the time? Every event must occur between 0% and 100% of the time. For every event E,

$$0\% \le p(E) \le 100\%$$

$$0 \le p(E) \le 1 \qquad \text{converting from percents to decimals}$$

This means that *if you ever get a negative answer or an answer greater than 1 when you calculate a probability, go back and find your error.*

What type of event occurs 100% of the time? That is, what type of event has a probability of 1? Such an event must include *all possible* outcomes; if any possible outcome is left out, the event could not occur 100% of the time. An event that has a probability of 1 must be the sample space, because the sample space is the set of all possible outcomes. As we discussed earlier, such an event is called a *certain event.*

What type of event occurs 0% of the time? That is, what type of event has a probability of 0? Such an event must not include any possible outcome; if a possible outcome is included, the event would not have a probability of 0. An event with a probability of 0 must be the null set. As we discussed earlier, such an event is called an *impossible event.*

PROBABILITY RULES

Rule 1	$p(\varnothing) = 0$	The probability of the null set is 0.
Rule 2	$p(S) = 1$	The probability of the sample space is 1.
Rule 3	$0 \le p(E) \le 1$	Probabilities are between 0 and 1 (inclusive).

Probability Rules 1, 2, and 3 can be formally verified as follows:

Rule 1: $$p(\varnothing) = \frac{n(\varnothing)}{n(S)} = \frac{0}{n(S)} = 0$$

Rule 2: $$p(S) = \frac{n(S)}{n(S)} = 1$$

Rule 3: E is a subset of S; therefore,

$$0 \le n(E) \le n(S)$$

$$\frac{0}{n(S)} \le \frac{n(E)}{n(S)} \le \frac{n(S)}{n(S)} \qquad \text{dividing by } n(S)$$

$$0 \le p(E) \le 1$$

EXAMPLE 1

ILLUSTRATING RULES 1, 2, AND 3 A single die is rolled once. Find the probability of:

a. event E, "a 15 is rolled"
b. event F, "a number between 1 and 6 (inclusive) is rolled"
c. event G, "a 3 is rolled"

SOLUTION

The sample space is $S = \{1, 2, 3, 4, 5, 6\}$, and $n(S) = 6$.

a. It is *wrong* to say that $E = \{15\}$. Remember, we are talking about rolling a single die. Rolling a 15 is not one of the possible outcomes. That is, $15 \notin S$. There are no possible outcomes that result in rolling a 15, so the number of outcomes in event E is $n(E) = 0$. Thus,

$$p(E) = n(E)/n(S) = \frac{0}{6} = 0$$

The number 15 will be rolled 0% of the time. Event E is an impossible event.

Mathematically, we say that event E consists of no possible outcomes, so $E = \varnothing$. This agrees with rule 1, since

$$p(E) = p(\varnothing) = 0$$

b. $F = \{1, 2, 3, 4, 5, 6\}$, so $n(F) = 6$. Thus,

$$p(F) = n(F)/n(S) = \frac{6}{6} = 1$$

A number between 1 and 6 (inclusive) will be rolled 100% of the time. Event F is a certain event.

Mathematically, we say that event F consists of every possible outcome, so $F = S$. This agrees with rule 2, since

$$p(F) = p(S) = 1$$

c. $G = \{3\}$, so $n(G) = 1$. Thus

$$p(G) = n(G)/n(S) = 1/6$$

This agrees with rule 3, since

$$0 \le p(G) \le 1$$

Mutually Exclusive Events

Two events that cannot both occur at the same time are called **mutually exclusive.** In other words, E and F are mutually exclusive if and only if $E \cap F = \varnothing$.

EXAMPLE 2

DETERMINING WHETHER TWO EVENTS ARE MUTUALLY EXCLUSIVE A die is rolled. Let E be the event "an even number comes up," F the event "a number greater than 3 comes up," and G the event "an odd number comes up."

a. Are E and F mutually exclusive?
b. Are E and G mutually exclusive?

SOLUTION

a. $E = \{2, 4, 6\}$, $F = \{4, 5, 6\}$, and $E \cap F = \{4, 6\} \neq \varnothing$ (see Figure 3.11). Therefore, E and F are *not* mutually exclusive; the number that comes up could be *both* even *and* greater than 3. In particular, it could be 4 or 6.
b. $E = \{2, 4, 6\}$, $G = \{1, 3, 5\}$, and $E \cap G = \varnothing$. Therefore, E and G *are* mutually exclusive; the number that comes up could *not* be both even and odd.

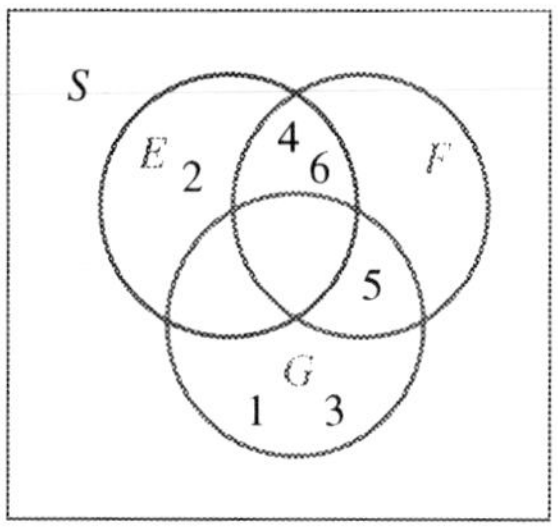

FIGURE 3.11 A Venn diagram for Example 2.

EXAMPLE 3

DETERMINING WHETHER TWO EVENTS ARE MUTUALLY EXCLUSIVE Let M be the event "being a mother," F the event "being a father," and D the event "being a daughter."

a. Are events M and D mutually exclusive?
b. Are events M and F mutually exclusive?

SOLUTION

a. M and D are mutually exclusive if $M \cap D = \varnothing$. $M \cap D$ is the set of all people who are both mothers and daughters, and that set is not empty. A person can be a mother and a daughter at the same time. M and D are not mutually exclusive because being a mother does not exclude being a daughter.

b. M and F are mutually exclusive if $M \cap F = \varnothing$. $M \cap F$ is the set of all people who are both mothers and fathers, and that set is empty. A person cannot be a mother and a father at the same time. M and F are mutually exclusive because being a mother does exclude the possibility of being a father.

Pair-of-Dice Probabilities

To find probabilities involving the rolling of a pair of dice, we must first determine the sample space. The *sum* can be anything from 2 to 12, but we will not use {2, 3, 4, 5, 6, 7, 8, 9, 10, 11, 12} as the sample space because those outcomes are not equally likely. Compare a sum of 2 with a sum of 7. There is only one way in which the sum can be 2, and that is if each die shows a 1. There are many ways in which the sum can be a 7, including:

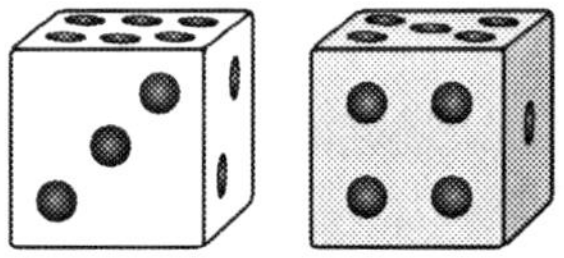

The event (3, 4).

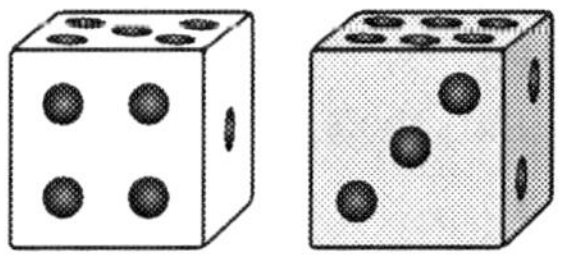

The event (4, 3).

- a 1 and a 6
- a 2 and a 5
- a 3 and a 4

What about a 4 and a 3? Is that the same as a 3 and a 4? They certainly *appear* to be the same. But if one die were blue and the other were white, it would be quite easy to tell them apart. So there are other ways in which the sum can be a 7:

- a 4 and a 3
- a 5 and a 2
- a 6 and a 1

Altogether, there are six ways in which the sum can be a 7. There is only one way in which the sum can be a 2. The outcomes "the sum is 2" and "the sum is 7" are not equally likely.

To have equally likely outcomes, we must use outcomes such as "rolling a 4 and a 3" and "rolling a 3 and a 4." We will use ordered pairs as a way of abbreviating these longer descriptions. So we will denote the outcome "rolling a 4 and a 3" with the ordered pair (4, 3), and we will denote the outcome "rolling a 3 and a 4" with the ordered pair (3, 4). Figure 3.12 lists all possible outcomes and the resulting sums. Notice that $n(S) = 6 \cdot 6 = 36$.

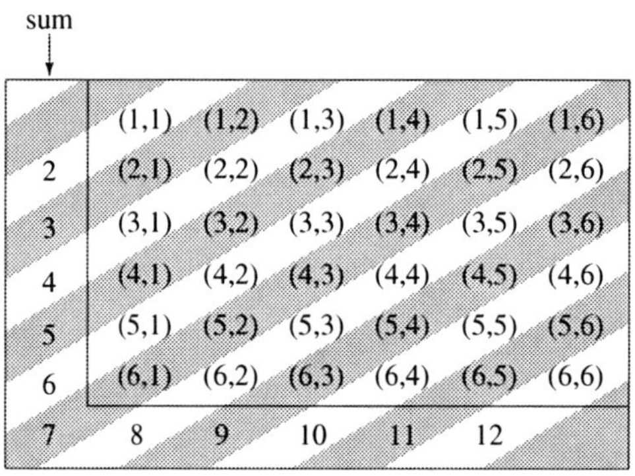

FIGURE 3.12 Outcomes of rolling two dice.

EXAMPLE 4

FINDING PAIR-OF-DICE PROBABILITIES A pair of dice is rolled. Find the probability of each of the following events.

a. The sum is 7. **b.** The sum is greater than 9. **c.** The sum is even.

SOLUTION

a. To find the probability that the sum is 7, let D be the event "the sum is 7." From Figure 3.12, $D = \{(1, 6), (2, 5), (3, 4), (4, 3), (5, 2), (6, 1)\}$, so $n(D) = 6$; therefore,

$$p(D) = \frac{n(D)}{n(S)} = \frac{6}{36} = \frac{1}{6}$$

This means that if we were to roll a pair of dice a large number of times, we should expect to get a sum of 7 approximately one-sixth of the time.

Notice that $p(D) = \frac{1}{6}$ is between 0 and 1, as are all probabilities.

b. To find the probability that the sum is greater than 9, let E be the event "the sum is greater than 9."

$E = \{(4, 6), (5, 5), (6, 4), (5, 6), (6, 5), (6, 6)\}$, so $n(E) = 6$; therefore,

$$p(E) = \frac{n(E)}{n(S)} = \frac{6}{36} = \frac{1}{6}$$

This means that if we were to roll a pair of dice a large number of times, we should expect to get a sum greater than 9 approximately one-sixth of the time.

c. Let F be the event "the sum is even."

$F = \{(1, 1), (1, 3), (2, 2), (3, 1), \ldots, (6, 6)\}$, so $n(F) = 18$ (refer to Figure 3.12); therefore,

$$p(F) = \frac{n(F)}{n(S)} = \frac{18}{36} = \frac{1}{2}$$

This means that if we were to roll a pair of dice a large number of times, we should expect to get an even sum approximately half of the time.

A Roman painting on marble of the daughters of Niobe using knucklebones as dice. This painting was found in the ruins of Herculaneum, a city that was destroyed along with Pompeii by the eruption of Vesuvius.

EXAMPLE 5

USING THE CARDINAL NUMBER FORMULAS TO FIND PROBABILITIES
A pair of dice is rolled. Use the Cardinal Number Formulas (where appropriate) and the results of Example 4 to find the probabilities of the following events.

a. the sum is not greater than 9.
b. the sum is greater than 9 and even.
c. the sum is greater than 9 or even.

SOLUTION

a. We could find the probability that the sum is not greater than 9 by counting, as in Example 4, but the counting would be rather excessive. It is easier to use one of the Cardinal Number Formulas from Chapter 2 on sets. The event "the sum is not greater than 9" is the complement of event E ("the sum is greater than 9") and can be expressed as E'.

$$n(E') = n(U) - n(E) \quad \text{Complement Formula}$$
$$= n(S) - n(E) \quad \text{"universal set" and "sample space" represent the same idea.}$$
$$= 36 - 6 = 30$$
$$p(E') = \frac{n(E')}{n(S)} = \frac{30}{36} = \frac{5}{6}$$

This means that if we were to roll a pair of dice a large number of times, we should expect to get a sum that's not greater than 9 approximately five-sixths of the time.

b. The event "the sum is greater than 9 and even" can be expressed as the event $E \cap F$. $E \cap F = \{(4, 6), (5, 5), (6, 4), (6, 6)\}$, so $n(E \cap F) = 4$; therefore,

$$p(E \cap F) = \frac{n(E \cap F)}{n(S)} = \frac{4}{36} = \frac{1}{9}$$

This means that if we were to roll a pair of dice a large number of times, we should expect to get a sum that's both greater than 9 and even approximately one-ninth of the time.

c. Finding the probability that the sum is greater than 9 or even by counting would require an excessive amount of counting. It is easier to use one of the Cardinal Number Formulas from Chapter 2. The event "the sum is greater than 9 or even" can be expressed as the event $E \cup F$.

$$n(E \cup F) = n(E) + n(F) - n(E \cap F) \quad \text{Union/Intersection Formula}$$
$$= 6 + 18 - 4 \quad \text{from part (e), and Example 4}$$
$$= 20$$
$$p(E \cup F) = \frac{n(E \cup F)}{n(S)} = \frac{20}{36} = \frac{5}{9}$$

This means that if we were to roll a pair of dice a large number of times, we should expect to get a sum that's either greater than 9 or even approximately five-ninths of the time.

More Probability Rules

In part (b) of Example 4, we found that $p(E) = \frac{1}{6}$, and in part (a) of Example 5, we found that $p(E') = \frac{5}{6}$. Notice that $p(E) + p(E') = \frac{1}{6} + \frac{5}{6} = 1$. This should make sense to you. It just means that if E happens one-sixth of the time, then E' has to happen the other five-sixths of the time. This always happens—for any event E, $p(E) + p(E') = 1$.

As we will see in Exercise 79, the fact that $p(E) + p(E') = 1$ is closely related to the Cardinal Number Formula $n(E) + n(E') = n(S)$. The main difference is that one is expressed in the language of probability theory and the other is expressed in the language of set theory. In fact, the following three rules are all set theory rules (from Chapter 2) rephrased so that they use the language of probability theory rather than the language of set theory.

MORE PROBABILITY RULES

Rule 4	**The Union/Intersection Rule**	$p(E \cup F) = p(E) + p(F) - p(E \cap F)$
Rule 5	**The Mutually Exclusive Rule**	$p(E \cup F) = p(E) + p(F)$, if E and F are mutually exclusive
Rule 6	**The Complement Rule**	$p(E) + p(E') = 1$ or, equivalently, $p(E') = 1 - p(E)$

In Example 5, we used some Cardinal Number Formulas from Chapter 2 to avoid excessive counting. Some find it easier to use Probability Rules to calculate probabilities, rather than Cardinal Number Formulas.

EXAMPLE **6**

USING RULES 4, 5, AND 6 Use a Probability Rule rather than a Cardinal Number Formula to find:

a. the probability that the sum is not greater than 9
b. the probability that the sum is greater than 9 or even

SOLUTION

We will use the results of Example 4.

a. $p(E') = 1 - p(E)$ the Complement Rule

$= 1 - \frac{1}{6}$ from Example 4

$= \frac{5}{6}$

b. $p(E \cup F) = p(E) + p(F) - p(E \cap F)$ the Union/Intersection Rule

$= \frac{1}{6} + \frac{1}{2} - \frac{1}{9}$ from parts (b) and (c) of Example 4, and part (b) of Example 5

$= \frac{3}{18} + \frac{9}{18} - \frac{2}{18} = \frac{5}{9}$ getting a common denominator

Probabilities and Venn Diagrams

Venn diagrams can be used to illustrate probabilities in the same way in which they are used in set theory. In this case, we label each region with its probability rather than its cardinal number.

EXAMPLE **7**

USING VENN DIAGRAMS Zaptronics manufactures compact discs and their cases for several major labels. A recent sampling of Zaptronics' products has indicated that 5% have a defective case, 3% have a defective disc, and 7% have at least one of the two defects.

a. Find the probability that a Zaptronics product has both defects.
b. Draw a Venn diagram that shows the probabilities of each of the basic regions.

c. Use the Venn diagram to find the probability that a Zaptronics product has neither defect.
d. Find the probability that a Zaptronics product has neither defect, using probability rules rather than a Venn diagram.

SOLUTION

a. Let C be the event that the case is defective, and let D be the event that the disc is defective. We are given that $p(C) = 5\% = 0.05$, $p(D) = 3\% = 0.03$, and $p(C \cup D) = 7\% = 0.07$, and we are asked to find $p(C \cap D)$. To do this, substitute into Probability Rule 4.

$$p(C \cup D) = p(C) + p(D) - p(C \cap D) \quad \text{the Union/Intersection Rule}$$
$$0.07 = 0.05 + 0.03 - p(C \cap D) \quad \text{substituting}$$
$$0.07 = 0.08 - p(C \cap D) \quad \text{adding}$$
$$p(C \cap D) = 0.01 = 1\% \quad \text{solving}$$

This means that 1% of Zaptronics' products have a defective case *and* a defective disc.

b. The Venn diagram for this is shown in Figure 3.13. The 0.93 probability in the lower right corner is obtained by first finding the sum of the probabilities of the other three basic regions:

$$0.04 + 0.01 + 0.02 = 0.07$$

The rectangle itself corresponds to the sample space, and its probability is 1. Thus, the probability of the outer region, the region outside of the C and D circles, is

$$1 - 0.07 = 0.93$$

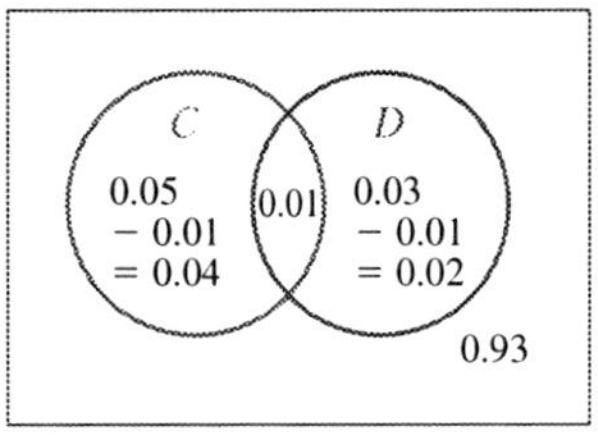

FIGURE 3.13
A Venn diagram for Example 6.

c. The only region that involves no defects is the outer region, the one whose probability is 0.93. This means that 93% of Zaptronics' products have neither defect.

d. If we are to not use the Venn diagram, we must first rephrase the event "a Zaptronics product has neither defect" as "a product does not have either defect." The "does not" part of this means *complement,* and the "either defect" means *either C or D.* This means that

- "has either defect" translates to $C \cup D$
- "does not have either defect" translates to $(C \cup D)'$

So we are asked to find $P((C \cup D)')$.

$$p((C \cup D)') = 1 - p(C \cup D) \quad \text{the Complement Rule}$$
$$= 1 - 0.07 \quad \text{substituting}$$
$$= 0.93 = 93\% \quad \text{subtracting}$$

This means that 93% of Zaptronics' products are defect-free.

3.3 Exercises

In Exercises 1–10, determine whether E and F are mutually exclusive. Write a sentence justifying your answer.

1. E is the event "being a doctor," and F is the event "being a woman."
2. E is the event "it is raining," and F is the event "it is sunny."
3. E is the event "being single," and F is the event "being married."
4. E is the event "having naturally blond hair," and F is the event "having naturally black hair."
5. E is the event "having brown hair," and F is the event "having gray hair."
6. E is the event "being a plumber," and F is the event "being a stamp collector."
7. E is the event "wearing boots," and F is the event "wearing sandals."
8. E is the event "wearing shoes," and F is the event "wearing socks."
9. If a die is rolled once, E is the event "getting a four," and F is the event "getting an odd number."

10. If a die is rolled once, E is the event "getting a four," and F is the event "getting an even number."

In Exercises 11–18, a card is dealt from a complete deck of fifty-two playing cards (no jokers). Use probability rules (when appropriate) to find the probability that the card is as stated. (Count an ace as high.)

11. **a.** a jack and red **b.** a jack or red
c. not a red jack

12. **a.** a jack and a heart **b.** a jack or a heart
c. not a jack of hearts

13. **a.** a 10 and a spade **b.** a 10 or a spade
c. not a 10 of spades

14. **a.** a 5 and black **b.** a 5 or black
c. not a black 5

15. **a.** under a 4
b. above a 9
c. both under a 4 and above a 9
d. either under a 4 or above a 9

16. **a.** above a jack
b. below a 3
c. both above a jack and below a 3
d. either above a jack or below a 3

17. **a.** above a 5
b. below a 10
c. both above a 5 and below a 10
d. either above a 5 or below a 10

18. **a.** above a 7
b. below a queen
c. both above a 7 and below a queen
d. either above a 7 or below a queen

In Exercises 19–26, use complements to find the probability that a card dealt from a full deck (no jokers) is as stated. (Count an ace as high.)

19. not a queen
20. not a 7
21. not a face card
22. not a heart
23. above a 3
24. below a queen
25. below a jack
26. above a 5

27. If $o(E) = 5{:}9$, find $o(E')$.
28. If $o(E) = 1{:}6$, find $o(E')$.
29. If $p(E) = \frac{2}{7}$, find $o(E)$ and $o(E')$.
30. If $p(E) = \frac{3}{8}$, find $o(E)$ and $o(E')$.
31. If $o(E) = a{:}b$, find $o(E')$.
32. If $p(E) = \frac{a}{b}$, find $o(E')$.

HINT: Use Exercise 49 from Section 3.2 and Exercise 31 above.

In Exercises 33–38, use Exercise 32 to find the odds that a card dealt from a full deck (no jokers) is as stated.

33. not a king **34.** not an 8
35. not a face card **36.** not a club
37. above a 4 **38.** below a king

In Exercises 39–42, use the following information: To determine the effect their salespeople have on purchases, a department store polled 700 shoppers regarding whether or not they made a purchase and whether or not they were pleased with the service they received. Of those who made a purchase, 151 were happy with the service and 133 were not. Of those who made no purchase, 201 were happy with the service and 215 were not. Use probability rules (when appropriate) to find the probability of the event stated.

39. **a.** A shopper made a purchase.
b. A shopper did not make a purchase.

40. **a.** A shopper was happy with the service received.
b. A shopper was unhappy with the service received.

41. **a.** A shopper made a purchase and was happy with the service.
b. A shopper made a purchase or was happy with the service.

42. **a.** A shopper made no purchase and was unhappy with the service.
b. A shopper made no purchase or was unhappy with the service.

In Exercises 43–46, use the following information: A supermarket polled 1,000 customers regarding the size of their bill. The results are given in Figure 3.14.

Size of Bill	Number of Customers
below \$20.00	208
\$20.00–\$39.99	112
\$40.00–\$59.99	183
\$60.00–\$79.99	177
\$80.00–\$99.99	198
\$100.00 or above	122

FIGURE 3.14 Supermarket bills.

Use probability rules (when appropriate) to find the relative frequency with which a customer's bill is as stated.

43. **a.** less than \$40.00
b. \$40.00 or more

44. **a.** less than \$80.00
b. \$80.00 or more

45. **a.** between \$40.00 and \$79.99
b. not between \$40.00 and \$79.99

46. **a.** between \$20.00 and \$79.99
b. not between \$20.00 and \$79.99

In Exercises 47–54, find the probability that the sum is as stated when a pair of dice is rolled.

47. **a.** 7 **b.** 9 **c.** 11
48. **a.** 2 **b.** 4 **c.** 6
49. **a.** 7 or 11 **b.** 7 or 11 or doubles
50. **a.** 8 or 10 **b.** 8 or 10 or doubles
51. **a.** odd and greater than 7
b. odd or greater than 7
52. **a.** even and less than 5
b. even or less than 5
53. **a.** even and doubles
b. even or doubles
54. **a.** odd and doubles
b. odd or doubles

In Exercises 55–58, use the following information: After examining their clients' records, Crashorama Auto Insurance calculated the probabilities in Figure 3.15.

	Miles/Year < 10,000	10,000 ≤ Miles/Year < 20,000	20,000 ≤ Miles/Year
Accident	0.05	0.10	0.3
No accident	0.20	0.15	0.2

FIGURE 3.15 Crashorama's Accident Incidence.

55. **a.** Find the probability that a client drives 20,000 miles/year or more or has an accident.
b. Find the probability that a client drives 20,000 miles/year or more and has an accident.
c. Find the probability that the client does not drive 20,000 miles/year or more and does not have an accident.
56. **a.** Find the probability that the client drives less than 10,000 miles/year and has an accident.
b. Find the probability that a client drives less than 10,000 miles/year or has an accident.
c. Find the probability that a client does not drive less than 10,000 miles per year and does not have an accident.
57. Are the probabilities in Figure 3.15 theoretical probabilities or relative frequencies? Why?
58. **a.** Find the probability that a client drives either less than 10,000 miles/year or 20,000 miles/year or more.
b. Find the probability that a client both has an accident and drives either less than 10,000 miles/year or 20,000 miles/year or more.

In Exercises 59–62, use the following information: The results of CNN's 2008 presidential election poll are given in Figure 3.16.

	Obama	McCain	Other/No Answer
Male Voters	23.0%	22.6%	1.4%
Female Voters	29.7%	22.8%	0.5%

FIGURE 3.16 Exit poll results, by gender. *Source:* CNN.

59. **a.** Find the probability that a polled voter voted for Obama or is male.
b. Find the probability that a polled voter voted for Obama and is male.
c. Find the probability that a polled voter didn't vote for Obama and is not male.
60. **a.** Find the probability that a polled voter voted for McCain or is female.
b. Find the probability that a polled voter voted for McCain and is female.
c. Find the probability that a polled voter didn't vote for McCain and is not female.
61. Are the probabilities in Figure 3.16 theoretical probabilities or relative frequencies? Why?
62. **a.** Find the probability that a polled voter didn't vote for Obama.
b. Find the probability that a polled voter is female.

63. Fried Foods hosted a group of twenty-five consumers at a tasting session. The consumers were asked to taste two new products and state whether or not they would be interested in buying them. Fifteen consumers said that they would be interested in buying Snackolas, twelve said that they would be interested in buying Chippers, and ten said that they would be interested in buying both.
a. What is the probability that one of the tasters is interested in buying at least one of the two new products?
b. What is the probability that one of the tasters is not interested in buying Snackolas but is interested in buying Chippers?
64. Ink Inc., a publishing firm, offers its 899 employees a cafeteria approach to benefits, in which employees can enroll in the benefit plan of their choice. Seven hundred thirteen employees have health insurance, 523 have dental insurance, and 489 have both health and dental insurance.
a. What is the probability that one of the employees has either health or dental insurance?
b. What is the probability that one of the employees has health insurance but not dental insurance?
65. ComCorp is considering offering city-wide Wi-Fi connections to the Web. They asked 3,000 customers about their use of the Web at home, and they received

1,451 responses. They found that 1,230 customers are currently connecting to the Web with a cable modem, 726 are interested in city-wide Wi-Fi, and 514 are interested in switching from a cable modem to city-wide Wi-Fi.

a. What is the probability that one of the respondents is interested in city-wide Wi-Fi and is not currently using a cable modem?

b. What is the probability that one of the respondents is interested in city-wide Wi-Fi and is currently using a cable modem?

66. The Video Emporium rents DVDs and blue-rays only. They surveyed their 1,167 rental receipts for the last two weeks. Eight hundred thirty-two customers rented DVDs, and 692 rented blue-rays.

a. What is the probability that a customer rents DVDs only?

b. What is the probability that a customer rents DVDs and blue-rays?

67. *Termiyak Magazine* conducted a poll of its readers, asking them about their telephones. Six hundred ninety-six readers responded. Five hundred seventy-two said they have a cell phone. Six hundred twelve said that they have a land line (a traditional, nonwireless telephone) at their home. Everyone has either a cell phone or a land line.

a. What is the probability that one of the readers has only a cell phone?

b. What is the probability that one of the readers has only a land line?

68. Which of the probabilities in Exercises 63–67 are theoretical probabilities and which are relative frequencies? Why?

69. Use probability rules to find the probability that a child will either have Tay-Sachs disease or be a carrier if (a) each parent is a Tay-Sachs carrier, (b) one parent is a Tay-Sachs carrier and the other parent has no Tay-Sachs gene, (c) one parent has Tay-Sachs and the other parent has no Tay-Sachs gene.

70. Use probability rules to find the probability that a child will have either sickle-cell anemia or sickle-cell trait if (a) each parent has sickle-cell trait, (b) one parent has sickle-cell trait and the other parent has no sickle-cell gene, (c) one parent has sickle-cell anemia and the other parent has no sickle-cell gene.

71. Use probability rules to find the probability that a child will neither have Tay-Sachs disease nor be a carrier if (a) each parent is a Tay-Sachs carrier, (b) one parent is a Tay-Sachs carrier and the other parent has no Tay-Sachs gene, (c) one parent has Tay-Sachs and the other parent has no Tay-Sachs gene.

72. Use probability rules to find the probability that a child will have neither sickle-cell anemia nor sickle-cell trait if (a) each parent has sickle-cell trait, (b) one parent has sickle-cell trait and the other parent has no sickle-cell gene, (c) one parent has sickle-cell anemia and the other parent has no sickle-cell gene.

73. Mary is taking two courses, photography and economics. Student records indicate that the probability of passing photography is 0.75, that of failing economics is 0.65, and that of passing at least one of the two courses is 0.85. Find the probability of the following.

a. Mary will pass economics.

b. Mary will pass both courses.

c. Mary will fail both courses.

d. Mary will pass exactly one course.

74. Alex is taking two courses; algebra and U.S. history. Student records indicate that the probability of passing algebra is 0.35, that of failing U.S. history is 0.35, and that of passing at least one of the two courses is 0.80. Find the probability of the following.

a. Alex will pass history.

b. Alex will pass both courses.

c. Alex will fail both courses.

d. Alex will pass exactly one course.

75. Of all the flashlights in a large shipment, 15% have a defective bulb, 10% have a defective battery, and 5% have both defects. If you purchase one of the flashlights in this shipment, find the probability that it has the following.

a. a defective bulb or a defective battery

b. a good bulb or a good battery

c. a good bulb and a good battery

76. Of all the DVDs in a large shipment, 20% have a defective disc, 15% have a defective case, and 10% have both defects. If you purchase one of the DVDs in this shipment, find the probability that it has the following.

a. a defective disc or a defective case

b. a good disc or a good case

c. a good disc and a good case

77. Verify the Union/Intersection Rule.

HINT: Divide the Cardinal Number Formula for the Union of Sets from Section 2.1 by $n(S)$.

78. Verify the Mutually Exclusive Rule.

HINT: Start with the Union/Intersection Rule. Then use the fact that E and F are mutually exclusive.

79. Verify the Complement Rule.

HINT: Are E and E' mutually exclusive?

Answer the following questions using complete sentences and your own words.

• CONCEPT QUESTIONS

80. What is the complement of a certain event? Justify your answer.

81. Compare and contrast mutually exclusive events and impossible events. Be sure to discuss both the similarities and the differences between these two concepts.

82. Explain why it is necessary to subtract $p(E \cap F)$ in Probability Rule 4. In other words, explain why Probability Rule 5 is not true for all events.

Fractions on a Graphing Calculator

Some graphing calculators—including the TI-83, TI-84, and Casio—will add, subtract, multiply, and divide fractions and will give answers in reduced fractional form.

Reducing Fractions

The fraction 42/70 reduces to 3/5. To do this on your calculator, you must make your screen read "42/70 ▶ Frac" ("42 ⌋70" on a Casio). The way that you do this varies.

TI-83/84	• Type 42 [÷] 70, but do not press [ENTER]. This causes "42/70" to appear on the screen. • Press the [MATH] button. • Highlight option 1, "▶Frac." (Option 1 is automatically highlighted. If we were selecting a different option, we would use the [▲] and [▼] buttons to highlight it.) • Press [ENTER]. This causes "42/70▶Frac" to appear on the screen. • Press [ENTER]. This causes 3/5 to appear on the screen.
CASIO	• Type 42 [a b/c] 70. • Type [EXE], and your display will read "3⌋5," which should be interpreted as "3/5."

EXAMPLE 8

Use your calculator to compute

$$\frac{1}{6} + \frac{1}{2} - \frac{1}{9}$$

and give your answer in reduced fractional form.

SOLUTION

On a TI, make your screen read "1/6 + 1/2 − 1/9 ▶Frac" by typing

1 [÷] 6 [+] 1 [÷] 2 [−] 1 [÷] 9

and then inserting the "▶Frac" command, as described above. Once you press [ENTER], the screen will read "5/9."

On a Casio, make your screen read "1⌟6 + 1⌟2 − 1⌟9" by typing

1 [$a^b/_c$] 6 + 1 [$a^b/_c$] 2 − 1 [$a^b/_c$] 9

Once you press EXE, the screen will read "5⌟9," which should be interpreted as "5/9."

EXERCISES

In Exercises 83–84, reduce the given fractions to lowest terms, both (a) by hand and (b) with a calculator. Check your work by comparing the two answers.

83. $\frac{18}{33}$ **84.** $-\frac{42}{72}$

In Exercises 85–90, perform the indicated operations, and reduce the answers to lowest terms, both (a) by hand and (b) with a calculator. Check your work by comparing the two answers.

85. $\frac{6}{15} \cdot \frac{10}{21}$

86. $\frac{6}{15} \div \frac{10}{21}$

87. $\frac{6}{15} + \frac{10}{21}$

88. $\frac{6}{15} - \frac{10}{21}$

89. $\frac{7}{6} - \frac{5}{7} + \frac{9}{14}$

90. $\frac{-8}{5} - \left(\frac{-3}{28} + \frac{5}{21}\right)$

91. How could you use your calculator to get decimal answers to the above exercises rather than fractional answers?

3.4 Combinatorics and Probability

OBJECTIVES

- Apply the concepts of permutations and combinations to probability calculations
- Use probabilities to analyze games of chance such as the lottery.

You would probably guess that it's pretty unlikely that two or more people in your math class share a birthday. It turns out, however, that it's surprisingly likely. In this section, we'll look at why that is.

Finding a probability involves finding the number of outcomes in an event and the number of outcomes in the sample space. So far, we have used probability rules as an alternative to excessive counting. Another alternative is combinatorics—that is, permutations, combinations, and the Fundamental Principle of Counting, as covered in Sections 2.3 and 2.4. The flowchart used in Chapter 2 is summarized below.

WHICH COMBINATORICS METHOD?

1. If the selection is done with replacement, use the Fundamental Principle of Counting.
2. If the selection is done without replacement and there is only one category, use:
   a. permutations if the order of selection does matter:

   $$_nP_r = \frac{n!}{(n-r)!}$$

   b. combinations if the order of selection does not matter:

   $$_nC_r = \frac{n!}{(n-r)!r!}$$

3. If there is more than one category, use the Fundamental Principle of Counting with one box per category. Use (2) above to determine the numbers that go in the boxes.

EXAMPLE 1

A BIRTHDAY PROBABILITY A group of three people is selected at random. What is the probability that at least two of them will have the same birthday?

SOLUTION

We will assume that all birthdays are equally likely, and for the sake of simplicity, we will ignore leap-year day (February 29).

- *The experiment* is to ask three people their birthdays. One possible outcome is (May 1, May 3, August 23).
- *The sample space* S is the set of all possible lists of three birthdays.
- *Finding* $n(S)$ by counting the elements in S is impractical, so we will use combinatorics as described in "Which Combinatorics Method" above.
  - The selected items are birthdays. They are selected with replacement because people can share the same birthday. This tells us to use *the Fundamental Principle of Counting:*

first person's birthday	second person's birthday	third person's birthday
$\square$	$\square$	$\square$

  - Each birthday can be any one of the 365 days in a year
  - $n(S) = \boxed{365} \cdot \boxed{365} \cdot \boxed{365} = 365^3$
- *The event* E is the set of all possible lists of three birthdays in which some birthdays are the same. It is difficult to compute $n(E)$ directly; instead, we will compute $n(E')$ and use the Complement Rule. E' is the set of all possible lists of three birthdays in which no birthdays are the same.
- *Finding* $n(E')$ using "Which Combinatorics Method":
  - The birthdays are selected without replacement, because no birthdays are the same.
  - There is only one category: birthdays.

- The order of selection does matter—(May 1, May 3, August 23) is a different list from (May 3, August 23, May 1)—so use permutations
- $n(E') = {}_{365}P_3$

- *Finding p(E′) and p(E):*

$$p(E') = \frac{n(E')}{n(S)} = \frac{{}_{365}P_3}{365^3} \quad \text{using the above results}$$

$$p(E) = 1 - p(E') \quad \text{the Complement Rule}$$

$$= 1 - \frac{{}_{365}P_3}{365^3} \quad \text{using the above result}$$

$$= 0.008204\ldots$$

$$\approx 0.8\% \quad \text{moving the decimal point left two places}$$

This result is not at all surprising. It means that two or more people in a group of three share a birthday slightly less than 1% of the time. In other words, this situation is extremely unlikely. However, we will see in Exercise 1 that it is quite likely that two or more people in a group of thirty share a birthday.

Lotteries

We're going to look at one of the most common forms of gambling in the United States: the lottery. Many people play the lottery, figuring that "someone's got to win—why not me?" Is this is a reasonable approach to lottery games?

EXAMPLE 2

WINNING A LOTTERY Arizona, Connecticut, Missouri, and Tennessee operate 6/44 lotteries. In this game, a gambler selects any six of the numbers from 1 to 44. If his or her six selections match the six winning numbers, the player wins first prize. If his or her selections include five of the winning numbers, the player wins second prize. Find the probability of:

a. the event E, winning first prize
b. the event F, winning second prize

SOLUTION

a. *The sample space S* is the set of all possible lottery tickets. That is, it is the set of all possible choices of six numbers selected from the numbers 1 through 44.

- *Finding n(S)* by counting the elements in S is impractical, so we will use combinatorics as described in "Which Combinatorics Method" above.
  - The selected is done without replacement, because you can't select the same lottery number twice.
  - Order does not matter, because the gambler can choose the six numbers in any order. Use combinations.
  - $n(S) = {}_{44}C_6$
- *Finding n(E)* is easy, because there is only one winning combination of numbers:

$$n(E) = 1$$

- *Finding p(E):*

$$p(E) = \frac{n(E)}{n(S)} = \frac{1}{{}_{44}C_6} = \frac{1}{7{,}059{,}052} \quad \text{using the above results}$$

This means that only one out of approximately *seven million* combinations is the first-prize-winning combinations. This is an incredibly unlikely event. It is less probable than dying in one year by being hit by lightning (see Figure 3.10 on page 154.). You probably don't worry about dying from lightning, because it essentially doesn't happen. Sure, it happens to somebody sometime, but not to anyone you've ever known. Winning the lottery is similar. The next time that you get tempted to buy a lottery ticket, remember that you should feel less certain of winning first prize that you should of dying from lightning.

b. To find the probability of event F, winning second prize, we need to find $n(F)$. We already found $n(S)$ in part a:

$$n(S) = {}_{44}C_6$$

- *Finding n(F):*
  - To win second prize, we must select five winning numbers and one losing number. This tells us to use *the Fundamental Principle of Counting:*

    winning numbers $\square$ losing numbers $\square$

  - We will use combinations in each box, for the same reasons that we used combinations to find $n(S)$.
  - The state selects six winning numbers, and the second-prize-winner must select five of them, so there are ${}_6C_5$ ways of selecting the five winning numbers.
  - There are $44 - 6 = 38$ losing numbers, of which the player selects 1, so there are ${}_{38}C_{51}$ ways of selecting one losing number.
  - $n(F) = \boxed{{}_6C_5} \cdot \boxed{{}_{38}C_1}$ using the above results
- *Finding p(F):*

- $p(F) = \dfrac{n(F)}{n(S)} = \dfrac{{}_6C_5 \cdot {}_{38}C_1}{{}_{44}C_6} = 0.000032\ldots$

How do we make sense of a decimal with so many zeros in front of it? One way is to round it off at the first nonzero digit and convert the result to a fraction:

$$p(F) = 0.00003\ldots \quad \text{rounding to the first nonzero digit}$$
$$= 3/100{,}000 \quad \text{converting to a fraction}$$

This means that if you buy a lot of 6/44 lottery tickets, you will win second prize approximately three times out of every 100,000 times you play. It also means that in any given game, there are about three second-prize ticket for every 100,000 ticket purchases.

To check your work in Example 2 part (b), notice that in the event, there is a distinction between two categories (winning numbers and losing numbers); in the sample space, there is no such distinction. Thus, the numerator of

$$p(E) = \frac{{}_6C_5 \cdot {}_{38}C_1}{{}_{44}C_6}$$

has two parts (one for each category), and the denominator has one part. Also, the numbers in front of the Cs add correctly (6 winning numbers + 38 losing numbers = 44 total numbers to choose from), and the numbers after the Cs add correctly (5 winning numbers + 1 losing number = 6 total numbers to select).

HOW TO WRITE A PROBABILITY

A probability can be written as a fraction, a percentage, or a decimal. Our goal is to write a probability in a form that is intuitively understandable.

- If $n(E)$ and $n(S)$ are both small numbers, such as $n(E) = 3$ and $n(S) = 12$, then write $p(E)$ as a **reduced fraction:**

  $$p(E) = \frac{n(E)}{n(S)} = \frac{3}{12} = \frac{1}{4}$$

  It's intuitively understandable to say that something happens about one out of every four times.
- If $n(E)$ and $n(S)$ are not small numbers and $p(E)$ is a decimal with at most two zeros after the decimal point, then write $p(E)$ as a **percentage,** as in Example 1:

  $$\begin{aligned} p(E) &= 0.008204\ldots && \text{two zeros after the decimal point} \\ &\approx 0.8\% && \text{writing as a percentage} \end{aligned}$$

  A probability of 0.008204 . . . is not easily understandable. But it is understandable to say that something happens slightly less than 1% of the time—that is, fewer than one out of every one hundred times.
- If $n(E)$ and $n(S)$ are not small numbers and $p(E)$ is a decimal with three or more zeros after the decimal point, then round off at the first nonzero digit and convert the result to a fraction, as in Example 2:

  $$\begin{aligned} p(F) &= \frac{n(F)}{n(S)} = \frac{6 \cdot 38}{7{,}059{,}059} = 0.000032298\ldots \\ &\approx 0.00003 && \text{rounding at the first nonzero digit} \\ &= 3/100{,}000 && \text{converting to a fraction} \end{aligned}$$

  A probability of $\frac{6.38}{7{,}059{,}059}$ or 0.000032298 . . . or 0.003% is not so easily understood. It's more understandable to say that something happens about three times out of every 100,000 times.

EXAMPLE 3

WINNING POWERBALL Powerball is played in thirty states, the District of Columbia, and the U.S. Virgin Islands. It involves selecting any five of the numbers from 1 to 59, plus a "powerball number," which is any one of the numbers from 1 to 39. Find the probability of winning first prize.

Multi-State Lottery Association

SOLUTION

The sample space S is the set of all possible lottery tickets. That is, it is the set of all possible choices of five numbers from 1 to 59 plus a powerball number from 1 to 39.

- *Finding n(S)*:
  - The numbers are selected without replacement, because the gambler can't select the same number twice.

- There are two categories—regular numbers and the powerball number—so we use the Fundamental Principle of Counting with two boxes.

regular powerball

☐ ☐

- In each box, we will use combinations, because the player can choose the numbers in any order.
- The state selects five of fifty-nine regular numbers, so there are ${}_{59}C_5$ possible selections.
- The state selects one of thirty-nine powerball numbers, so there are ${}_{39}C_1$ possible powerball selections.
- The number of different lottery tickets is

$$n(S) = \boxed{{}_{59}C_5} \cdot \boxed{{}_{39}C_1} \quad \text{using the above results}$$

- *Finding $n(E)$*: Only one of these combinations is the winning lottery ticket, so $n(E) = 1$.
- *Finding $p(E)$*:

$$\begin{aligned} p(E) &= \frac{n(E)}{n(S)} = \frac{1}{{}_{59}C_5 \cdot {}_{39}C_1} && \text{using the above results} \\ &= 0.00000000512\ldots && \text{more than three zeros after the decimal point} \\ &\approx 0.000000005 && \text{rounding to the first nonzero digit} \\ &= 5/1{,}000{,}000{,}000 && \text{converting to a fraction} \\ &= 1/200{,}000{,}000 && \text{reducing} \end{aligned}$$

This means that if you play powerball a lot, you will win first prize about once every 200 million games. It also means that in any given game, there is about one first prize ticket for every 200 million ticket purchases.

Keno

The game of keno is a casino version of the lottery. In this game, the casino has a container filled with balls numbered from 1 to 80. The player buys a keno ticket, with which he or she selects anywhere from 1 to 15 (usually 6, 8, 9, or 10) of those 80 numbers; the player's selections are called "spots." The casino chooses 20 winning numbers, using a mechanical device to ensure a fair game. If a sufficient number of the player's spots are winning numbers, the player receives an appropriate payoff.

EXAMPLE 4

WINNING AT KENO In the game of keno, if eight spots are marked, the player wins if five or more of his or her spots are selected. Find the probability of having five winning spots.

SOLUTION

The sample space S is the set of all ways in which a player can select eight numbers from the eighty numbers in the game.

- *Finding $n(S)$*:
  - Selection is done without replacement.
  - Order doesn't matter, so use combinations.
  - $n(S) = {}_{80}C_8$
- *The event E* is the set of all ways in which an eight-spot player can select five winning numbers and three losing numbers.
- *Finding $n(E)$*:
  - There are two categories—winning numbers and losing numbers—so use the Fundamental Principle of Counting.

TOPIC X THE BUSINESS OF GAMBLING: *PROBABILITIES IN THE REAL WORLD*

It used to be that legal commercial gambling was not common. Nevada made casino gambling legal in 1931, and for more than thirty years, it was the only place in the United States that had legal commercial gambling. Then in 1964, New Hampshire instituted the first lottery in the United States since 1894. In 1978, New Jersey became the second state to legalize casino gambling. Now, state-sponsored lotteries are common, Native American tribes have casinos in more than half the states, and some states have casinos in selected cities or on riverboats.

It is very likely that you will be exposed to gambling if you have not been already. You should approach gambling with an educated perspective. If you are considering gambling, know what you are up against. The casinos all use probabilities and combinatorics in designing their games *to ensure that they make a consistent profit.* Learn this mathematics so that they do not take advantage of you.

Public lotteries have a long history in the United States. The settlement of Jamestown was financed in part by an English lottery. George Washington managed a lottery that paid for a road through the Cumberland Mountains. Benjamin Franklin used lotteries to finance cannons for the Revolutionary War, and John Hancock used lotteries to rebuild Faneuil Hall in Boston. Several universities, including Harvard, Dartmouth, Yale, and Columbia, were partly financed by lotteries. The U.S. Congress operated a lottery to help finance the Revolutionary War.

Today, public lotteries are a very big business. In 2008, Americans spent almost $61 billion on lottery tickets. Lotteries are quite lucrative for the forty-two states that offer them; on the average, 30% of the money went back into government budgets. Fewer than half of the states dedicate the proceeds to education. Frequently, this money goes into the general fund. The states' cuts vary quite a bit. In Oregon in 2008, 54% of the money went to the state, and the remaining 46% went to prizes and administrative costs. In Rhode Island, the state took only 15%.

Most lottery sales come from a relatively small number of people. In Pennsylvania, for example, 29% of the players accounted for 79% of the spending on the lottery in 2008.However, many people who don't normally play go berserk when the jackpots accumulate, partially because of the amazingly large winnings but also because of a lack of understanding of how unlikely it is that they will actually win. The largest cumulative jackpot was $390 million, which was spilt by winners in Georgia and New Jersey in 2007. The largest single-winner jackpot was $315 million in West Virginia in 2002. The winner opted to take a lump sum payment of $114 million, instead of receiving twenty years of regular payments that would have added up to $315 million.

In this section, we will discuss probabilities and gambling. Specifically, we will explore lotteries, keno, and card games. See Examples 2–7, and Exercises 5–22 and 35.

In Section 3.5, we will discuss how much money you can expect to win or lose when gambling. We will continue to discuss lotteries, keno, and card games, but we will also discuss roulette and raffles. See Exercises 1–12, 28–34, and 39–48 in that section.

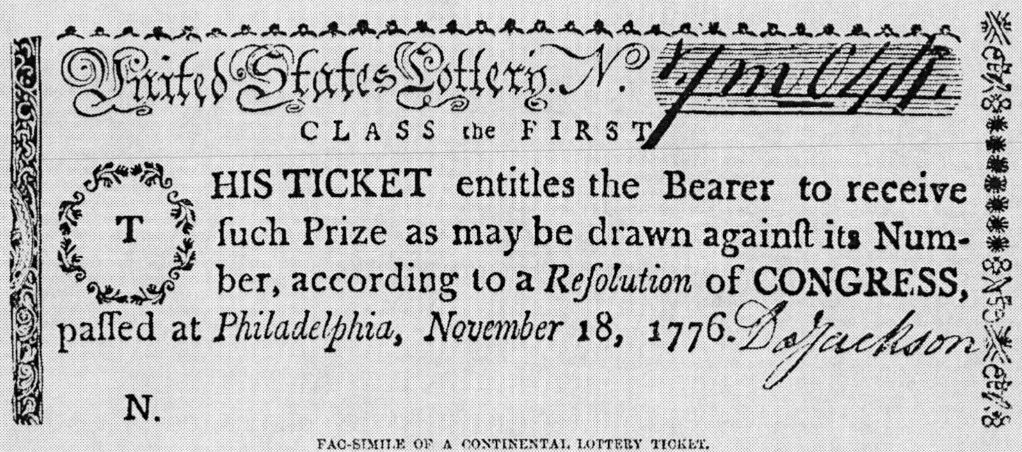
United States Lottery. No.
CLASS the FIRST
THIS TICKET entitles the Bearer to receive ſuch Prize as may be drawn againſt its Number, according to a *Reſolution* of CONGRESS, paſſed at *Philadelphia, November* 18, 1776.
N.

FAC-SIMILE OF A CONTINENTAL LOTTERY TICKET.

- Use combinations in each category, as with $n(S)$.
- The casino selects twenty winning numbers, from which the gambler is to select five winning spots. There are ${}_{20}C_5$ different ways of doing this.
- The casino selects $80 - 20 = 60$ losing numbers, from which the gambler is to select $8 - 5 = 3$ losing spots. There are ${}_{60}C_3$ different ways of doing this.
- $n(E) = {}_{20}C_5 \cdot {}_{60}C_3$ using the above results

- *Finding p(E):*

$$p(E) = \frac{{}_{20}C_5 \cdot {}_{60}C_3}{{}_{80}C_8} = 0.0183\ldots \quad \text{one zero after the decimal point}$$

$$\approx 1.8\% \quad \text{writing as a percentage}$$

This means that if you play eight-spot keno a lot, you will have five winning spots about 1.8% of the time. It also means that in any given game, about 1.8% of the players will have five winning spots.

Cards

One common form of poker is five-card draw, in which each player is dealt five cards. The order in which the cards are dealt is unimportant, so we compute probabilities with combinations rather than permutations.

EXAMPLE 5 GETTING FOUR ACES Find the probability of being dealt four aces.

SOLUTION *The sample space* consists of all possible five-card hands that can be dealt from a deck of fifty-two cards

- *Finding n(S):*
  - Selection is done without replacement.
  - Order does not matter, so use combinations.
  - $n(S) = {}_{52}C_5$
- *The event E* consists of all possible five-card hands that include four aces and one non-ace.
  - There are two categories—aces and non-aces—so use the Fundamental Principle of Counting.
  - Use combinations as with $n(S)$.
  - The gambler is to be dealt four of four aces. This can happen in ${}_4C_4$ ways.
  - The gambler is to be dealt one of $52 - 4 = 48$ non-aces. This can be done in ${}_{48}C_1$ ways.
  - $n(E) = {}_4C_4 \cdot {}_{48}C_1$
- *Finding p(E):*

$$p(E) = \frac{{}_4C_4 \cdot {}_{48}C_1}{{}_{52}C_5} = 0.000018\ldots \quad \text{more than two zeros after the decimal point}$$

$$\approx 0.00002 \quad \text{rounding to the first nonzero digit}$$

$$= 2/100{,}000 \quad \text{rewriting as a fraction}$$

$$= 1/50{,}000 \quad \text{reducing}$$

This means that if you play cards a lot, you will be dealt four aces about once every 50,000 deals.

In the event, there is a distinction between two categories (aces and non-aces); in the sample space, there is no such distinction. Thus, the numerator of

$$p(E) = \frac{{}_4C_4 \cdot {}_{48}C_1}{{}_{52}C_5}$$

has two parts (one for each category), and the denominator has one part. Also, the numbers in front of the Cs add correctly (4 aces + 48 non-aces = 52 cards to choose from), and the numbers after the Cs add correctly (4 aces + 1 non-ace = 5 cards to select).

EXAMPLE 6

GETTING FOUR OF A KIND Find the probability of being dealt four of a kind.

SOLUTION

The sample space is the same as that in Example 5. The event "being dealt four of a kind" means "being dealt four 2's or being dealt four 3's or being dealt four 4's . . . or being dealt four kings or being dealt four aces." These latter events ("four 2's," "four 3's," etc.) are all mutually exclusive, so we can use the Mutually Exclusive Rule.

$$\begin{aligned} &p\text{ (four of a kind)} \\ &\quad = p(\text{four 2's} \cup \text{four 3's} \cup \cdots \cup \text{four kings} \cup \text{four aces}) \\ &\quad = p(\text{four 2's}) + p(\text{four 3's}) + \cdots + p(\text{four kings}) + p(\text{four aces}) \end{aligned}$$

using the Mutually Exclusive Rule

Furthermore, these probabilities are all the same:

$$p(\text{four 2's}) = p(\text{four 3's}) = \cdots = p(\text{four aces}) = \frac{{}_4C_4 \cdot {}_{48}C_1}{{}_{52}C_5}$$

This means that the probability of being dealt four of a kind is

$$\begin{aligned} p(\text{four of a kind}) &= p(\text{four 2's}) + p(\text{four 3's}) + \cdots + p(\text{four kings}) \\ &\quad + p(\text{four aces}) \\ &= \frac{{}_4C_4 \cdot {}_{48}C_1}{{}_{52}C_5} + \frac{{}_4C_4 \cdot {}_{48}C_1}{{}_{52}C_5} + \cdots + \frac{{}_4C_4 \cdot {}_{48}C_1}{{}_{52}C_5} + \frac{{}_4C_4 \cdot {}_{48}C_1}{{}_{52}C_5} \\ &= 13 \cdot \frac{{}_4C_4 \cdot {}_{48}C_1}{{}_{52}C_5} && \text{there are thirteen denominations (2 through ace)} \\ &= 13 \cdot \frac{1 \cdot 48}{2{,}598{,}960} && \text{from Example 4} \\ &= \frac{624}{2{,}598{,}960} \\ &= 0.0002400\ldots && \text{three zeros after the decimal point} \\ &\approx 0.0002 && \text{rounding to the first nonzero digit} \\ &= 2/10{,}000 && \text{converting to a fraction} \\ &= 1/5{,}000 && \text{reducing} \end{aligned}$$

This means that if you play cards a lot, you will be dealt four of a kind about once every 5,000 deals.

See Example 8 in Section 2.4.

EXAMPLE 7 **GETTING FIVE HEARTS** Find the probability of being dealt five hearts.

SOLUTION

The sample space is the same as in Example 5. The event consists of all possible five-card hands that include five hearts and no non-hearts. This involves two categories (hearts and non-hearts), so we will use the Fundamental Counting Principle and multiply the number of ways of getting five hearts and the number of ways of getting no non-hearts. There are

$$_{13}C_5 = \frac{13!}{5! \cdot 8!} = 1{,}287$$

ways of getting five hearts, and there is

$$_{39}C_0 = \frac{39!}{0! \cdot 39!} = 1$$

ways of getting no non-hearts. Thus, the probability of being dealt five hearts is

$$p(E) = \frac{_{13}C_5 \cdot {_{39}C_0}}{_{52}C_5} = \frac{1287 \cdot 1}{2{,}598{,}960} \approx 0.000495198 \approx 0.0005 = 1/2000$$

In the event, there is a distinction between two categories (hearts and non-hearts); in the sample space, there is no such distinction. Thus, the numerator of

$$p(E) = \frac{_{13}C_5 \cdot {_{39}C_0}}{_{52}C_5}$$

has two parts (one for each category), and the denominator has one part. Also, the numbers in front of the Cs add correctly (13 hearts + 39 non-hearts = 52 cards to choose from), and the numbers after the Cs add correctly (5 hearts + 0 non-hearts = 5 cards to select).

Notice that in Example 7, we could argue that since we're selecting only hearts, we can disregard the non-hearts. This would lead to the answer obtained in Example 7:

$$p(E) = \frac{_{13}C_5}{_{52}C_5} = \frac{1287}{2{,}598{,}960} \approx 0.000495198 \approx 0.0005 = 1/2000$$

However, this approach would not allow us to check our work in the manner described above; the numbers in front of the Cs don't add correctly, nor do the numbers after the Cs.

3.4 EXERCISES

1. A group of thirty people is selected at random. What is the probability that at least two of them will have the same birthday?
2. A group of sixty people is selected at random. What is the probability that at least two of them will have the same birthday?
3. How many people would you have to have in a group so that there is a probability of at least 0.5 that at least two of them will have the same birthday?
4. How many people would you have to have in a group so that there is a probability of at least 0.9 that at least two of them will have the same birthday?

5. In 1990, California switched from a 6/49 lottery to a 6/53 lottery. Later, the state switched again, to a 6/51 lottery.

a. Find the probability of winning first prize in a 6/49 lottery.

b. Find the probability of winning first prize in a 6/53 lottery.

c. Find the probability of winning first prize in a 6/51 lottery.

d. How much more probable is it that one will win the 6/49 lottery than the 6/53 lottery?

e. Why do you think California switched from a 6/49 lottery to a 6/53 lottery? And why do you think the state then switched to a 6/51 lottery?
(Answer using complete sentences.)

6. Find the probability of winning second prize—that is, picking five of the six winning numbers—with a 6/53 lottery.

7. Currently, the most common multinumber game is the 5/39 lottery. It is played in California, Georgia, Illinois, Michigan, New York, North Carolina, Ohio, Tennessee, and Washington.

a. Find the probability of winning first prize.

b. Find the probability of winning second prize.

8. Currently, the second most common multinumber game is the 6/49 lottery. It is played in Massachusetts, New Jersey, Ohio, Pennsylvania, Washington, and Wisconsin.

a. Find the probability of winning first prize.

b. Find the probability of winning second prize.

9. "Cash 5" is a 5/35 lottery. It is played in Arizona, Connecticut, Iowa, Massachusetts and South Dakota.

a. Find the probability of winning first prize.

b. Find the probability of winning second prize.

10. The 6/44 lottery is played in Arizona, Connecticut, and New Jersey.

a. Find the probability of winning first prize.

b. Find the probability of winning second prize.

11. Games like "Mega Millions" are played in thirty-nine states and the District of Columbia. It involves selecting any five of the numbers from 1 to 56, plus a number from 1 to 46.

Find the probability of winning first prize.

12. "Hot Lotto" is played in Delaware, Idaho, Iowa, Kansas, Minnesota, Montana, New Hampshire, New Mexico, North Dakota, Oklahoma, South Dakota, West Virginia, and the District of Columbia. It involves selecting any five of the numbers from 1 to 39 plus a number from 1 to 19. Find the probability of winning first prize.

Multi-State Lottery Association

13. "Wild Card 2" is played in Idaho, Montana, North Dakota, and South Dakota. It involves selecting any five of the numbers from 1 to 31, plus a card that's either a jack, queen, king or ace of any one of the four suits. Find the probability of winning first prize.

Multi-State Lottery Association

14. "2 by 2" is played in Kansas, Nebraska, and North Dakota. It involves selecting two red numbers from 1 to 26 and two white numbers from 1 to 26. Find the probability of winning first prize.

Multi-State Lottery Association

15. There is an amazing variety of multinumber lotteries played in the United States. Currently, the following lotteries are played: 4/26, 4/77, 5/30, 5/31, 5/32, 5/34, 5/35, 5/36, 5/37, 5/38, 5/39, 5/40, 5/43, 5/47, 5/50, 6/25, 6/35, 6/39, 6/40, 6/42, 6/43, 6/44, 6/46, 6/47, 6/48, 6/49, 6/52, 6/53, and 6/54. Which is the easiest to win? Which is the hardest to win? Explain your reasoning.

HINT: It isn't necessary to compute every single probability.

16. In the game of keno, if six spots are marked, the player wins if four or more of his or her spots are selected. Complete the chart in Figure 3.17.

Outcome	Probability
6 winning spots	
5 winning spots	
4 winning spots	
3 winning spots	
fewer than 3 winning spots	

FIGURE 3.17 Chart for Exercise 16.

17. In the game of keno, if eight spots are marked, the player wins if five or more of his or her spots are selected. Complete the chart in Figure 3.18.

Outcome	Probability
8 winning spots	
7 winning spots	
6 winning spots	
5 winning spots	
4 winning spots	
fewer than 4 winning spots	

FIGURE 3.18 Chart for Exercise 17.

18. In the game of keno, if nine spots are marked, the player wins if six or more of his or her spots are selected. Complete the chart in Figure 3.19.

Outcome	Probability
9 winning spots	
8 winning spots	
7 winning spots	
6 winning spots	
5 winning spots	
fewer than 5 winning spots	

FIGURE 3.19 Chart for Exercise 18.

19. "Pick three" games are played in thirty-six states. In this game, the player selects a three-digit number, such as 157. Also, the state selects a three-digit winning number.

a. How many different three-digit numbers are there? Explain your reasoning.

b. If the player opts for "straight play," she wins if her selecting matches the winning number *in exact order.* For example, if she selected the number 157, she wins only if the winning number is 157. How many different winning straight play numbers are there?

c. Find the probability of winning with straight play.

20. If a pick three player (see Exercise 19) opts for "box play," the player wins if his selection matches the winning number *in any order.* For example, if he selected the number 157, he wins if the winning number is 715 or any other reordering of 157.

a. How many different winning box play numbers are there if the three digits are different? That is, how many different ways are there to reorder a number such as 157?

b. Find the probability of winning with box play if the three digits are different.

c. How many different winning box play numbers are there if two digits are the same? That is, how many different ways are there to reorder a number such as 266?

d. Find the probability of winning with box play if two digits are the same.

21. **a.** Find the probability of being dealt five spades when playing five-card draw poker.

b. Find the probability of being dealt five cards of the same suit when playing five-card draw poker.

c. When you are dealt five cards of the same suit, you have either a *flush* (if the cards are not in sequence) or a *straight flush* (if the cards are in sequence). For each suit, there are ten possible straight flushes ("ace, two, three, four, five," through "ten, jack, queen, king, ace"). Find the probability of being dealt a straight flush.

d. Find the probability of being dealt a flush.

22. **a.** Find the probability of being dealt an "aces over kings" full house (three aces and two kings).

b. Why are there $13 \cdot 12$ different types of full houses?

c. Find the probability of being dealt a full house. (Round each answer off to six decimal places.)

You order twelve burritos to go from a Mexican restaurant, five with hot peppers and seven without. However, the restaurant forgot to label them. If you pick three burritos at random, find the probability of each event in Exercises 23–30.

23. All have hot peppers.

24. None has hot peppers.

25. Exactly one has hot peppers.

26. Exactly two have hot peppers.

27. At most one has hot peppers.

28. At least one has hot peppers.

29. At least two have hot peppers.
30. At most two have hot peppers.
31. Two hundred people apply for two jobs. Sixty of the applicants are women.
   a. If two people are selected at random, what is the probability that both are women?
   b. If two people are selected at random, what is the probability that only one is a woman?
   c. If two people are selected at random, what is the probability that both are men?
   d. If you were an applicant and the two selected people were not of your gender, do you think that the above probabilities would indicate the presence or absence of gender discrimination in the hiring process? Why or why not?
32. Two hundred people apply for three jobs. Sixty of the applicants are women.
   a. If three people are selected at random, what is the probability that all are women?
   b. If three people are selected at random, what is the probability that two are women?
   c. If three people are selected at random, what is the probability that one is a woman?
   d. If three people are selected at random, what is the probability that none is a woman?
   e. If you were an applicant and the three selected people were not of your gender, should the above probabilities have an impact on your situation? Why or why not?
33. In Example 2, $n(E) = 1$ because only one of the 7,059,052 possible lottery tickets is the first prize winner. Use combinations to show that $n(E) = 1$.

Answer the following questions using complete sentences and your own words.

• CONCEPT QUESTIONS

34. Explain why, in Example 6, $p(\text{four 2's}) = p(\text{four 3's}) = \cdots = p(\text{four kings}) = p(\text{four aces})$.
35. Do you think a state lottery is a good thing for the state's citizens? Why or why not? Be certain to include a discussion of both the advantages and disadvantages of a state lottery to its citizens.
36. Why are probabilities for most games of chance calculated with combinations rather than permutations?
37. Suppose a friend or relative of yours regularly spends (and loses) a good deal of money on lotteries. How would you explain to this person why he or she loses so frequently?
38. In Example 2, $n(E) = 1$ because only one of the 7,059,052 possible lottery tickets is the first prize winner. Does this mean that it is impossible for two people to each buy a first-prize-winning lottery ticket? Explain.

• HISTORY QUESTION

39. Are public lotteries relative newcomers to the American scene? Explain.

3.5 Expected Value

OBJECTIVES

- Understand how expected values take both probabilities and winnings into account
- Use expected values to analyze games of chance
- Use expected values to make decisions

Suppose you are playing roulette, concentrating on the \$1 single-number bet. At one point, you were \$10 ahead, but now you are \$14 behind. How much should you expect to win or lose, on the average, if you place the bet many times?

The probability of winning a single-number bet is $\frac{1}{38}$, because there are thirty-eight numbers on the roulette wheel and only one of them is the subject of the bet. This means that if you place the bet a large number of times, it is most likely that you will win once for every thirty-eight times you place the bet (and lose the other thirty-seven times). When you win, you win \$35, because the house odds are 35 to 1. When you lose, you lose \$1. Your average profit would be

$$\frac{\$35 + 37 \cdot (-\$1)}{38} = \frac{-\$2}{38} \approx -\$0.053$$

per game. This is called the *expected value* of a \$1 single-number bet, because you should expect to lose about a nickel for every dollar you bet if you play the game a long time. If you play a few times, anything could happen—you could win every single bet (though it is not likely). The house makes the bet so many times that it can be certain that its profit will be \$0.053 per dollar bet.

The standard way to find the **expected value** of an experiment is to multiply the value of each outcome of the experiment by the probability of that outcome and add the results. Here the experiment is placing a \$1 single-number bet in roulette. The outcomes are winning the bet and losing the bet. The values of the outcomes are +\$35 (if you win) and −\$1 (if you lose); the probabilities of the outcomes are $\frac{1}{38}$ and $\frac{37}{38}$, respectively (see Figure 3.20). The expected value would then be

$$35 \cdot \frac{1}{38} + (-1) \cdot \frac{37}{38} \approx -\$0.053$$

It is easy to see that this calculation is algebraically equivalent to the calculation done above.

Outcome	Value	Probability
winning	35	$\frac{1}{38}$
losing	−1	$\frac{37}{38}$

FIGURE 3.20 Finding the expected value.

Finding an expected value of a bet is very similar to finding your average test score in a class. Suppose that you are a student in a class in which you have taken four tests. If your scores were 80%, 76%, 90%, and 90%, your average test score would be

$$\frac{80 + 76 + 2 \cdot 90}{4} = 84\%$$

or, equivalently,

$$80 \cdot \frac{1}{4} + 76 \cdot \frac{1}{4} + 90 \cdot \frac{2}{4}$$

The difference between finding an average test score and finding the expected value of a bet is that with the average test score you are summarizing what *has* happened, whereas with a bet you are using probabilities to project what *will* happen.

EXPECTED VALUE

To find the **expected value** (or "long-term average") of an experiment, multiply the value of each outcome of the experiment by its probability and add the results.

Roulette, the oldest casino game played today, has been popular since it was introduced to Paris in 1765. Does this game have any good bets?

EXAMPLE 1

COMPUTING AN EXPECTED VALUE By analyzing her sales records, a saleswoman has found that her weekly commissions have the probabilities in Figure 3.21. Find the saleswoman's expected commission.

Commission	0	\$100	\$200	\$300	\$400
Probability	0.05	0.15	0.25	0.45	0.1

FIGURE 3.21 Commission data for Example 1.

SOLUTION

To find the expected commission, we multiply each possible commission by its probability and add the results. Therefore,

$$\begin{aligned}\text{expected commission} &= (0)(0.05) + (100)(0.15) + (200)(0.25) \\ &\quad + (300)(0.45) + (400)(0.1) \\ &= 240\end{aligned}$$

On the basis of her history, the saleswoman should expect to average \$240 per week in future commissions. Certainly, anything can happen in the future—she could receive a \$700 commission (it's not likely, though, because it has never happened before).

Why the House Wins

Four of the "best" bets that can be made in a casino game of chance are the pass, don't pass, come, and don't come bets in craps. They all have the same expected value, −\$0.014. In the long run, *there isn't a single bet in any game of chance with which you can expect to break even, let alone make a profit.* After all, the casinos are out to make money. The expected values for \$1 bets in the more common games are shown in Figure 3.22.

Game	Expected Value of \$1 Bet
baccarat	−\$0.014
blackjack	−\$0.06 to +\$0.10 (varies with strategies)
craps	−\$0.014 for pass, don't pass, come, don't come bets *only*
slot machines	−\$0.13 to ? (varies)
keno (eight-spot ticket)	−\$0.29
average state lottery	−\$0.48

FIGURE 3.22 Expected values of common games of chance.

It is possible to achieve a positive expected value in blackjack. To do this, the player must keep a running count of the cards dealt, following a system that assigns a value to each card. Casinos use their pit bosses and video surveillance to watch for gamblers who use this tactic, and casinos will harass or kick such gamblers out when they find them. There is an application available for the iPhone and iPod Touch that will count cards. The casinos do everything in their power to eliminate its use. Some casinos use four decks at once and shuffle frequently to minimize the impact of counting.

Decision Theory

Which is the better bet: a \$1 single-number bet in roulette or a lottery ticket? Each costs \$1. The roulette bet pays \$35, but the lottery ticket might pay several million dollars. Lotteries are successful in part because the possibility of winning a large amount of money distracts people from the fact that winning is extremely unlikely. In Example 2 of Section 3.4, we found that the probability of winning first prize in many state lotteries is $\frac{1}{7{,}059{,}052} \approx 0.00000014$. At the beginning of this section, we found that the probability of winning the roulette bet is $\frac{1}{38} \approx 0.03$.

A more informed decision would take into account not only the potential winnings and losses but also their probabilities. The expected value of a bet does just that, since its calculation involves both the value and the probability of each outcome. We found that the expected value of a \$1 single-number bet in roulette is about −\$0.053. The expected value of the average state lottery is −\$0.48 (see Figure 3.22). The roulette bet is a much better bet than is the lottery. (Of course, there is a third option, which has an even better expected value of \$0.00: not gambling!)

A decision always involves choosing between various alternatives. If you compare the expected values of the alternatives, then you are taking into account

FEATURED IN THE NEWS

VIRGINIA LOTTERY HEDGES ON SYNDICATE'S BIG WIN

RICHMOND, VA.—Virginia lottery officials confirmed yesterday that an Australian gambling syndicate won last month's record $27 million jackpot after executing a massive block-buying operation that tried to cover all 7 million possible ticket combinations.

But lottery director Kenneth Thorson said the jackpot may not be awarded because the winning ticket may have been bought in violation of lottery rules.

The rules say tickets must be paid for at the same location where they are issued. The Australian syndicate, International Lotto Fund, paid for many of its tickets at the corporate offices of Farm Fresh Inc. grocery stores, rather than at the Farm Fresh store in Chesapeake where the winning ticket was issued, Thorson said.

"We have to validate who bought the ticket, where the purchase was made and how the purchase was made," Thorson said. "It's just as likely that we will honor the ticket as we won't honor the ticket." He said he may not decide until the end of next week.

Two Australians representing the fund, Joseph Franck and Robert Hans Roos, appeared at lottery headquarters yesterday to claim the prize.

The group succeeded in buying about 5 million of the more than 7 million possible numerical combinations before the February 15 drawing. The tactic is not illegal, although lottery officials announced new rules earlier this week aimed at making such block purchases more difficult.

The Australian fund was started last year and raised about $13 million from an estimated 2,500 shareholders who each paid a minimum of $4,000, according to Tim Phillipps of the Australian Securities Commission.

Half the money went for management expenses, much of that to Pacific Financial Resources, a firm controlled by Stefan Mandel, who won fame when he covered all the numbers in a 1986 Sydney lottery. Roos owns 10 percent of Pacific Financial Resources.

Australian Securities Commission officials said last week that the fund is under investigation for possible violations of Australian financial laws.

the alternatives' potential advantages and disadvantages as well as their probabilities. This form of decision making is called **decision theory.**

EXAMPLE 2

USING EXPECTED VALUES TO MAKE A DECISION The saleswoman in Example 1 has been offered a new job that has a fixed weekly salary of $290. Financially, which is the better job?

SOLUTION

In Example 1, we found that her expected weekly commission was $240. The new job has a guaranteed weekly salary of $290. Financial considerations indicate that she should take the new job.

Betting Strategies

One very old betting strategy is to "cover all the numbers." In 1729, the French philosopher and writer François Voltaire organized a group that successfully implemented this strategy to win the Parisian city lottery by buying most if not all of the tickets. Their strategy was successful because, owing to a series of poor financial decisions by the city of Paris, the total value of the prizes was greater than the combined price of all of the tickets! Furthermore, there were not a great number of tickets to buy. This strategy is still being used. (See the above newspaper article on its use in 1992 in Virginia.)

A **martingale** is a gambling strategy in which the gambler doubles his or her bet after each loss. A person using this strategy in roulette, concentrating on the black numbers bet (which has 1-to-1 house odds), might lose three times before winning. This strategy would result in a net gain, as illustrated in Figure 3.23 below.

This seems to be a great strategy. Sooner or later, the player will win a bet, and because each bet is larger than the player's total losses, he or she has to come out ahead! We will examine this strategy further in Exercises 47 and 48.

Bet Number	Bet	Result	Total Winnings/Losses
1	\$1	lose	−\$1
2	\$2	lose	−\$3
3	\$4	lose	−\$7
4	\$8	win	+\$1

FIGURE 3.23 Analyzing the martingale strategy.

3.5 EXERCISES

In Exercises 1–10, (a) find the expected value of each \$1 bet in roulette and (b) use the Law of Large Numbers to interpret it.

1. the two-number bet
2. the three-number bet
3. the four-number bet
4. the five-number bet
5. the six-number bet
6. the twelve-number bet
7. the low-number bet
8. the even-number bet
9. the red-number bet
10. the black-number bet
11. Using the expected values obtained in the text and in the preceding odd-numbered exercises, determine a casino's expected net income from a 24-hour period at a single roulette table if the casino's total overhead for the table is \$50 per hour and if customers place a total of \$7,000 on single-number bets, \$4,000 on two-number bets, \$4,000 on four-number bets, \$3,000 on six-number bets, \$7,000 on low-number bets, and \$8,000 on red-number bets.
12. Using the expected values obtained in the text and in the preceding even-numbered exercises, determine a casino's expected net income from a 24-hour period at a single roulette table if the casino's total overhead for the table is \$50 per hour and if customers place a total of \$8,000 on single-number bets, \$3,000 on three-number bets, \$4,000 on five-number bets, \$4,000 on twelve-number bets, \$8,000 on even-number bets, and \$9,000 on black-number bets.
13. On the basis of his previous experience, the public librarian at Smallville knows that the number of books checked out by a person visiting the library has the probabilities shown in Figure 3.24.

Number of Books	0	1	2	3	4	5
Probability	0.15	0.35	0.25	0.15	0.05	0.05

FIGURE 3.24 Probabilities for Exercise 13.

Find the expected number of books checked out by a person visiting this library.

14. On the basis of his sale records, a salesman knows that his weekly commissions have the probabilities shown in Figure 3.25.

Commission	0	\$1,000	\$2,000	\$3,000	\$4,000
Probability	0.15	0.2	0.45	0.1	0.1

FIGURE 3.25 Probabilities for Exercise 14.

Find the salesman's expected commission.

15. Of all workers at a certain factory, the proportions earning certain hourly wages are as shown in Figure 3.26.

Hourly Wage	$8.50	$9.00	$9.50	$10.00	$12.50	$15.00
Proportion	20%	15%	25%	20%	15%	5%

FIGURE 3.26 Data for Exercise 15.

Find the expected hourly wage that a worker at this factory makes.

16. Of all students at the University of Metropolis, the proportions taking certain numbers of units are as shown in Figure 3.27. Find the expected number of units that a student at U.M. takes.

Units	3	4	5	6	7	8
Proportion	3%	4%	5%	6%	5%	4%

Units	9	10	11	12	13	14
Proportion	8%	12%	13%	13%	15%	12%

FIGURE 3.27 Data for Exercise 16.

17. You have been asked to play a dice game. It works like this:

- If you roll a 1 or 2, you win $50.
- If you roll a 3, you lose $20.
- If you roll a 4, 5, or 6, you lose $30.

Should you play the game? Use expected values to justify your answer.

18. You have been asked to play a dice game. It works like this:

- If you roll a 1, 2, 3, or 4, you win $50.
- If you roll a 5 or 6, you lose $80.

Should you play the game? Use expected values to justify your answer.

19. You are on a TV show. You have been asked to either play a dice game ten times or accept a $100 bill. The dice game works like this:

- If you roll a 1 or 2, you win $50.
- If you roll a 3, you win $20.
- If you roll a 4, 5, or 6, you lose $30.

Should you play the game? Use expected values and decision theory to justify your answer.

20. You are on a TV show. You have been asked to either play a dice game five times or accept a $50 bill. The dice game works like this:

- If you roll a 1, 2, or 3, you win $50.
- If you roll a 4 or 5, you lose $25.
- If you roll a 6, you lose $90.

Should you play the game? Use expected values and decision theory to justify your answer.

21. Show why the calculation at the top of page 182 is algebraically equivalent to the calculation in the middle of the same page.

22. In Example 1, the saleswoman's most likely weekly commission was $300. With her new job (in Example 2), she will always make $290 per week. This implies that she would be better off with the old job. Is this reasoning more or less valid than that used in Example 2? Why?

23. Maria just inherited $10,000. Her bank has a savings account that pays 4.1% interest per year. Some of her friends recommended a new mutual fund, which has been in business for three years. During its first year, the fund went up in value by 10%; during the second year, it went down by 19%; and during its third year, it went up by 14%. Maria is attracted by the mutual fund's potential for relatively high earnings but concerned by the possibility of actually losing some of her inheritance. The bank's rate is low, but it is insured by the federal government. Use decision theory to find the best investment. (Assume that the fund's past behavior predicts its future behavior.)

24. Trang has saved $8,000. It is currently in a bank savings account that pays 3.9% interest per year. He is considering putting the money into a speculative investment that would either earn 20% in one year if the investment succeeds or lose 18% in one year if it fails. At what probability of success would the speculative investment be the better choice?

25. Erica has her savings in a bank account that pays 4.5% interest per year. She is considering buying stock in a pharmaceuticals company that is developing a cure for cellulite. Her research indicates that she could earn 50% in one year if the cure is successful or lose 60% in one year if it is not. At what probability of success would the pharmaceuticals stock be the better choice?

26. Debra is buying prizes for a game at her school's fundraiser. The game has three levels of prizes, and she has already bought the second and third prizes. She wants the first prize to be nice enough to attract people to the game. The game's manufacturer has supplied her with the probabilities of winning first, second, and third prizes. Tickets cost $3 each, and she wants the school to profit an average of $1 per ticket. How much should she spend on each first prize?

Prize	Cost of Prize	Probability
1st	?	.15
2nd	$1.25	.30
3rd	$0.75	.45

FIGURE 3.28 Data for Exercise 26.

27. Few students manage to complete their schooling without taking a standardized admissions test such as the Scholastic Achievement Test, or S.A.T. (used for admission to college); the Law School Admissions Test, or L.S.A.T.; and the Graduate Record Exam, or G.R.E. (used for admission to graduate school). Sometimes, these multiple-choice tests discourage guessing by subtracting points for wrong answers. In particular, a correct answer will be worth +1 point, and an incorrect answer on a question with five listed answers (a through e) will be worth $-\frac{1}{4}$ point.

a. Find the expected value of a random guess.

b. Find the expected value of eliminating one answer and guessing among the remaining four possible answers.

c. Find the expected value of eliminating three answers and guessing between the remaining two possible answers.

d. Use decision theory and your answers to parts (a), (b), and (c) to create a guessing strategy for standardized tests such as the S.A.T.

28. Find the expected value of a $1 bet in six-spot keno if three winning spots pays $1 (but you pay $1 to play, so you actually break even), four winning spots pays $3 (but you pay $1 to play, so your profit is $2), five pays $100, and six pays $2,600. (You might want to use the probabilities computed in Exercise 16 of Section 3.4.)

29. Find the expected value of a $1 bet in eight-spot keno if four winning spots pays $1 (but you pay $1 to play, so you actually break even), five winning spots pays $5 (but you pay $1 to play, so your profit is $4), six pays $100, seven pays $1,480, and eight pays $19,000. (You might want to use the probabilities computed in Exercise 17 of Section 3.4.)

30. Find the expected value of a $1 bet in nine-spot keno if five winning spots pays $1 (but you pay $1 to play, so you actually break even), six winning spots pays $50 (but you pay $1 to play, so your profit is $49), seven pays $390, eight pays $6,000, and nine pays $25,000. (You might want to use the probabilities computed in Exercise 18 of Section 3.4.)

31. Arizona's "Cash 4" is a 4/26 lottery. It differs from many other state lotteries in that its payouts are set; they do not vary with sales. If you match all four of the winning numbers, you win $10,000 (but you pay $1 to play, so your profit is $9,999). If you match three of the winning numbers, you win $25, and if you match two of the winning numbers, you win $2. Otherwise, you lose your $1.

a. Find the probabilities of winning first prize, second prize, and third prize.

b. Use the results of parts a and the Complement Rule to find the probability of losing.

c. Use the results of parts a and b to find the expected value of Cash 4.

32. New York's "Pick 10" is a 10/80 lottery. Its payouts are set; they do not vary with sales. If you match all ten winning numbers, you win $500,000 (but you pay $1 to play, so your profit is $499,999). If you match nine winning numbers, you win $6000. If you match eight, seven, or six you win $300, $40, or $10, respectively. If you match no winning numbers, you win $4. Otherwise, you lose your $1.

a. Find the probabilities of winning first prize, second prize, and third prize.

b. Use the results of part a and the Complement Rule to find the probability of losing.

c. Use the results of parts a and b to find the expected value of Pick 10.

33. Arizona and New York have Pick 3 games (New York's is called "Numbers"), as described in Exercises 19 and 20 of Section 3.4. If the player opts for "straight play" and wins, she wins $500 (but she pays $1 to play, so her profit is $499). If the player opts for "box play" with three different digits and wins, he wins $80. If the player opts for "box play" with two of the same digits and wins, he wins $160.

Courtesy of the New York Lottery

a. Use the probabilities from Exercises 19 and 20 of Section 3.4 to find the probability of losing.

b. Use the probabilities from part a and from Exercises 19 and 20 of Section 3.4 to find the expected value of the game.

34. Write a paragraph in which you compare the states' fiscal policies concerning their lotteries with the casinos' fiscal policies concerning their keno games. Assume that the expected value of a $1 lottery bet described in Exercises 31, 32, and/or 33 is representative of that of the other states' lotteries, and assume that the expected value of a $1 keno bet described in Exercises 28, 29, and/or 30 is representative of other keno bets.

35. Trustworthy Insurance Co. estimates that a certain home has a 1% chance of burning down in any one year. They calculate that it would cost $120,000 to rebuild that home. Use expected values to determine the annual insurance premium.

36. Mr. and Mrs. Trump have applied to the Trustworthy Insurance Co. for insurance on Mrs. Trump's diamond tiara. The tiara is valued at $97,500. Trustworthy estimates that the jewelry has a 2.3% chance of being stolen in any one year. Use expected values to determine the annual insurance premium.

37. The Black Gold Oil Co. is considering drilling either in Jed Clampett's back yard or his front yard. After thorough testing and analysis, they estimate that there is a 30% chance of striking oil in the back yard and a 40% chance in the front yard. They also estimate that the back yard site would either net \$60 million (if oil is found) or lose \$6 million (if oil is not found), and the front yard site would either net \$40 million or lose \$6 million. Use decision theory to determine where they should drill.

38. If in Exercise 37, Jed Clampett rejected the use of decision theory, where would he drill if he were an optimist? What would he do if he were a pessimist?

39. A community youth group is having a raffle to raise funds. Several community businesses have donated prizes. The prizes and their retail value are listed in Figure 3.29. Each prize will be given away, regardless of the number of raffle tickets sold. Tickets are sold for \$15 each. Determine the expected value of a ticket, and discuss whether it would be to your financial advantage to buy a ticket under the given circumstances.

a. 1000 tickets are sold.
b. 2000 tickets are sold.
c. 3000 tickets are sold.

Prize	Retail Value	Number of These Prizes to Be Given Away
new car	\$21,580	1
a cell phone and a one-year subscription	\$940	1
a one-year subscription to an Internet service provider	\$500	2
dinner for two at Spiedini's restaurant	\$100	2
a one-year subscription to the local newspaper	\$180	20

FIGURE 3.29 Data for Exercise 39.

40. The Centerville High School PTA is having a raffle to raise funds. Several community businesses have donated prizes. The prizes and their retail value are listed in Figure 3.30. Each prize will be given away, regardless of the number of raffle tickets sold. Tickets are sold for \$30 each. Determine the expected value of a ticket, and discuss whether it would be to your financial advantage to buy a ticket under the given circumstances.

a. 100 tickets are sold.
b. 200 tickets are sold.
c. 300 tickets are sold.

Prize	Retail Value	Number of These Prizes to Be Given Away
one week in a condo in Hawaii, and airfare for two	\$2,575	1
tennis lessons for two	\$500	1
a one-year subscription to an Internet service provider	\$500	2
dinner for two at Haute Stuff restaurant	\$120	2
a one-year subscription at the Centerville Skate Park	\$240	5
a copy of *Centerville Cooks,* a cookbook containing PTA members' favorite recipes	\$10	20

FIGURE 3.30 Data for Exercise 40.

41. The Central State University Young Republicans Club is having a raffle to raise funds. Several community businesses have donated prizes. The prizes and their retail value are listed in Figure 3.31. Each prize will be given away, regardless of the number of raffle tickets sold. Tickets are sold for \$5 each. Determine the expected value of a ticket, and discuss whether it would be to your financial advantage to buy a ticket under the given circumstances.

a. 1000 tickets are sold.
b. 2000 tickets are sold.
c. 3000 tickets are sold.

Prize	Retail Value	Number of These Prizes to Be Given Away
laptop computer	\$2,325	1
MP3 player	\$425	2
20 CDs at Einstein Entertainment	\$320	3
a giant pizza and your choice of beverage at Freddie's Pizza	\$23	4
a one-year subscription to the *Young Republican Journal*	\$20	15

FIGURE 3.31 Data for Exercise 41.

42. The Southern State University Ecology Club is having a raffle to raise funds. Several community businesses have donated prizes. The prizes and their retail value are listed in Figure 3.32. Each prize will be given away, regardless of the number of raffle tickets sold. Tickets are sold for $20 each. Determine the expected value of a ticket, and discuss whether it would be to your financial advantage to buy a ticket under the given circumstances.

a. 100 tickets are sold.

b. 200 tickets are sold.

c. 300 tickets are sold.

Prize	Retail Value	Number of These Prizes to Be Given Away
one-week ecovacation to Costa Rica, including airfare for two	$2,990	1
Sierra Designs tent	$750	1
REI backpack	$355	2
Jansport daypack	$75	5
fleece jacket with Ecology Club logo	$50	10
Ecology Club T-shirt	$18	20

FIGURE 3.32 Data for Exercise 42.

43. Find the expected value of the International Lotto Fund's application of the "cover all of the numbers" strategy from the newspaper article on page 185. Assume that $5 million was spent on lottery tickets, that half of the $13 million raised went for management expenses, and that the balance was never spent. Also assume that Virginia honors the winning ticket.

44. One application of the "cover all the numbers" strategy would be to bet $1 on every single number in roulette.

a. Find the results of this strategy.

b. How could you use the expected value of the $1 single-number bet $\left(\frac{-\$2}{38}\right)$ to answer part (a)?

45. The application of the "cover all the numbers" strategy to a modern state lottery would involve the purchase of a large number of tickets.

a. How many tickets would have to be purchased if you were in a state that has a 6/49 lottery (the player selects 6 out of 49 numbers)?

b. How much would it cost to purchase these tickets if each costs $1?

c. If you organized a group of 100 people to purchase these tickets and it takes one minute to purchase each ticket, how many days would it take to purchase the required number of tickets?

46. The application of the "cover all the numbers" strategy to Keno would involve the purchase of a large number of tickets.

a. How many tickets would have to be purchased if you were playing eight-spot keno?

b. How much would it cost to purchase these tickets if each costs $5?

c. If it takes five seconds to purchase one ticket and the average keno game lasts twenty minutes, how many people would it take to purchase the required number of tickets?

47. If you had $100 and were applying the martingale strategy to the black-number bet in roulette and you started with a $1 bet, how many successive losses could you afford? How large would your net profit be if you lost each bet except for the last one?

48. If you had $10,000 and were applying the martingale strategy to the black-number bet in roulette and you started with a $1 bet, how many successive losses could you afford? How large would your net profit be if you lost each bet except for the last one?

Answer the following questions using complete sentences and your own words.

• CONCEPT QUESTIONS

49. Discuss the meaningfulness of the concept of expected value to three different people: an occasional gambler, a regular gambler, and a casino owner. To whom is the concept most meaningful? To whom is it least meaningful? Why?

50. Discuss the advantages and disadvantages of decision theory. Consider the application of decision theory to a nonrecurring situation and to a recurring situation. Also consider Jed Clampett and the Black Gold Oil Co. in Exercises 37 and 38.

• PROJECTS

51. Design a game of chance. Use probabilities and expected values to set the house odds so that the house will make a profit. However, be certain that your game is not so obviously pro-house that no one would be willing to play. Your project should include the following:

- a complete description of how the game is played
- a detailed mathematical analysis of the expected value of the game (or of each separate bet in the game, whichever is appropriate)
- a complete description of the bet(s) and the house odds

3.6 Conditional Probability

OBJECTIVES

- Use conditional probabilities to focus on one or two groups rather than the entire sample space
- Understand the relationship between intersections of events and products of probabilities
- Use tree diagrams to combine probabilities

Public opinion polls, such as those found in newspapers and magazines and on television, frequently categorize the respondents by such groups as sex, age, race, or level of education. This is done so that the reader or listener can make comparisons and observe trends, such as "people over 40 are more likely to support the Social Security system than are people under 40." The tool that enables us to observe such trends is conditional probability.

Probabilities and Polls

In a newspaper poll concerning violence on television, 600 people were asked, "What is your opinion of the amount of violence on prime-time television—is there too much violence on television?" Their responses are indicated in Figure 3.33.

	Yes	No	Don't Know	Total
Men	162	95	23	280
Women	256	45	19	320
Total	418	140	42	600

FIGURE 3.33 Results of "Violence on Television" poll.

Six hundred people were surveyed in this poll; that is, the sample space consists of 600 responses. Of these, 418 said they thought there was too much violence on television, so the probability of a "yes" response is $\frac{418}{600}$, or about $0.70 = 70\%$. The probability of a "no" response is $\frac{140}{600}$, or about $0.23 = 23\%$.

If we are asked to find the probability that a *woman* responded yes, we do not consider all 600 responses but instead limit the sample space to only the responses from women. See Figure 3.34.

	Yes	No	Don't Know	Total
Women	256	45	19	320

FIGURE 3.34 Women and violence on television.

The probability that a woman responded yes is $\frac{256}{320} = 0.80 = 80\%$.

Is there too much violence on TV?

Suppose we label the events in the following manner: W is the event that a response is from a woman, M is the event that a response is from a man, Y is the event that a response is yes, and N is the event that a response is no. Then the event that a woman responded yes would be written as

$$Y \mid W$$

The vertical bar stands for the phrase "given that"; the event $Y \mid W$ is read "a response is yes, given that the response is from a woman." The probability of this event is called a *conditional probability:*

$$p(Y \mid W) = \frac{256}{320} = \frac{4}{5} = 0.80 = 80\%$$

The numerator of this probability, 256, is the number of responses that are yes and are from women; that is, $n(Y \cap W) = 256$. The denominator, 320, is the number of responses that are from women; that is, $n(W) = 320$. A **conditional probability** is a probability whose sample space has been limited to only those outcomes that fulfill a certain condition. Because an event is a subset of the sample space, the event must also fulfill that condition. The numerator of $p(Y \mid W)$ is 256 rather than 418 even though there were 418 "yes" responses, because many of those 418 responses were made by men; we are interested only in the probability that a woman responded yes.

CONDITIONAL PROBABILITY DEFINITION

The **conditional probability** of event A, given event B, is

$$p(A \mid B) = \frac{n(A \cap B)}{n(B)}$$

EXAMPLE 1

COMPUTING CONDITIONAL PROBABILITIES Using the data in Figure 3.33, find the following.

a. the probability that a response is yes, given that the response is from a man.
b. the probability that a response is from a man, given that the response is yes.
c. the probability that a response is yes and is from a man.

SOLUTION

a. *Finding* $p(Y \mid M)$: We are told to consider only the male responses—that is, to limit our sample space to men. See Figure 3.35.

	Yes	No	Don't Know	Total
Men	162	95	23	280

FIGURE 3.35 Men and violence on television.

$$p(Y \mid M) = \frac{n(Y \cap M)}{n(M)} = \frac{162}{280} \approx 0.58 = 58\%$$

In other words, approximately 58% of the men responded yes. (Recall that 80% of the women responded yes. This poll indicates that men and women do not have the same opinion regarding violence on television and, in particular, that a woman is more likely to oppose the violence.)

b. *Finding* $p(M \mid Y)$: We are told to consider only the "yes" responses shown in Figure 3.36.

	Yes
Men	162
Women	256
Total	418

FIGURE 3.36
Limiting our sample space to "yes" responses.

$$p(M \mid Y) = \frac{n(M \cap Y)}{n(Y)} = \frac{162}{418} \approx 0.39 = 39\%$$

Therefore, of those who responded yes, approximately 39% were male.

c. *Finding* $p(Y \cap M)$: This is *not* a conditional probability (there is no vertical bar), so we do *not* limit our sample space.

$$p(Y \cap M) = \frac{n(Y \cap M)}{n(S)} = \frac{162}{600} = 0.27 = 27\%$$

Therefore, of all those polled, 27% were men who responded yes.

Each of the above three probabilities has the *same numerator,* $n(Y \cap M) = 162$. This is the number of responses that are yes and are from men.

But the three probabilities have *different denominators*. This means that we are comparing the group of men who said yes with three different larger groups. See Figure 3.37.

In the probability:	the denominator is:	so we're comparing the group of men who said yes with:
$p(Y \mid M)$	$n(M)$, the number of male responses	all of the men
$p(M \mid Y)$	$n(Y)$, the number of yes responses	all of the yes responses
$p(Y \cap M)$	$n(S)$, the number of responses	all of the people polled

FIGURE 3.37 The impact of the different denominators.

The Product Rule

If two cards are dealt from a full deck (no jokers), how would you find the probability that both are hearts? The probability that the first card is a heart is easy to find—it is $\frac{13}{52}$, because there are fifty-two cards in the deck and thirteen of them are hearts. The probability that the second card is a heart is more difficult to find. There are only fifty-one cards left in the deck (one was already dealt), but how many of these are hearts? The number of hearts left in the deck depends on the first card that was dealt. If it was a heart, then there are twelve hearts left in the deck; if it was not a heart, then there are thirteen hearts left. We could certainly say that the probability that the second card is a heart, *given that the first card was a heart,* is $\frac{12}{51}$.

Therefore, the probability that the first card is a heart is $\frac{13}{52}$, and the probability that the second card is a heart, given that the first was a heart, is $\frac{12}{51}$. How do we put these two probabilities together to find the probability that *both* the first and the second cards are hearts? Should we add them? Subtract them? Multiply them? Divide them?

The answer is obtained by algebraically rewriting the Conditional Probability Definition to obtain what is called the *Product Rule:*

$$p(A \mid B) = \frac{n(A \cap B)}{n(B)} \quad \text{Conditional Probability Definition}$$

$$p(A \mid B) \cdot n(B) = n(A \cap B) \quad \text{multiplying by } n(B)$$

$$\frac{p(A \mid B) \cdot n(B)}{n(S)} = \frac{n(A \cap B)}{n(S)} \quad \text{dividing by } n(S)$$

$$\frac{p(A \mid B)}{1} \cdot \frac{n(B)}{n(S)} = \frac{n(A \cap B)}{n(S)} \quad \text{since } 1 \cdot n(S) = n(S)$$

$$p(A \mid B) \cdot p(B) = p(A \cap B) \quad \text{definition of probability}$$

PRODUCT RULE

For any events A and B, the probability of A and B is

$$p(A \cap B) = p(A \mid B) \cdot p(B)$$

EXAMPLE 2

USING THE PRODUCT RULE If two cards are dealt from a full deck, find the probability that both are hearts.

SOLUTION

$$
\begin{aligned}
p(A \cap B) &= p(A \mid B) \cdot p(B) \\
p(\text{2nd heart and 1st heart}) &= p(\text{2nd heart} \mid \text{1st heart}) \cdot p(\text{1st heart}) \\
&= \frac{12}{51} \cdot \frac{13}{52} \\
&= \frac{4}{17} \cdot \frac{1}{4} \\
&= \frac{1}{17} \approx 0.06 = 6\%
\end{aligned}
$$

Therefore, there is a 6% probability that both cards are hearts.

Tree Diagrams

Many people find that a *tree diagram* helps them understand problems like the one in Example 2, in which an experiment is performed in stages over time. Figure 3.38 shows the tree diagram for Example 2. The first column gives a list of the possible outcomes of the first stage of the experiment; in Example 2, the first stage is dealing the first card, and its outcomes are "heart," and "not a heart." The branches leading to those outcomes represent their probabilities. The second column gives a list of the possible outcomes of the second stage of the experiment; in Example 2, the second stage is dealing the second card. A branch leading from a first-stage outcome to a second-stage outcome is the conditional probability $p(\text{2nd stage outcome} \mid \text{1st stage outcome})$.

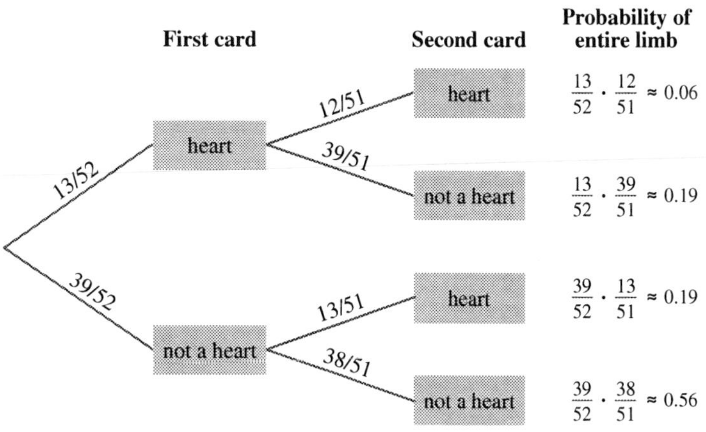

FIGURE 3.38 A tree diagram of dealing two hearts.

Looking at the top pair of branches, we see that the first branch stops at "first card is a heart" and the probability is $p(\text{1st heart}) = \frac{13}{52}$. The second branch starts at "first card is a heart" and stops at "second card is a heart" and gives the conditional probability $p(\text{2nd heart} \mid \text{1st heart}) = \frac{12}{51}$. The probability that we were asked to calculate in Example 2, p(1st heart and 2nd heart), is that of the top limb:

$$p(\text{1st heart and 2nd heart}) = p(\text{2nd heart} \mid \text{1st heart}) \cdot p(\text{1st heart}) = \frac{12}{51} \cdot \frac{13}{52}$$

(We use the word *limb* to refer to a sequence of branches that starts at the beginning of the tree.) Notice that the sum of the probabilities of the four limbs is 1.00. Because the four limbs are the only four possible outcomes of the experiment, they must add up to 1.

Conditional probabilities always start at their condition, never at the beginning of the tree. For example, p(2nd heart | 1st heart) is a conditional probability; its condition is that the first card is a heart. Thus, its branch starts at the box "first card is a heart." However, p(1st heart) is not a conditional probability, so it starts at the beginning of the tree. Similarly, p(1st heart and 2nd heart) is not a conditional probability, so it too starts at the beginning of the tree. The product rule tells us that

$$p(\text{1st heart and 2nd heart}) = p(\text{2nd heart} \mid \text{1st heart}) \cdot p(\text{1st heart})$$

That is, the product rule tells us to multiply the branches that make up the top horizontal limb. In fact, "Multiply when moving horizontally across a limb" is a restatement of the product rule.

EXAMPLE 3

USING TREE DIAGRAMS AND THE MUTUALLY EXCLUSIVE RULE
Two cards are drawn from a full deck. Use the tree diagram in Figure 3.38 on page 195 to find the probability that the second card is a heart.

SOLUTION

The second card can be a heart if the first card is a heart *or* if it is not. The event "the second card is a heart" is the union of the following two mutually exclusive events:

$$E = \text{1st heart and 2nd heart}$$
$$F = \text{1st not heart and 2nd heart}$$

We previously used the tree diagram to find that

$$p(E) = \frac{13}{52} \cdot \frac{12}{51}$$

Similarly,

$$p(F) = \frac{39}{52} \cdot \frac{13}{51}$$

Thus, we add the probabilities of limbs that result in the second card being a heart:

$$\begin{aligned} p(\text{2nd heart}) &= p(E \cup F) \\ &= p(E) + p(F) && \text{the Mutually Exclusive Rule} \\ &= \frac{13}{52} \cdot \frac{12}{51} + \frac{39}{52} \cdot \frac{13}{51} = 0.25 \end{aligned}$$

In Example 3, the first and third limbs represent the only two ways in which the second card can be a heart. These two limbs represent mutually exclusive events, so we used Probability Rule 5 [$p(E \cup F) = p(E) + p(F)$] to add their probabilities. In fact, "add when moving vertically from limb to limb" is a good restatement of Probability Rule 5.

TREE DIAGRAM SUMMARY

- Conditional probabilities start at their condition.
- Nonconditional probabilities start at the beginning of the tree.
- Multiply when moving horizontally across a limb.
- Add when moving vertically from limb to limb.

EXAMPLE 4

USING TREE DIAGRAMS, THE PRODUCT RULE, AND THE UNION/INTERSECTION RULE Big Fun Bicycles manufactures its product at two plants, one in Korea and one in Peoria. The Korea plant manufactures 60% of the bicycles; 4% of the Korean bikes are defective; and 5% of the Peorian bikes are defective.

a. draw a tree diagram that shows this information.
b. use the tree diagram to find the probability that a bike is defective and came from Korea.
c. use the tree diagram to find the probability that a bike is defective.
d. use the tree diagram to find the probability that a bike is defect-free.

SOLUTION

a. First, we need to determine which probabilities have been given and find their complements, as shown in Figure 3.39.

Probabilities Given	Complements of These Probabilities
$p(\text{Korea}) = 60\% = 0.60$	$p(\text{Peoria}) = p(\text{not Korea}) = 1 - 0.60 = 0.40$
$p(\text{defective} \mid \text{Korea}) = 4\% = 0.04$	$p(\text{not defective} \mid \text{Korea}) = 1 - 0.04 = 0.96$
$p(\text{defective} \mid \text{Peoria}) = 5\% = 0.05$	$p(\text{not defective} \mid \text{Peoria}) = 1 - 0.05 = 0.95$

FIGURE 3.39 Probabilities for Example 4.

The first two of these probabilities [$p(\text{Korea})$ and $p(\text{Peoria})$] are not conditional, so they start at the beginning of the tree. The next two probabilities [$p(\text{defective} \mid \text{Korea})$ and $p(\text{not defective} \mid \text{Korea})$] are conditional, so they start at their condition (Korea). Similarly, the last two probabilities are conditional, so they start at their condition (Peoria). This placement of the probabilities yields the tree diagram in Figure 3.40.

b. The probability that a bike is defective and came from Korea is a nonconditional probability, so it starts at the beginning of the tree. Do not confuse it with the conditional probability that a bike is defective, *given that* it came from Korea, which starts at its condition (Korea). The former is the limb that goes through "Korea" and stops at "defective"; the latter is one branch of that limb. We use the product rule to multiply when moving horizontally across a limb:

$$p(\text{defective and Korea}) = p(\text{defective} \mid \text{Korea}) \cdot p(\text{Korea})$$
$$= 0.04 \cdot 0.60 = 0.024 \qquad \text{Product Rule}$$

This means that 2.4% of all of Big Fun's bikes are defective bikes manufactured in Korea.

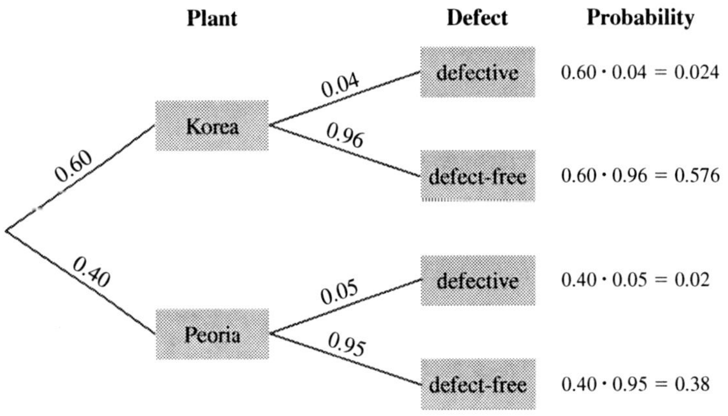

FIGURE 3.40 A tree diagram for Example 4.

c. The event that a bike is defective is the union of two mutually exclusive events:

The bike is defective and came from Korea.

The bike is defective and came from Peoria.

These two events are represented by the first and third limbs of the tree. We use the Union/Intersection Rule to add when moving vertically from limb to limb:

$$\begin{aligned} p(\text{defective}) &= p(\text{defective and Korea} \cup \text{defective and Peoria}) \\ &= p(\text{defective and Korea}) + p(\text{defective and Peoria}) \\ &= 0.024 + 0.02 = 0.044 \end{aligned}$$

This means that 4.4% of Big Fun's bicycles are defective.

d. The probability that a bike is defect-free is the complement of part (c).

$$p(\text{defect-free}) = p(\text{not defective}) = 1 - 0.044 = 0.956$$

Alternatively, we can find the sum of all the limbs that stop at "defect-free":

$$p(\text{defect-free}) = 0.576 + 0.38 = 0.956$$

This means that 95.6% of Big Fun's bicycles are defect-free.

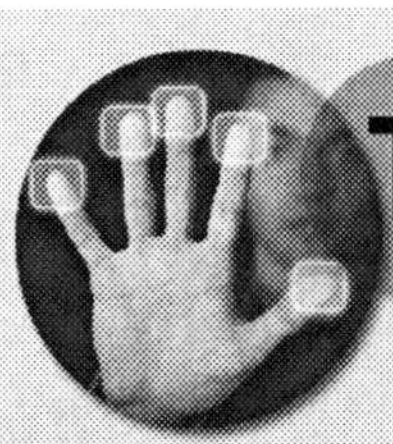

TOPIC X HIV/AIDS: PROBABILITIES IN THE REAL WORLD

The human immunodeficiency virus (or HIV) is the virus that causes AIDS. The Centers for Disease Control estimates that in 2005, 950,000 Americans had HIV/AIDS. They are of both genders, all ages and sexual orientations. World-wide, nearly half of the 38 million people living with HIV/AIDS are women between 15 and 24 years old.

Most experts agree that the HIV/AIDS epidemic is in its early stages and that a vaccine is not on the immediate horizon. The only current hope of stemming the infection lies in education and prevention. Education is particularly important, because an infected person can be symptom-free for eight years or more. You may well know many people who are unaware that they are infected, because they have no symptoms.

Health organizations such as the Centers for Disease Control and Prevention routinely use probabilities to determine whether men or women are more likely to get HIV, which age groups are more likely to develop AIDS, and the different sources of exposure to HIV. See Exercises 49 and 50. In Section 3.7, we will investigate the accuracy of some HIV/AIDS tests. See Exercises 33 and 34 in Section 3.7.

3.6 EXERCISES

In Example 1, we wrote, "the probability that a response is yes, given that it is from a man" as $p(Y \mid M)$, and we wrote, "the probability that a response is yes and is from a man" as $p(Y \cap M)$. In Exercises 1–4, write the given probabilities in a similar manner. Also, identify which are conditional and which are not conditional.

1. Let H be the event that a job candidate is hired, and let Q be the event that a job candidate is well qualified. Use the symbols H, Q, $|$, and $\cap$ to write the following probabilities.

a. The probability that a job candidate is hired given that the candidate is well qualified.

b. The probability that a job candidate is hired and the candidate is well qualified.

2. Let W be the event that a gambler wins a bet, and let L be the event that a gambler is feeling lucky. Use the symbols W, L, $|$, and $\cap$ to write the following probabilities.

a. The probability that a gambler wins a bet and is feeling lucky.

b. The probability that a gambler wins a bet given that the gambler is feeling lucky.

3. Let S be the event that a cell phone user switches carriers, and let D be the event that a cell phone user gets dropped a lot. Use the symbols S, D, $|$, and $\cap$ to write the following probabilities.

a. The probability that a cell phone user switches carriers given that she gets dropped a lot.

b. The probability that a cell phone user switches carriers and gets dropped a lot.

c. The probability that a cell phone user gets dropped a lot given that she switches carriers.

4. Let P be the event that a student passes the course, and S be the event that a student studies hard. Use the symbols P, S, $|$, and $\cap$ to write the following probabilities.

a. The probability that a student passes the course given that the student studies hard.

b. The probability that a student studies hard given that the student passes the course.

c. The probability that a student passes the course and studies hard.

In Exercises 5–8, use Figure 3.41.

5. **a.** Find $p(B \mid A)$
b. Find $p(B \cap A)$

6. **a.** Find $p(B' \mid A)$
b. Find $p(B' \cap A)$

7. **a.** Find $p(B \mid A')$
b. Find $p(B \cap A')$

8. **a.** Find $p(B' \mid A')$
b. Find $p(B' \cap A')$

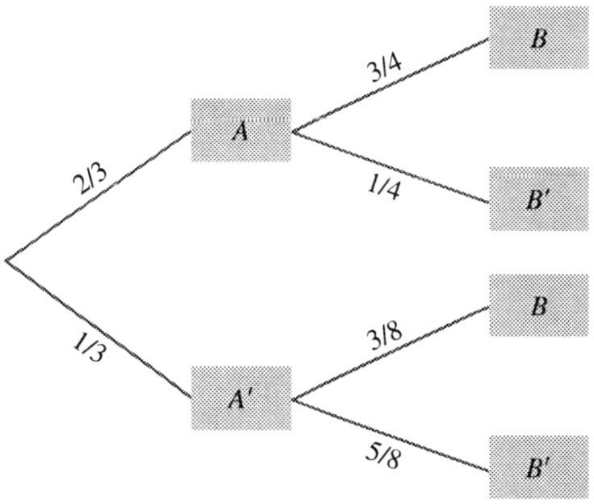

FIGURE 3.41 A tree diagram for Exercises 5–8.

9. Use the data in Figure 3.33 on page 191 to find the given probabilities. Also, write a sentence explaining what each means.

a. $p(N)$ **b.** $p(W)$ **c.** $p(N \mid W)$
d. $p(W \mid N)$ **e.** $p(N \cap W)$ **f.** $p(W \cap N)$

10. Use the data in Figure 3.33 on page 191 to find the given probabilities. Also, write a sentence explaining what each means.

a. $p(N)$ **b.** $p(M)$ **c.** $p(N \mid M)$
d. $p(M \mid N)$ **e.** $p(N \cap M)$ **f.** $p(M \cap N)$

Use the information in Figure 3.10 on page 154 to answer Exercises 11–14. Round your answers off to the nearest hundredth. Also interpret each of your answers using percentages and everyday English.

11. **a.** Find and interpret the probability that a U.S. resident dies of a pedestrian transportation accident, given that the person dies of a transportation accident in one year.

b. Find and interpret the probability that a U.S. resident dies of a pedestrian transportation accident, given that the person dies of a transportation accident, in a lifetime.

c. Find and interpret the probability that a U.S. resident dies of a pedestrian transportation accident, given that the person dies of a non-transportation accident, in a lifetime.

12. **a.** Find and interpret the probability that a U.S. resident dies from lightning, given that the person dies of a non-transportation accident, in one year.

b. Find and interpret the probability that a U.S. resident dies from lightning, given that the person dies of a non-transportation accident, in a lifetime.

c. Find and interpret the probability that a U.S. resident dies from lightning, given that the person dies of a transportation accident, in a lifetime.

13. **a.** Find and interpret the probability that a U.S. resident dies from an earthquake, given that the person dies from a non-transportation accident, in a lifetime.

b. Find and interpret the probability that a U.S. resident dies from an earthquake, given that the person dies from a non-transportation accident, in one year.

c. Find and interpret the probability that a U.S. resident dies from an earthquake, given that the person dies from an external cause, in one year.

14. **a.** Find and interpret the probability that a U.S. resident dies from a motorcyclist transportation accident, given that the person dies from a transportation accident, in a lifetime.

b. Find and interpret the probability that a U.S. resident dies from a motorcyclist transportation accident, given that the person dies from a transportation accident, in one year.

c. Find and interpret the probability that a U.S. resident dies from a motorcyclist transportation accident, given that the person dies from an external cause, in one year.

In Exercises 15–18, cards are dealt from a full deck of 52. Find the probabilities of the given events.

15. **a.** The first card is a club.

b. The second card is a club, given that the first was a club.

c. The first and second cards are both clubs.

d. Draw a tree diagram illustrating this.

16. **a.** The first card is a king.

b. The second card is a king, given that the first was a king.

c. The first and second cards are both kings.

d. Draw a tree diagram illustrating this. (Your diagram need not be a complete tree. It should have all the branches referred to in parts (a), (b), and (c), but it does not need other branches.)

17. **a.** The first card is a diamond.

b. The second card is a spade, given that the first was a diamond.

c. The first card is a diamond and the second is a spade.

d. Draw a tree diagram illustrating this. (Your diagram need not be a complete tree. It should have all the branches referred to in parts (a), (b), and (c), but it does not need other branches.)

18. **a.** The first card is a jack.

b. The second card is an ace, given that the first card was a jack.

c. The first card is a jack and the second is an ace.

d. Draw a tree diagram illustrating this.

*In Exercises 19 and 20, determine which probability the indicated branch in Figure 3.42 refers to. For example, the branch labeled * refers to the probability p(A).*

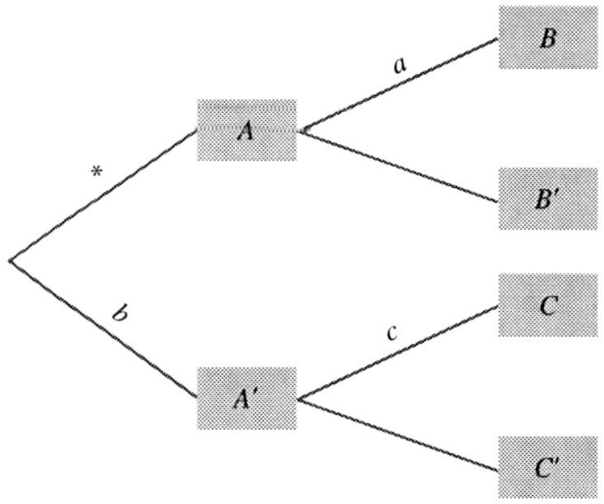

FIGURE 3.42 A tree diagram for Exercises 19 and 20.

19. **a.** the branch labeled a

b. the branch labeled b

c. the branch labeled c

20. **a.** Should probabilities (*) and (a) be added or multiplied? What rule tells us that? What is the result of combining them?

b. Should probabilities (b) and (c) be added or multiplied? What rule tells us that? What is the result of combining them?

c. Should the probabilities that result from parts (a) and (b) of this exercise be added or multiplied? What rule tells us that? What is the result of combining them?

In Exercises 21 and 22, a single die is rolled. Find the probabilities of the given events.

21. **a.** rolling a 6

b. rolling a 6, given that the number rolled is even

c. rolling a 6, given that the number rolled is odd

d. rolling an even number, given that a 6 was rolled

22. **a.** rolling a 5

b. rolling a 5, given that the number rolled is even

c. rolling a 5, given that the number rolled is odd

d. rolling an odd number, given that a 5 was rolled

In Exercises 23–26, a pair of dice is rolled. Find the probabilities of the given events.

23. **a.** The sum is 6.

b. The sum is 6, given that the sum is even.

c. The sum is 6, given that the sum is odd.

d. The sum is even, given that the sum is 6.

24. **a.** The sum is 12.

b. The sum is 12, given that the sum is even.

c. The sum is 12, given that the sum is odd.

d. The sum is even, given that the sum was 12.

25. a. The sum is 4.
b. The sum is 4, given that the sum is less than 6.
c. The sum is less than 6, given that the sum is 4.

26. a. The sum is 11.
b. The sum is 11, given that the sum is greater than 10.
c. The sum is greater than 10, given that the sum is 11.

27. A single die is rolled. Determine which of the following events is least likely and which is most likely. Do so without making any calculations. Explain your reasoning.
E_1 is the event "rolling a 4."
E_2 is the event "rolling a 4, given that the number rolled is even."
E_3 is the event "rolling a 4, given that the number rolled is odd."

28. A pair of dice is rolled. Determine which of the following events is least likely and which is most likely. Do so without making any calculations. Explain your reasoning.
E_1 is the event "rolling a 7."
E_2 is the event "rolling a 7, given that the number rolled is even."
E_3 is the event "rolling a 7, given that the number rolled is odd."

In Exercises 29 and 30, use the following information. To determine what effect the salespeople had on purchases, a department store polled 700 shoppers as to whether or not they had made a purchase and whether or not they were pleased with the service. Of those who had made a purchase, 125 were happy with the service and 111 were not. Of those who had made no purchase, 148 were happy with the service and 316 were not.

29. Find the probability that a shopper who was happy with the service had made a purchase (round off to the nearest hundredth). What can you conclude?

30. Find the probability that a shopper who was unhappy with the service had not made a purchase. (Round off to the nearest hundredth.) What can you conclude?

In Exercises 31–34, five cards are dealt from a full deck. Find the probabilities of the given events. (Round off to four decimal places.)

31. All are spades.

32. The fifth is a spade, given that the first four were spades.

33. The last four are spades, given that the first was a spade.

34. All are the same suit.

In Exercises 35–40, round off to the nearest hundredth.

35. If three cards are dealt from a full deck, use a tree diagram to find the probability that exactly two are spades.

36. If three cards are dealt from a full deck, use a tree diagram to find the probability that exactly one is a spade.

37. If three cards are dealt from a full deck, use a tree diagram to find the probability that exactly one is an ace.

38. If three cards are dealt from a full deck, use a tree diagram to find the probability that exactly two are aces.

39. If a pair of dice is rolled three times, use a tree diagram to find the probability that exactly two throws result in 7's.

40. If a pair of dice is rolled three times, use a tree diagram to find the probability that all three throws result in 7's.

In Exercises 41–44, use the following information: A personal computer manufacturer buys 38% of its chips from Japan and the rest from the United States. Of the Japanese chips, 1.7% are defective, and 1.1% of the American chips are defective.

41. Find the probability that a chip is defective and made in Japan.

42. Find the probability that a chip is defective and made in the United States.

43. Find the probability that a chip is defective.

44. Find the probability that a chip is defect-free.

45. The results of CNN's 2008 presidential election poll are given in Figure 3.43.

	Obama	McCain	Other/No Answer
Male Voters	23.0%	22.6%	1.4%
Female Voters	29.7%	22.8%	0.5%

FIGURE 3.43 Exit poll results, by gender. *Source:* CNN.

a. Find the probability that a voter voted for Obama, given that the voter is male.
b. Find the probability that a voter voted for Obama, given that the voter is female.
c. What observations can you make?

46. Use the information in Figure 3.43 to answer the following questions.
a. Find the probability that a voter is male, given that the voter voted for Obama.
b. Find the probability that a voter is female, given that the voter voted for Obama.
c. What observations can you make?

47. The results of CNN's 2008 presidential election poll are given in Figure 3.44 on page 202.
a. Find that probability that a voter voted for McCain, given that the voter is under 45.
b. Find the probability that a voter voted for McCain, given that the voter is 45 or over.
c. What observations can you make?

Age of Voter	Obama	McCain	Other/No Answer
18–29	11.9%	5.8%	0.4%
30–44	15.1%	13.3%	0.6%
45–64	18.5%	18.1%	0.4%
65 and Older	7.2%	8.5%	0.3%

FIGURE 3.44 Exit poll results, by age. *Source:* CNN.

48. Use the information in Figure 3.44 to answer the following questions.

a. Find the probability that a voter is under 45, given that the voter voted for McCain.

b. Find the probability that a voter is 45 or over, given that the voter voted for McCain.

c. What observations can you make?

49. Figure 3.45 gives the estimated number of diagnoses of AIDS among adults and adolescents in the United States by transmission category through the year 2007.

Transmission Category	Estimated Number of AIDS Cases, through 2007		
	Adult and Adolescent Male	Adult and Adolescent Female	Total
Male-to-male sexual contact	487,695	—	487,695
Injection drug use	175,704	80,155	255,859
Male-to-male sexual contact and injection drug use	71,242	—	71,242
High-risk heterosexual contact	63,927	112,230	176,157
Other*	12,108	6,158	18,266

FIGURE 3.45 AIDS sources.
*Includes hemophilia, blood transfusion, perinatal exposure, and risk not reported or not identified. *Source:* Centers for Disease Control and Prevention.

a. Find the probability that a U.S. adult or adolescent diagnosed with AIDS is male and the probability that a U.S. adult or adolescent diagnosed with AIDS is female.

b. Find the probability that a male U.S. adult or adolescent diagnosed with AIDS was exposed by injection drug use (possibly combined with male-to-male sexual contact) and the probability that a female U.S. adult or adolescent diagnosed with AIDS was exposed by injection drug use.

c. Find the probability that a U.S. adult or adolescent diagnosed with AIDS is male and was exposed by injection drug use (possibly combined with male-to-male sexual contact) and the probability that a U.S. adult or adolescent diagnosed with AIDS is female and was exposed by injection drug use.

d. Find the probability that a male U.S. adult or adolescent diagnosed with AIDS was exposed by heterosexual contact and the probability that a female U.S. adult or adolescent diagnosed with AIDS was exposed by heterosexual contact.

e. Find the probability that a U.S. adult or adolescent diagnosed with AIDS is male and was exposed by heterosexual contact and the probability that a U.S. adult or adolescent diagnosed with AIDS is female and was exposed by heterosexual contact.

f. Explain the difference between parts (b) and (c) and the difference between parts (d) and (e).

50. Figure 3.46 gives the estimated number of diagnoses of AIDS in the United States by age at the time of diagnosis, through 2007. Also note that in 2007, there were 301,621,157 residents of the United States, according to the U.S. Census Bureau.

Age (Years)	Cumulative Number of AIDS Cases
Under 13	9,209
13–14	1,169
15–24	44,264
25–34	322,370
35–44	396,851
45–54	176,304
55–64	52,409
65 or older	15,853

FIGURE 3.46 AIDS by age. *Source:* Centers for Disease Control and Prevention.

a. Find the probability that a U.S. resident diagnosed with AIDS was 15 to 24 years old at the time of diagnosis.

b. Find the probability that a U.S. resident was diagnosed with AIDS and was 15 to 24 years old at the time of diagnosis.

c. Find the probability that a U.S. resident diagnosed with AIDS was 25 to 34 years old at the time of diagnosis.

d. Find the probability that a U.S. resident was diagnosed with AIDS and was 25 to 34 years old at the time of diagnosis.

e. Explain in words the difference between parts (c) and (d).

51. In November 2007, the National Center for Health Statistics published a document entitled "Obesity Among Adults in the United States." According to that document, 32% of adult men and 35% of adult women in the United States were obese in 2006. At that time, there were 148 million adult men and 152 million adult women in the country.

Source: National Center for Health Statistics, C. Ogden et al: Obesity Among Adults in the United States.

a. Find the probability that an adult man was obese.
b. Find the probability that an adult was an obese man.
c. Find the probability that an adult woman was obese.
d. Find the probability that an adult was an obese woman.
e. Find the probability that an adult was obese.
f. Explain in words the difference between parts (c) and (d).

52. The document mentioned in Exercise 51 also says that 23.2% of adult men and 29.5% of adult women in the United States were overweight.

a. Find the probability that an adult man was overweight.
b. Find the probability that an adult was an overweight man.
c. Find the probability that an adult woman was overweight.
d. Find the probability that an adult was an overweight woman.
e. Find the probability that an adult was overweight.

53. In 1981 a study on race and the death penalty was released. The data in Figure 3.47 are from that study.

Death penalty imposed?	Victim's race	Defendant's race	Frequency
Yes	White	White	19
Yes	White	Black	11
Yes	Black	White	0
Yes	Black	Black	6
No	White	White	132
No	White	Black	152
No	Black	White	9
No	Black	Black	97

FIGURE 3.47 Race and the death penalty. *Source:* M. Radelet (1981) "Racial Characteristics and the Imposition of the Death Penalty." American Sociological Review, 46, 918–927.

a. Find p(death penalty imposed | victim white and defendant white).
b. Find p(death penalty imposed | victim white and defendant black).
c. Find p(death penalty imposed | victim black and defendant white).
d. Find p(death penalty imposed | victim black and defendant black).
e. What can you conclude from parts (a) through (d)?
f. Determine what other conditional probabilities would affect this issue, and calculate those probabilities.
g. Discuss your results.

54. The information in Exercise 53 is rather dated. Use the following information to determine whether things have improved since then. In April 2003, Amnesty International issued a study on race and the death penalty. The following quote is from that study:

"The population of the USA is approximately 75 percent white and 12 percent black. Since 1976, blacks have been six to seven times more likely to be murdered than whites, with the result that blacks and whites are the victims of murder in about equal numbers. Yet, 80 percent of the more than 840 people put to death in the USA since 1976 were convicted of crimes involving white victims, compared to the 13 percent who were convicted of killing blacks. Less than four percent of the executions carried out since 1977 in the USA were for crimes involving Hispanic victims. Hispanics represent about 12 percent of the US population. Between 1993 and 1999, the recorded murder rate for Hispanics was more than 40 percent higher than the national homicide rate."

(*Source:* Amnesty International. "United States of America: Death by discrimination—the continuing role of race in capital cases." AMR 51/046/2003)

a. In the above quote, what conditional probability is 80%? That is, find events A and B such that $p(A \mid B) = 80\%$.
b. In the above quote, what conditional probability is 13%? That is, find events C and D such that $p(C \mid D) = 13\%$.
c. Use the information in Exercise 53 to compute $p(A \mid B)$ for 1981.
d. Use the information in Exercise 53 to compute $p(C \mid D)$ for 1981.
e. Have things improved since 1981? Justify your answer.

55. A man and a woman have a child. Both parents have sickle-cell trait. They know that their child does not have sickle-cell anemia because he shows no symptoms, but they are concerned that he might be a carrier. Find the probability that he is a carrier.

56. A man and a woman have a child. Both parents are Tay-Sachs carriers. They know that their child does not have Tay-Sachs disease because she shows no symptoms, but they are concerned that she might be a carrier. Find the probability that she is a carrier.

In Exercises 57–60, use the following information. The University of Metropolis requires its students to pass an examination in college-level mathematics before they can graduate. The students are given three chances to pass the exam; 61% pass it on their first attempt, 63% of those that take it a second time pass it then, and 42% of those that take it a third time pass it then.

57. What percentage of the students pass the exam?

58. What percentage of the students are not allowed to graduate because of their performance on the exam?

59. What percentage of the students take the exam at least twice?

60. What percent of the students take the test three times?

61. In the game of blackjack, if the first two cards dealt to a player are an ace and either a ten, jack, queen, or king, then the player has a blackjack, and he or she wins. Find the probability that a player is dealt a blackjack out of a full deck (no jokers).

62. In the game of blackjack, the dealer's first card is dealt face up. If that card is an ace, then the player has the option of "taking insurance." "Insurance" is a side bet. If the dealer has blackjack, the player wins the insurance bet and is paid 2 to 1 odds. If the dealer does not have a blackjack, the player loses the insurance bet. Find the probability that the dealer is dealt a blackjack if his or her first card is an ace.

63. Use the data in Figure 3.33 to find the following probabilities, where N is the event "saying no," and W is the event "being a woman":

a. $p(N' \mid W)$ **b.** $p(N \mid W')$ **c.** $p(N' \mid W')$

d. Which event, $N' \mid W$, $N \mid W'$, or $N' \mid W'$, is the complement of the event $N \mid W$? Why?

64. Use the data in Figure 3.33 to find the following probabilities, where Y is the event "saying yes," and M is the event "being a man.":

a. $p(Y' \mid M)$ **b.** $p(Y \mid M')$ **c.** $p(Y' \mid M')$

d. Which event, $Y' \mid M$, $Y \mid M'$, or $Y' \mid M'$, is the complement of the event $Y \mid M$? Why?

65. If A and B are arbitrary events, what is the complement of the event $A \mid B$?

66. Show that $p(A \mid B) = \dfrac{p(A \cap B)}{p(B)}$.

HINT: Divide the numerator and denominator of the Conditional Probability Definition by $n(S)$.

67. Use Exercise 66 and appropriate answers from Exercise 9 to find $P(N \mid W)$.

68. Use Exercise 66 and appropriate answers from Exercise 10 to find $P(Y \mid M)$.

Answer the following questions using complete sentences and your own words.

• CONCEPT QUESTIONS

69. Which must be true for any events A and B?

- $P(A \mid B)$ is always greater than or equal to $P(A)$.
- $P(A \mid B)$ is always less than or equal to $P(A)$.
- Sometimes $P(A \mid B)$ is greater than or equal to $P(A)$, and sometimes $P(A \mid B)$ is less than or equal to $P(A)$, depending on the events A and B.

Answer this without making any calculations. Explain your reasoning.

70. Compare and contrast the events A, $A \mid B$, $B \mid A$, and $A \cap B$. Be sure to discuss both the similarities and the differences between these events.

WEB PROJECTS

71. There are many different blood type systems, but the ABO and Rh systems are the most important systems for blood donation purposes. These two systems generate eight different blood types: A+, A−, B+, B−, AB+, AB−, O+, and O−.

a. For each of these eight blood types, use the web to determine the percentage of U.S. residents that have that blood type. Interpret these percentages as probabilities. Use language like "the probability that a randomly-selected U.S. resident . . . ."

b. You can always give blood to someone with the same blood type. In some cases, you can give blood to someone with a different blood type, or receive blood from someone with a different blood type. This depends on the donor's blood type and the receiver's blood type. Use the web to complete the chart in Figure 3.48.

c. For each of the eight blood types, use the web to determine the percentage of U.S. residents that can donate blood to a person of that blood type. Interpret these percentages as conditional probabilities.

d. For each of the eight blood types, use the web to determine the percentage of U.S. residents that can receive blood from a person of that blood type. Interpret these percentages as conditional probabilities.

Blood group	A+	A−	B+	B−	AB+	AB−	O+	O−
Can donate blood to								
Can receive blood from								

FIGURE 3.48 Blood types.

e. Are the probabilities in parts (a), (c), and (d) theoretical probabilities or relative frequencies? Why?

Some useful links for this web project are listed on the text web site:
www.cengage.com/math/johnson

72. According to the U.S. Department of Transportation's National Highway Traffic Safety Administration, "rollovers are dangerous incidents and have a higher fatality rate than other kinds of crashes. Of the nearly 11 million passenger car, SUV, pickup, and van crashes in 2002, only 3% involved a rollover. However, rollovers accounted for nearly 33% of all deaths from passenger vehicle crashes."

(*Source:* **http://www. safercar.gov/Rollover/pages/RolloCharFat.htm**)

Use probabilities and the web to investigate rollovers. How likely is a rollover if you are driving a sedan, an SUV, or a van? Which models have the highest probability of a rollover? Which models have the lowest probability? Wherever possible, give specific conditional probabilities.

Some useful links for this web project are listed on the text web site:
www.cengage.com/math/johnson

PROJECTS

73. In 1973, the University of California at Berkeley admitted 1,494 of 4,321 female applicants for graduate study and 3,738 of 8,442 male applicants.

(*Source:* P.J. Bickel, E.A. Hammel, and J.W. O'Connell, "Sex Bias in Graduate Admissions: Data from Berkeley," *Science*, vol. 187, 7 February 1975.)

a. Find the probability that:
- an applicant was admitted
- an applicant was admitted, given that he was male
- an applicant was admitted, given that she was female

Discuss whether these data indicate a possible bias against women.

b. Berkeley's graduate students are admitted by the department to which they apply rather than by a campuswide admissions panel. When $P(\text{admission} \mid \text{male})$ and $P(\text{admission} \mid \text{female})$ were computed for each of the school's more than 100 departments, it was found that in four departments, $P(\text{admission} \mid \text{male})$ was greater than $P(\text{admission} \mid \text{female})$ by a significant amount and that in six departments, $P(\text{admission} \mid \text{male})$ was less than $P(\text{admission} \mid \text{female})$ by a significant amount. Discuss whether this information indicates a possible bias against women and whether it is consistent with that in part (a).

c. The authors of "Sex Bias in Graduate Admissions: Data from Berkeley" attempt to explain the paradox by discussing an imaginary school with only two departments: "machismatics" and "social wafare." Machismatics admitted 200 of 400 male applicants for graduate study and 100 of 200 female applicants, while social warfare admitted 50 of 150 male applicants for graduate study and 150 of 450 female applicants. For *the school as a whole and for each of the two departments*, find the probability that:
- an applicant was admitted
- an applicant was admitted, given that he was male
- an applicant was admitted, given that she was female

Discuss whether these data indicate a possible bias against women.

d. Explain the paradox illustrated in parts (a)–(c).

e. What conclusions would you make, and what further information would you obtain, if you were an affirmative action officer at Berkeley?

74. a. Use the data in Figure 3.49 to find the probability that:
- a New York City resident died from tuberculosis
- a Caucasian New York City resident died from tuberculosis
- a non-Caucasian New York City resident died from tuberculosis
- a Richmond resident died from tuberculosis
- a Caucasian Richmond resident died from tuberculosis
- a non-Caucasian Richmond resident died from tuberculosis

b. What conclusions would you make, and what further information would you obtain, if you were a public health official?

	New York City		Richmond, Virginia	
	Population	TB deaths	Population	TB deaths
Caucasian	4,675,000	8400	81,000	130
Non-Caucasian	92,000	500	47,000	160

FIGURE 3.49 Tuberculosis deaths by race and location, 1910.
Source: Morris R. Cohen and Ernest Nagel, *An Introduction to Logic and Scientific Method* (New York: Harcourt Brace & Co, 1934.)

3.7 Independence; Trees in Genetics

OBJECTIVES

- Understand the difference between dependent and independent events
- Know the effect of independence on the Product Rule
- Be able to apply tree diagrams in medical and genetic situations

Dependent and Independent Events

Consider the dealing of two cards from a full deck. An observer who saw that the first card was a heart would be better able to predict whether the second card will be a heart than would another observer who did not see the first card. If the first card was a heart, there is one fewer heart in the deck, so it is slightly less likely that the second card will be a heart. In particular,

$$p(\text{2nd heart} \mid \text{1st heart}) = \frac{12}{51} \approx 0.24$$

whereas, as we saw in Example 3 of Section 3.6,

$$p(\text{2nd heart}) = 0.25$$

These two probabilities are different because of the effect the first card drawn has on the second. We say that the two events "first card is a heart" and "second card is a heart" are *dependent;* the result of dealing the second card depends, to some extent, on the result of dealing the first card. In general, two events E and F are **dependent** if $p(E \mid F) \neq p(E)$.

Consider two successive tosses of a single die. An observer who saw that the first toss resulted in a three would be *no better able* to predict whether the second toss will result in a three than another observer who did not observe the first toss. In particular,

$$p(\text{2nd toss is a three}) = \frac{1}{6}$$

and

$$p(\text{2nd toss is a three} \mid \text{1st toss was a three}) = \frac{1}{6}$$

These two probabilities are the same, because the first toss has no effect on the second toss. We say that the two events "first toss is a three" and "second toss is a three" are *independent;* the result of the second toss does *not* depend on the result of the first toss. In general, two events E and F are **independent** if $p(E \mid F) = p(E)$.

INDEPENDENCE/DEPENDENCE DEFINITIONS

Two events E and F are **independent** if $p(E \mid F) = p(E)$.
(Think: Knowing F *does not affect* E*'s probability.)*
Two events E and F are **dependent** if $p(E \mid F) \neq p(E)$.
(Think: Knowing F *does affect* E*'s probability.)*

Many people have difficulty distinguishing between *independent* and *mutually exclusive*. (Recall that two events E and F are mutually exclusive if $E \cap F = \emptyset$, that is, if one event excludes the other.) This is probably because the relationship between mutually exclusive events and the relationship between independent events both could be described, in a very loose sort of way, by saying that "the two events have nothing to do with each other." *Never think this way;* mentally replacing "mutually exclusive" or "independent" with "having nothing to do with each other" only obscures the distinction between these two concepts. E and F are independent if knowing that F has occurred *does not* affect the probability that E will occur. E and F are dependent if knowing that F has occurred *does* affect the probability that E will occur. E and F are mutually exclusive if E and F cannot occur simultaneously.

EXAMPLE 1

INDEPENDENT EVENTS AND MUTUALLY EXCLUSIVE EVENTS Let F be the event "a person has freckles," and let R be the event "a person has red hair."

a. Are F and R independent?
b. Are F and R mutually exclusive?

SOLUTION

a. F and R are independent if $p(F|R) = p(F)$. With $p(F|R)$, we are given that a person has red hair; with $p(F)$, we are not given that information. Does knowing that a person has red hair affect the probability that the person has freckles? Yes, it does; $p(F \mid R) > p(F)$. Therefore, F and R are not independent; they are dependent.
b. F and R are mutually exclusive if $F \cap R = \emptyset$. Many people have both freckles and red hair, so $F \cap R \neq \emptyset$, and F and R are not mutually exclusive. In other words, having freckles does not exclude the possibility of having red hair; freckles and red hair can occur simultaneously.

EXAMPLE 2

INDEPENDENT EVENTS AND MUTUALLY EXCLUSIVE EVENTS Let T be the event "a person is tall," and let R be the event "a person has red hair."

a. Are T and R independent?
b. Are T and R mutually exclusive?

SOLUTION

a. T and R are independent if $p(T|R) = p(T)$. With $p(T|R)$, we are given that a person has red hair; with $p(T)$, we are not given that information. Does knowing that a person has red hair affect the probability that the person is tall? No, it does not; $p(T \mid R) = p(T)$, so T and R are independent.
b. T and R are mutually exclusive if $T \cap R = \emptyset$. $T \cap R$ is the event "a person is tall and has red hair." There are tall people who have red hair, so $T \cap R \neq \emptyset$, and T and R are not mutually exclusive. In other words, being tall does not exclude the possibility of having red hair; being tall and having red hair can occur simultaneously.

In Examples 1 and 2, we had to rely on our personal experience in concluding that knowledge that a person has red hair does affect the probability that he or she has freckles and does not affect the probability that he or she is tall. It may be the case that you have seen only one red-haired person and she was short and without freckles. Independence is better determined by computing the appropriate probabilities than by relying on one's own personal experiences. This is especially crucial in determining the effectiveness of an experimental drug. *Double-blind* experiments, in which neither the patient nor the doctor knows whether the given medication is the experimental drug or an inert substance, are often done to ensure reliable, unbiased results.

Independence is an important tool in determining whether an experimental drug is an effective vaccine. Let D be the event that the experimental drug was administered to a patient, and let R be the event that the patient recovered. It is hoped that $p(R \mid D) > p(R)$, that is, that the rate of recovery is greater among those who were given the drug. In this case, R and D are dependent. Independence is also an important tool in determining whether an advertisement effectively promotes a product. An ad is effective if p(consumer purchases product | consumer saw ad) > p(consumer purchases product).

EXAMPLE 3

DETERMINING INDEPENDENCE Use probabilities to determine whether the events "thinking there is too much violence in television" and "being a man" in Example 1 of Section 3.6 are independent.

SOLUTION

Two events E and F are independent if $p(E \mid F) = p(E)$. The events "responding yes to the question on violence in television" and "being a man" are independent if $p(Y \mid M) = p(Y)$. We need to compute these two probabilities and compare them.

In Example 1 of Section 3.6, we found $p(Y \mid M) \approx 0.58$. We can use the data from the poll in Figure 3.33 to find $p(Y)$.

$$p(Y) = \frac{418}{600} \approx 0.70$$

$$p(Y \mid M) \neq p(Y)$$

The events "responding yes to the question on violence in television" and "being a man" are dependent. According to the poll, men are less likely to think that there is too much violence on television.

In Example 3, what should we conclude if we found that $p(Y) = 0.69$ and $p(Y \mid M) = 0.67$? Should we conclude that $p(Y \mid M) \neq p(Y)$ and that the events "thinking that there is too much violence on television" and "being a man" are dependent? Or should we conclude that $p(Y \mid M) \approx p(Y)$ and that the events are (probably) independent? In this particular case, the probabilities are relative frequencies rather than theoretical probabilities, and relative frequencies can vary. A group of 600 people was polled to determine the opinions of the entire viewing public; if the same question was asked of a different group, a somewhat different set of relative frequencies could result. While it would be reasonable to conclude that the events are (probably) independent, it would be more appropriate to include more people in the poll and make a new comparison.

Product Rule for Independent Events

The product rule says that $p(A \cap B) = p(A \mid B) \cdot p(B)$. If A and B are independent, then $p(A \mid B) = p(A)$. Combining these two equations, we get the following rule:

PRODUCT RULE FOR INDEPENDENT EVENTS

If A and B are independent events, then the probability of A and B is

$$p(A \cap B) = p(A) \cdot p(B)$$

A common error that is made in computing probabilities is using the formula $p(A \cap B) = p(A) \cdot p(B)$ without verifying that A and B are independent. In fact, the Federal Aviation Administration (FAA) has stated that this is the most frequently encountered error in probabilistic analysis of airplane component failures. If it is not known that A and B are independent, you must use the Product Rule $p(A \cap B) = p(A \mid B) \cdot p(B)$.

EXAMPLE 4

USING THE PRODUCT RULE FOR INDEPENDENT EVENTS If a pair of dice is tossed twice, find the probability that each toss results in a seven.

SOLUTION

In Example 4 of Section 3.3, we found that the probability of a seven is $\frac{1}{6}$. The two rolls are independent (one roll has no influence on the next), so the probability of a seven is $\frac{1}{6}$ regardless of what might have happened on an earlier roll; we can use the Product Rule for Independent Events:

$$\begin{aligned} p(A \cap B) &= p(A) \cdot p(B) \\ p(\text{1st is 7 and 2nd is 7}) &= p(\text{1st is 7}) \cdot p(\text{2nd is 7}) \\ &= \frac{1}{6} \cdot \frac{1}{6} \\ &= \frac{1}{36} \end{aligned}$$

See Figure 3.50. The thicker branch of the tree diagram starts at the event "1st roll is 7"and ends at the event "2nd roll is 7," so it is the conditional probability $p(\text{2nd is 7} \mid \text{1st is 7})$. However, the two rolls are independent, so $p(\text{2nd is 7} \mid \text{1st is 7}) = p(\text{2nd is 7})$. We are free to label this branch as either of these two equivalent probabilities.

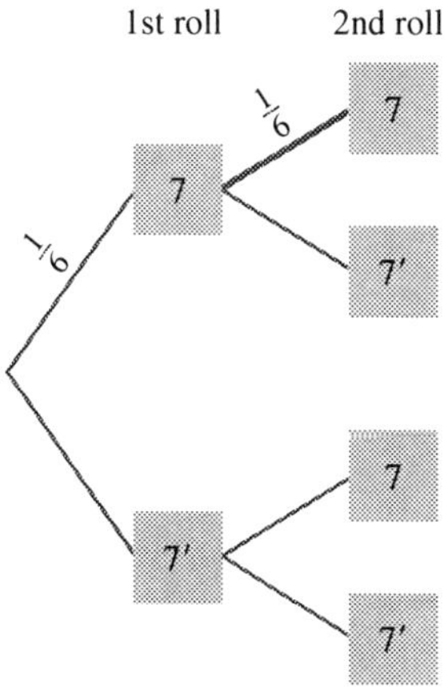

FIGURE 3.50
A tree diagram for Example 4.

Trees in Medicine and Genetics

Usually, medical diagnostic tests are not 100% accurate. A test might indicate the presence of a disease when the patient is in fact healthy (this is called a **false positive**), or it might indicate the absence of a disease when the patient does in fact have the disease (a **false negative**). Probability trees can be used to determine the probability that a person whose test results were positive actually has the disease.

EXAMPLE 5

ANALYZING THE EFFECTIVENESS OF DIAGNOSTIC TESTS Medical researchers have recently devised a diagnostic test for "white lung" (an imaginary disease caused by the inhalation of chalk dust). Teachers are particularly susceptible to this disease; studies have shown that half of all teachers are afflicted with it. The test correctly diagnoses the presence of white lung in 99% of the people who have it and correctly diagnoses its absence in 98% of the people who do not have it. Find the probability that a teacher whose test results are positive actually has white lung and the probability that a teacher whose test results are negative does not have white lung.

SOLUTION

First, we determine which probabilities have been given and find their complements, as shown in Figure 3.51. We use + to denote the event that a person receives a positive diagnosis and − to denote the event that a person receives a negative diagnosis.

Probabilities Given	Complements of Those Probabilities
$p(\text{ill}) = 0.50$	$p(\text{healthy}) = p(\text{not ill}) = 1 - 0.50 = 0.50$
$p(- \mid \text{healthy}) = 98\% = 0.98$	$p(+ \mid \text{healthy}) = 1 - 0.98 = 0.02$
$p(+ \mid \text{ill}) = 99\% = 0.99$	$p(- \mid \text{ill}) = 1 - 0.99 = 0.01$

FIGURE 3.51 Data from Example 5.

The first two of these probabilities [p(ill) and p(healthy)] are not conditional, so they start at the beginning of the tree. The next two probabilities [$p(- \mid \text{healthy})$

and $p(+ \mid \text{healthy})$] are conditional, so they start at their condition (healthy). Similarly, the last two probabilities are conditional, so they start at their condition (ill). This placement of the probabilities yields the tree diagram in Figure 3.52.

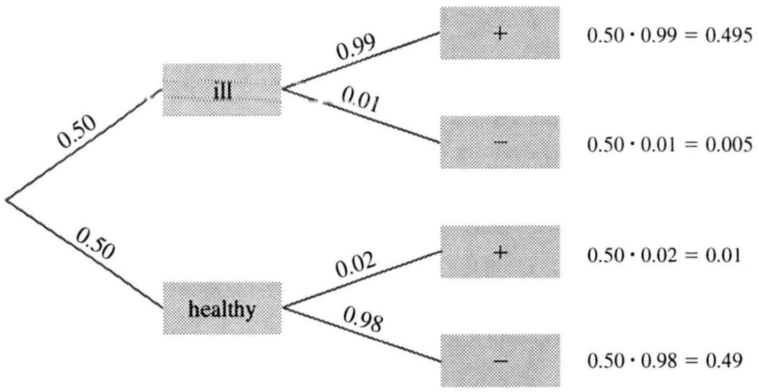

FIGURE 3.52 A tree diagram for Example 5.

The four probabilities to the right of the tree are

$$p(\text{ill and } +) = 0.495$$
$$p(\text{ill and } -) = 0.005$$
$$p(\text{healthy and } +) = 0.01$$
$$p(\text{healthy and } -) = 0.49$$

We are to find the probability that a teacher whose test results are positive actually has white lung; that is, we are to find $p(\text{ill} \mid +)$. Thus, we are given that the test results are positive, and we need only consider branches that involve positive test results: $p(\text{ill and } +) = 0.495$ and $p(\text{healthy and } +) = 0.01$. The probability that a teacher whose test results are positive actually has white lung is

$$p(\text{ill} \mid +) = \frac{0.495}{0.495 + 0.01} = 0.9801\ldots \approx 98\%$$

The probability that a teacher whose test results are negative does not have white lung is

$$p(\text{healthy} \mid -) = \frac{0.49}{0.49 + 0.005} = 0.98989\ldots \approx 99\%$$

The probabilities show that this diagnostic test works well in determining whether a teacher actually has white lung. In the exercises, we will see how well it would work with schoolchildren.

EXAMPLE 6

FINDING THE PROBABILITY THAT YOUR CHILD INHERITS A DISEASE Mr. and Mrs. Smith each had a sibling who died of cystic fibrosis. The Smiths are considering having a child. They have not been tested to determine whether they are carriers. What is the probability that their child would have cystic fibrosis?

SOLUTION

In our previous examples involving trees, we were given probabilities; the conditional or nonconditional status of those probabilities helped us to determine the physical layout of the tree. In this example, we are not given any probabilities. To determine the physical layout of the tree, we have to separate what we know to be true from what is only possible or probable. The tree focuses on what is possible or probable. We know that both Mr. and Mrs. Smith had a sibling who died of cystic fibrosis, and it is possible that their child would inherit that disease. The tree's branches will represent the series of possible events that could result in the Smith child inheriting cystic fibrosis.

What events must take place if the Smith child is to inherit the disease? First, the grandparents would have to have had cystic fibrosis genes. Next, the Smiths themselves would have to have inherited those genes from their parents. Finally, the Smith child would have to inherit those genes from his or her parents.

Cystic fibrosis is recessive, which means that a person can inherit it only if he or she receives two cystic fibrosis genes, one from each parent. Mr. and Mrs. Smith each had a sibling who had cystic fibrosis, so each of the four grandparents must have been a carrier. They could not have actually had the disease because the Smiths would have known, and we would have been told.

We now know the physical layout of our tree. The four grandparents were definitely carriers. Mr. and Mrs. Smith were possibly carriers. The Smith child will possibly inherit the disease. The first set of branches will deal with Mr. and Mrs. Smith, and the second set will deal with the child.

The Smith child will not have cystic fibrosis unless Mr. and Mrs. Smith are both carriers. Figure 3.53 shows the Punnett square for Mr. and Mrs. Smith's possible genetic configuration.

	N	c	← one grandparent's genes
N	NN	cN	← possible offspring
c	Nc	cc	← possible offspring

↑
other grandparent's genes

FIGURE 3.53 A Punnett square for Example 6.

Neither Mr. Smith nor Mrs. Smith has the disease, so we can eliminate the cc possibility. Thus, the probability that Mr. Smith is a carrier is $\frac{2}{3}$, as is the probability that Mrs. Smith is a carrier. Furthermore, these two events are independent, since the Smiths are (presumably) unrelated.

Using the Product Rule for Independent Events, we have

$$p(\text{Mr. S is a carrier and Mrs. S is a carrier})$$

$$= p(\text{Mr. S is a carrier}) \cdot p(\text{Mrs. S is a carrier}) = \frac{2}{3} \cdot \frac{2}{3} = \frac{4}{9}$$

The same Punnett square tells us that the probability that their child will have cystic fibrosis, given that the Smiths are both carriers, is $\frac{1}{4}$. Letting B be the event that both parents are carriers and letting F be the event that the child has cystic fibrosis, we obtain the tree diagram in Figure 3.54.

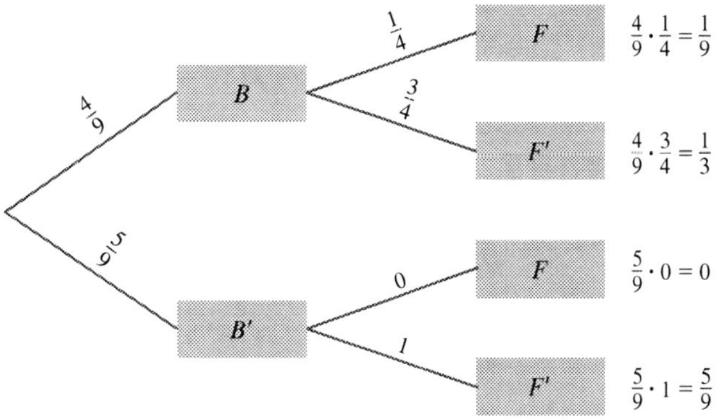

FIGURE 3.54 A tree diagram for Example 6.

The probability that the Smiths' child would have cystic fibrosis is $\frac{1}{9}$.

Notice that the tree in Figure 3.54 could have been drawn differently, as shown in Figure 3.55. It is not necessary to draw a branch going from the B' box to the F box, since the child cannot have cystic fibrosis if both parents are not carriers.

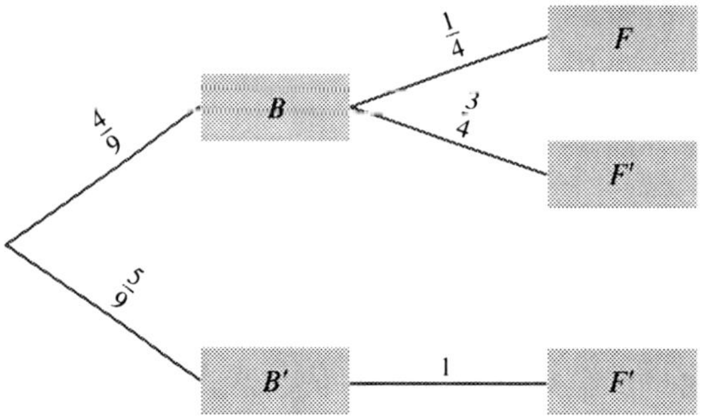

FIGURE 3.55 An alternative to the tree in Figure 3.54.

Hair Color

Like cystic fibrosis, hair color is inherited, but the method by which it is transmitted is more complicated than the method by which an inherited disease is transmitted. This more complicated method of transmission allows for the possibility that a child might have hair that is colored differently from that of other family members.

Hair color is determined by two pairs of genes: one pair that determines placement on a blond/brown/black spectrum and one pair that determines the presence or absence of red pigment. These two pairs of genes are independent.

Melanin is a brown pigment that affects the color of hair (as well as the colors of eyes and skin). The pair of genes that controls the brown hair colors does so by determining the amount of melanin in the hair. This gene has three forms, each traditionally labeled with an M (for melanin): M^{Bd}, or blond (a light melanin deposit); M^{Bw}, or brown (a medium melanin deposit); and M^{Bk}, or black (a heavy melanin deposit). Everyone has two of these genes, and their combination determines the brown aspect of hair color, as illustrated in Figure 3.56.

Genes	Hair Color
$M^{Bd}M^{Bd}$	blond
$M^{Bd}M^{Bw}$	light brown
$M^{Bd}M^{Bk}$ $M^{Bw}M^{Bw}$	medium brown
$M^{Bw}M^{Bk}$	dark brown
$M^{Bk}M^{Bk}$	black

FIGURE 3.56 Melanin.

The hair colors in Figure 3.56 are altered by the presence of red pigment, which is determined by another pair of genes. This gene has two forms: R^-, or no red pigment, and R^+, or red pigment. Because everyone has two of these genes, there are three possibilities for the amount of red pigment: R^-R^-, R^+R^-, and R^+R^+. The amount of red pigment in a person's hair is independent of the brownness of his or her hair.

The actual color of a person's hair is determined by the interaction of these two pairs of genes, as shown in Figure 3.57.

Genes	Blond ($M^{Bd}M^{Bd}$)	Light Brown ($M^{Bd}M^{Bw}$)	Medium Brown ($M^{Bd}M^{Bk}$, $M^{Bw}M^{Bw}$)	Dark Brown ($M^{Bw}M^{Bk}$)	Black ($M^{Bk}M^{Bk}$)
R^-R^-	blond	light brown	medium brown	dark brown	black
R^+R^-	strawberry blond	reddish brown	chestnut	shiny dark brown	shiny black
R^+R^+	bright red	dark red	auburn	glossy dark brown	glossy black

FIGURE 3.57 Melanin and red pigment.

EXAMPLE 7

PREDICTING A CHILD'S HAIR COLOR The Rosses are going to have a child. Mr. Ross has blond hair, and Mrs. Ross has reddish brown hair. Find their child's possible hair colors and the probabilities of each possibility.

SOLUTION

The parent with blond hair has genes $M^{Bd}M^{Bd}$ and R^-R^-. The parent with reddish brown hair has genes $M^{Bd}M^{Bw}$ and R^+R^-. We need to use two Punnett squares, one for the brownness of the hair (Figure 3.58) and one for the presence of red pigment (Figure 3.59).

	M^{Bd}	M^{Bd}
M^{Bd}	$M^{Bd}M^{Bd}$	$M^{Bd}M^{Bd}$
M^{Bw}	$M^{Bd}M^{Bw}$	$M^{Bd}M^{Bw}$

FIGURE 3.58 Brownness.

	R^-	R^-
R^+	R^+R^-	R^+R^-
R^-	R^-R^-	R^-R^-

FIGURE 3.59 Red pigment.

$$p(M^{Bd}M^{Bd}) = \frac{2}{4} = \frac{1}{2} \qquad p(R^+R^-) = \frac{2}{4} = \frac{1}{2}$$

$$p(M^{Bd}M^{Bw}) = \frac{2}{4} = \frac{1}{2} \qquad p(R^-R^-) = \frac{2}{4} = \frac{1}{2}$$

We will use a tree diagram to determine the possible hair colors and their probabilities (see Figure 3.60).

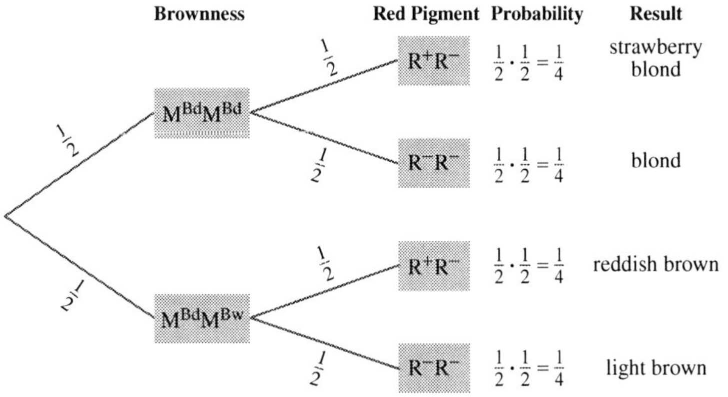

FIGURE 3.60 A tree diagram for Example 7.

The Rosses' child could have strawberry blond, blond, reddish brown, or light brown hair. The probability of each color is $\frac{1}{4}$. Notice that there is a 50% probability that the child will have hair that is colored differently from that of either parent.

3.7 EXERCISES

In Exercises 1–8, use your own personal experience with the events described to determine whether (a) E and F are independent and (b) E and F are mutually exclusive. (Where appropriate, these events are meant to be simultaneous; for example, in Exercise 2, "E and F" would mean "it's simultaneously raining and sunny.") Write a sentence justifying each of your answers.

1. E is the event "being a doctor," and F is the event "being a woman."
2. E is the event "it's raining," and F is the event "it's sunny."
3. E is the event "being single," and F is the event "being married."
4. E is the event "having naturally blond hair," and F is the event "having naturally black hair."
5. E is the event "having brown hair," and F is the event "having gray hair."

6. E is the event "being a plumber," and F is the event "being a stamp collector."
7. E is the event "wearing shoes," and F is the event "wearing sandals."
8. E is the event "wearing shoes," and F is the event "wearing socks."

In Exercises 9–10, use probabilities, rather than your own personal experience, to determine whether (a) E and F are independent and (b) E and F are mutually exclusive. (c) Interpret your answers to parts (a) and (b).

9. If a die is rolled once, E is the event "getting a 4," and F is the event "getting an odd number."
10. If a die is rolled once, E is the event "getting a 4," and F is the event "getting an even number."

11. Determine whether the events "responding yes to the question on violence in television" and "being a woman" in Example 1 of Section 3.6 are independent.
12. Determine whether the events "having a defect" and "being manufactured in Peoria" in Example 4 of Section 3.6 are independent.
13. A single die is rolled once.
    a. Find the probability of rolling a 5.
    b. Find the probability of rolling a 5 given that the number rolled is even.
    c. Are the events "rolling a 5" and "rolling an even number" independent? Why?
    d. Are the events "rolling a 5" and "rolling an even number" mutually exclusive? Why?
    e. Interpret the results of parts (c) and (d).
14. A pair of dice is rolled once.
    a. Find the probability of rolling a 6.
    b. Find the probability of rolling a 6 given that the number rolled is even.
    c. Are the events "rolling a 6" and "rolling an even number" independent? Why?
    d. Are the events "rolling a 6" and "rolling an even number" mutually exclusive? Why?
    e. Interpret the results of parts (c) and (d).
15. A card is dealt from a full deck (no jokers).
    a. Find the probability of being dealt a jack.
    b. Find the probability of being dealt a jack given that you were dealt a red card.
    c. Are the events "being dealt a jack" and "being dealt a red card" independent? Why?
    d. Are the events "being dealt a jack" and "being dealt a red card" mutually exclusive? Why?
    e. Interpret the results of parts (c) and (d).
16. A card is dealt from a full deck (no jokers).
    a. Find the probability of being dealt a jack.
    b. Find the probability of being dealt a jack given that you were dealt a card above a 7 (count aces high).
    c. Are the events "being dealt a jack" and "being dealt a card above a 7" independent? Why?
    d. Are the events "being dealt a jack" and "being dealt a card above a 7" mutually exclusive? Why?
    e. Interpret the results of parts (c) and (d).

In Exercises 17 and 18, use the following information: To determine what effect salespeople had on purchases, a department store polled 700 shoppers as to whether or not they made a purchase and whether or not they were pleased with the service. Of those who made a purchase, 125 were happy with the service and 111 were not. Of those who made no purchase, 148 were happy with the service and 316 were not.

17. Are the events "being happy with the service" and "making a purchase" independent? What conclusion can you make?
18. Are the events "being unhappy with the service" and "not making a purchase" independent? What conclusion can you make?

19. A personal computer manufacturer buys 38% of its chips from Japan and the rest from the United States. Of the Japanese chips, 1.7% are defective, whereas 1.1% of the U.S. chips are defective. Are the events "defective" and "Japanese-made" independent? What conclusion can you draw? (See Exercises 41–44 in Section 3.6.)
20. A skateboard manufacturer buys 23% of its ball bearings from a supplier in Akron, 38% from one in Atlanta, and the rest from one in Los Angeles. Of the ball bearings from Akron, 4% are defective; 6.5% of those from Atlanta are defective; and 8.1% of those from Los Angeles are defective.
    a. Find the probability that a ball bearing is defective.
    b. Are the events "defective" and "from the Los Angeles supplier" independent?
    c. Are the events "defective" and "from the Atlanta supplier" independent?
    d. What conclusion can you draw?
21. Over the years, a group of nutritionists have observed that their vegetarian clients tend to have fewer health problems. To determine whether their observation is accurate, they collected the following data:
    - They had 365 clients.
    - 281 clients are healthy.
    - Of the healthy clients, 189 are vegetarians.
    - Of the unhealthy clients, 36 are vegetarians.

    a. Use the data to determine whether the events "being a vegetarian" and "being healthy" are independent.
    b. Use the data to determine whether the events "being a vegetarian" and "being healthy" are mutually exclusive.
    c. Interpret the results of parts (a) and (b).

HINT: Start by organizing the data in a chart, similar to that used in the "Violence on Television" poll in Figure 3.33 in Section 3.6.

22. Over the years, a group of exercise physiologists have observed that their clients who exercise solely by running tend to have more ankle problems than do their clients who vary between running and other forms of exercise. To determine whether their observation is accurate, they collected the following data:

- They had 422 clients.
- 276 clients are have ankle problems.
- Of the clients with ankle problems, 191 only run.
- Of the clients without ankle problems, 22 only run.

a. Use the data to determine whether the events "running only" and "having ankle problems" are independent.

b. Use the data to determine whether the events "running only" and "having ankle problems" are mutually exclusive.

c. Interpret the results of parts (a) and (b).

HINT: Start by organizing the data in a chart, similar to that used in the "Violence on Television" poll in Figure 3.33 in Section 3.6.

23. The Venn diagram in Figure 3.61 contains the results of a survey. Event *A* is "supports proposition 3," and event *B* is "lives in Bishop."

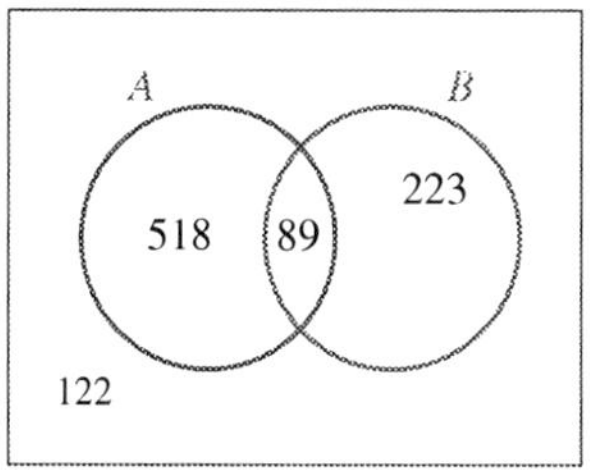

FIGURE 3.61 Venn diagram for Exercise 23.

a. Are the events "supports proposition 3" and "lives in Bishop" independent?

b. Are the events "supports proposition 3" and "lives in Bishop" mutually exclusive?

c. Interpret the results of parts (a) and (b).

24. The Venn diagram in Figure 3.62 contains the results of a survey. Event *A* is "uses Ipana toothpaste," and event *B* is "has good dental checkups."

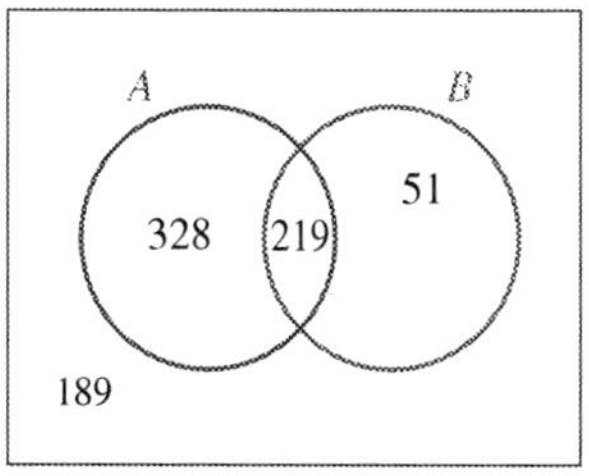

FIGURE 3.62 Venn diagram for Exercise 24.

a. Are the events "uses Ipana toothpaste" and "has good dental checkups" independent?

b. Are the events "uses Ipana toothpaste" and "has good dental checkups" mutually exclusive?

c. Interpret the results of parts (a) and (b).

In Exercises 25–28, you may wish to use Exercise 66 in Section 3.6.

25. Sixty percent of all computers sold last year were HAL computers, 4% of all computer users quit using computers, and 3% of all computer users used HAL computers and then quit using computers. Are the events "using a HAL computer" and "quit using computers" independent? What conclusion can you make?

26. Ten percent of all computers sold last year were Peach computers, 4% of all computer users quit using computers, and 0.3% of all computer users used Peach computers and then quit using computers. Are the events "using a Peach computer" and "quit using computers" independent? What conclusion can you make?

27. Forty percent of the country used SmellSoGood deodorant, 10% of the country quit using deodorant, and 4% of the country used SmellSoGood deodorant and then quit using deodorant. Are the events "used SmellSoGood" and "quit using deodorant" independent? What conclusion can you make?

28. Nationwide, 20% of all TV viewers use VideoLink cable, 8% of all TV viewers switched from cable to a satellite service, and 7% of all TV viewers used VideoLink cable and then switched from cable to a satellite service. Are the events "used VideoLink cable" and "switched from cable to satellite" independent? What conclusion can you make?

29. Suppose that the space shuttle has three separate computer control systems: the main system and two backup duplicates of it. The first backup would monitor the main system and kick in if the main system failed. Similarly, the second backup would monitor the first. We can assume that a failure of one system is independent of a failure of another system, since the systems are separate. The probability of failure for any one system on any one mission is known to be 0.01.

a. Find the probability that the shuttle is left with no computer control system on a mission.

b. How many backup systems does the space shuttle need if the probability that the shuttle is left with no computer control system on a mission must be $\frac{1}{1 \text{ billion}}$?

30. Use the information in Example 5 to find $p(\text{healthy} \mid +)$, the probability that a teacher receives a false positive.

31. Use the information in Example 5 to find $p(\text{ill} \mid -)$, the probability that a teacher receives a false negative.

32. Overwhelmed with their success in diagnosing white lung in teachers, public health officials decided to administer the test to all schoolchildren, even though only one child in 1,000 has contracted the disease. Recall from Example 5 that the test correctly diagnoses the presence of white lung in 99% of the people who have it and correctly diagnoses its absence in 98% of the people who do not have it.

a. Find the probability that a schoolchild whose test results are positive actually has white lung.

b. Find the probability that a schoolchild whose test results are negative does not have white lung.

c. Find the probability that a schoolchild whose test results are positive does not have white lung.

d. Find the probability that a schoolchild whose test results are negative actually has white lung.

e. Which of these events is a false positive? Which is a false negative?

f. Which of these probabilities would you be interested in if you or one of your family members tested positive?

g. Discuss the usefulness of this diagnostic test, both for teachers (as in Example 5) and for schoolchildren.

33. In 2004, the Centers for Disease Control and Prevention estimated that 1,000,000 of the 287,000,000 residents of the United States are HIV-positive. (HIV is the virus that is believed to cause AIDS.) The SUDS diagnostic test correctly diagnoses the presence of AIDS/HIV 99.9% of the time and correctly diagnoses its absence 99.6% of the time.

a. Find the probability that a person whose test results are positive actually has HIV.

b. Find the probability that a person whose test results are negative does not have HIV.

c. Find the probability that a person whose test results are positive does not have HIV.

d. Find the probability that a person whose test results are negative actually has HIV.

e. Which of the probabilities would you be interested in if you or someone close to you tested positive?

f. Which of these events is a false positive? Which is a false negative?

g. Discuss the usefulness of this diagnostic test.

h. It has been proposed that all immigrants to the United States should be tested for HIV before being allowed into the country. Discuss this proposal.

34. Assuming that the SUDS test cannot be made more accurate, what changes in the circumstances described in Exercise 33 would increase the usefulness of the SUDS diagnostic test? Give a specific example of this change in circumstances, and demonstrate how it would increase the test's usefulness by computing appropriate probabilities.

35. Compare and contrast the circumstances and the probabilities in Example 5 and in Exercises 32, 33, and 34. Discuss the difficulties in using a diagnostic test when that test is not 100% accurate.

36. Use the information in Example 6 to find the following:

a. the probability that the Smiths' child is a cystic fibrosis carrier

b. the probability that the Smiths' child is healthy (i.e., has no symptoms)

c. the probability that the Smiths' child is healthy and not a carrier

HINT: You must consider three possibilities: Both of the Smiths are carriers, only one of the Smiths is a carrier, and neither of the Smiths is a carrier.

37. a. Two first cousins marry. Their mutual grandfather's sister died of cystic fibrosis. They have not been tested to determine whether they are carriers. Find the probability that their child will have cystic fibrosis. (Assume that no other grandparents are carriers.)

b. Two unrelated people marry. Each had a grandparent whose sister died of cystic fibrosis. They have not been tested to determine whether they are carriers. Find the probability that their child will have cystic fibrosis. (Assume that no other grandparents are carriers.)

38. It is estimated that one in twenty-five Americans is a cystic fibrosis carrier. Find the probability that a randomly selected American couple's child will have cystic fibrosis (assuming that they are unrelated).

39. In 1989, researchers announced a new carrier-detection test for cystic fibrosis. However, it was discovered in 1990 that the test will detect the presence of the cystic fibrosis gene in only 85% of cystic fibrosis carriers. If two unrelated people are in fact carriers of cystic fibrosis, find the probability that they both test positive. Find the probability that they don't both test positive.

40. Ramon del Rosario's mother's father died of Huntington's disease. His mother died in childbirth, before symptoms of the disease would have appeared in her. Find the probability that Ramon will have the disease. (Assume that Ramon's grandfather had one bad gene, and that there was no other source of Huntington's in the family.) (Huntington's disease is discussed on page 149 in Section 3.2.)

41. Albinism is a recessive disorder that blocks the normal production of pigmentation. The typical albino has white hair, white skin, and pink eyes and is visually impaired. Mr. Jones is an albino, and although Ms. Jones is normally pigmented, her brother is an albino. Neither of Ms. Jones' parents is an albino. Find the probability that their child will be an albino.

42. Find the probability that the Joneses' child will not be an albino but will be a carrier. (See Exercise 41.)

43. If the Joneses' first child is an albino, find the probability that their second child will be an albino. (See Exercise 41.)

44. The Donohues are going to have a child. She has shiny black hair, and he has bright red hair. Find their child's possible hair colors and the probabilities of each possibility.

45. The Yorks are going to have a child. She has black hair, and he has dark red hair. Find their child's possible hair colors and the probabilities of each possibility.

46. The Eastwoods are going to have a child. She has chestnut hair ($M^{Bd}M^{Bk}$), and he has dark brown hair. Find their child's possible hair colors and the probabilities of each possibility.

47. The Wilsons are going to have a child. She has strawberry blond hair, and he has shiny dark brown hair. Find their child's possible hair colors and the probabilities of each possibility.

48. The Breuners are going to have a child. She has blond hair, and he has glossy dark brown hair. Find their child's possible hair colors and the probabilities of each possibility.

49. The Landres are going to have a child. She has chestnut hair ($M^{Bd}M^{Bk}$), and he has shiny dark brown hair. Find their child's possible hair colors and the probabilities of each possibility.

50. The Hills are going to have a child. She has reddish brown hair, and he has strawberry blond hair. Find their child's possible hair colors and the probabilities of each possibility.

51. Recall from Section 3.1 that Antoine Gombauld, the Chevalier de Méré, had made money over the years by betting with even odds that he could roll at least one 6 in four rolls of a single die. This problem finds the probability of winning that bet and the expected value of the bet.

a. Find the probability of rolling a 6 in one roll of one die.

b. Find the probability of not rolling a 6 in one roll of one die.

c. Find the probability of never rolling a 6 in four rolls of a die.

HINT: This would mean that the first roll is not a 6 *and* the second roll is not a 6 *and* the third is not a 6 *and* the fourth is not a 6.

d. Find the probability of rolling at least one 6 in four rolls of a die.

HINT: Use complements.

e. Find the expected value of this bet if $1 is wagered.

52. Recall from Section 3.1 that Antoine Gombauld, the Chevalier de Méré, had lost money over the years by betting with even odds that he could roll at least one pair of 6's in twenty-four rolls of a pair of dice and that he could not understand why. This problem finds the probability of winning that bet and the expected value of the bet.

a. Find the probability of rolling a double 6 in one roll of a pair of dice.

b. Find the probability of not rolling a double 6 in one roll of a pair of dice.

c. Find the probability of never rolling a double 6 in twenty-four rolls of a pair of dice.

d. Find the probability of rolling at least one double 6 in twenty-four rolls of a pair of dice.

e. Find the expected value of this bet if $1 is wagered.

53. Probability theory began when Antoine Gombauld, the Chevalier de Méré, asked his friend Blaise Pascal why he had made money over the years betting that he could roll at least one 6 in four rolls of a single die but he had lost money betting that he could roll at least one pair of 6's in twenty-four rolls of a pair of dice. Use decision theory and the results of Exercises 51 and 52 to answer Gombauld.

54. Use Exercise 66 of Section 3.6 to explain the calculations of $p(\text{ill} \mid +)$ and $p(\text{healthy} \mid -)$ in Example 5 of this section.

55. Dr. Wellby's patient exhibits symptoms associated with acute neural toxemia (an imaginary disease), but the symptoms can have other, innocuous causes. Studies show that only 25% of those who exhibit the symptoms actually have acute neural toxemia (ANT). A diagnostic test correctly diagnoses the presence of ANT in 88% of the persons who have it and correctly diagnoses its absence in 92% of the persons who do not have it. ANT can be successfully treated, but the treatment causes side effects in 2% of the patients. If left untreated, 90% of those with ANT die; the rest recover fully. Dr. Wellby is considering ordering a diagnostic test for her patient and treating the patient if test results are positive, but she is concerned about the treatment's side effects.

a. Dr. Wellby could choose to test her patient and administer the treatment if the results are positive. Find the probability that her patient's good health will return under this plan.

b. Dr. Wellby could choose to avoid the treatment's side effects by not administering the treatment. (This also implies not testing the patient.) Find the probability that her patient's good health will return under this plan.

c. Find the probability that the patient's good health will return if he undergoes treatment regardless of the test's outcome.

d. On the basis of the probabilities, should Dr. Wellby order the test and treat the patient if test results are positive?

Answer the following questions using complete sentences and your own words.

• CONCEPT QUESTIONS

56. In Example 5, we are given that $p(- \mid \text{healthy}) = 98\%$ and $p(+ \mid \text{ill}) = 99\%$, and we computed that $p(\text{ill} \mid +) \approx 98\%$ and that $p(\text{healthy} \mid -) \approx 99\%$. Which of these probabilities would be most important to a teacher who was diagnosed as having white lung? Why?

57. Are the melanin hair color genes (M^{Bd}, M^{Bw}, and M^{Bk}) dominant, recessive, or codominant? Are the redness hair color genes (R^{+} and R^{-}) dominant, or recessive, or codominant? Why?

58. Compare and contrast the concepts of independence and mutual exclusivity. Be sure to discuss both the similarities and the differences between these two concepts.

3 CHAPTER REVIEW

TERMS

carrier
certain event
codominant gene
conditional probability
decision theory
dependent events
dominant gene
event
expected value
experiment
false negative
false positive
genes
impossible event
independent events
Law of Large Numbers
mutually exclusive events
odds
outcome
probability
Punnett square
recessive gene
relative frequency
sample space

PROBABILITY RULES

1. The probability of the null set is 0: $p(\emptyset) = 0$
2. The probability of the sample space is 1: $p(S) = 1$
3. Probabilities are between 0 and 1 (inclusive): $0 \le p(S) \le 1$
4. $p(E \cup F) = p(E) + p(F) - p(E \cap F)$
5. $p(E \cup F) = p(E) + p(F)$ if E and F are mutually exclusive
6. $p(E) + p(E') = 1$

FORMULAS

If outcomes are equally likely, then:

- the **probability** of an event E is $p(E) = n(E)/n(S)$
- the **odds** of an event E are $o(E) = n(E):n(E')$

To find the **expected value** of an experiment, multiply the value of each outcome by its probability and add the results.

The **conditional probability** of A given B is

$p(A \mid B) = n(A \cap B)/n(B)$ if outcomes are equally likely

The **product rule:**

$p(A \cap B) = p(A \mid B) \cdot p(B)$ for any two events A and B

$p(A \cap B) = p(A) \cdot p(B)$ if A and B are independent

Review Exercises

In Exercises 1–6, a card is dealt from a well-shuffled deck of fifty-two cards.

1. Describe the experiment and the sample space.
2. Find and interpret the probability and the odds of being dealt a queen.
3. Find and interpret the probability and the odds of being dealt a club.
4. Find and interpret the probability and the odds of being dealt the queen of clubs.
5. Find and interpret the probability and the odds of being dealt a queen or a club.
6. Find and interpret the probability and the odds of being dealt something other than a queen.

In Exercises 7–12, three coins are tossed.

7. Find the experiment and the sample space.
8. Find the event E that exactly two are tails.
9. Find the event F that two or more are tails.
10. Find and interpret the probability of E and the odds of E.
11. Find and interpret the probability of F and the odds of F.
12. Find and interpret the probability of E' and the odds of E'.

In Exercises 13–18, a pair of dice is tossed. Find and interpret the probability of rolling each of the following.

13. a 7
14. an 11
15. a 7, an 11, or doubles
16. a number that is both odd and greater than 8
17. a number that is either odd or greater than 8
18. a number that is neither odd nor greater than 8

In Exercises 19–24, three cards are dealt from a deck of fifty-two cards. Find the probability of each of the following.

19. All three are hearts.
20. Exactly two are hearts.
21. At least two are hearts.
22. The first is an ace of hearts, the second is a 2 of hearts, and the third is a 3 of hearts.
23. The second is a heart, given that the first is a heart.
24. The second is a heart, and the first is a heart.

In Exercises 25–30, a pair of dice is rolled three times. Find the probability of each of the following.

25. All three are 7's.
26. Exactly two are 7's.
27. At least two are 7's.
28. None are 7's.
29. The second roll is a 7, given that the first roll is a 7.
30. The second roll is a 7 and the first roll is a 7.

In Exercises 31–32, use the following information. A long-stemmed pea is dominant over a short-stemmed one. A pea with one long-stemmed gene and one short-stemmed gene is crossed with a pea with two short-stemmed genes.

31. Find and interpret the probability that the offspring will be long-stemmed.
32. Find and interpret the probability that the offspring will be short-stemmed.

In Exercises 33–35, use the following information: Cystic fibrosis is caused by a recessive gene. Two cystic fibrosis carriers produce a child.

33. Find the probability that that child will have the disease.
34. Find the probability that that child will be a carrier.
35. Find the probability that that child will neither have the disease nor be a carrier.

In Exercises 36–38, use the following information: Sickle-cell anemia is caused by a codominant gene. A couple, each of whom has sickle-cell trait, produce a child. (Sickle-cell trait involves having one bad gene. Sickle-cell anemia involves having two bad genes.)

36. Find the probability that that child will have the disease.
37. Find the probability that that child will have sickle-cell trait.
38. Find the probability that that child will have neither sickle-cell disease nor sickle-cell trait.

In Exercises 39–41, use the following information: Huntington's disease is caused by a dominant gene. One parent has Huntington's disease. This parent has a single gene for Huntington's disease.

39. Find the probability that that child will have the disease.
40. Find the probability that that child will be a carrier.
41. Find the probability that that child will neither have the disease nor be a carrier.
42. Find the probability of being dealt a pair of 10's (and no other 10's) when playing five-card draw poker.
43. Find the probability of being dealt a pair of tens and three jacks when playing five-card draw poker.
44. Find the probability of being dealt a pair of tens and a pair of jacks (and no other tens or jacks) when playing five-card draw poker.
45. In nine-spot keno, five winning spots breaks even, six winning spots pays \$50, seven pays \$390, eight pays \$6000, and nine pays \$25,000.
    a. Find the probability of each of these events.
    b. Find the expected value of a \$1 bet.

46. Some $1 bets in craps (specifically, the pass, don't pass, come, and don't come bets) have expected values of −$0.014. Use decision theory to compare these bets with a $1 bet in nine-spot keno. (See Exercise 45.)
47. At a certain office, three people make $6.50 per hour, three make $7, four make $8.50, four make $10, four make $13.50, and two make $25. Find the expected hourly wage at that office.

In Exercises 48–55, use the following information: Jock O'Neill, a sportscaster, and Trudy Bell, a member of the state assembly, are both running for governor of the state of Erehwon. A recent telephone poll asked 800 randomly selected voters whom they planned on voting for. The results of this poll are shown in Figure 3.63.

	Jock O'Neill	Trudy Bell	Undecided
Urban residents	266	184	22
Rural residents	131	181	16

FIGURE 3.63 Data for Exercises 48–55.

48. Find the probability that an urban resident supports O'Neill and the probability that an urban resident supports Bell.
49. Find the probability that a rural resident supports O'Neill and the probability that a rural resident supports Bell.
50. Find the probability that an O'Neill supporter lives in an urban area and the probability that an O'Neill supporter lives in a rural area.
51. Find the probability that a Bell supporter lives in an urban area and the probability that a Bell supporter lives in a rural area.
52. Where are O'Neill supporters more likely to live? Where are Bell supporters more likely to live?
53. Which candidate do the urban residents tend to prefer? Which candidate do the rural residents tend to prefer?
54. Are the events "supporting O'Neill" and "living in an urban area" independent or dependent? What can you conclude?
55. On the basis of the poll, who is ahead in the gubernatorial race?

Are the following events independent or dependent? Are they mutually exclusive?

56. "It is summer" and "it is sunny." (Use your own personal experience.)
57. "It is summer" and "it is Monday." (Use your own personal experience.)
58. "It is summer" and "it is autumn." (Use your own personal experience.)
59. "The first card dealt is an ace" and "the second card dealt is an ace." (Do not use personal experience.)
60. "The first roll of the dice results in a 7" and "the second roll results in a 7" (Do not use personal experience.)

In Exercises 61–66, use the following information: Gregor's Garden Corner buys 40% of their plants from the Green Growery and the balance from Herb's Herbs. Twenty percent of the plants from the Green Growery must be returned, and 10% of those from Herb's Herbs must be returned.

61. What percent of all plants are returned?
62. What percent of all plants are not returned?
63. What percent of all plants are from the Green Growery and are returned?
64. What percent of the returned plants are from the Green Growery?
65. What percent of the returned plants are from Herb's Herbs?
66. Are the events "a plant must be returned" and "a plant was from Herb's Herbs" independent? What conclusion can you make?

In Exercises 67–73, use the following information: The Nissota Automobile Company buys emergency flashers from two different manufacturers, one in Arkansas and one in Nevada. Thirty-nine percent of its turn-signal indicators are purchased from the Arkansas manufacturer, and the rest are purchased from the Nevada manufacturer. Two percent of the Arkansas turn-signal indicators are defective, and 1.7% of the Nevada indicators are defective.

67. What percent of the indicators are defective and made in Arkansas?
68. What percentage of the indicators are defective and made in Nevada?
69. What percentage of the indicators are defective?
70. What percentage of the indicators are not defective?
71. What percentage of the defective indicators are made in Arkansas?
72. What percentage of the defective indicators are made in Nevada?
73. Are the events "made in Arkansas" and "a defective" independent? What conclusion can you make?

Answer the following questions with complete sentences and your own words.

• CONCEPT QUESTIONS

74. What is a conditional probability?
75. Give two events that are independent, and explain why they are independent.
76. Give two events that are dependent, and explain why they are dependent.
77. Give two events that are mutually exclusive, and explain why they are mutually exclusive.

78. Give two events that are not mutually exclusive, and explain why they are not mutually exclusive.

79. Why are probabilities always between 0 and 1, inclusive?

80. Give an example of a permutation and a similar example of a combination.

81. Give an example of two events that are mutually exclusive, and explain why they are mutually exclusive.

82. Give an example of two events that are not mutually exclusive, and explain why they are not mutually exclusive.

83. How is set theory used in probability theory?

• HISTORY

84. What two mathematicians invented probability theory? Why?

85. Why was probability theory not considered to be a serious branch of mathematics? What changed that viewpoint? Give a specific example of something that helped change that viewpoint.

86. What role did the following people play in the development of probability theory?

Jacob Bernoulli

Gerolamo Cardano

Pierre de Fermat

Antoine Gombauld, the Chevalier de Méré

John Graunt

Pierre-Simon, the Marquis de Laplace

Gregor Mendel

Blaise Pascal

4 STATISTICS

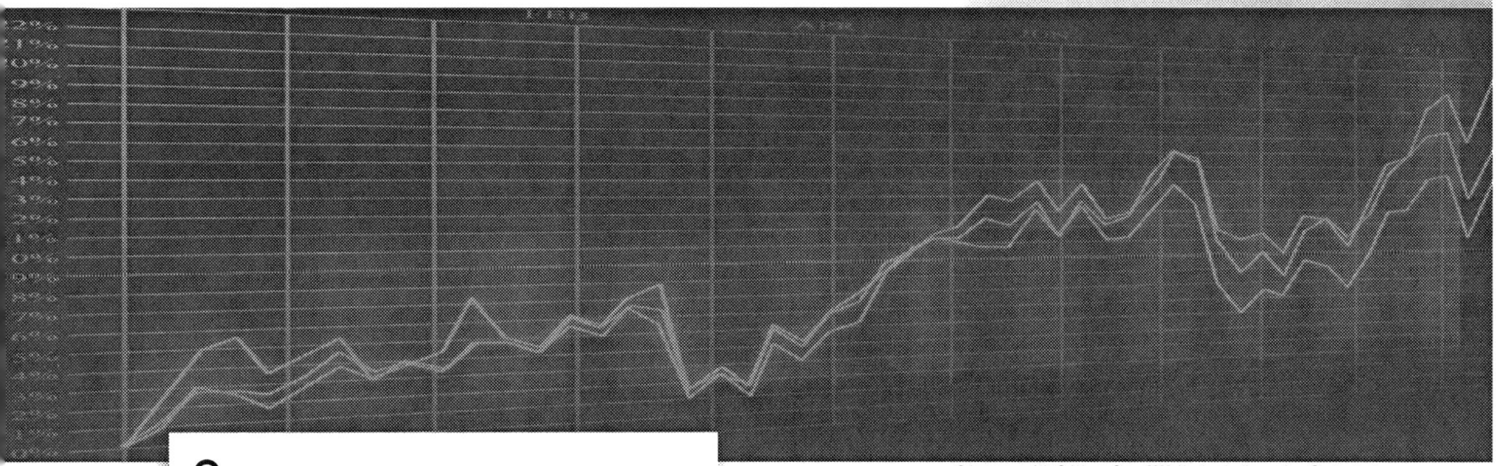

Statistics are everywhere. The news, whether reported in a newspaper, on television, through the Internet, or over the radio, includes statistics of every kind. When shopping for a new car, you will certainly examine the statistics (average miles per gallon, acceleration times, braking distances, and so on) of the various makes and models you are considering. Statistics abound in government studies, and the interpretation of these statistics affects us all. Industry is driven by statistics; they are essential to the direction of quality control, marketing research, productivity, and many other factors. Sporting events are laden with statistics concerning the past performance of the teams and players.

A person who understands the nature of statistics is equipped to see beyond short-term and individual perspectives.

continued

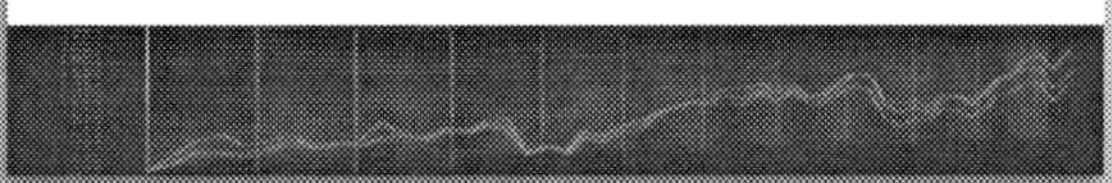

WHAT WE WILL DO IN THIS CHAPTER

WE'LL EXPLORE DIFFERENT TYPES OF STATISTICS, BOTH DESCRIPTIVE AND INFERENTIAL:

- Once data have been collected from a sample, how should they be organized and presented?
- Once they have been organized, how should data be summarized?
- Once data have been summarized, how are conclusions drawn and predictions made?

WE'LL ANALYZE AND EXPLORE DISTRIBUTIONS OF DATA THAT EXHIBIT SPECIFIC TRENDS OR PATTERNS:

- It is not uncommon for data to be clustered around a central value. In many instances, this type of pattern can be represented as a "bell-shaped" curve (high in the middle, low at each end).
- All bell curves share common features. What are these features and how can they be applied in specific situations?

continued

WHAT WE WILL DO IN THIS CHAPTER — *continued*

WE'LL ANALYZE AND INTERPRET SURVEYS AND OPINION POLLS:

- Once the opinions of a specific sample of people have been collected, how can they be used to predict the overall opinion of a large population?
- How accurate or reliable are opinion polls?

WE'LL EXAMINE RELATIONSHIPS BETWEEN TWO SETS OF DATA:

- Are the values of one variable related to the values of another variable?
- If two variables are related, how can the relationship be expressed so that predictions can be made?

He or she is also better prepared to deal with those who use statistics in misleading ways. To many people, the word *statistics* conjures up an image of an endless list of facts and figures. Where do statistics come from? What do they mean? In this chapter, you will learn to handle basic statistical problems and expand your knowledge of the meanings, uses, and misuses of statistics.

4.1 Population, Sample, and Data

OBJECTIVES

- Construct a frequency distribution
- Construct a histogram
- Construct a pie chart

The field of **statistics** can be defined as the science of collecting, organizing, and summarizing data in such a way that valid conclusions and meaningful predictions can be drawn from them. The first part of this definition, "collecting, organizing, and summarizing data," applies to **descriptive statistics.** The second part, "drawing valid conclusions and making meaningful predictions," describes **inferential statistics.**

Population versus Sample

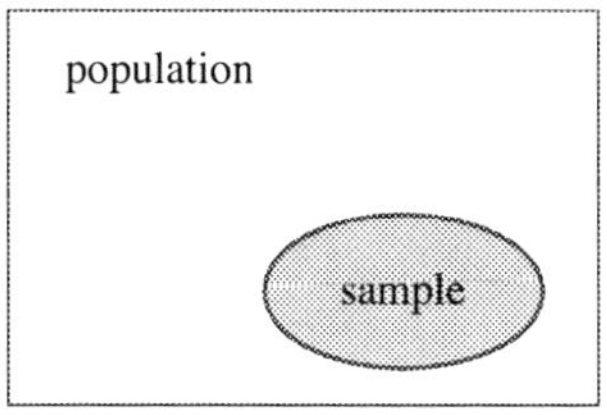

FIGURE 4.1
Population versus sample.

Who will become the next president of the United States? During election years, political analysts spend a lot of time and money trying to determine what percent of the vote each candidate will receive. However, because there are over 175 million registered voters in the United States, it would be virtually impossible to contact each and every one of them and ask, "Whom do you plan on voting for?" Consequently, analysts select a smaller group of people, determine their intended voting patterns, and project their results onto the entire body of all voters.

Because of time and money constraints, it is very common for researchers to study the characteristics of a small group in order to estimate the characteristics of a larger group. In this context, the set of all objects under study is called the **population,** and any subset of the population is called a **sample** (see Figure 4.1).

When we are studying a large population, we might not be able to collect data from every member of the population, so we collect data from a smaller, more manageable sample. Once we have collected these data, we can summarize by calculating various descriptive statistics, such as the average value. Inferential statistics, then, deals with drawing conclusions (hopefully, valid ones!) about the population, based on the descriptive statistics of the sample data.

Sample data are collected and summarized to help us draw conclusions about the population. A good sample is representative of the population from which it was taken. Obviously, if the sample is not representative, the conclusions concerning the population might not be valid. The most difficult aspect of inferential statistics is obtaining a representative sample. Remember that conclusions are only as reliable as the sampling process and that information will usually change from sample to sample.

What person who lived in the twentieth century do you admire most? The top ten responses in a Gallup poll taken on Dec. 20–21, 1999, were as follows: (1) Mother Teresa, (2) Martin Luther King, Jr., (3) John F. Kennedy, (4) Albert Einstein, (5) Helen Keller, (6) Franklin D. Roosevelt, (7) Billy Graham, (8) Pope John Paul II, (9) Eleanor Roosevelt, and (10) Winston Churchill.

Frequency Distributions

The first phase of any statistical study is the collection of data. Each element in a set of data is referred to as a **data point.** When data are first collected, the data points might show no apparent patterns or trends. To summarize the data and detect any trends, we must organize the data. This is the second phase of descriptive statistics. The most common way to organize raw data is to create a **frequency distribution,** a table that lists each data point along with the number of times it occurs (its **frequency**).

The composition of a frequency distribution is often easier to see if the frequencies are converted to percents, especially if large amounts of data are being summarized. The **relative frequency** of a data point is the frequency of the data point expressed as a percent of the total number of data points (that is, made *relative* to the total). The relative frequency of a data point is found by dividing its frequency by the total number of data points in the data set. Besides listing the frequency of each data point, a frequency distribution should also contain a column that gives the relative frequencies.

EXAMPLE 1

CREATING A FREQUENCY DISTRIBUTION: SINGLE VALUES While bargaining for their new contract, the employees of 2 Dye 4 Clothing asked their employers to provide daycare service as an employee benefit. Examining the personnel files of the company's fifty employees, the management recorded the number of children under six years of age that each employee was caring for. The following results were obtained:

0	2	1	0	3	2	0	1	1	0
0	1	1	2	4	1	0	1	1	0
2	1	0	0	3	0	0	1	2	1
0	0	2	4	1	1	0	1	2	0
1	1	0	3	5	1	2	1	3	2

Organize the data by creating a frequency distribution.

SOLUTION

First, we list each different number in a column, putting them in order from smallest to largest (or vice versa). Then we use tally marks to count the number of times each data point occurs. The frequency of each data point is shown in the third column of Figure 4.2.

Number of Children under Six	Tally	Frequency	Relative Frequency
0	𝍸 𝍸 𝍸 |	16	$\frac{16}{50} = 0.32 = 32\%$
1	𝍸 𝍸 𝍸 |||	18	$\frac{18}{50} = 0.36 = 36\%$
2	𝍸 ||||	9	$\frac{9}{50} = 0.18 = 18\%$
3	||||	4	$\frac{4}{50} = 0.08 = 8\%$
4	||	2	$\frac{2}{50} = 0.04 = 4\%$
5	|	1	$\frac{1}{50} - 0.02 = 2\%$
		$n = 50$	total = 100%

FIGURE 4.2 Frequency distribution of data.

To get the relative frequencies, we divide each frequency by 50 (the total number of data points) and change the resulting decimal to a percent, as shown in the fourth column of Figure 4.2.

Adding the frequencies, we see that there is a total of $n = 50$ data points in the distribution. This is a good way to monitor the tally process.

The raw data have now been organized and summarized. At this point, we can see that about one-third of the employees have no need for child care (32%), while the remaining two-thirds (68%) have at least one child under 6 years of age who would benefit from company-sponsored daycare. The most common trend (that is, the data point with the highest relative frequency for the fifty employees) is having one child (36%).

Grouped Data

When raw data consist of only a few distinct values (for instance, the data in Example 1, which consisted of only the numbers 0, 1, 2, 3, 4, and 5), we can easily organize the data and determine any trends by listing each data point along with its frequency and relative frequency. However, when the raw data consist of many nonrepeated data points, listing each one separately does not help us to see any trends the data set might contain. In such cases, it is useful to group the data into intervals or classes and then determine the frequency and relative frequency of each group rather than of each data point.

EXAMPLE 2

CREATING A FREQUENCY DISTRIBUTION: GROUPED DATA Keith Reed is an instructor for an acting class offered through a local arts academy. The class is open to anyone who is at least 16 years old. Forty-two people are enrolled; their ages are as follows:

26	16	21	34	45	18	41	38	22
48	27	22	30	39	62	25	25	38
29	31	28	20	56	60	24	61	28
32	33	18	23	27	46	30	34	62
49	59	19	20	23	24			

Organize the data by creating a frequency distribution.

SOLUTION

This example is quite different from Example 1. Example 1 had only six different data values, whereas this example has many. Listing each distinct data point and its frequency might not summarize the data well enough for us to draw conclusions. Instead, we will work with grouped data.

First, we find the largest and smallest values (62 and 16). Subtracting, we find the range of ages to be $62 - 16 = 46$ years. In working with grouped data, it is customary to create between four and eight groups of data points. We arbitrarily choose six groups, the first group beginning at the smallest data point, 16. To find the beginning of the second group (and hence the end of the first group), divide the range by the number of groups, round off this answer to be consistent with the data, and then add the result to the smallest data point:

$46 \div 6 = 7.6666666\ldots \approx 8$ This is the width of each group.

The beginning of the second group is $16 + 8 = 24$, so the first group consists of people from 16 up to (but not including) 24 years of age.

In a similar manner, the second group consists of people from 24 up to (but not including) 32 $(24 + 8 = 32)$ years of age. The remaining groups are formed and the ages tallied in the same way. The frequency distribution is shown in Figure 4.3.

x = Age (years)	Tally	Frequency	Relative Frequency
$16 \le x < 24$	~~||||~~ ~~||||~~ |	11	$\frac{11}{42} \approx 26\%$
$24 \le x < 32$	~~||||~~ ~~||||~~ |||	13	$\frac{13}{42} \approx 31\%$
$32 \le x < 40$	~~||||~~ ||	7	$\frac{7}{42} \approx 17\%$
$40 \le x < 48$	|||	3	$\frac{3}{42} \approx 7\%$
$48 \le x < 56$	||	2	$\frac{2}{42} \approx 5\%$
$56 \le x < 64$	~~||||~~ |	6	$\frac{6}{42} \approx 14\%$
		$n = 42$	total = 100%

FIGURE 4.3 Frequency distribution of grouped data.

Now that the data have been organized, we can observe various trends: Ages from 24 to 32 are most common (31% is the highest relative frequency), and ages from 48 to 56 are least common (5% is the lowest). Also, over half the people enrolled (57%) are from 16 to 32 years old.

When we are working with grouped data, we can choose the groups in any desired fashion. The method used in Example 2 might not be appropriate in all situations. For example, we used the smallest data point as the beginning of the first group, but we could have begun the first group at an even smaller number. The following box gives a general method for constructing a frequency distribution.

CONSTRUCTING A FREQUENCY DISTRIBUTION

1. If the raw data consist of many different values, create intervals and work with grouped data. If not, list each distinct data point. [When working with grouped data, choose from four to eight intervals. Divide the range (high minus low) by the desired number of intervals, round off this answer to be consistent with the data, and then add the result to the lowest data point to find the beginning of the second group.]
2. Tally the number of data points in each interval or the number of times each individual data point occurs.
3. List the frequency of each interval or each individual data point.
4. Find the relative frequency by dividing the frequency of each interval or each individual data point by the total number of data points in the distribution. The resulting decimal can be expressed as a percent.

Histograms

When data are grouped in intervals, they can be depicted by a **histogram,** a bar chart that shows how the data are distributed in each interval. To construct a histogram, mark off the class limits on a horizontal axis. If each interval has equal width, we draw two vertical axes; the axis on the left exhibits the frequency of an interval, and the axis on the right gives the corresponding relative frequency. We then draw a rectangle above each interval; the height of the rectangle corresponds to the number of data points contained in the interval. The vertical scale on the right gives the percentage of data contained in each interval. The histogram depicting the distribution of the ages of the people in Keith Reed's acting class (Example 2) is shown in Figure 4.4.

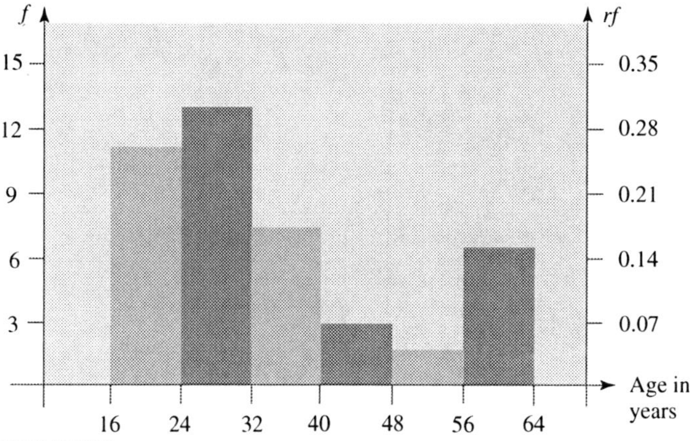

FIGURE 4.4 Ages of the people in Keith Reed's acting class.

What happens if the intervals do not have equal width? For instance, suppose the ages of the people in Keith Reed's acting class are those given in the frequency distribution shown in Figure 4.5.

x = Age (years)	Frequency	Relative Frequency	Class Width
$15 \le x < 20$	4	$\frac{4}{42} \approx 10\%$	5
$20 \le x < 25$	11	$\frac{11}{42} \approx 26\%$	5
$25 \le x < 30$	6	$\frac{6}{42} \approx 14\%$	5
$30 \le x < 45$	11	$\frac{11}{42} \approx 26\%$	15
$45 \le x < 65$	10	$\frac{10}{42} \approx 24\%$	20
	$n = 42$	Total = 100%	

FIGURE 4.5 Frequency distribution of age.

With frequency and relative frequency as the vertical scales, the histogram depicting this new distribution is given in Figure 4.6.

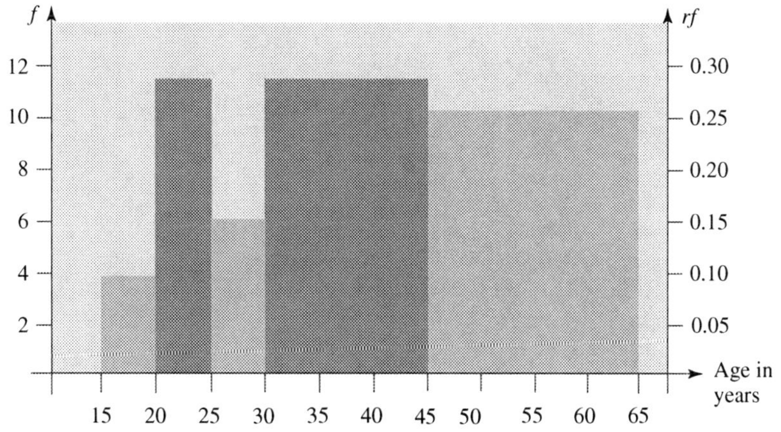

FIGURE 4.6 Why is this histogram misleading?

Does the histogram in Figure 4.6 give a truthful representation of the distribution? No; the rectangle over the interval from 45 to 65 appears to be larger than the rectangle over the interval from 30 to 45, yet the interval from 45 to 65 contains fewer data than the interval from 30 to 45. This is misleading; rather than comparing the heights of the rectangles, our eyes naturally compare the areas of the rectangles. Therefore, to make an accurate comparison, *the areas of the rectangles must correspond to the relative frequencies of the intervals.* This is accomplished by utilizing the **density** of each interval.

Histograms and Relative Frequency Density

Density is a ratio. In science, density is used to determine the concentration of weight in a given volume: Density = weight/volume. For example, the density of water is 62.4 pounds per cubic foot. In statistics, density is used to determine the concentration of data in a given interval: Density = (percent of total data)/(size of an interval). Because relative frequency is a measure of the percentage of data within an interval, we shall calculate the relative frequency density of an interval to determine the concentration of data within the interval.

For example, if the interval $20 \le x < 25$ contains eleven out of forty-two data points, then the relative frequency density of the interval is

$$rfd = \frac{f/n}{\Delta x} = \frac{11/42}{5} = 0.052380952\ldots$$

DEFINITION OF RELATIVE FREQUENCY DENSITY

Given a set of n data points, if an interval contains f data points, then the **relative frequency density** (rfd) of the interval is

$$rfd = \frac{f/n}{\Delta x}$$

where Δx is the width of the interval.

If a histogram is constructed using relative frequency density as the vertical scale, the area of a rectangle will correspond to the relative frequency of the interval, as shown in Figure 4.7.

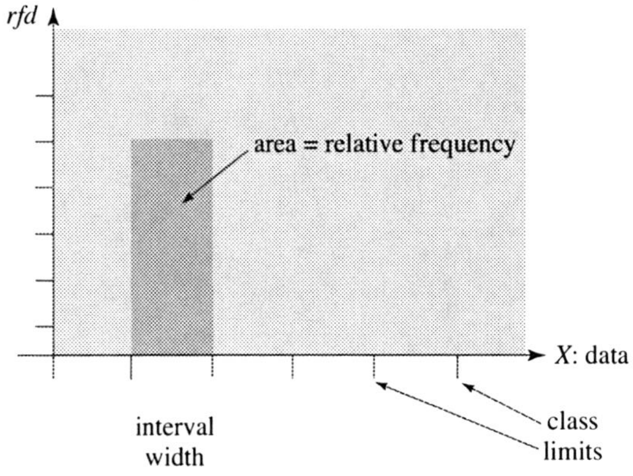

$$\begin{aligned} \text{area} &= \text{base} \cdot \text{height} \\ &= \Delta x \cdot rfd \\ &= \Delta x \cdot \frac{f/n}{\Delta x} \\ &= f/n \\ &= \text{relative frequency} \end{aligned}$$

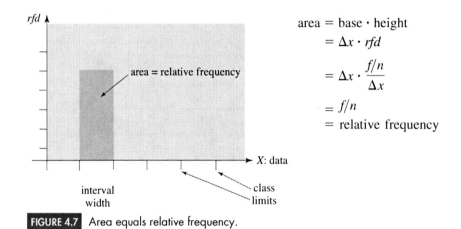

FIGURE 4.7 Area equals relative frequency.

Adding a new column to the frequency distribution given in Figure 4.5, we obtain the relative frequency densities shown in Figure 4.8.

x = Age (years)	Relative Frequency	Class Frequency	Relative Width	Frequency Density
$15 \le x < 20$	4	$\frac{4}{42} \approx 10\%$	5	$\frac{4}{42} \div 5 \approx 0.020$
$20 \le x < 25$	11	$\frac{11}{42} \approx 26\%$	5	$\frac{11}{42} \div 5 \approx 0.052$
$25 \le x < 30$	6	$\frac{6}{42} \approx 14\%$	5	$\frac{6}{42} \div 5 \approx 0.028$
$30 \le x < 45$	11	$\frac{11}{42} \approx 26\%$	15	$\frac{11}{42} \div 15 \approx 0.017$
$45 \le x < 65$	10	$\frac{10}{42} \approx 24\%$	20	$\frac{10}{42} \div 20 \approx 0.012$
	$n = 42$	Total = 100%		

FIGURE 4.8 Calculating relative frequency density.

We now construct a histogram using relative frequency density as the vertical scale. The histogram depicting the distribution of the ages of the people in Keith Reed's acting class (using the frequency distribution in Figure 4.8) is shown in Figure 4.9.

Comparing the histograms in Figures 4.6 and 4.9, we see that using relative frequency density as the vertical scale (rather than frequency) gives a more truthful representation of a distribution when the interval widths are unequal.

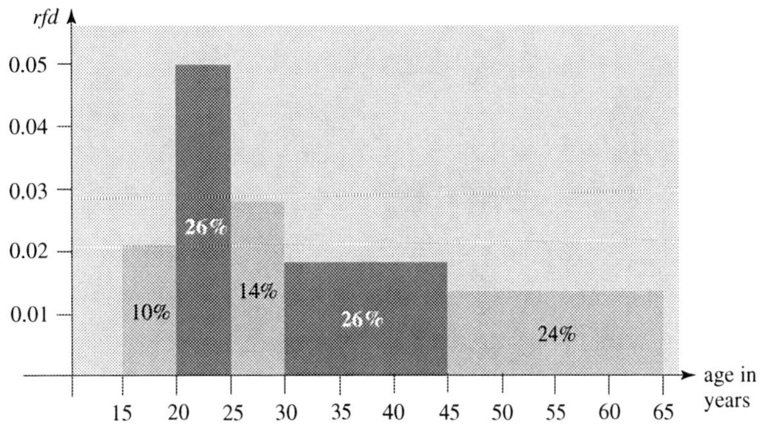

FIGURE 4.9 Ages of the people in Keith Reed's acting class.

EXAMPLE 3

CONSTRUCTING A HISTOGRAM: GROUPED DATA To study the output of a machine that fills bags with corn chips, a quality control engineer randomly selected and weighed a sample of 200 bags of chips. The frequency distribution in Figure 4.10 summarizes the data. Construct a histogram for the weights of the bags of corn chips.

x = Weight (ounces)	f = Number of Bags
$15.3 \le x < 15.5$	10
$15.5 \le x < 15.7$	24
$15.7 \le x < 15.9$	36
$15.9 \le x < 16.1$	58
$16.1 \le x < 16.3$	40
$16.3 \le x < 16.5$	20
$16.5 \le x < 16.7$	12

FIGURE 4.10 Weights of bags of corn chips.

SOLUTION

Because each interval has the same width ($\Delta x = 0.2$), we construct a combined frequency and relative frequency histogram. The relative frequencies are given in Figure 4.11.

x	f	$rf = f/n$
$15.3 \le x < 15.5$	10	0.05
$15.5 \le x < 15.7$	24	0.12
$15.7 \le x < 15.9$	36	0.18
$15.9 \le x < 16.1$	58	0.29
$16.1 \le x < 16.3$	40	0.20
$16.3 \le x < 16.5$	20	0.10
$16.5 \le x < 16.7$	12	0.06
	$n = 200$	Sum = 1.00

FIGURE 4.11 Relative frequencies.

We now draw coordinate axes with appropriate scales and rectangles (Figure 4.12). Notice the (near) symmetry of the histogram. We will study this type of distribution in more detail in Section 4.4.

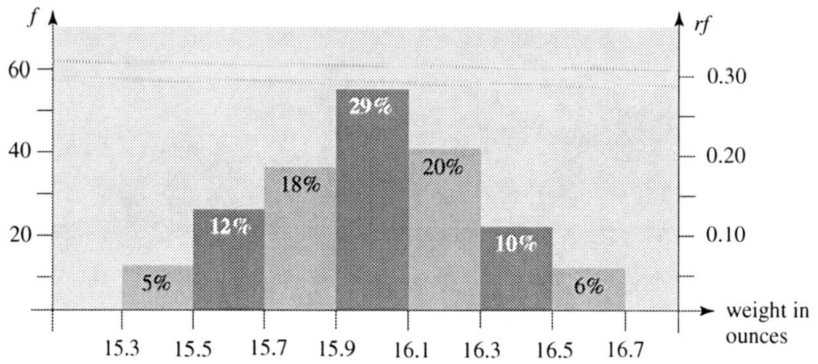

FIGURE 4.12 Weights of bags of corn chips.

Histograms and Single-Valued Classes

The histograms that we have constructed so far have all utilized intervals of grouped data. For instance, the first group of ages in Keith Reed's acting class was from 16 to 24 years old ($16 \le x < 24$). However, if a set of data consists of only a few distinct values, it may be advantageous to consider each distinct value to be a "class" of data; that is, we utilize single-valued classes of data.

EXAMPLE 4

CONSTRUCTING A HISTOGRAM: SINGLE VALUES A sample of high school seniors was asked, "How many television sets are in your house?" The frequency distribution in Figure 4.13 summarizes the data. Construct a histogram using single-valued classes of data.

Number of Television Sets	Frequency
0	2
1	13
2	18
3	11
4	5
5	1

FIGURE 4.13 Frequency distribution.

SOLUTION

Rather than using intervals of grouped data, we use a single value to represent each class. Because each class has the same width ($\Delta x = 1$), we construct a combined frequency and relative frequency histogram. The relative frequencies are given in Figure 4.14.

We now draw coordinate axes with appropriate scales and rectangles. In working with single valued classes of data, it is common to write the single value at the midpoint of the base of each rectangle as shown in Figure 4.15.

Number of Television Sets	Frequency	Relative Frequency
0	2	2/50 = 4%
1	13	13/50 = 26%
2	18	18/50 = 36%
3	11	11/50 = 22%
4	5	5/50 = 10%
5	1	1/50 = 2%
	$n = 50$	Total = 100%

FIGURE 4.14 Calculating relative frequency.

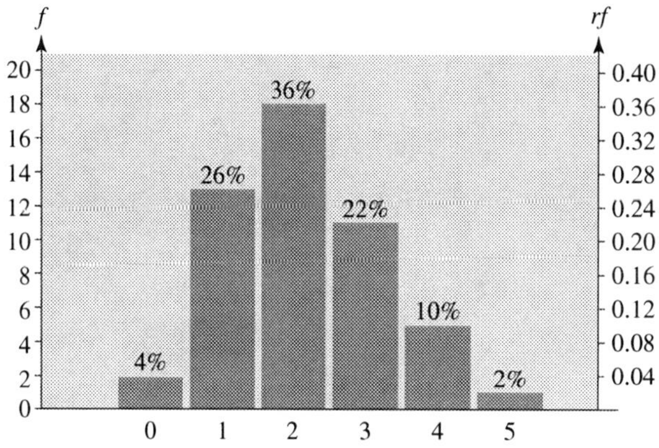

FIGURE 4.15 Number of television sets.

Pie Charts

Many statistical studies involve **categorical data**—that which is grouped according to some common feature or quality. One of the easiest ways to summarize categorical data is through the use of a **pie chart.** A pie chart shows how various categories of a set of data account for certain proportions of the whole. Financial incomes and expenditures are invariably shown as pie charts, as in Figure 4.16.

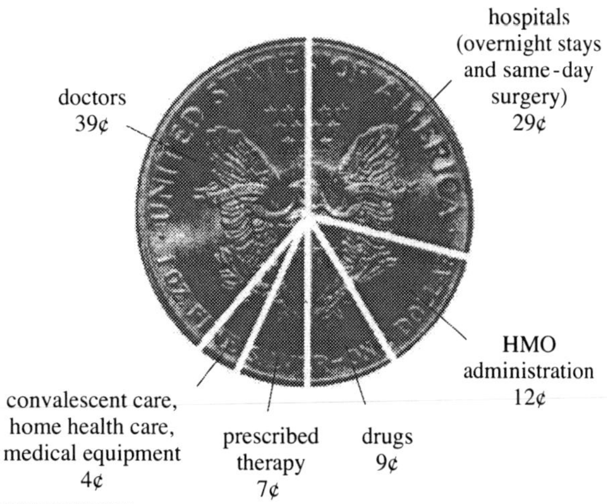

FIGURE 4.16 How a typical medical dollar is spent.

To draw the "slice" of the pie representing the relative frequency (percentage) of the category, the appropriate central angle must be calculated. Since a complete circle comprises 360 degrees, we obtain the required angle by multiplying 360° times the relative frequency of the category.

EXAMPLE 5

CONSTRUCTING A PIE CHART What type of academic degree do you hope to earn? The different types of degrees and the number of each type of degree conferred in the United States during the 2006–07 academic year is given in Figure 4.17. Construct a pie chart to summarize the data.

Type of Degree	Frequency (thousands)
Associate's	728
Bachelor's	1,524
Master's	605
Doctorate	61
Total	2,918

FIGURE 4.17 Academic degrees conferred, 2006–07. *Source:* National Center for Education Statistics.

SOLUTION

Find the relative frequency of each category and multiply it by 360° to determine the appropriate central angle. The necessary calculations are shown in Figure 4.18.

Type of Degree	Frequency (thousands)	Relative Frequency	Central Angle
Associate's	728	$\frac{728}{2,918} \approx 0.249$	$0.249 \times 360° = 89.64°$
Bachelor's	1,524	$\frac{1,524}{2,918} \approx 0.523$	$0.523 \times 360° = 188.28°$
Master's	605	$\frac{605}{2,918} \approx 0.207$	$0.207 \times 360° = 74.52°$
Doctorate	61	$\frac{61}{2,918} \approx 0.021$	$0.021 \times 360° = 7.56°$
	$n = 2,918$	Sum = 1.000	Total = 360°

FIGURE 4.18 Calculating relative frequency and central angles.

Now use a protractor to lay out the angles and draw the "slices." The name of each category can be written directly on the slice, or, if the names are too long, a legend consisting of various shadings may be used. Each slice of the pie should contain its relative frequency, expressed as a percent. Remember, the whole reason for constructing a pie chart is to convey information visually; pie charts should enable the reader to instantly compare the relative proportions of categorical data. See Figure 4.19.

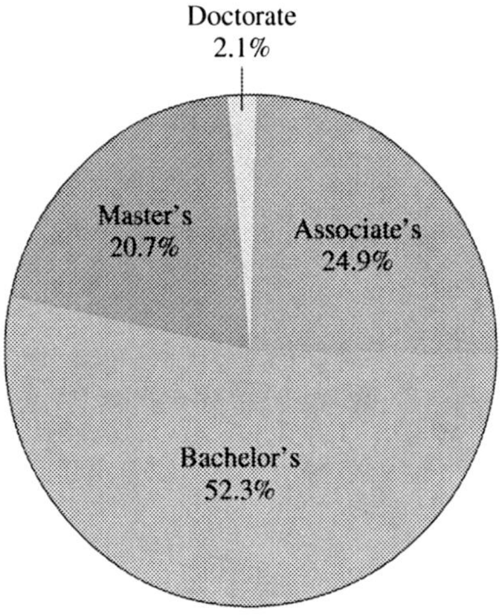

FIGURE 4.19 Academic degrees conferred, 2006–07.

4.1 EXERCISES

1. To study the library habits of students at a local college, thirty randomly selected students were surveyed to determine the number of times they had been to the library during the last week. The following results were obtained:

1	5	2	1	1	4	2	1	5	4
5	2	5	1	2	3	4	1	1	2
3	5	4	1	2	2	4	5	1	2

 a. Organize the given data by creating a frequency distribution.
 b. Construct a pie chart to represent the data.
 c. Construct a histogram using single-valued classes of data.

2. To study the eating habits of students at a local college, thirty randomly selected students were surveyed to determine the number of times they had purchased food at the school cafeteria during the last week. The following results were obtained:

2	3	1	2	2	3	2	1	2	3
1	1	5	5	4	3	5	2	2	2
3	2	3	4	1	4	1	2	3	3

 a. Organize the given data by creating a frequency distribution.
 b. Construct a pie chart to represent the data.
 c. Construct a histogram using single-valued classes of data.

3. To study the composition of families in Manistee, Michigan, forty randomly selected married couples were surveyed to determine the number of children in each family. The following results were obtained:

2	1	3	0	1	0	2	5	1	2	0	2	2	1
4	3	1	1	3	4	1	1	0	2	0	0	2	2
1	0	3	1	1	3	4	2	1	3	0	1		

 a. Organize the given data by creating a frequency distribution.
 b. Construct a pie chart to represent the data.
 c. Construct a histogram using single-valued classes of data.

4. To study the spending habits of shoppers in Orlando, Florida, fifty randomly selected shoppers at a mall were surveyed to determine the number of credit cards they carried. The following results were obtained:

2	5	0	4	2	1	0	6	3	5	4	3	4	0
5	2	5	2	0	2	5	0	2	5	2	5	4	3
5	6	1	0	6	3	5	3	4	0	5	2	2	5
2	0	2	0	4	2	1	0						

 a. Organize the given data by creating a frequency distribution.
 b. Construct a pie chart to represent the data.
 c. Construct a histogram using single-valued classes of data.

5. The speeds, in miles per hour, of forty randomly monitored cars on Interstate 40 near Winona, Arizona, were as follows:

66	71	76	61	73	78	74	67	80	63	69
78	66	70	77	60	72	58	65	70	64	75
80	75	62	67	72	59	74	65	54	69	73
79	64	68	57	51	68	79				

 a. Organize the given data by creating a frequency distribution. (Group the data into six intervals.)
 b. Construct a histogram to represent the data.

6. The weights, in pounds, of thirty-five packages of ground beef at the Cut Above Market were as follows:

1.0	1.9	2.5	1.2	2.0	0.7	1.3	2.4	1.1
3.3	2.4	0.8	2.3	1.7	1.0	2.8	1.4	3.0
0.9	1.1	1.4	2.2	1.5	3.2	2.1	2.7	1.8
1.6	2.3	2.6	1.3	2.9	1.9	1.2	0.5	

 a. Organize the given data by creating a frequency distribution. (Group the data into six intervals.)
 b. Construct a histogram to represent the data.

7. To examine the effects of a new registration system, a campus newspaper asked freshmen how long they had to wait in a registration line. The frequency distribution in Figure 4.20 summarizes the responses. Construct a histogram to represent the data.

x = Time (minutes)	Number of Freshmen
$0 \le x < 10$	32
$10 \le x < 20$	47
$20 \le x < 30$	36
$30 \le x < 40$	22
$40 \le x < 50$	13
$50 \le x < 60$	10

FIGURE 4.20 Time waiting in line.

8. The frequency distribution shown in Figure 4.21 lists the annual salaries of the managers at Universal Manufacturing of Melonville. Construct a histogram to represent the data.

x = Salary (thousands of dollars)	Number of Managers
$30 \le x < 40$	6
$40 \le x < 50$	12
$50 \le x < 60$	10
$60 \le x < 70$	5
$70 \le x < 80$	7
$80 \le x < 90$	3

FIGURE 4.21 Annual salaries.

9. The frequency distribution shown in Figure 4.22 lists the hourly wages of the workers at Universal Manufacturing of Melonville. Construct a histogram to represent the data.

x = Hourly Wage (dollars)	Number of Employees
$8.00 \le x < 9.50$	21
$9.50 \le x < 11.00$	35
$11.00 \le x < 12.50$	42
$12.50 \le x < 14.00$	27
$14.00 \le x < 15.50$	18
$15.50 \le x < 17.00$	9

FIGURE 4.22 Hourly wages.

10. To study the output of a machine that fills boxes with cereal, a quality control engineer weighed 150 boxes of Brand X cereal. The frequency distribution in Figure 4.23 summarizes her findings. Construct a histogram to represent the data.

x = Weight (ounces)	Number of Boxes
$15.3 \le x < 15.6$	13
$15.6 \le x < 15.9$	24
$15.9 \le x < 16.2$	84
$16.2 \le x < 16.5$	19
$16.5 \le x < 16.8$	10

FIGURE 4.23 Weights of boxes of cereal.

11. The ages of the nearly 4 million women who gave birth in the United States in 1997 are given in Figure 4.24. Construct a histogram to represent the data.

Age (years)	Number of Women
$15 \le x < 20$	486,000
$20 \le x < 25$	948,000
$25 \le x < 30$	1,075,000
$30 \le x < 35$	891,000
$35 \le x < 40$	410,000
$40 \le x < 45$	77,000
$45 \le x < 50$	4,000

FIGURE 4.24 Ages of women giving birth in 1997. *Source:* U.S. Bureau of the Census.

12. The age composition of the population of the United States in the year 2000 is given in Figure 4.25. Replace the interval "85 and over" with the interval $85 \le x \le 100$ and construct a histogram to represent the data.

Age (years)	Number of People (thousands)
$0 < x < 5$	19,176
$5 \le x < 10$	20,550
$10 \le x < 15$	20,528
$15 \le x < 25$	39,184
$25 \le x < 35$	39,892
$35 \le x < 45$	44,149
$45 \le x < 55$	37,678
$55 \le x < 65$	24,275
$65 \le x < 85$	30,752
85 and over	4,240

FIGURE 4.25 Age composition of the population of the United States in the year 2000. *Source:* U.S. Bureau of the Census.

In Exercises 13 and 14, use the age composition of the 14,980,000 students enrolled in institutions of higher education in the United States during 2000, as given in Figure 4.26.

13. Using the data in Figure 4.26, replace the interval "35 and over" with the interval $35 \le x \le 60$ and construct a histogram to represent the male data.

Age of Males (years)	Number of Students
$14 \le x < 18$	94,000
$18 \le x < 20$	1,551,000
$20 \le x < 22$	1,420,000
$22 \le x < 25$	1,091,000
$25 \le x < 30$	865,000
$30 \le x < 35$	521,000
35 and over	997,000
Total	6,539,000

Age of Females (years)	Number of Students
$14 \le x < 18$	78,000
$18 \le x < 20$	1,907,000
$20 \le x < 22$	1,597,000
$22 \le x < 25$	1,305,000
$25 \le x < 30$	1,002,000
$30 \le x < 35$	664,000
35 and over	1,888,000
Total	8,441,000

FIGURE 4.26 Age composition of students in higher education. *Source:* U.S. National Center for Education Statistics.

14. Using the data in Figure 4.26, replace the interval "35 and over" with the interval $35 \le x \le 60$ and construct a histogram to represent the female data.

15. The frequency distribution shown in Figure 4.27 lists the ages of 200 randomly selected students who received a bachelor's degree at State University last year. Where possible, determine what percent of the graduates had the following ages:

a. less than 23
b. at least 31
c. at most 20
d. not less than 19
e. at least 19 but less than 27
f. not between 23 and 35

Age (years)	Number of Students
$10 \le x < 15$	1
$15 \le x < 19$	4
$19 \le x < 23$	52
$23 \le x < 27$	48
$27 \le x < 31$	31
$31 \le x < 35$	16
$35 \le x < 39$	29
39 and over	19

FIGURE 4.27 Age of students.

16. The frequency distribution shown in Figure 4.28 lists the number of hours per day a randomly selected sample of teenagers spent watching television. Where possible, determine what percent of the teenagers spent the following number of hours watching television:

a. less than 4 hours
b. at least 5 hours
c. at least 1 hour
d. less than 2 hours
e. at least 2 hours but less than 4 hours
f. more than 3.5 hours

Hours per Day	Number of Teenagers
$0 \le x < 1$	18
$1 \le x < 2$	31
$2 \le x < 3$	24
$3 \le x < 4$	38
$4 \le x < 5$	27
$5 \le x < 6$	12
$6 \le x < 7$	15

FIGURE 4.28 Time watching television.

17. Figure 4.29 lists the top five reasons given by patients for emergency room visits in 2006. Construct a pie chart to represent the data.

Reason	Number of Patients (thousands)
Stomach pain	8,057
Chest pain	6,392
Fever	4,485
Headache	3,354
Shortness of breath	3,007

FIGURE 4.29 Reasons for emergency room visits, 2006. *Source:* National Center for Health Statistics.

18. Figure 4.30 lists the world's top six countries as tourist destinations in 2007. Construct a pie chart to represent the data.

Country	Number of Arrivals (millions)
France	81.9
Spain	59.2
United States	56.0
China	54.7
Italy	43.7
United Kingdom	30.7

FIGURE 4.30 World's top tourist destinations, 2007. *Source:* World Tourism Organization.

19. Figure 4.31 lists the race of new AIDS cases in the United States in 2005.

Race	Male	Female
White	10,027	1,747
Black	13,048	7,093
Hispanic	5,949	1,714
Asian	389	92
Native American	137	45

FIGURE 4.31 New AIDS cases in the United States, 2005. *Source:* National Center for Health Statistics.

a. Construct a pie chart to represent the male data.
b. Construct a pie chart to represent the female data.
c. Compare the results of parts (a) and (b). What conclusions can you make?
d. Construct a pie chart to represent the total data.

20. Figure 4.32 lists the types of accidental deaths in the United States in 2006. Construct a pie chart to represent the data.

21. Figure 4.33 lists some common specialties of physicians in the United States in 2005.

a. Construct a pie chart to represent the male data.
b. Construct a pie chart to represent the female data.
c. Compare the results of parts (a) and (b). What conclusions can you make?
d. Construct a pie chart to represent the total data.

Type of Accident	Number of Deaths
Motor vehicle	44,700
Poison	25,300
Falls	21,200
Suffocation	4,100
Drowning	3,800
Fire	2,800
Firearms	680

FIGURE 4.32 Types of accidental deaths in the United States, 2006. *Source:* National Safety Council.

Specialty	Male	Female
Family practice	54,022	26,305
General surgery	32,329	5,173
Internal medicine	104,688	46,245
Obstetric/gynecology	24,801	17,258
Pediatrics	33,515	36,636
Psychiatry	27,213	13,079

FIGURE 4.33 Physicians by Gender and Specialty, 2005. *Source:* American Medical Association.

22. Figure 4.34 lists the major metropolitan areas on intended residence for immigrants admitted to the United States in 2002. Construct a pie chart to represent the data.

Metropolitan Area	Number of Immigrants
Los Angeles/ Long Beach, CA	100,397
New York, NY	86,898
Chicago, IL	41,616
Miami, FL	39,712
Washington DC	36,371

FIGURE 4.34 Immigrants and areas of residence, 2002. *Source:* U.S. Immigration and Naturalization Service.

Answer the following questions using complete sentences and your own words.

• CONCEPT QUESTIONS

23. Explain the meanings of the terms *population* and *sample*.

24. The cholesterol levels of the 800 residents of Land-o-Lakes, Wisconsin, were recently collected and organized in a frequency distribution. Do these data represent a sample or a population? Explain your answer.

25. Explain the difference between frequency, relative frequency, and relative frequency density. What does each measure?

26. When is relative frequency density used as the vertical scale in constructing a histogram? Why?

27. In some frequency distributions, data are grouped in intervals; in others, they are not.
    a. When should data be grouped in intervals?
    b. What are the advantages and disadvantages of using grouped data?

HISTOGRAMS ON A GRAPHING CALCULATOR

In Example 2 of this section, we created a frequency distribution and a histogram for the ages of the students in an acting class. Much of this work can be done on a computer or a graphing calculator.

TECHNOLOGY AND STATISTICAL GRAPHS

Entering the Data

To enter the data from Example 2, do the following.

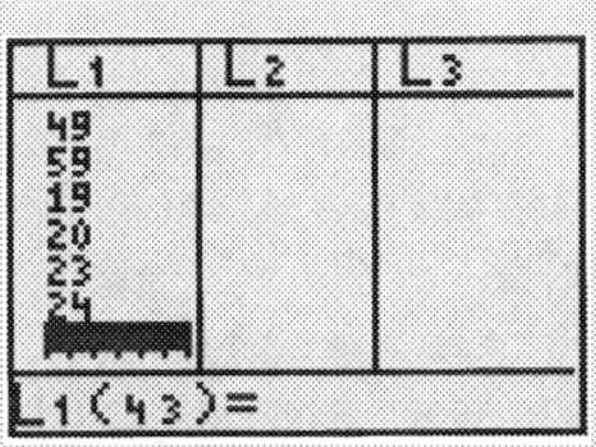

FIGURE 4.35
A TI-83/84's list screen.

ENTERING THE DATA ON A TI-83/84:

- *Put the calculator into statistics mode* by pressing STAT.
- *Set the calculator up for entering the data* by selecting "Edit" from the "EDIT" menu, and the "list screen" appears, as shown in Figure 4.35. If data already appear in a list (as they do in list L_1 in Figure 4.35), use the arrow buttons to highlight the name of the list (i.e., "L_1" or "xStat") and press CLEAR ENTER.
- *Enter the students' ages* in list L_1, in any order, using the arrow buttons and the ENTER button. When this has been completed, your screen should look similar to that in Figure 4.35. Notice the "$L_1(43)=$" at the bottom of the screen; this indicates that 42 entries have been made, and the calculator is ready to receive the 43rd. This allows you to check whether you have left any entries out.

  Note: If some data points frequently recur, you can enter the data points in list L_1 and their frequencies in list L_2 rather than reentering a data point each time it recurs.
- Press 2nd QUIT.

ENTERING THE DATA ON A CASIO:

- *Put the calculator into statistics mode* by pressing MENU, highlighting STAT, and pressing EXE.

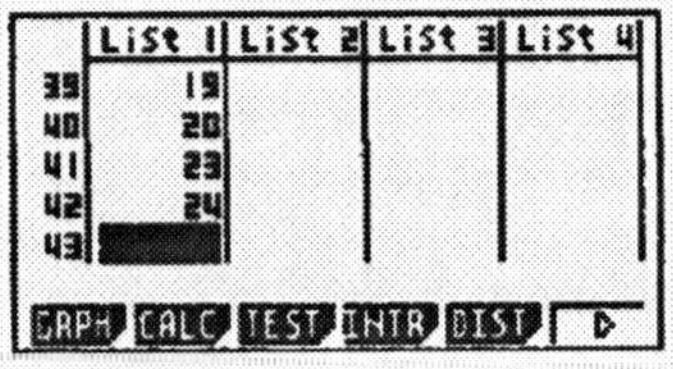

FIGURE 4.36
A Casio's list screen.

- *Enter the data* in List 1 in any order, using the arrow buttons and the EXE button. See Figure 4.36. If data already appear in a list and you want to clear the list, use the arrow keys to move to that list, press F6 and then F4 (i.e., DEL-A, which stands for "delete all"), and then press F1 (i.e., YES). Notice that the entries are numbered; this allows you to check whether you have left any entries out.

Drawing a Histogram

Once the data have been entered, you can draw a histogram.

DRAWING A HISTOGRAM ON A TI-83/84:

- Press Y= and clear any functions that may appear.
- *Enter the group boundaries* by pressing WINDOW, entering the left boundary of the first group as xmin (16 for this problem), the right boundary of the last group plus 1 as xmax ($64 + 1 = 65$ for this problem), and the group width as xscl (8 for this problem). (The calculator will create histograms only with equal group widths.) Enter 0 for ymin, and the largest frequency for ymax. (You may guess; it's easy to change it later if you guess wrong.)
- *Set the calculator up to draw a histogram* by pressing 2nd STAT PLOT and selecting "Plot 1." Turn the plot on and select the histogram icon.
- Tell the calculator to put the data entered in list L_1 on the x-axis by selecting "L_1" for "Xlist," and to consider each entered data point as having a frequency of 1 by selecting "1" for "Freq."

  Note: If some data points frequently recur and you entered their frequencies in list L_2, then select "L_2" rather than "1" for "Freq" by typing 2nd L_2.
- *Draw a histogram* by pressing GRAPH. If some of the bars are too long or too short for the screen, alter ymin accordingly.
- Press TRACE to find out the left and right boundaries and the frequency of the bars, as shown in Figure 4.37. Use the arrow buttons to move from bar to bar.
- Press 2nd STAT PLOT, select "Plot 1" and turn the plot off, or else the histogram will appear on future graphs.

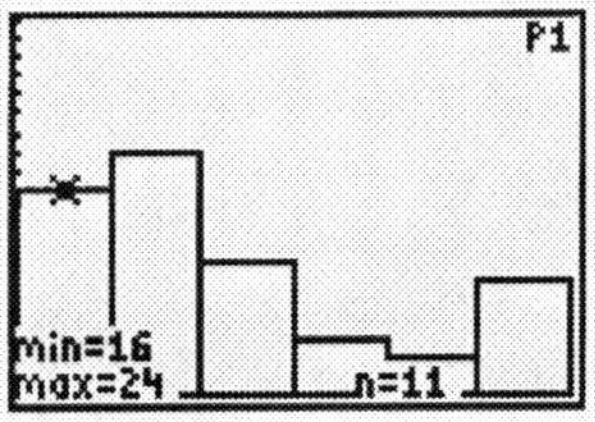

FIGURE 4.37
The first bar's boundaries are 16 and 24; its frequency is 11.

DRAWING A HISTOGRAM ON A CASIO:

- Press GRPH (i.e., F1).
- Press SET (i.e., F6).
- Make the resulting screen, which is labeled "StatGraph1," read as follows:

Graph Type	:Hist
Xlist	:List1
Frequency	: 1

  To make the screen read as described above, do the following:
  - Use the down arrow button to scroll down to "Graph Type."
  - Press F6.
  - Press HIST (i.e., F1).
  - In a similar manner, change "Xlist" and "Frequency" if necessary.

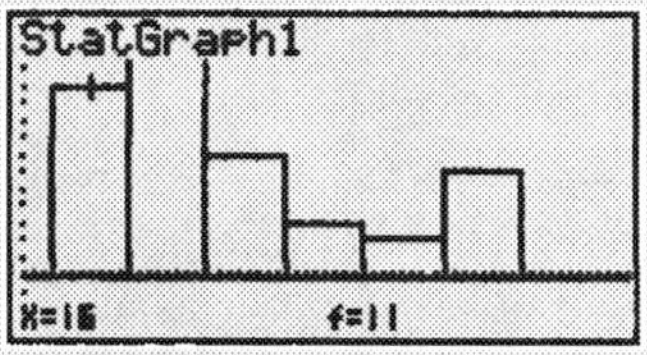

FIGURE 4.38

The first bar's left boundary is 16; its frequency is 11.

- Press EXE and return to List1.
- Press GPH1 (i.e., F1) and make the resulting screen, which is labeled "Set Interval," read as follows:

  Start: 16 "Start" refers to the beginning of the first group

  Pitch: 8 "Pitch" refers to the width of each group

- Press DRAW (i.e., F6), and the calculator will display the histogram.
- Press SHIFT and then TRCE (i.e., F1) to find out the left boundaries and the frequency of the bars. Use the arrow buttons to move from bar to bar. See Figure 4.38.

Histograms and Pie Charts on a Computerized Spreadsheet

A **spreadsheet** is a large piece of paper marked off in rows and columns. Accountants use spreadsheets to organize numerical data and perform computations. A **computerized spreadsheet,** such as Microsoft Excel, is a computer program that mimics the appearance of a paper spreadsheet. It frees the user from performing any computations; instead, it allows the user merely to give instructions on how to perform those computations. The instructions in this subsection were specifically written for Microsoft Excel; however, all computerized spreadsheets work somewhat similarly.

When you start a computerized spreadsheet, you see something that looks like a table waiting to be filled in. The rows are labeled with numbers and the columns with letters, as shown in Figure 4.39.

◇	A	B	C	D
1				
2				
3				
4				
5				

FIGURE 4.39 A blank spreadsheet.

The individual boxes are called **cells.** The cell in column A row 1 is called cell A1; the cell below it is called cell A2, because it is in column A row 2.

A computerized spreadsheet is an ideal tool to use in creating a histogram or a pie chart. We will illustrate this process by preparing both a histogram and a pie chart for the ages of the students in Keith Reed's acting class, as discussed in Example 2.

Entering the Data

1. *Label the columns.* Use the mouse and/or the arrow buttons to move to cell A1, type in "age of student" and press "return" or "enter." (If there were other data, we could

enter it in other columns. For example, if the students' names were included, we could type "name of student" in cell A1, and "age of student" in cell B1.)

2. *Enter the students' ages in column A.* Move to cell A2, type in "26" and press "return" or "enter." Move to cell A3, type in "16" and press "return" or "enter." In a similar manner, enter all of the ages. You can enter the ages in any order. After you complete this step, your spreadsheet should look like that in Figure 4.40 (except that it should go down a lot further).

◇	A	B	C	D
1	age of student			
2	26			
3	16			
4	21			
5	34			

FIGURE 4.40 The spreadsheet after entering the students' ages.

3. *Save the spreadsheet.* Use your mouse to select "File" at the very top of the screen. Then pull your mouse down until "Save As" is highlighted, and let go. Your instructor may give you further instructions on where and how to save your spreadsheet.

Preparing the Data for a Chart

◇	A	B
1	age of student	bin numbers
2	26	23
3	16	31
4	21	39
5	34	47
6	45	55
7	18	63
8	41	

FIGURE 4.41

The spreadsheet after entering the bin numbers.

1. *Enter the group boundaries.* Excel uses "bin numbers" rather than group boundaries. A group's bin number is the highest number that should be included in that group. In Example 2, the first group was $16 \leq x < 24$. The highest age that should be included in this group is 23, so the group's bin number is 23. Be careful; this group's bin number is not 24, since people who are 24 years old should be included in the second group.
   - If necessary, use the up arrow in the upper-right corner of the spreadsheet to scroll to the top 6of the spreadsheet.
   - Type "bin numbers" in cell B1.
   - Determine each group's bin number, and enter them in column B.

   After you complete this step, your spreadsheet should look like that in Figure 4.41. You might need to adjust the width of column A. Click on the right edge of the "A" label at the top of the first column, and move it.

2. *Have Excel determine the frequencies.*
   - Use your mouse to select "Tools" at the very top of the screen. Then pull your mouse down until "Data Analysis" is highlighted, and let go.
   - If "Data Analysis" is not listed under "Tools," then
     - Select "Tools" at the top of the screen, pull down until "Add-Ins" is highlighted, and let go.
     - Select "Analysis ToolPak-VBA."
     - Use your mouse to press the "OK" button.
     - Select "Tools" at the top of the screen, pull down until "Data Analysis" is highlighted, and let go.
   - In the "Data Analysis" box that appears, use your mouse to highlight "Histogram" and press the "OK" button.
   - In the "Histogram" box that appears, use your mouse to click on the white rectangle that follows "Input Range," and then use your mouse to draw a box around all of the ages. (To draw the box, move your mouse to cell A2, press the mouse button, and move the mouse down until all of the entered ages are enclosed in a box.) This should cause "$A$2:$A$43" to appear in the Input Range rectangle.

- Use your mouse to click on the white rectangle that follows "Bin Range," and then use your mouse to draw a box around all of the bin numbers. This should cause "\$B\$2:\$B\$7" to appear in the Bin Range rectangle.
- Be certain that there is a dot in the button to the left of "Output Range." Then click on the white rectangle that follows "Output Range," and use your mouse to click on cell C1. This should cause "\$C\$1" to appear in the Output Range rectangle.
- Use your mouse to press the "OK" button.

After you complete this step, your spreadsheet should look like that in Figure 4.42.

◇	A	B	C	D
1	age of student	bin numbers	*Bin*	*Frequency*
2	26	23	23	11
3	16	31	31	13
4	21	39	39	7
5	34	47	47	3
6	45	55	55	2
7	18	63	63	6
8	41		More	0
9	38			

FIGURE 4.42 The spreadsheet after Excel determines the frequencies.

3. *Prepare labels for the chart.*
   - In column E, list the group boundaries.
   - In column F, list the frequencies.

   After you complete this step, your spreadsheet should look like that in Figure 4.43 (the first few columns are not shown).

C	D	E	F
Bin	*Frequency*	group boundaries	frequency
23	11	16≤x<24	11
31	13	24≤x<32	13
39	7	32≤x<40	7
47	3	40≤x<48	3
55	2	48≤x<56	2
63	6	56≤x<64	6
More	0		

FIGURE 4.43 The spreadsheet after preparing labels.

Drawing a Histogram

1. *Use the Chart Wizard to draw a bar chart.*
   - Use your mouse to press the "Chart Wizard" button at the top of the spreadsheet. (It looks like a histogram and it might have a magic wand.)
   - Select "Column" and then an appropriate style.

- Press the "Next" button.
- Use your mouse to click on the white rectangle that follows "Data range," and then use your mouse to draw a box around all of the group boundaries and frequencies from step 3 in "Preparing the data for a chart". This should cause "=Sheet1!$E$2:$F$7" to appear in the Data range rectangle.
- Press the "Next" button.
- Under "Chart Title" type an appropriate title, such as "Acting class ages."
- After "Category (X)" type an appropriate title for the *x*-axis, such as "Students' ages."
- After "Value (Y)" type an appropriate title for the *y*-axis, such as "frequencies."
- Press "Legend" at the top of the chart options box, and then remove the check mark next to "show legend."
- Press the "Finish" button and the bar chart will appear. See Figure 4.44.

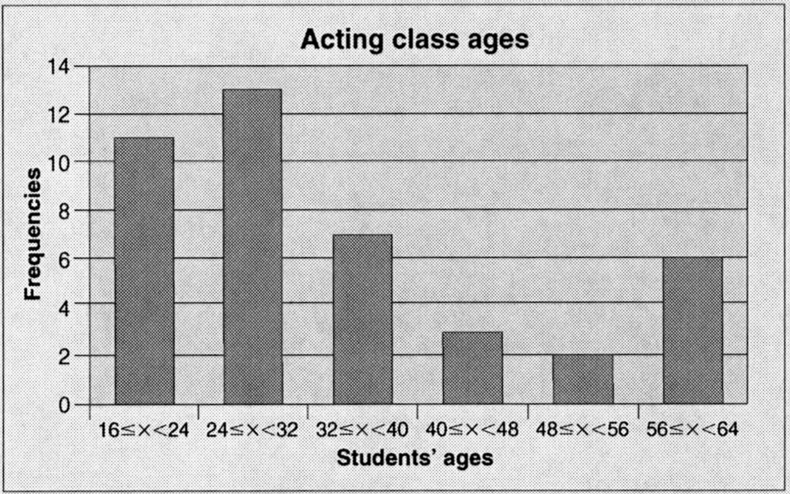

FIGURE 4.44 The bar chart.

2. *Save the spreadsheet.* Use your mouse to select "File" at the very top of the screen. Then pull your mouse down until "Save" is highlighted, and let go.
3. *Convert the bar graph to a histogram.* The graph is not a histogram because of the spaces between the bars.
   - Double click or right click on the bar graph and a "Format Data Series" box will appear.
   - Press the "Options" tab.
   - Remove the spaces between the bars by changing the "Gap width" to 0.
   - Press OK.
   - Save the spreadsheet. See Figure 4.45.
4. *Print the histogram.*
   - Click on the histogram, and it will be surrounded by a thicker border than before.
   - Use your mouse to select "File" at the very top of the screen. Then pull your mouse down until "Print" is highlighted, and let go.
   - Respond appropriately to the "Print" box that appears.

Drawing a Pie Chart

1. *Use the Chart Wizard to draw the pie chart.*
   - Use your mouse to press the "Chart Wizard" button at the top of the spreadsheet. (It looks like a histogram and it might have a magic wand.)

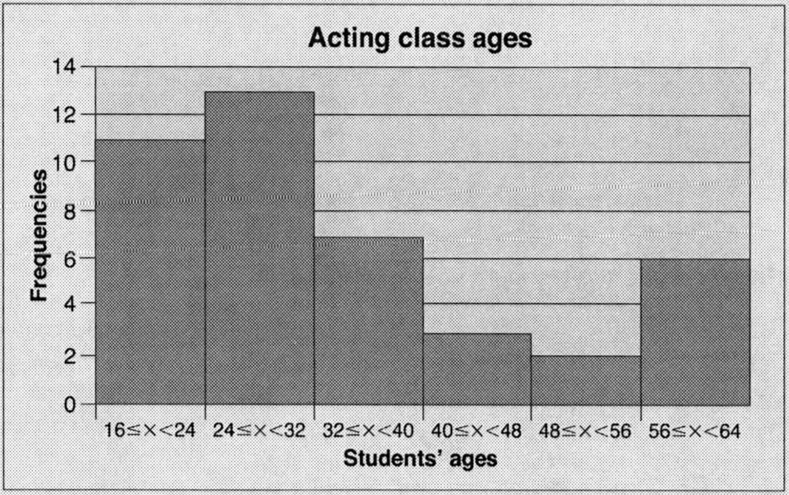

FIGURE 4.45 The histogram.

- Select "Pie" and then an appropriate style.
- Press the "Next" button.
- Use your mouse to click on the white rectangle that follows "Data range," and then use your mouse to draw a box around all of the group boundaries and frequencies from step 3 in "Preparing the data for a chart." This should cause "=Sheet1!$E$2:$F$7" to appear in the Data range rectangle.
- Press the "Next" button.
- Under "Chart Title" type an appropriate title.
- Press the "Finish" button and the pie chart will appear. See Figure 4.46.

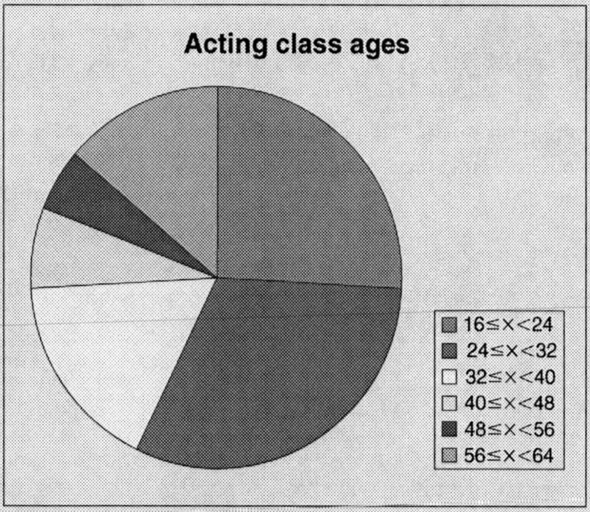

FIGURE 4.46 The pie chart.

2. *Save the spreadsheet.* Use your mouse to select "File" at the very top of the screen. Then pull your mouse down until "Save" is highlighted, and let go.
3. *Print the pie chart.*
   - Click on the chart, and it will be surrounded by a thicker border than before.
   - Use your mouse to select "File" at the very top of the screen. Then pull your mouse down until "Print" is highlighted, and let go.
   - Respond appropriately to the "Print" box that appears.

EXERCISES

Use a graphing calculator or Excel on the following exercises. Answers will vary depending on your choice of groups.

Graphing calculator instructions: *By hand, copy a graphing calculator's histogram onto paper for submission.*

Excel instructions: *Print out a histogram for submission.*

28. Do Double Stuf Oreos really contain twice as much "stuf"? In 1999, Marie Revak and Jihan William's "interest was piqued by the unusual spelling of the word stuff and the bold statement 'twice the filling' on the package." So they conducted an experiment in which they weighed the amount of filling in a sample of traditional Oreos and Double Stuf Oreos. The data from that experiment are in Figure 4.47 (weight in grams). An unformatted spreadsheet containing these data can be downloaded from the text web site: **www.cengage.com/math/johnson**

Forsling Research

a. Create one histogram for Double Stuf Oreos and one for Traditional Oreos. Use the same categories for each histogram.

b. Compare the results of part (a). What do they tell you about the two cookies?

29. Every year, CNNMoney.com lists "the best places to live." They compare the "best small towns" on several bases, one of which is the "quality of life." See Figure 4.48. An unformatted spreadsheet containing these data can be downloaded from the text web site: **www.cengage.com/math/johnson**

a. Create a histogram to represent the data for the air quality index.

b. What do the results say about these towns?

30. Every year, CNNMoney.com lists "the best places to live." They compare the "best small towns" on several bases, one of which is the "quality of life." See Figure 4.48. An unformatted spreadsheet containing these data can be downloaded from the text web site: **www.cengage.com/math/johnson**

a. Create a histogram to represent the data for the mean commute time.

b. What do the results say about these towns?

31. Every year, CNNMoney.com lists "the best places to live." They compare the "best small towns" on several bases, one of which is the "quality of life." See Figure 4.48. An unformatted spreadsheet containing these data can be downloaded from the text web site: **www.cengage.com/math/johnson**

a. Create a histogram to represent the data for the property crime levels.

b. What do the results say about these towns?

Double Stuf Oreos:																		
4.7	6.5	5.5	5.6	5.1	5.3	5.4	5.4	3.5	5.5	6.5	5.9	5.4	4.9	5.6	5.7	5.3	6.9	6.5
6.3	4.8	3.3	6.4	5.0	5.3	5.5	5.0	6.0	5.7	6.3	6.0	6.3	6.1	6.0	5.8	5.8	5.9	6.2
5.9	6.5	6.5	6.1	5.8	6.0	6.2	6.2	6.0	6.8	6.2	5.4	6.6	6.2					
Traditional Oreos:																		
2.9	2.8	2.6	3.5	3.0	2.4	2.7	2.4	2.5	2.2	2.6	2.6	2.9	2.6	2.6	3.1	2.9	2.4	2.8
3.8	3.1	2.9	3.0	2.1	3.8	3.0	3.0	2.8	2.9	2.7	3.2	2.8	3.1	2.7	2.8	2.6	2.6	3.0
2.8	3.5	3.3	3.3	2.8	3.1	2.6	3.5	3.5	3.1	3.1								

FIGURE 4.47 Amount of Oreo filling. *Source:* Revak, Marie A. and Jihan G. Williams, "Sharing Teaching Ideas: The Double Stuf Dilemma," *The Mathematics Teacher,* November 1999, Volume 92, Issue 8, page 674.

Rank	City	Air quality index (% days AQI ranked good)	Property crime—incidents per 1000	Median commute time (mins)
1	Louisville, CO	74.0%	15	19.5
2	Chanhassen, MN	N.A.	16	23.1
3	Papillion, NE	92.0%	16	19.3
4	Middleton, WI	76.0%	27	16.3
5	Milton, MA	81.0%	9	27.7
6	Warren, NJ	N.A.	21	28.7
7	Keller, TX	65.0%	14	29.4
8	Peachtree City, GA	84.0%	12	29.3
9	Lake St. Louis, MO	79.0%	16	26.8
10	Mukilteo, WA	89.0%	27	22.3

FIGURE 4.48 The best places to live.

32. Figure 4.49 gives EPA fuel efficiency ratings for 2010 family sedans with an automatic transmission and the smallest available engine. An unformatted spreadsheet containing these data can be downloaded from the text web site: **www.cengage.com/math/johnson**

a. For each of the two manufacturing regions, create a histogram for city driving mileage. Use the same categories for the two regions.

b. What conclusions can you draw?

33. Figure 4.49 gives EPA fuel efficiency ratings for 2010 family sedans with an automatic transmission and the smallest available engine. An unformatted spreadsheet containing these data can be downloaded from the text web site: **www.cengage.com/math/johnson**

a. For each of the two manufacturing regions, create a histogram for highway driving mileage. Use the same categories for the two regions.

b. What conclusions can you draw?

American cars	city mpg	highway mpg	Asian cars	city mpg	highway mpg
Ford Fusion hybrid	41	36	Toyota Prius	51	48
Chevrolet Malibu hybrid	26	34	Nissan Altima hybrid	35	33
Ford Fusion fwd	24	34	Toyota Camry Hybrid	33	34
Mercury Milan	23	34	Kia Forte	27	36
Chevrolet Malibu	22	33	Hyundai Elantra	26	35
Saturn Aura	22	33	Nissan Altima	23	32
Dodge Avenger	21	30	Subaru Legacy awd	23	31
Buick Lacrosse	17	27	Toyota Camry	22	33
			Hyundai Sonata	22	32
			Kia Optima	22	32
			Honda Accord	22	31
			Mazda 6	21	30
			Mitsubishi Galant	21	30
			Hyunda Azera	18	26

FIGURE 4.49 Fuel efficiency. Source: www.fueleconomy.gov

4.2 Measures of Central Tendency

Objectives

- Find the mean
- Find the median
- Find the mode

Who is the best running back in professional football? How much does a typical house cost? What is the most popular television program? The answers to questions like these have one thing in common: They are based on averages. To compare the capabilities of athletes, we compute their average performances. This computation usually involves the ratio of two totals, such as (total yards gained)/(total number of carries) = average gain per carry. In real estate, the average-price house is found by listing the prices of all houses for sale (from lowest to highest) and selecting the price in the middle. Television programs are rated by the average number of households tuned in to each particular program.

Rather than listing every data point in a large distribution of numbers, people tend to summarize the data by selecting a representative number, calling it the average. Three figures—the *mean,* the *median,* and the *mode*—describe the "average" or "center" of a distribution of numbers. These averages are known collectively as the **measures of central tendency.**

The Mean

© Reuters/Corbis

In 2001, Barry Bonds of the San Francisco Giants hit 73 home runs and set an all-time record in professional baseball. On average, how many home runs per season did Bonds hit? See Exercise 13.

The **mean** is the average people are most familiar with; it can also be the most misleading. Given a collection of n data points, $x_1, x_2, \ldots, x_n$, the mean is found by adding up the data and dividing by the number of data points:

$$\text{mean of } n \text{ data points} = \frac{x_1 + x_2 + \cdots + x_n}{n}$$

If the data are collected from a sample, then the mean is denoted by $\bar{x}$ (read "x bar"); if the data are collected from an entire population, then the mean is denoted by μ (lowercase Greek letter "mu"). Unless stated otherwise, we will assume that the data represent a sample, so the mean will be symbolized by $\bar{x}$.

Mathematicians have developed a type of shorthand, called **summation notation,** to represent the sum of a collection of numbers. The Greek letter Σ ("sigma") corresponds to the letter S and represents the word *sum.* Given a group of data points $x_1, x_2, \ldots, x_n$, we use the symbol Σx to represent their sum; that is, $\Sigma x = x_1 + x_2 + \cdots + x_n$.

DEFINITION OF THE MEAN

Given a sample of n data points, $x_1, x_2, \ldots, x_n$, the **mean,** denoted by $\bar{x}$, is

$$\bar{x} = \frac{\Sigma x}{n} \quad \text{or} \quad \bar{x} = \frac{\text{the sum of the data points}}{\text{the number of data points}}$$

Many scientific calculators have statistical functions built into them. These functions allow you to enter the data points and press the "x-bar" button to obtain the mean. Consult your manual for specific instructions.

EXAMPLE 1

FINDING THE MEAN: SINGLE VALUES AND MIXED UNITS In 2005, Lance Armstrong won his seventh consecutive Tour de France bicycle race. No one in the 100-year history of the race has won so many times. (Miguel Indurain of Spain won five times from 1991 to 1995.) Lance's winning times are given in Figure 4.50. Find the mean winning time of Lance Armstrong's Tour de France victory rides.

Year	1999	2000	2001	2002	2003	2004	2005
Time (h:m:s)	91:32:16	92:33:08	86:17:28	82:05:12	83:41:12	83:36:02	86:15:02

FIGURE 4.50 Lance Armstrong's Tour de France winning times. *Source: San Francisco Chronicle.*

SOLUTION

To find the mean, we must add up the data and divide by the number of data points. However, before we can sum the data, they must be converted to a common unit, say, minutes. Therefore, we multiply the number of hours by 60 (minutes per hour), divide the number of seconds by 60 (seconds per minute), and add each result to the number of minutes. Lance Armstrong's 1999 winning time of 91 hours, 32 minutes, 16 seconds is converted to minutes as follows:

$$\left(91\text{ hours} \times \frac{60\text{ minutes}}{1\text{ hour}}\right) + 32\text{ minutes} + \left(16\text{ seconds} \times \frac{1\text{ minute}}{60\text{ seconds}}\right)$$

$$= 5{,}460\text{ minutes} + 32\text{ minutes} + 0.266666\ldots\text{ minutes}$$

$$= 5{,}492.267\text{ minutes} \quad \text{rounding off to three decimal places}$$

In a similar fashion, all of Lance Armstrong's winning times are converted to minutes, and the results are given in Figure 4.51.

Year	1999	2000	2001	2002	2003	2004	2005
Time (min)	5492.267	5553.133	5177.467	4925.200	5021.200	5016.033	5175.033

FIGURE 4.51 Winning time in minutes.

$$\bar{x} = \frac{\Sigma x}{n}$$

$$= \frac{5492.267 + 5553.133 + 5177.467 + 4925.200 + 5021.200 + 5016.033 + 5175.033}{7}$$

$$= \frac{36360.333}{7}$$

$$= 5194.333\text{ minutes}$$

Converting back to hours, minutes, and seconds, we find that Lance Armstrong's mean winning time is 86 hours, 34 minutes, 20 seconds.

THE FAR SIDE® By GARY LARSON

"Bob and Ruth! Come on in. ... Have you met Russell and Bill, our 1.5 children?"

If a sample of 460 families have 690 children altogether, then the mean number of children per family is $\bar{x} = 1.5$.

EXAMPLE 2

FINDING THE MEAN: GROUPED DATA In 2008, the U.S. Bureau of Labor Statistics tabulated a survey of workers' ages and wages. The frequency distribution in Figure 4.52 summarizes the age distribution of workers who received minimum wage ($6.55 per hour). Find the mean age of a worker receiving minimum wage.

y = Age (years)	Number of Workers
$16 \le y < 20$	108,000
$20 \le y < 25$	53,000
$25 \le y < 35$	41,000
$35 \le y < 45$	23,000
$45 \le y < 55$	29,000
$55 \le y < 65$	23,000
	$n = 277{,}000$

FIGURE 4.52 Age distribution of workers receiving minimum wage.
Source: Bureau of Labor Statistics, U.S. Department of Labor.

SOLUTION

To find the mean age of the workers, we must add up the ages of all the workers and divide by 277,000. However, because we are given grouped data, the ages of the individual workers are unknown to us. In this situation, we use the midpoint of each interval as the representative of the interval; consequently, our answer is an approximation.

To find the midpoint of an interval, add the endpoints and divide by 2. For instance, the midpoint of the interval $16 \le y < 20$ is

$$\frac{16 + 20}{2} = 18$$

We can then say that each of the 108,000 people in the first interval is approximately eighteen years old. Adding up these workers' ages, we obtain

$$\begin{aligned} &18 + 18 + 18 + \cdots + 18 \quad \text{(one hundred eight thousand times)} \\ &= (108{,}000)(18) \\ &= 1{,}944{,}000 \text{ years} \end{aligned}$$

If we let f = frequency and x = the midpoint of an interval, the product $f \cdot x$ gives us the total age of the workers in an interval. The results of this procedure for all the workers are shown in Figure 4.53.

y = Age (years)	f = Frequency	x = Midpoint	$f \cdot x$
$16 \le y < 20$	108,000	18	1,944,000
$20 \le y < 25$	53,000	22.5	1,192,500
$25 \le y < 35$	41,000	30	1,230,000
$35 \le y < 45$	23,000	40	920,000
$45 \le y < 55$	29,000	50	1,450,000
$55 \le y < 65$	23,000	60	1,380,000
	$n = 277{,}000$		$\Sigma(f \cdot x) = 8{,}116{,}500$

FIGURE 4.53 Using midpoints of grouped data.

Because $f \cdot x$ gives the sum of the ages of the workers in an interval, the symbol $\Sigma(f \cdot x)$ represents the sum of all the $(f \cdot x)$; that is, $\Sigma(f \cdot x)$ represents the sum of the ages of *all* workers.

The mean age is found by dividing the sum of the ages of all the workers by the number of workers:

$$\begin{aligned} \bar{x} &= \frac{\Sigma(f \cdot x)}{n} \\ &= \frac{8{,}116{,}500}{277{,}000} \\ &= 29.30144 \text{ years} \end{aligned}$$

The mean age of the workers earning minimum wage is approximately 29.3 years.

> One common mistake that is made in working with grouped data is to forget to multiply the midpoint of an interval by the frequency of the interval. Another common mistake is to divide by the number of intervals instead of by the total number of data points.

The procedure for calculating the mean when working with grouped data (as illustrated in Example 2) is summarized in the following box.

CALCULATING THE MEAN: GROUPED DATA

Given a frequency distribution containing several groups of data, the mean $\bar{x}$ can be found by using the following formula:

$\bar{x} = \frac{\Sigma(f \cdot x)}{n}$ where x = the midpoint of a group, f = the frequency of the group, and $n = \Sigma f$

EXAMPLE 3

FINDING THE MEAN: REPEATED DATA Ten college students were comparing their wages earned at part-time jobs. Nine earned \$10.00 per hour working at jobs ranging from waiting on tables to working in a bookstore. The tenth student earned \$200.00 per hour modeling for a major fashion magazine. Find the mean wage of the ten students.

SOLUTION

The data point \$10.00 is repeated nine times, so we multiply it by its frequency. Thus, the mean is as follows:

$$\begin{aligned}\bar{x} &= \frac{\Sigma(f \cdot x)}{n} \\ &= \frac{(9 \cdot 10) + (1 \cdot 200)}{10} \\ &= \frac{290}{10} \\ &= 29\end{aligned}$$

The mean wage of the students is \$29.00 per hour.

Example 3 seems to indicate that the average wage of the ten students is \$29.00 per hour. Is that a reasonable figure? If nine out of ten students earn \$10.00 per hour, can we justify saying that their average wage is \$29.00? Of course not! Even though the mean wage *is* \$29.00, it is not a convincing "average" for this specific group of data. The mean is inflated because one student made \$200.00 per hour. This wage is called an **outlier** (or **extreme value**) because it is significantly different from the rest of the data. Whenever a collection of data has extreme values, the mean can be greatly affected and might not be an accurate measure of the average.

The Median

The **median** is the "middle value" of a distribution of numbers. To find it, we first put the data in numerical order. (If a number appears more than once, we include it as many times as it occurs.) If there is an odd number of data points, the median is the middle data point; if there is an even number of data points, the median is defined to be the mean of the two middle values. In either case, the median separates the distribution into two equal parts. Thus, the median can be viewed as an "average." (The word *median* is also used to describe the strip that runs down the middle of a freeway; half the freeway is on one side, and half is on the other. This common usage is in keeping with the statistical meaning.)

EXAMPLE 4

FINDING THE MEDIAN: ODD VERSUS EVEN NUMBERS OF DATA

Find the median of the following sets of raw data.

a. 2 8 3 12 6 2 11
b. 2 8 3 12 6 2 11 8

SOLUTION

a. First, we put the data in order from smallest to largest. Because there is an odd number of data points, ($n = 7$), we pick the middle one:

2 2 3 6 8 11 12
↑
middle value

The median is 6. (Notice that 6 is the fourth number in the list.)

b. We arrange the data first. Because there is an even number of data points ($n = 8$), we pick the two middle values and find their mean:

2 2 3 6 8 8 11 12
↑ ↑

$$\frac{(6 + 8)}{2} = 7$$

Therefore, the median is 7. (Notice that 7 is halfway between the fourth and fifth numbers in the list.)

In Example 4, we saw that when $n = 7$, the median was the fourth number in the list, and that when $n = 8$, the median was halfway between the fourth and fifth numbers in the list. Consequently, the *location* of the median depends on n, the number of numbers in the set of data. The formula

$$L = \frac{n + 1}{2}$$

can be used to find the location, L, of the median; when $n = 7$,

$$L = \frac{7 + 1}{2} = 4$$

(the median is the fourth number in the list), and when $n = 8$,

$$L = \frac{8 + 1}{2} = 4.5$$

(the median is halfway between the fourth and fifth numbers in the list).

LOCATION OF THE MEDIAN

Given a sample of n data points, the location of the median can be found by using the following formula:

$$L = \frac{n + 1}{2}$$

Once the data have been arranged from smallest to largest, the median is the Lth number in the list.

EXAMPLE 5

FINDING THE LOCATION AND VALUE OF THE MEDIAN Find the median wage for the ten students in Example 3.

SOLUTION

First, we put the ten wages in order:

10 10 10 10 10 10 10 10 10 200

Because there are $n = 10$ data points, the location of the median is

$$L = \frac{10 + 1}{2} = 5.5$$

That is, the median is halfway between the fifth and sixth numbers. To find the median, we add the fifth and sixth numbers and divide by 2. Now, the fifth number is 10 and the sixth number is 10, so the median is

$$\frac{10 + 10}{2} = 10$$

Therefore, the median wage is $10.00. This is a much more meaningful "average" than the mean of $29.00.

If a collection of data contains extreme values, the median, rather than the mean, is a better indicator of the "average" value. For instance, in discussions of real estate, the median is usually used to express the "average" price of a house. (Why?) In a similar manner, when the incomes of professionals are compared, the median is a more meaningful representation. The discrepancy between mean (average) income and median income is illustrated in the following news article. Although the article is dated (it's from 1992), it is very informative as to the differences between the mean and the median.

FEATURED IN THE NEWS

DOCTORS' AVERAGE INCOME IN U.S. IS NOW $177,000

Washington—The average income of the nation's physicians rose to $177,400 in 1992, up 4 percent from the year before, the American Medical Association said yesterday. . . .

The average income figures, compiled annually by the AMA and based on a telephone survey of more than 4,100 physicians, ranged from a low of $111,800 for general practitioners and family practice doctors to a high of $253,300 for radiologists.

Physicians' median income was $148,000 in 1992, or 6.5 percent more than a year earlier. Half the physicians earned more than that and half earned less.

The average is pulled higher than the median by the earnings of the highest paid surgeons, anesthesiologists and other specialists at the top end of the scale. . . .

Here are the AMA's average and median net income figures by specialty for 1992:

General/family practice:
$111,800, $100,000.
Internal medicine:
$159,300, $130,000.
Surgery: $244,600, $207,000.
Pediatrics: $121,700, $112,000.
Obstetrics/gynecology:
$215,100, $190,000.
Radiology: $253,300, $240,000.
Psychiatry: $130,700, $120,000.
Anesthesiology:
$228,500, $220,000.
Pathology: $189,800, $170,000.
Other: $165,400, $150,000.

EXAMPLE 6

FINDING THE LOCATION AND VALUE OF THE MEDIAN The students in Ms. Kahlo's art class were asked how many siblings they had. The frequency distribution in Figure 4.54 summarizes the responses. Find the median number of siblings.

Number of Siblings	Number of Responses
0	2
1	8
2	5
3	6

FIGURE 4.54 Frequency distribution of data.

SOLUTION

The frequency distribution indicates that two students had no (0) siblings, eight students had one (1) sibling, five students had two (2) siblings, and six students had three (3) siblings. Therefore, there were $n = 2 + 8 + 5 + 6 = 21$ students in the class; consequently, there are twenty-one data points. Listing the data in order, we have

$$0, 0, 1, \ldots 1, 2, \ldots 2, 3, \ldots 3$$

Because there are $n = 21$ data points, the location of the median is

$$L = \frac{21 + 1}{2} = 11$$

That is, the median is the eleventh number. Because the first ten numbers are 0s and 1s ($2 + 8 = 10$), the eleventh number is a 2. Consequently, the median number of siblings is 2.

The Mode

The third measure of central tendency is the **mode.** The mode is the most frequent number in a collection of data; that is, it is the data point with the highest frequency. Because it represents the most common number, the mode can be viewed as an average. A distribution of data can have more than one mode or none at all.

EXAMPLE 7

FINDING THE MODE Find the mode(s) of the following sets of raw data:

a. 4 10 1 8 5 10 5 10
b. 4 9 1 10 1 10 4 9
c. 9 6 1 8 3 10 3 9

SOLUTION

a. The mode is 10, because it has the highest frequency (3).
b. There is no mode, because each number has the same frequency (2).
c. The distribution has two modes—namely, 3 and 9—each with a frequency of 2. A distribution that has two modes is called *bimodal.*

In summarizing a distribution of numbers, it is most informative to list all three measures of central tendency. It helps to avoid any confusion or misunderstanding in situations in which the word *average* is used. In the hands of someone with questionable intentions, numbers can be manipulated to mislead people. In his book *How to Lie with Statistics,* Darrell Huff states, "The secret language of statistics, so appealing in a fact-minded culture, is employed to sensationalize, inflate, confuse, and oversimplify. Statistical methods and statistical terms are necessary in reporting the mass data of social and economic trends, business conditions, 'opinion' polls, the census. But without writers who use the words with honesty and understanding, and readers who know what they mean, the result can only be semantic nonsense." An educated public should always be on the alert for oversimplification of data via statistics. The next time someone mentions an "average," ask "Which one?" Although people might not intentionally try to mislead you, their findings can be misinterpreted if you do not know the meaning of their statistics and the method by which the statistics were calculated.

4.2 EXERCISES

In Exercises 1–4, find the mean, median, and mode of the given set of raw data.

1. 9 12 8 10 9 11 12
15 20 9 14 15 21 10

2. 20 25 18 30 21 25 32 27
32 35 19 26 38 31 20 23

3. 1.2 1.8 0.7 1.5 1.0 0.7 1.9 1.7 1.2
0.8 1.7 1.3 2.3 0.9 2.0 1.7 1.5 2.2

4. 0.07 0.02 0.09 0.04 0.10 0.08 0.07 0.13
0.05 0.04 0.10 0.07 0.04 0.01 0.11 0.08

5. Find the mean, median, and mode of each set of data.
a. 9 9 10 11 12 15
b. 9 9 10 11 12 102
c. How do your answers for parts (a) and (b) differ (or agree)? Why?

6. Find the mean, median, and mode for each set of data.
a. 80 90 100 110 110 140
b. 10 90 100 110 110 210
c. How do your answers for parts (a) and (b) differ (or agree)? Why?

7. Find the mean, median, and mode of each set of data.
a. 2 4 6 8 10 12
b. 102 104 106 108 110 112
c. How are the data in part (b) related to the data in part (a)?
d. How do your answers for parts (a) and (b) compare?

8. Find the mean, median, and mode of each set of data.
a. 12 16 20 24 28 32
b. 600 800 1,000 1,200 1,400 1,600
c. How are the data in part (b) related to the data in part (a)?
d. How do your answers for parts (a) and (b) compare?

9. Kaitlin Mowry is a member of the local 4-H club and has six mini Rex rabbits that she enters in regional rabbit competition shows. The weights of the rabbits are given in Figure 4.55. Find the mean, median, and mode of the rabbits' weights.

Weight (lb:oz)	3:12	4:03	3:06	3:15	3:12	4:02

FIGURE 4.55 Weights of rabbits.

10. As was stated in Example 1, Lance Armstrong won the Tour de France bicycle race every year from 1999 to 2005. His margins of victory (time difference of the second place finisher) are given in Figure 4.56. Find the mean, median, and mode of Lance Armstrong's victory margins.

Year	1999	2000	2001	2002	2003	2004	2005
Margin (m:s)	7:37	6:02	6:44	7:17	1:01	6:19	4:40

FIGURE 4.56 Lance Armstrong's Tour de France victory margins. *Source: San Francisco Chronicle.*

11. Jerry Rice holds the all-time record in professional football for scoring touchdowns. The number of touchdown receptions (TDs) for each of his seasons is given in Figure 4.57. Find the mean, median, and mode of the number of touchdown receptions per year by Rice.

Year	TDs	Year	TDs
1985	3	1995	15
1986	15	1996	8
1987	22	1997	1
1988	9	1998	9
1989	17	1999	5
1990	13	2000	7
1991	14	2001	9
1992	10	2002	7
1993	15	2003	2
1994	13	2004	3

FIGURE 4.57 Touchdown receptions for Jerry Rice.
Source: http://sportsillustrated.cnn.com/football/nfl/players/.

12. Wayne Gretzky, known as "The Great One," holds the all-time record in professional hockey for scoring goals. The number of goals for each of his seasons is given in Figure 4.58. Find the mean, median, and mode of the number of goals per season by Gretzky.

Season	Goals	Season	Goals
1979–80	51	1989–90	40
1980–81	55	1990–91	41
1981–82	92	1991–92	31
1982–83	71	1992–93	16
1983–84	87	1993–94	38
1984–85	73	1994–95	11
1985–86	52	1995–96	23
1986–87	62	1996–97	25
1987–88	40	1997–98	23
1988–89	54	1998–99	9

FIGURE 4.58 Goals made by Wayne Gretzky.
Source: The World Almanac.

13. Barry Bonds of the San Francisco Giants set an all-time record in professional baseball by hitting 73 home runs in one season (2001). The number of home runs (HR) for each of his seasons in professional baseball is given in Figure 4.59. Find the mean, median, and mode of the number of home runs hit per year by Bonds.

Year	HR	Year	HR
1986	16	1997	40
1987	25	1998	37
1988	24	1999	34
1989	19	2000	49
1990	33	2001	73
1991	25	2002	46
1992	34	2003	45
1993	46	2004	45
1994	37	2005	5
1995	33	2006	26
1996	42	2007	28

FIGURE 4.59 Home runs hit by Barry Bonds.
Source: http://sportsillustrated.cnn.com/baseball/mlb/players/.

14. Michael Jordan has been recognized as an extraordinary player in professional basketball, especially in terms of the number of points per game he has scored. Jordan's average number of points per game (PPG) for each of his seasons is given in Figure 4.60. Find the mean, median, and mode of the average points per game made per season by Jordan.

Season	PPG	Season	PPG
1984–85	28.2	1992–93	32.6
1985–86	22.7	1994–95	26.9
1986–87	37.1	1995–96	30.4
1987–88	35.0	1996–97	29.6
1988–89	32.5	1997–98	28.7
1989–90	33.6	2001–02	22.9
1990–91	31.5	2002–03	20.0
1991–92	30.1		

FIGURE 4.60 Points per game made by Michael Jordan.
Source: http://www.nba.com/playerfile/.

15. The frequency distribution in Figure 4.61 lists the results of a quiz given in Professor Gilbert's statistics class. Find the mean, median, and mode of the scores.

Score	Number of Students
10	3
9	10
8	9
7	8
6	10
5	2

FIGURE 4.61 Quiz scores in Professor Gilbert's statistics class.

16. Todd Booth, an avid jogger, kept detailed records of the number of miles he ran per week during the past year. The frequency distribution in Figure 4.62 summarizes his records. Find the mean, median, and mode of the number of miles per week that Todd ran.

Miles Run per Week	Number of Weeks
0	5
1	4
2	10
3	9
4	10
5	7
6	3
7	4

FIGURE 4.62 Miles run by Todd Booth.

17. To study the output of a machine that fills boxes with cereal, a quality control engineer weighed 150 boxes of Brand X cereal. The frequency distribution in Figure 4.63 summarizes his findings. Find the mean weight of the boxes of cereal.

x = Weight (ounces)	Number of Boxes
$15.3 \le x < 15.6$	13
$15.6 \le x < 15.9$	24
$15.9 \le x < 16.2$	84
$16.2 \le x < 16.5$	19
$16.5 \le x \le 16.8$	10

FIGURE 4.63 Amount of Brand X cereal per box.

18. To study the efficiency of its new price-scanning equipment, a local supermarket monitored the amount of time its customers had to wait in line. The frequency distribution in Figure 4.64 summarizes the findings. Find the mean amount of time spent in line.

x = Time (minutes)	Number of Customers
$0 \le x < 1$	79
$1 \le x < 2$	58
$2 \le x < 3$	64
$3 \le x < 4$	40
$4 \le x \le 5$	35

FIGURE 4.64 Time spent waiting in a supermarket checkout line.

19. Katrina must take five exams in a math class. If her scores on the first four exams are 71, 69, 85, and 83, what score does she need on the fifth exam for her overall mean to be

a. at least 70? **b.** at least 80?

c. at least 90?

20. Eugene must take four exams in a geography class. If his scores on the first three exams are 91, 67, and 83, what score does he need on the fourth exam for his overall mean to be

a. at least 70? **b.** at least 80?

c. at least 90?

21. The mean salary of twelve men is $58,000, and the mean salary of eight women is $42,000. Find the mean salary of all twenty people.

22. The mean salary of twelve men is $52,000, and the mean salary of four women is $84,000. Find the mean salary of all sixteen people.

23. Maria drove from Chicago, Illinois, to Milwaukee, Wisconsin, a distance of ninety miles, at a mean speed of 60 miles per hour. On her return trip, the traffic was much heavier, and her mean speed was 45 miles per hour. Find Maria's mean speed for the round trip.

HINT: Divide the total distance by the total time.

24. Sully drove from Atlanta, Georgia, to Birmingham, Alabama, a distance of 150 miles, at a mean speed of 50 miles per hour. On his return trip, the traffic was much lighter, and his mean speed was 60 miles per hour. Find Sully's mean speed for the round trip.

HINT: Divide the total distance by the total time.

25. The mean age of a class of twenty-five students is 23.4 years. How old would a twenty-sixth student have to be for the mean age of the class to be 24.0 years?

26. The mean age of a class of fifteen students is 18.2 years. How old would a sixteenth student have to be for the mean age of the class to be 21.0 years?

27. The mean salary of eight employees is $40,000, and the median is $42,000. The highest-paid employee gets a $6,000 raise.

a. What is the new mean salary of the eight employees?

b. What is the new median salary of the eight employees?

28. The mean salary of ten employees is $32,000, and the median is $30,000. The highest-paid employee gets a $5,000 raise.

a. What is the new mean salary of the ten employees?

b. What is the new median salary of the ten employees?

29. The number of civilians holding government jobs in various federal departments and their monthly payrolls for September 2007 are given in Figure 4.65.

Department	Number of Civilian Workers	Monthly Payroll (thousands of dollars)
Department of Education	4,201	44,497
Department of Health and Human Services	62,502	576,747
Environmental Protection Agency	18,119	195,320
Department of Homeland Security	159,447	1,251,754
Department of the Navy	175,722	712,172
Department of the Air Force	157,182	635,901
Department of the Army	245,599	678,044

FIGURE 4.65 Monthly earnings for civilian jobs (September 2007). *Source:* U.S. Office of Personnel Management.

a. Which department or agency has the highest mean monthly earnings? What is the mean monthly earnings for this department or agency?

b. Which department or agency has the lowest mean monthly earnings? What is the mean monthly earnings for this department or agency?

c. Find the mean monthly earnings of all civilians employed by the Department of Education, the Department of Health and Human Services, and the Environmental Protection Agency.

d. Find the mean monthly earnings of all civilians employed by the Department of Homeland Security, the Department of the Navy, the Department of the Air Force, and the Department of the Army.

30. The ages of the nearly 4 million women who gave birth in the United States in 2001 are given in Figure 4.66. Find the mean age of these women.

Age (years)	Number of Women
$15 \le x < 20$	549,000
$20 \le x < 25$	872,000
$25 \le x < 30$	897,000
$30 \le x < 35$	859,000
$35 \le x < 40$	452,000
$40 \le x < 45$	137,000

FIGURE 4.66 Ages of women giving birth in 2001. *Source:* U.S. Bureau of the Census.

31. The age composition of the population of the United States in the year 2000 is given in Figure 4.67.

a. Find the mean age of all people in the United States under the age of 85.

b. Replace the interval "85 and over" with the interval $85 \le X \le 100$ and find the mean age of all people in the United States.

Age (years)	Number of People (thousands)
$0 < x < 5$	19,176
$5 \le x < 10$	20,550
$10 \le x < 15$	20,528
$15 \le x < 25$	39,184
$25 \le x < 35$	39,892
$35 \le x < 45$	44,149
$45 \le x < 55$	37,678
$55 \le x < 65$	24,275
$65 \le x < 85$	30,752
85 and over	4,240

FIGURE 4.67 Age composition of the population of the United States in the year 2000. *Source:* U.S. Bureau of the Census.

In Exercises 32 and 33, use the age composition of the 14,980,000 students enrolled in institutions of higher education in the United States during 2000, as given in Figure 4.68.

Age of Males (years)	Number of Students
$14 \le x < 18$	94,000
$18 \le x < 20$	1,551,000
$20 \le x < 22$	1,420,000
$22 \le x < 25$	1,091,000
$25 \le x < 30$	865,000
$30 \le x < 35$	521,000
35 and over	997,000
total	6,539,000

FIGURE 4.68 Age composition of students in higher education. *Source:* U.S. National Center for Education Statistics.

Age of Females (years)	Number of Students
$14 \le x < 18$	78,000
$18 \le x < 20$	1,907,000
$20 \le x < 22$	1,597,000
$22 \le x < 25$	1,305,000
$25 \le x < 30$	1,002,000
$30 \le x < 35$	664,000
35 and over	1,888,000
total	8,441,000

FIGURE 4.68 *Continued.*

32. **a.** Find the mean age of all male students in higher education under 35.
b. Replace the interval "35 and over" with the interval $35 \le x \le 60$ and find the mean age of all male students in higher education.

33. **a.** Find the mean age of all female students in higher education under 35.
b. Replace the interval "35 and over" with the interval $35 \le x \le 60$ and find the mean age of all female students in higher education.

Answer the following questions using complete sentences and your own words.

• CONCEPT QUESTIONS

34. What are the three measures of central tendency? Briefly explain the meaning of each.

35. Suppose the mean of Group I is A and the mean of Group II is B. We combine Groups I and II to form Group III. Is the mean of Group III equal to $\frac{A+B}{2}$? Explain.

36. Why do we use the midpoint of an interval when calculating the mean of grouped data?

WEB PROJECT

37. What were last month's "average" high and low temperatures in your favorite city? Pick a city that interests you and obtain the high and low temperatures for each day last month. Print the data, and submit them as evidence in answering the following.
a. Find the mean, median, and mode of the daily high temperatures.
b. How does the mean daily high temperature last month compare to the mean seasonal high temperature for last month? That is, were the high temperatures last month above or below the normal high temperatures for the month?
c. Find the mean, median, and mode of the daily low temperatures.
d. How does the mean daily low temperature last month compare to the mean seasonal low temperature for last month? That is, were the low temperatures last month above or below the normal low temperatures for the month?

Some useful links for this web project are listed on the text web site:
www.cengage.com/math/johnson

4.3 Measures of Dispersion

OBJECTIVES

- Find the standard deviation of a data set
- Examine the dispersion of data relative to the mean

To settle an argument over who was the better bowler, George and Danny agreed to bowl six games, and whoever had the highest "average" would be considered best. Their scores were as shown in Figure 4.69.

George	185	135	200	185	250	155
Danny	182	185	188	185	180	190

FIGURE 4.69 Bowling scores.

Each bowler then arranged his scores from lowest to highest and computed the mean, median, and mode:

George

135 155 185 185 200 250

$$\text{mean} = \frac{\text{sum of scores}}{6} = \frac{1{,}110}{6} = 185$$

$$\text{median} = \text{middle score} = \frac{185 + 185}{2} = 185$$

$$\text{mode} = \text{most common score} = 185$$

Danny

180 182 185 185 188 190

$$\text{mean} = \frac{1{,}110}{6} = 185$$

$$\text{median} = \frac{185 + 185}{2} = 185$$

$$\text{mode} = 185$$

Much to their surprise, George's mean, median, and mode were exactly the same as Danny's! Using the measures of central tendency alone to summarize their performances, the bowlers appear identical. Even though their averages were identical, however, their performances were not; George was very erratic, while Danny was very consistent. Who is the better bowler? On the basis of high score, George is better. On the basis of consistency, Danny is better.

George and Danny's situation points out a fundamental weakness in using only the measures of central tendency to summarize data. In addition to finding the averages of a set of data, the consistency, or spread, of the data should also be taken into account. This is accomplished by using **measures of dispersion,** which determine how the data points differ from the average.

Deviations

It is clear from George and Danny's bowling scores that it is sometimes desirable to measure the relative consistency of a set of data. Are the numbers consistently bunched up? Are they erratically spread out? To measure the dispersion of a set of data, we need to identify an average or typical distance between the data points and the mean. The difference between a single data point x and the mean $\bar{x}$ is called the **deviation from the mean** (or simply the **deviation**) and is given by $(x - \bar{x})$. A data point that is close to the mean will have a small deviation, whereas data points far from the mean will have large deviations, as shown in Figure 4.70.

To find the typical deviation of the data points, you might be tempted to add up all the deviations and divide by the total number of data points, thus finding the "average" deviation. Unfortunately, this process leads nowhere. To see why, we will find the mean of the deviations of George's bowling scores.

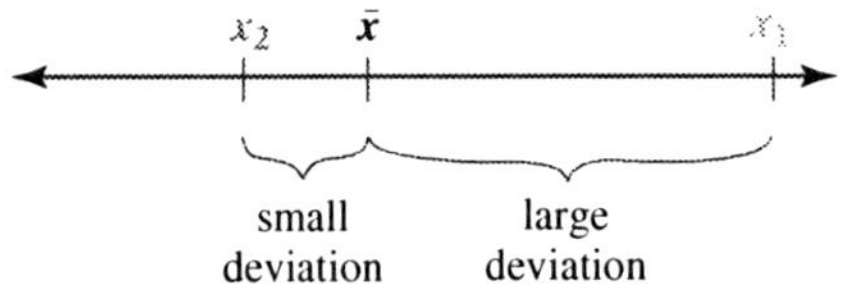

FIGURE 4.70 Large versus small deviations.

EXAMPLE 1

FINDING THE DEVIATIONS OF A DATA SET George bowled six games, and his scores were 185, 135, 200, 185, 250, and 155. Find the mean of the scores, the deviation of each score, and the mean of the deviations.

SOLUTION

$$\bar{x} = \frac{\text{sum of scores}}{6} = \frac{1{,}110}{6} = 185$$

The mean score is 185.

To find the deviations, subtract the mean from each score, as shown in Figure 4.71.

$$\text{mean of the deviations} = \frac{\text{sum of deviations}}{6} = \frac{0}{6} = 0$$

The mean of the deviations is zero.

Score (x)	Deviation ($x - 185$)
135	−50
155	−30
185	0
185	0
200	15
250	65
	Sum = 0

FIGURE 4.71 Deviations of George's score.

In Example 1, the sum of the deviations of the data is zero. This *always* happens; that is, $\Sigma\,(x - \bar{x}) = 0$ for any set of data. The negative deviations are the "culprits"—they will always cancel out the positive deviations. Therefore, to use deviations to study the spread of the data, we must modify our approach and convert the negatives into positives. We do this by squaring each deviation.

Variance and Standard Deviation

Before proceeding, we must be reminded of the difference between a population and a sample. A population is the universal set of all possible items under study; a sample is any group or subset of items selected from the population. (Samples are used to study populations.) In this context, George's six bowling scores represent a sample, not a population; because we do not know the scores of *all* the games George has ever bowled, we are limited to a sample. Unless otherwise specified, we will consider any given set of data to represent a sample, not an entire population.

To measure the typical deviation contained within a set of data points, we must first find the **variance** of the data. Given a sample of n data points, the variance of the data is found by squaring each deviation, adding the squares, and then dividing the sum by the number $(n - 1)$.*

Because we are working with n data points, you might wonder why we divide by $n - 1$ rather than by n. The answer lies in the study of inferential statistics. Recall that inferential statistics deal with the drawing of conclusions concerning the nature of a population based on observations made within a sample. Hence, the variance of a sample can be viewed as an estimate of the variance of the population. However, because the population will vary more than the sample (a population has more data points), dividing the sum of the squares of the sample deviations by n would underestimate the true variance of the entire population.

*If the n data points represent the entire population, the population variance, denoted by σ^2, is found by squaring each deviation, adding the squares, and then dividing the sum by n.

SAMPLE VARIANCE DEVIATION

Given a sample of n data points, $x_1, x_2, \ldots, x_n$, the **variance** of the data, denoted by s^2, is

$$s^2 = \frac{\Sigma(x - \bar{x})^2}{n - 1}$$

The variance of a sample is found by dividing the sum of the squares of the deviations by $n - 1$. The symbol s^2 is a reminder that the deviations have been squared.

To compensate for this underestimation, statisticians have determined that dividing the sum of the squares of the deviations by $n - 1$ rather than by n produces the best estimate of the true population variance.

Variance is the tool with which we can obtain a measure of the typical deviation contained within a set of data. However, because the deviations have been squared, we must perform one more operation to obtain the desired result: We must take the square root. The square root of variance is called the **standard deviation** of the data.

STANDARD DEVIATION DEFINITION

Given a sample of n data points, $x_1, x_2, \ldots, x_n$, the **standard deviation** of the data, denoted by s, is

$$s = \sqrt{\text{variance}}$$

To find the standard deviation of a set of data, first find the variance and then take the square root of the variance.

EXAMPLE 2

FINDING THE VARIANCE AND STANDARD DEVIATION: SINGLE VALUES George bowled six games, and his scores were 185, 135, 200, 185, 250, and 155. Find the standard deviation of his scores.

SOLUTION

To find the standard deviation, we must first find the variance. The mean of the six data points is 185. The necessary calculations for finding variance are shown in Figure 4.72.

Data (x)	Deviation ($x - 185$)	Deviation Squared $(x - 185)^2$
135	−50	$(-50)^2 = 2{,}500$
155	−30	$(-30)^2 = 900$
185	0	$(0)^2 = 0$
185	0	$(0)^2 = 0$
200	15	$(15)^2 = 225$
250	65	$(65)^2 = 4{,}225$
		Sum = 7,850

FIGURE 4.72 Finding variance.

$$\text{variance} = \frac{\text{sum of the squares of the deviations}}{n - 1}$$

$$s^2 = \frac{7{,}850}{6 - 1}$$

$$= \frac{7{,}850}{5}$$

$$= 1{,}570$$

The variance is $s^2 = 1{,}570$. Taking the square root, we have

$$s = \sqrt{1{,}570}$$

$$= 39.62322551 \ldots$$

It is customary to round off s to one place more than the original data. Hence, the standard deviation of George's bowling scores is $s = 39.6$ points.

Because they give us information concerning the spread of data, variance and standard deviation are called **measures of dispersion.** Standard deviation (and variance) is a relative measure of the dispersion of a set of data; the larger the standard deviation, the more spread out the data. Consider George's standard deviation of 39.6. This appears to be high, but what exactly constitutes a "high" standard deviation? Unfortunately, because it is a relative measure, there is no hard-and-fast distinction between a "high" and a "low" standard deviation.

By itself, the standard deviation of a set of data might not be very informative, but standard deviations are very useful in comparing the relative consistencies of two sets of data. Given two groups of numbers of the same type (for example, two sets of bowling scores, two sets of heights, or two sets of prices), the set with the lower standard deviation contains data that are more consistent, whereas the data with the higher standard deviation are more spread out. Calculating the standard deviation of Danny's six bowling scores, we find $s = 3.7$. Since Danny's standard deviation is less than George's, we infer that Danny is more consistent. If George's standard deviation is less than 39.6 the next time he bowls six games, we would infer that his game has become more consistent (the scores would not be spread out as far).

Alternative Methods for Finding Variance

The procedure for calculating variance is very direct: First find the mean of the data, then find the deviation of each data point, and finally divide the sum of the squares of the deviations by $(n - 1)$. However, using the definition of sample variance to find the variance can be rather tedious. Fortunately, many scientific calculators are programmed to find the variance (and standard deviation) if you just push a few buttons. Consult your manual to utilize the statistical capabilities of your calculator.

If your calculator does not have built-in statistical functions, you might still be able to take a shortcut in calculating variance. Instead of using the definition of variance (as in Example 2), we can use an alternative formula that contains the two sums Σx and Σx^2, where Σx represents the sum of the data and Σx^2 represents the sum of the squares of the data, as shown in the following box.

ALTERNATIVE FORMULA FOR SAMPLE VARIANCE

Given a sample of n data points, $x_1, x_2, \ldots, x_n$, the **variance** of the data, denoted by s^2, can be found by

$$s^2 = \frac{1}{(n-1)}\left[\Sigma x^2 - \frac{(\Sigma x)^2}{n}\right]$$

Note: Σx^2 means "square each data point, then add"; $(\Sigma x)^2$ means "add the data points, then square."

Although we will not prove it, this Alternative Formula for Sample Variance is algebraically equivalent to the sample variance definition; given any set of data, either method will produce the same answer. At first glance, the Alternative Formula might appear to be more difficult to use than the definition. Do not be fooled by its appearance! As we will see, the Alternative Formula is relatively quick and easy to apply.

EXAMPLE 3

FINDING THE VARIANCE AND STANDARD DEVIATION: ALTERNATIVE FORMULA Using the Alternative Formula for Sample Variance, find the standard deviation of George's bowling scores as given in Example 2.

SOLUTION

Recall that George's scores were 185, 135, 200, 185, 250, and 155. To find the standard deviation, we must first find the variance.

The Alternative Formula for Sample Variance requires that we find the sum of the data and the sum of the squares of the data. These calculations are shown in Figure 4.73. Applying the Alternative Formula for Sample Variance, we have

x	x^2
135	18,225
155	24,025
185	34,225
185	34,225
200	40,000
250	62,500
$\Sigma x = 1{,}110$	$\Sigma x^2 = 213{,}200$

FIGURE 4.73
Data and data squared.

$$\begin{aligned} s^2 &= \frac{1}{(n-1)}\left[\Sigma x^2 - \frac{(\Sigma x)^2}{n}\right] \\ &= \frac{1}{6-1}\left[213{,}200 - \frac{(1{,}110)^2}{6}\right] \\ &= \frac{1}{5}[213{,}200 - 205{,}350] \\ &= \frac{7{,}850}{5} = 1{,}570 \end{aligned}$$

The variance is $s^2 = 1{,}570$. (Note that this is the same as the variance calculated in Example 2 using the definition of sample variance.)

Taking the square root, we have

$$\begin{aligned} s &= \sqrt{1{,}570} \\ &= 39.62322551\ldots \end{aligned}$$

Rounded off, the standard deviation of George's bowling scores is $s = 39.6$ points.

When we are working with grouped data, the individual data points are unknown. In such cases, the midpoint of each interval should be used as the representative value of the interval.

EXAMPLE 4

FIND THE VARIANCE AND STANDARD DEVIATION: GROUPED DATA In 2008, the U.S. Bureau of Labor Statistics tabulated a survey of workers' ages and wages. The frequency distribution in Figure 4.74 summarizes the age distribution of workers who received minimum wage ($6.55 per hour). Find the standard deviation of the ages these workers.

y = Age (years)	Number of Workers
$16 \le y < 20$	108,000
$20 \le y < 25$	53,000
$25 \le y < 35$	41,000
$35 \le y < 45$	23,000
$45 \le y < 55$	29,000
$55 \le y < 65$	23,000
	$n = 277,000$

FIGURE 4.74 Age distribution of workers receiving minimum wage.
Source: Bureau of Labor Statistics, U.S. Department of Labor.

SOLUTION

Because we are given grouped data, the first step is to determine the midpoint of each interval. We do this by adding the endpoints and dividing by 2.

To utilize the Alternative Formula for Sample Variance, we must find the sum of the data and the sum of the squares of the data. The sum of the data is found by multiplying each midpoint by the frequency of the interval and adding the results; that is, $\Sigma(f \cdot x)$. The sum of the squares of the data is found by squaring each midpoint, multiplying by the corresponding frequency, and adding; that is, $\Sigma(f \cdot x^2)$. The calculations are shown in Figure 4.75.

y = Age (years)	f = Frequency	x = Midpoint	$f \cdot x$	$f \cdot x^2$
$16 \le y < 20$	108,000	18	1,944,000	34,992,000
$20 \le y < 25$	53,000	22.5	1,192,500	26,831,250
$25 \le y < 35$	41,000	30	1,230,000	36,900,000
$35 \le y < 45$	23,000	40	920,000	36,800,000
$45 \le y < 55$	29,000	50	1,450,000	72,500,000
$55 \le y < 65$	23,000	60	1,380,000	82,800,000
	$n = 277,000$		$\Sigma(f \cdot x) = 8,116,500$	$\Sigma(f \cdot x^2) = 290,823,250$

FIGURE 4.75 Finding variance of grouped data.

Applying the Alternative Formula for Sample Variance, we have

$$
\begin{aligned}
s^2 &= \frac{1}{n-1}\left[\Sigma(f\cdot x^2) - \frac{(\Sigma f\cdot x)^2}{n}\right] \\
&= \frac{1}{277{,}000 - 1}\left[290{,}823{,}250 - \frac{(8{,}116{,}500)^2}{277{,}000}\right] \\
&= \frac{1}{276{,}999}\left[290{,}823{,}250 - 237{,}825{,}170.6\right] \\
&= \frac{52{,}998{,}079.4}{276{,}999}
\end{aligned}
$$

The variance is $s^2 = 191.3294972$. Taking the square root, we have

$$
\begin{aligned}
s &= \sqrt{191.3294972} \\
&= 13.83219062\ldots
\end{aligned}
$$

Rounded off, the standard deviation of the ages of the workers receiving minimum wage is $s = 13.8$ years.

The procedure for calculating variance when working with grouped data (as illustrated in Example 4) is summarized in the following box.

ALTERNATIVE FORMULA FOR SAMPLE VARIANCE: GROUPED DATA

Given a frequency distribution containing several groups of data, the variance s^2 can be found by

$$s^2 = \frac{1}{(n-1)}\left[\Sigma(f\cdot x^2) - \frac{(\Sigma f\cdot x)^2}{n}\right]$$

where x = the midpoint of a group, f = the frequency of the group, and $n = \Sigma f$

To obtain the best analysis of a collection of data, we should use the measures of central tendency and the measures of dispersion in conjunction with each other. The most common way to combine these measures is to determine what percent of the data lies within a specified number of standard deviations of the mean. The phrase "one standard deviation of the mean" refers to all numbers within the interval $[\bar{x} - s, \bar{x} + s]$, that is, all numbers that differ from $\bar{x}$ by at most s. Likewise, "two standard deviations of the mean" refers to all numbers within the interval $[\bar{x} - 2s, \bar{x} + 2s]$. One, two, and three standard deviations of the mean are shown in Figure 4.76.

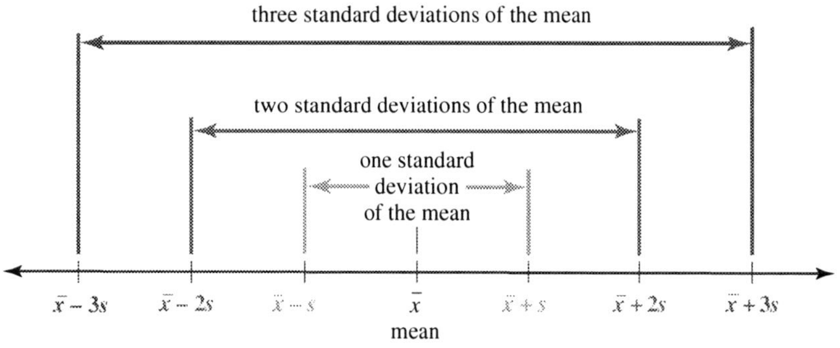

FIGURE 4.76 One, two, and three standard deviations of the mean.

EXAMPLE 5

FINDING THE PERCENTAGE OF DATA WITHIN A SPECIFIED INTERVAL Paki Mowry is a rabbit enthusiast and has eleven mini Rex rabbits that she enters in regional rabbit competition shows. The weights of the rabbits are given in Figure 4.77. What percent of the rabbits' weights lie within one standard deviation of the mean?

Weight (lb:oz)						
	3:10	4:02	3:06	3:15	3:12	4:01
	3:11	4:00	4:03	3:13	3:15	

FIGURE 4.77 Weights of rabbits in pounds and ounces.

SOLUTION

First, the weights must be converted to a common unit, say, ounces. Therefore, we multiply the number of pounds by 16 (ounces per pound) and add the given number of ounces. For instance, 3 pounds, 10 ounces is converted to ounces as follows:

$$\begin{aligned}\text{3 pounds, 10 ounces} &= \left(\text{3 pounds} \times \frac{\text{16 ounces}}{\text{pound}}\right) + \text{10 ounces}\\ &= \text{48 ounces} + \text{10 ounces}\\ &= \text{58 ounces}\end{aligned}$$

The converted weights are given in Figure 4.78.

Weight (ounces)						
	58	66	54	63	60	65
	59	64	67	61	63	

FIGURE 4.78 Weights of rabbits in ounces.

Now we find the mean and standard deviation.

Summing the eleven data points, we have $\Sigma x = 680$. Summing the squares of the data points, we have $\Sigma x^2 = 42{,}186$. The mean is

$$\bar{x} = \frac{680}{11} = 61.81818181\ldots = 61.8 \text{ ounces} \quad \text{(rounded to one decimal place)}$$

Using the Alternative Formula for Sample Variance, we have

$$\begin{aligned}s^2 &= \frac{1}{(n-1)}\left[\Sigma x^2 - \frac{(\Sigma x)^2}{n}\right]\\ &= \frac{1}{(11-1)}\left[42{,}186 - \frac{(680)^2}{11}\right]\\ &= \frac{1}{10}\left[\frac{1646}{11}\right] && \text{subtracting fractions with LCD} = 11\\ &= \frac{823}{55} && \text{reducing}\end{aligned}$$

The variance is $\frac{823}{55} = 14.963636363\ldots$. Taking the square root, we have

$$s = \sqrt{823/55} = 3.868285972\ldots$$

The standard deviation is 3.9 ounces (rounded to one decimal place).

To find one standard deviation of the mean, we add and subtract the standard deviation to and from the mean:

$$\begin{aligned}[\bar{x} - s, \bar{x} + s] &= [61.8 - 3.9, 61.8 + 3.9]\\ &= [57.9, 65.7]\end{aligned}$$

Arranging the data from smallest to largest, we see that eight of the eleven data points are between 57.9 and 65.7:

54 58 59 60 61 63 63 64 65 66 67

$\bar{x} - s$ 57.9 $\quad$ $\bar{x}$ 61.8 $\quad$ $\bar{x} + s$ 65.7

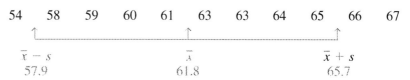

Therefore, $\frac{8}{11} = 0.7272727272\ldots$, or 72.7%, of the data lie within one standard deviation of the mean.

4.3 Exercises

1. Perform each task, given the following sample data:

3 8 5 3 10 13

a. Use the Sample Variance Definition to find the variance and standard deviation of the data.

b. Use the Alternative Formula for Sample Variance to find the variance and standard deviation of the data.

2. Perform each task, given the following sample data:

6 10 12 12 11 17 9

a. Use the Sample Variance Definition to find the variance and standard deviation of the data.

b. Use the Alternative Formula for Sample Variance to find the variance and standard deviation of the data.

3. Perform each task, given the following sample data:

10 10 10 10 10 10

a. Find the variance of the data.

b. Find the standard deviation of the data.

4. Find the mean and standard deviation of each set of data.

a. 2 4 6 8 10 12

b. 102 104 106 108 110 112

c. How are the data in (b) related to the data in (a)?

d. How do your answers for (a) and (b) compare?

5. Find the mean and standard deviation of each set of data.

a. 12 16 20 24 28 32

b. 600 800 1,000 1,200 1,400 1,600

c. How are the data in (b) related to the data in (a)?

d. How do your answers for (a) and (b) compare?

6. Find the mean and standard deviation of each set of data.

a. 50 50 50 50 50

b. 46 50 50 50 54

c. 5 50 50 50 95

d. How do your answers for (a), (b), and (c) compare?

7. Joey and Dee Dee bowled five games at the Rock 'n' Bowl Lanes. Their scores are given in Figure 4.79.

a. Find the mean score of each bowler. Who has the highest mean?

b. Find the standard deviation of each bowler's scores.

c. Who is the more consistent bowler? Why?

Joey	144	171	220	158	147
Dee Dee	182	165	187	142	159

FIGURE 4.79 Bowling scores.

8. Paki surveyed the price of unleaded gasoline (self-serve) at gas stations in Novato and Lafayette. The raw data, in dollars per gallon, are given in Figure 4.80.

a. Find the mean price in each city. Which city has the lowest mean?

b. Find the standard deviation of prices in each city.

c. Which city has more consistently priced gasoline? Why?

Novato	2.899	3.089	3.429	2.959	2.999	3.099
Lafayette	3.595	3.389	3.199	3.199	3.549	3.349

FIGURE 4.80 Price (in dollars) of one gallon of unleaded gasoline.

9. Kaitlin Mowry is a member of the local 4-H club and has six mini Rex rabbits that she enters in regional rabbit competition shows. The weights of the rabbits are given in Figure 4.81. Find the standard deviation of the rabbits' weights.

Weight (lb:oz)	4:12	5:03	4:06	3:15	4:12	4:08

FIGURE 4.81 Weights of rabbits.

10. As was stated in Example 1 of Section 4.2, Lance Armstrong won the Tour de France bicycle race every year from 1999 to 2005. His margins of victory (time difference of the second place finisher) are given in Figure 4.82. Find the standard deviation of Lance Armstrong's victory margins.

Year	1999	2000	2001	2002	2003	2004	2005
Margin	7:37	6:02	6:44	7:17	1:01	6:19	4:40

FIGURE 4.82 Lance Armstrong's Tour de France victory margins. *Source: San Francisco Chronicle.*

11. Barry Bonds of the San Francisco Giants set an all-time record in professional baseball by hitting 73 home runs in one season (2001). The number of home runs (HR) for each of his seasons in professional baseball is given in Figure 4.83. Find the standard deviation of the number of home runs hit per year by Bonds.

Year	HR
1986	16
1987	25
1988	24
1989	19
1990	33
1991	25
1992	34
1993	46
1994	37
1995	33
1996	42

Year	HR
1997	40
1998	37
1999	34
2000	49
2001	73
2002	46
2003	45
2004	45
2005	5
2006	26
2007	28

FIGURE 4.83 Barry Bonds's home runs.

12. Michael Jordan has been recognized as an extraordinary player in professional basketball, especially in terms of the number of points per game he has scored. Jordan's average number of points per game (PPG) for each of his seasons is given in Figure 4.84. Find the standard deviation of the average points per game made per season by Jordan.

Season	PPG
1984–85	28.2
1985–86	22.7
1986–87	37.1
1987–88	35.0
1988–89	32.5
1989–90	33.6
1990–91	31.5
1991–92	30.1

Season	PPG
1992–93	32.6
1994–95	26.9
1995–96	30.4
1996–97	29.6
1997–98	28.7
2001–02	22.9
2002–03	20.0

FIGURE 4.84 Michael Jordan's points per game.

13. The Truly Amazing Dudes are a group of comic acrobats. The heights (in inches) of the ten acrobats are as follows:

68 50 70 67 72 78 69 68 66 67

© Michael P. Gadomski/Photo Researchers, Inc.

Is your height or weight "average"? These characteristics can vary considerably within any specific group of people. The mean is used to represent the average, and the standard deviation is used to measure the "spread" of a collection of data.

a. Find the mean and standard deviation of the heights.
b. What percent of the data lies within one standard deviation of the mean?
c. What percent of the data lies within two standard deviations of the mean?

14. The weights (in pounds) of the ten Truly Amazing Dudes are as follows:

152 196 144 139 166 83 186 157 140 138

a. Find the mean and standard deviation of the weights.
b. What percent of the data lies within one standard deviation of the mean?
c. What percent of the data lies within two standard deviations of the mean?

15. The normal monthly rainfall in Seattle, Washington, is given in Figure 4.85.

Month	Jan.	Feb.	Mar.	Apr.	May	June
Inches	5.4	4.0	3.8	2.5	1.8	1.6

Month	July	Aug.	Sept.	Oct.	Nov.	Dec.
Inches	0.9	1.2	1.9	3.3	5.7	6.0

FIGURE 4.85 Monthly rainfall in Seattle, WA. *Source:* U.S. Department of Commerce.

a. Find the mean and standard deviation of the monthly rainfall in Seattle.
b. What percent of the year will the monthly rainfall be within one standard deviation of the mean?
c. What percent of the year will the monthly rainfall be within two standard deviations of the mean?

16. The normal monthly rainfall in Phoenix, Arizona, is given in Figure 4.86.

Month	Jan.	Feb.	Mar.	Apr.	May	June
Inches	0.7	0.7	0.9	0.2	0.1	0.1

Month	July	Aug.	Sept.	Oct.	Nov.	Dec.
Inches	0.8	1.0	0.9	0.7	0.7	1.0

FIGURE 4.86 Monthly rainfall in Phoeniz, AZ. *Source:* U.S. Department of Commerce.

a. Find the mean and standard deviation of the monthly rainfall in Phoenix.
b. What percent of the year will the monthly rainfall be within one standard deviation of the mean?
c. What percent of the year will the monthly rainfall be within two standard deviations of the mean?

17. The frequency distribution in Figure 4.87 lists the results of a quiz given in Professor Gilbert's statistics class.

Score	Number of Students
10	5
9	10
8	6

Score	Number of Students
7	8
6	3
5	2

FIGURE 4.87 Quiz scores in Professor Gilbert's statistics class.

a. Find the mean and standard deviation of the scores.
b. What percent of the data lies within one standard deviation of the mean?
c. What percent of the data lies within two standard deviations of the mean?
d. What percent of the data lies within three standard deviations of the mean?

18. Amy surveyed the prices for a quart of a certain brand of motor oil. The sample data, in dollars per quart, is summarized in Figure 4.88.

Price per Quart (dollars)	Number of Stores
1.99	2
2.09	5
2.19	10
2.29	13
2.39	9
2.49	3

FIGURE 4.88 Price for a quart of motor oil.

a. Find the mean and the standard deviation of the prices.
b. What percent of the data lies within one standard deviation of the mean?
c. What percent of the data lies within two standard deviations of the mean?
d. What percent of the data lies within three standard deviations of the mean?

19. To study the output of a machine that fills boxes with cereal, a quality control engineer weighed 150 boxes of Brand X cereal. The frequency distribution in Figure 4.89 summarizes his findings. Find the standard deviation of the weight of the boxes of cereal.

x = Weight (ounces)	Number of Boxes
$15.3 \le x < 15.6$	13
$15.6 \le x < 15.9$	24
$15.9 \le x < 16.2$	84
$16.2 \le x < 16.5$	19
$16.5 \le x < 16.8$	10

FIGURE 4.89 Amount of Brand X cereal per box.

20. To study the efficiency of its new price-scanning equipment, a local supermarket monitored the amount of time its customers had to wait in line. The frequency distribution in Figure 4.90 summarizes the findings. Find the standard deviation of the amount of time spent in line.

x = Time (minutes)	Number of Customers
$0 \le x < 1$	79
$1 \le x < 2$	58
$2 \le x < 3$	64
$3 \le x < 4$	40
$4 \le x \le 5$	35

FIGURE 4.90 Time spent waiting in a supermarket checkout line.

21. The ages of the nearly 4 million women who gave birth in the United States in 2001 are given in Figure 4.91. Find the standard deviation of the ages of these women.

Age (years)	Number of Women
$15 \le x < 20$	549,000
$20 \le x < 25$	872,000
$25 \le x < 30$	897,000
$30 \le x < 35$	859,000
$35 \le x < 40$	452,000
$40 \le x < 45$	137,000

FIGURE 4.91 Ages of women giving birth in 2001. *Source:* U.S. Bureau of the Census.

22. The age composition of the population of the United States in the year 2000 is given in Figure 4.92. Replace the interval "85 and over" with the interval $85 \le x \le 100$ and find the standard deviation of the ages of all people in the United States.

Age	Number of People (thousands)
$0 < x < 5$	19,176
$5 \le x < 10$	20,550
$10 \le x < 15$	20,528
$15 \le x < 25$	39,184
$25 \le x < 35$	39,892
$35 \le x < 45$	44,149
$45 \le x < 55$	37,678
$55 \le x < 65$	24,275
$65 \le x < 85$	30,752
85 and over	4,240

FIGURE 4.92 Age composition of the population of the United States in the year 2000. *Source:* U.S. Bureau of the Census.

Answer the following questions using complete sentences and your own words.

• CONCEPT QUESTIONS

23. **a.** When studying the dispersion of a set of data, why are the deviations from the mean squared?

b. What effect does squaring have on a deviation that is less than 1?

c. What effect does squaring have on a deviation that is greater than 1?

d. What effect does squaring have on the data's units?

e. Why is it necessary to take a square root when calculating standard deviation?

24. Why do we use the midpoint of an interval when calculating the standard deviation of grouped data?

WEB PROJECT

25. This project is a continuation of Exercise 37 in Section 4.2. How did last month's daily high and low temperatures vary in your favorite city? Pick a city that interests you, and obtain the high and low temperatures for each day last month. Print the data, and submit them as evidence in answering the following.

a. Find the standard deviation of the daily high temperatures.

b. Find the standard deviation of the daily low temperatures.

c. Comparing your answers to parts (a) and (b), what can you conclude?

Some useful links for this web project are listed on the text web site:

www.cengage.com/math/johnson

• PROJECTS

26. The purpose of this project is to explore the variation in the pricing of a common commodity. Go to several different stores that sell food (the more stores the better) and record the price of one gallon of whole milk.

a. Compute the mean, median, and mode of the data.

b. Compute the standard deviation of the data.

c. What percent of the data lie within one standard deviation of the mean?

d. What percent of the data lie within two standard deviations of the mean?

e. What percent of the data lie within three standard deviations of the mean?

27. The purpose of this project is to explore the variation in the pricing of a common commodity. Go to several different gas stations (the more the better) and record the price of one gallon of premium gasoline (91 octane).

a. Compute the mean, median, and mode of the data.

b. Compute the standard deviation of the data.

c. What percent of the data lie within one standard deviation of the mean?

d. What percent of the data lie within two standard deviations of the mean?

e. What percent of the data lie within three standard deviations of the mean?

MEASURES OF CENTRAL TENDENCY AND DISPERSION ON A GRAPHING CALCULATOR

In Examples 1, 2, and 3 of this section, we found the mean, median, mode, variance, and standard deviation of George's bowling scores. This work can be done quickly and easily on either a graphing calculator or Excel.

TECHNOLOGY AND MEASURES OF CENTRAL TENDENCY AND DISPERSION

Calculating the Mean, the Variance, and the Standard Deviation

ON A TI-83/84:

- Enter the data from Example 1 of this section in list L_1 as discussed in Section 4.1.
- Press STAT.
- Select "1-Var Stats" from the "CALC" menu.
- When "1-Var" appears on the screen, press 2nd L_1 ENTER.

ON A CASIO:

- Enter the data from Example 1 of this section as described in Section 4.1.
- Press CALC (i.e., F2).
- Press 1VAR (i.e., F1).

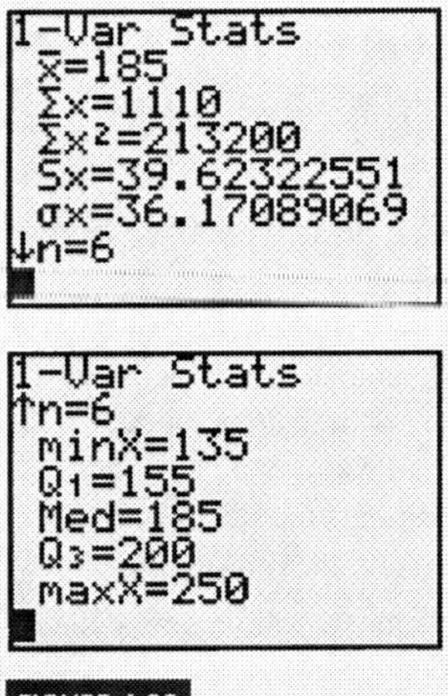

FIGURE 4.93

After computing the mean, standard deviation, and more on a TI graphing calculator.

The above steps will result in the first screen in Figure 4.93. This screen gives the mean, the sample standard deviation (S_x on a TI, $\chi\sigma$n-1 on a Casio), the population standard deviation (σ_x on a TI, $\chi\sigma$n on a Casio), and the number of data points (n). The second screen can be obtained by pressing the down arrow. It gives the minimum and maximum data points (minX and maxX, respectively) as well as the median (Med).

Calculating the Sample Variance

The above work does not yield the sample variance. To find it, follow these steps:

- Quit the statistics mode.
- Get S_x on the screen.
- Press VARS, select "Statistics," and then select "S_x" from the "X/Y" menu.
- Once S_x is on the screen, square it by pressing x^2 ENTER. The variance is 1,570.

Calculating the Mean, the Variance, and the Standard Deviation with Grouped Data

To calculate the mean and standard deviation from the frequency distribution that utilizes grouped data, follow the same steps except:

ON A TI-83/84:

- Enter the midpoints of the classes in list L_1.
- Enter the frequencies of those classes in list L_2. Each frequency must be less than 100.
- After "1-Var Stats" appears on the screen, press 2nd L_1 , 2nd L_2 ENTER.

Measures of Central Tendency and Dispersion on Excel

1. *Enter the data* from Example 1 of this section as discussed in Section 4.1. See column A of Figure 4.94.
2. *Have Excel compute the mean, standard deviation, etc.*
   - Use your mouse to select "Tools" at the very top of the screen. Then pull your mouse down until "Data Analysis" is highlighted, and let go. (If "Data Analysis" is not listed under "Tools," then follow the instructions given in Section 4.1.)
   - In the "Data Analysis" box that appears, use your mouse to highlight "Descriptive Statistics" and press the "OK" button.
   - In the "Descriptive Statistics" box that appears, use your mouse to click on the white rectangle that follows "Input Range," and then use your mouse to draw a box around all of the bowling scores. (To draw the box, move your mouse to cell A2, press the mouse button, and move the mouse down until all of the entered scores are enclosed in a box.) This should cause "\$A\$2 : \$A\$7" to appear in the Input Range rectangle.

◇	A	B	C
1	George's scores	*Column 1*	
2	185		
3	135	Mean	185
4	200	Standard Errc	16.1761141
5	185	Median	185
6	250	Mode	185
7	155	Standard Dev	39.6232255
8		Sample Varia	1570
9		Kurtosis	0.838675
10		Skewness	0.60763436
11		Range	115
12		Minimum	135
13		Maximum	250
14		Sum	1110
15		Count	6

FIGURE 4.94 After computing the mean, standard deviation, and more on Excel.

- Be certain that there is a dot in the button to the left of "Output Range." Then click on the white rectangle that follows "Output Range," and use your mouse to click on cell B1. This should cause "\$B\$1" to appear in the Output Range rectangle.
- Select "Summary Statistics."
- Use your mouse to press the "OK" button.

After you complete this step, your spreadsheet should look like that in Figure 4.94. To compute the population standard deviation, first type "pop std dev" in cell B16, and then click on cell C16. Press the *fx* button on the Excel ribbon at the top of the screen. In the "Paste Function" box that appears, click on "Statistical" under "Function category" and on "STDEVP" under "Function name." Then press the "OK" button. In the "STDEVP" box that appears, click on the white rectangle that follows "Number 1." Then use your mouse to draw a box around all of the bowling scores. Press the "OK" button and the population standard deviation will appear in cell C16.

EXERCISES

28. Do Double Stuf Oreos really contain twice as much "stuf"? In 1999, Marie Revak and Jihan William's "interest was piqued by the unusual spelling of the word stuff and the bold statement 'twice the filling' on the package." So they conducted an experiment in which they weighed the amount of filling in a sample of traditional Oreos and Double Stuf Oreos. The data from that experiment are in Figure 4.95 (weight in grams). An unformatted spreadsheet containing these data can be downloaded from the text web site: **www.cengage.com/math/johnson**

a. Find the mean of these data.

b. Find the standard deviation.

c. Did you choose the population or sample standard deviation? Why?

d. What do your results say about Double Stuf Oreos? Interpret the results of both parts (a) and (b).

Double Stuf Oreos:																		
4.7	6.5	5.5	5.6	5.1	5.3	5.4	5.4	3.5	5.5	6.5	5.9	5.4	4.9	5.6	5.7	5.3	6.9	6.5
6.3	4.8	3.3	6.4	5.0	5.3	5.5	5.0	6.0	5.7	6.3	6.0	6.3	6.1	6.0	5.8	5.8	5.9	6.2
5.9	6.5	6.5	6.1	5.8	6.0	6.2	6.2	6.0	6.8	6.2	5.4	6.6	6.2					
Traditional Oreos:																		
2.9	2.8	2.6	3.5	3.0	2.4	2.7	2.4	2.5	2.2	2.6	2.6	2.9	2.6	2.6	3.1	2.9	2.4	2.8
3.8	3.1	2.9	3.0	2.1	3.8	3.0	3.0	2.8	2.9	2.7	3.2	2.8	3.1	2.7	2.8	2.6	2.6	3.0
2.8	3.5	3.3	3.3	2.8	3.1	2.6	3.5	3.5	3.1	3.1								

FIGURE 4.95 Amount of Oreo filling. *Source:* Revak, Marie A. and Jihan G. Williams, "Sharing Teaching Ideas: The Double Stuf Dilemma," *The Mathematics Teacher,* November 1999, Volume 92, Issue 8, page 674.

Rank	City	Air Quality Index (% days AQI ranked good)	Property Crime (incidents per 1000)	Median Commute Time (in minutes)
1	Louisville, CO	74.0%	15	19.5
2	Chanhassen, MN	N.A.	16	23.1
3	Papillion, NE	92.0%	16	19.3
4	Middleton, WI	76.0%	27	16.3
5	Milton, MA	81.0%	9	27.7
6	Warren, NJ	N.A.	21	28.7
7	Keller, TX	65.0%	14	29.4
8	Peachtree City, GA	84.0%	12	29.3
9	Lake St. Louis, MO	79.0%	16	26.8
10	Mukilteo, WA	89.0%	27	22.3

FIGURE 4.96 The best places to live. *Source:* CNNMoney.com.

29. Every year, CNNMoney.com lists "the best places to live." They compare the "best small towns" on several bases, one of which is the "quality of life." See Figure 4.96. An unformatted spreadsheet containing these data can be downloaded from the text web site: **www.cengage.com/math/johnson**

a. Find the mean of the air quality indices.

b. Would the population standard deviation or the sample standard deviation be more appropriate here? Why?

c. Find the standard deviation from part (b).

d. What do your results say about the air quality index of all of the best places to live in the United States?

30. Every year, CNNMoney.com lists "the best places to live." They compare the "best small towns" on several bases, one of which is the "quality of life." See Figure 4.96. An unformatted spreadsheet containing these data can be downloaded from the text web site: **www.cengage.com/math/johnson**

a. Find the mean of the commute times.

b. Would the population standard deviation or the sample standard deviation be more appropriate here? Why or why not?

c. Find the standard deviation from part (b).

d. What do your results say about commuting in all of the best places to live in the United States?

31. Every year, CNNMoney.com lists "the best places to live." They compare the "best small towns" on several bases, one of which is the "quality of life." See Figure 4.96. An unformatted spreadsheet containing these data can be downloaded from the text web site: **www.cengage.com/math/johnson**

a. Find the mean of the property crime levels.

b. Would the population standard deviation or the sample standard deviation be more appropriate here? Why or why not?

c. Find the standard deviation from part (b).

d. What do your results say about property crime in all of the best places to live in the United States?

32. Figure 4.97 gives EPA fuel efficiency ratings for 2010 family sedans with an automatic transmission and the smallest available engine. An unformatted spreadsheet containing these data can be downloaded from the text web site: **www.cengage.com/math/johnson**

a. For each of the two manufacturing regions, find the mean and an appropriate standard deviation for city driving, and defend your choice of standard deviations.

b. What do the results of part (a) tell you about the two regions?

33. Figure 4.97 gives EPA fuel efficiency ratings for 2010 family sedans with an automatic transmission and the smallest available engine. An unformatted spreadsheet containing these data can be downloaded from the text web site: **www.cengage.com/math/johnson**

a. For each of the two manufacturing regions, find the mean and an appropriate standard deviation for highway driving, and defend your choice of standard deviations.

b. What do the results of part (a) tell you about the two regions?

American Cars	City mpg	Highway mpg	Asian Cars	City mpg	Highway mpg
Ford Fusion hybrid	41	36	Toyota Prius	51	48
Chevrolet Malibu hybrid	26	34	Nissan Altima hybrid	35	33
Ford Fusion fwd	24	34	Toyota Camry Hybrid	33	34
Mercury Milan	23	34	Kia Forte	27	36
Chevrolet Malibu	22	33	Hyundai Elantra	26	35
Saturn Aura	22	33	Nissan Altima	23	32
Dodge Avenger	21	30	Subaru Legacy awd	23	31
Buick Lacrosse	17	27	Toyota Camry	22	33
			Hyundai Sonata	22	32
			Kia Optima	22	32
			Honda Accord	22	31
			Mazda 6	21	30
			Mitsubishi Galant	21	30
			Hyunda Azera	18	26

FIGURE 4.97 Fuel efficiency. *Source:* www.fueleconomy.gov.

4.4 The Normal Distribution

OBJECTIVES

- Find probabilities of the standard normal distribution
- Find probabilities of a nonstandard normal distribution
- Find the value of a normally distributed variable that will produce a specific probability

Sets of data may exhibit various trends or patterns. Figure 4.98 shows a histogram of the weights of bags of corn chips. Notice that most of the data are near the "center" and that the data taper off at either end. Furthermore, the histogram is nearly symmetric; it is almost the same on both sides. This type of distribution (nearly symmetric, with most of the data in the middle) occurs quite often in many different situations. To study the composition of such distributions, statisticians have created an ideal **bell-shaped curve** describing a **normal distribution,** as shown in Figure 4.99.

Before we can study the characteristics and applications of a normal distribution, we must make a distinction between different types of variables.

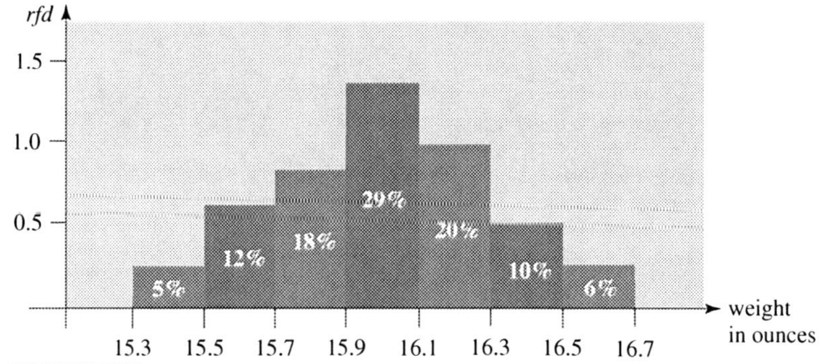

FIGURE 4.98 Weights of bags of corn chips.

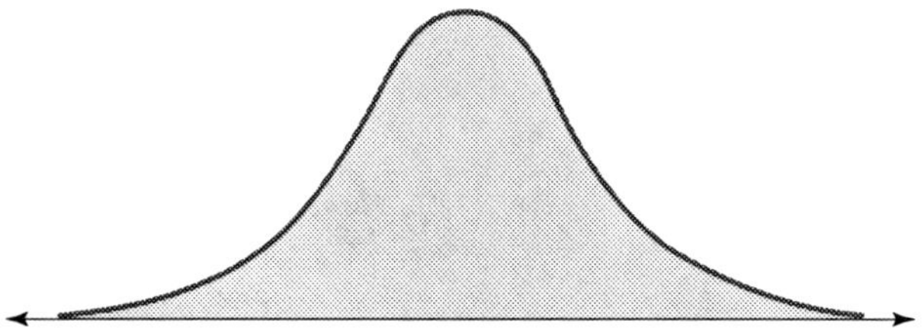

FIGURE 4.99 Normal distribution.

Discrete versus Continuous Variables

The number of children in a family is variable, because it varies from family to family. In listing the number of children, only whole numbers (0, 1, 2, and so on) can be used. In this respect, we are limited to a collection of discrete, or separate, values. A variable is **discrete** if there are "gaps" between the possible variable values. Consequently, any variable that involves counting is discrete.

On the other hand, a person's height or weight does not have such a restriction. When someone grows, he or she does not instantly go from 67 inches to 68 inches; a person grows continuously from 67 inches to 68 inches, attaining all possible values in between. For this reason, height is called a continuous variable. A variable is **continuous** if it can assume *any value* in an interval of real numbers (see Figure 4.100). Consequently, any variable that involves measurement is continuous; someone might claim to be 67 inches tall and to weigh 152 pounds, but the true values might be 67.13157 inches and 151.87352 pounds. Heights and weights are expressed (discretely) as whole numbers solely for convenience; most people do not have rulers or bathroom scales that allow them to obtain measurements that are accurate to ten or more decimal places!

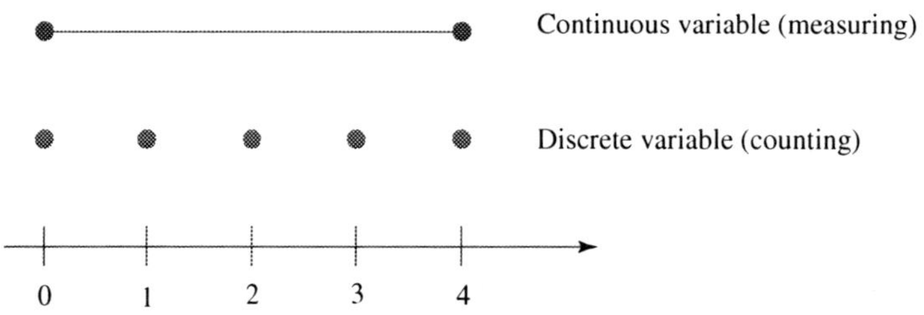

FIGURE 4.100 Discrete versus continuous variables.

Normal Distributions

The collection of all possible values that a discrete variable can assume forms a countable set. For instance, we can list all the possible numbers of children in a family. In contrast, a continuous variable will have an uncountable number of possibilities because it can assume any value in an interval. For instance, the weights (a continuous variable) of bags of corn chips could be *any* value x such that $15.3 \leq x \leq 16.7$.

When we sample a continuous variable, some values may occur more often than others. As we can see in Figure 4.98, the weights are "clustered" near the center of the histogram, with relatively few located at either end. If a continuous variable has a symmetric distribution such that the highest concentration of values is at the center and the lowest is at both extremes, the variable is said to have a **normal distribution** and is represented by a smooth, continuous, bell-shaped curve like that in Figure 4.99*.

The normal distribution, which is found in a wide variety of situations, has two main qualities: (1) the frequencies of the data points nearer the center or "average" are increasingly higher than the frequencies of data points far from the center, and (2) the distribution is symmetric (one side is a mirror image of the other). *Because of these two qualities, the mean, median, and mode of a normal distribution all coincide at the center of the distribution.*

Just like any other collection of numbers, the spread of normal distribution is measured by its standard deviation. It can be shown that for any normal distribution, slightly more than two-thirds of the data (68.26%) will lie within one standard deviation of the mean, 95.44% will lie within two standard deviations, and virtually all the data (99.74%) will lie within three standard deviations of the mean. Recall that μ (the Greek letter "mu") represents the mean of a population and σ (the Greek letter "sigma") represents the standard deviation of the population. The spread of a normal distribution, with μ and σ used to represent the mean and standard deviation, is shown in Figure 4.101.

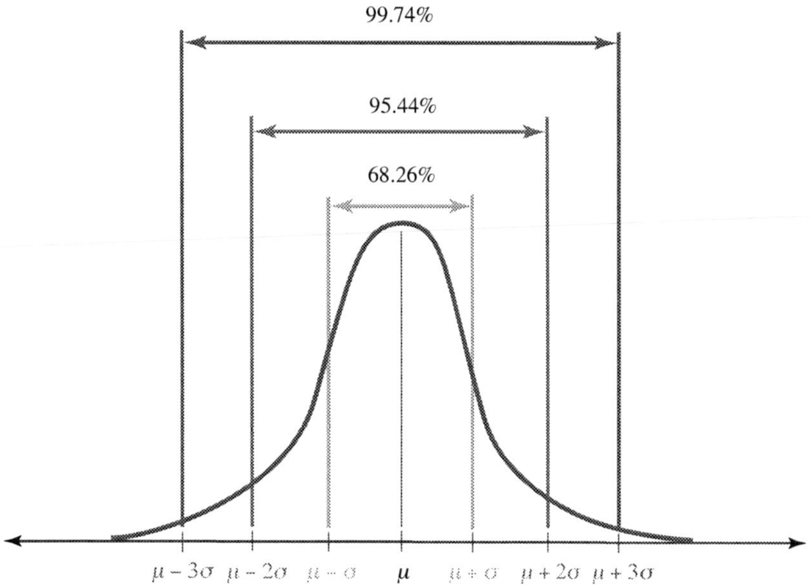

FIGURE 4.101 The spread of a normal distribution.

* This is an informal definition only. The formal definition of a normal distribution involves the number pi, the natural exponential e^x, and the mean and variance of the distribution.

EXAMPLE 1 ANALYZING THE DISPERSION OF A NORMAL DISTRIBUTION The heights of a large group of people are assumed to be normally distributed. Their mean height is 66.5 inches, and the standard deviation is 2.4 inches. Find and interpret the intervals representing one, two, and three standard deviations of the mean.

SOLUTION The mean is $\mu = 66.5$, and the standard deviation is $\sigma = 2.4$.

1. *One standard deviation of the mean:*

$$\begin{aligned} \mu \pm 1\sigma &= 66.5 \pm 1(2.4) \\ &= 66.5 \pm 2.4 \\ &= [64.1, 68.9] \end{aligned}$$

Therefore, approximately 68% of the people are between 64.1 and 68.9 inches tall.

2. *Two standard deviations of the mean:*

$$\begin{aligned} \mu \pm 2\sigma &= 66.5 \pm 2(2.4) \\ &= 66.5 \pm 4.8 \\ &= [61.7, 71.3] \end{aligned}$$

Therefore, approximately 95% of the people are between 61.7 and 71.3 inches tall.

3. *Three standard deviations of the mean:*

$$\begin{aligned} \mu \pm 3\sigma &= 66.5 \pm 3(2.4) \\ &= 66.5 \pm 7.2 \\ &= [59.3, 73.7] \end{aligned}$$

Nearly all of the people (99.74%) are between 59.3 and 73.7 inches tall.

In Example 1, we found that virtually all the people under study were between 59.3 and 73.7 inches tall. A clothing manufacturer might want to know what percent of these people are shorter than 66 inches or what percent are taller than 73 inches. Questions like these can be answered by using probability and a normal distribution. (We will do this in Example 6.)

Probability, Area, and Normal Distributions

In Chapter 3, we mentioned that relative frequency is really a type of probability. If 3 out of every 100 people have red hair, you could say that the relative frequency of red hair is $\frac{3}{100}$ (or 3%), or you could say that the probability of red hair $p(x = \text{red hair})$ is 0.03. Therefore, to find out what percent of the people in a population are taller than 73 inches, we need to find $p(x > 73)$, the probability that x is greater than 73, where x represents the height of a randomly selected person.

Recall that a sample space is the set S of all possible outcomes of a random experiment. Consequently, the probability of a sample space must always equal 1; that is, $p(S) = 1$ (or 100%). If the sample space S has a normal distribution, its outcomes and their respective probabilities can be represented by a bell curve.

Recall that when constructing a histogram, relative frequency density (*rfd*) was used to measure the heights of the rectangles. Consequently, the *area* of a rectangle gave the relative frequency (percent) of data contained in an interval. In a similar manner, we can imagine a bell curve being a histogram composed of infinitely many "skinny" rectangles, as in Figure 4.102.

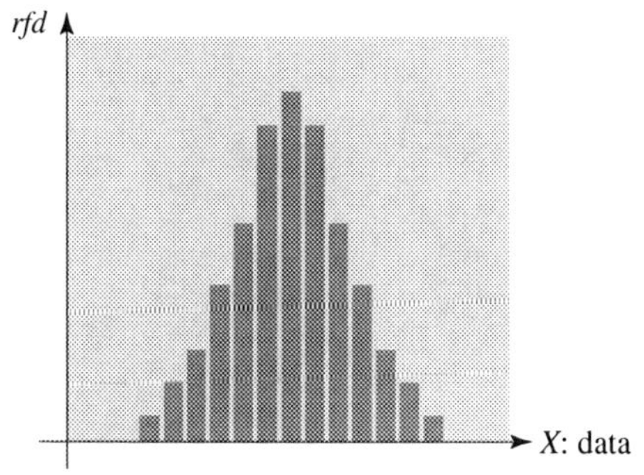

FIGURE 4.102 Symmetric, bell-shaped histogram.

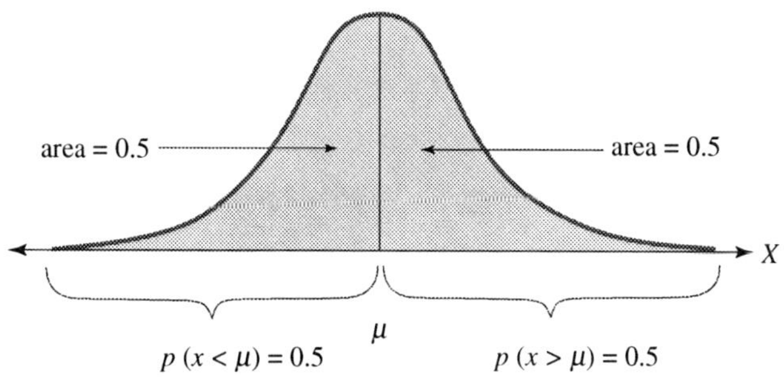

FIGURE 4.103 Total probability equals 1 (or 100%).

For a normal distribution, the outcomes nearer the center of the distribution occur more frequently than those at either end; the distribution is denser in the middle and sparser at the extremes. This difference in density is taken into account by consideration of the area under the bell curve; the center of the distribution is denser, contains more area, and has a higher probability of occurrence than the extremes. Consequently, we use the area under the bell curve to represent the probability of an outcome. Because $p(S) = 1$, we define the entire area under the bell curve to equal 1.

Because a normal distribution is symmetric, 50% of the data will be greater than the mean, and 50% will be less. (The mean and the median coincide in a symmetric distribution.) Therefore, the probability of randomly selecting a number x greater than the mean is $p(x > \mu) = 0.5$, and that of selecting a number x less than the mean is $p(x < \mu) = 0.5$, as shown in Figure 4.103.

To find the probability that a randomly selected number x is between two values (say a and b), we must determine the area under the curve from a to b; that is, $p(a < x < b) =$ area under the bell curve from $x = a$ to $x = b$, as shown in Figure 4.104(a). Likewise, the probability that x is greater than or less than any specific number is given by the area of the tail, as shown in Figure 4.104(b). To find probabilities involving data that are normally distributed, we must find the area of the appropriate region under the bell curve.

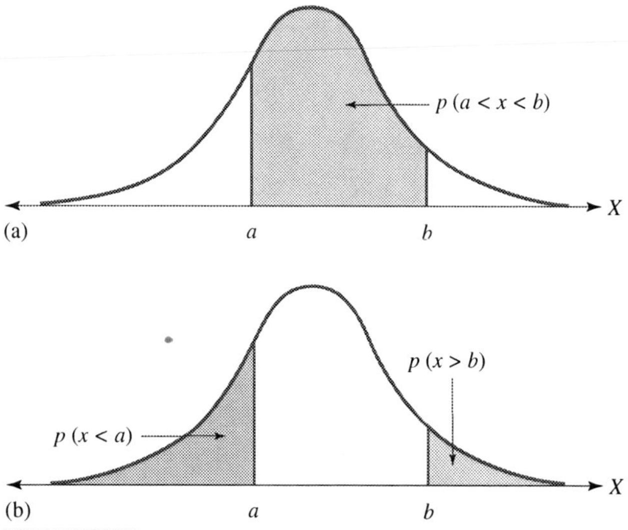

FIGURE 4.104 Regions under a bell curve.

HISTORICAL NOTE — CARL FRIEDRICH GAUSS, 1777–1855

© akg-images

Dubbed "the Prince of Mathematics," Carl Gauss is considered by many to be one of the greatest mathematicians of all time. At the age of three, Gauss is said to have discovered an arithmetic error in his father's bookkeeping. The child prodigy was encouraged by his teachers and excelled throughout his early schooling. When he was fourteen, Gauss was introduced to Ferdinand, the Duke of Brunswick. Impressed with the youth, the duke gave Gauss a yearly stipend and sponsored his education for many years.

In 1795, Gauss enrolled at Göttingen University, where he remained for three years. While at Göttingen, Gauss had complete academic freedom; he was not required to attend lectures, he had no required conferences with professors or tutors, and he did not take exams. Much of his time was spent studying independently in the library. For reasons unknown to us, Gauss left the university in 1798 without a diploma. Instead, he sent his dissertation to the University of Helmstedt and in 1799 was awarded his degree without the usual oral examination.

In 1796, Gauss began his famous mathematical diary. Discovered forty years after his death, the 146 sometimes cryptic entries exhibit the diverse range of topics that Gauss pondered and pioneered. The first entry was Gauss's discovery (at the age of nineteen) of a method for constructing a seventeen-sided polygon with a compass and a straightedge. Other entries include important results in number theory, algebra, calculus, analysis, astronomy, electricity, magnetism, the foundations of geometry, and probability.

At the dawn of the nineteenth century, Gauss began his lifelong study of astronomy. On January 1, 1801, the Italian astronomer Giuseppe Piazzi discovered Ceres, the first of the known planetoids (minor planets or asteroids). Piazzi and others observed Ceres for forty-one days, until it was lost behind the sun. Because of his interest in the mathematics of astronomy, Gauss turned his attention to Ceres. Working with a minimal amount of data, he successfully calculated the orbit of Ceres. At the end of the year, the planetoid was rediscovered in exactly the spot that Gauss had predicted!

To obtain the orbit of Ceres, Gauss utilized his method of least squares, a technique for dealing with experimental error. Letting x represent the error between an experimentally obtained value and the true value it represents, Gauss's theory involved minimizing x^2—that is, obtaining the least square of the error. Theorizing that the probability of a small error was higher than that of a large error, Gauss subsequently developed the normal distribution, or bell-shaped curve, to explain the probabilities of the random errors. Because of his pioneering efforts, some mathematicians refer to the normal distribution as the Gaussian distribution.

In 1807, Gauss became director of the newly constructed observatory at Göttingen. He held the position until his death some fifty years later.

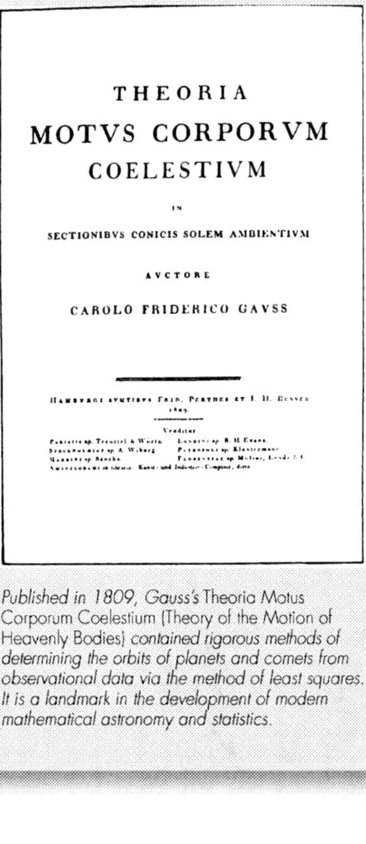
THEORIA
MOTVS CORPORVM
COELESTIVM
IN
SECTIONIBVS CONICIS SOLEM AMBIENTIVM
AVCTORE
CAROLO FRIDERICO GAVSS

© Stock Montage

Published in 1809, Gauss's Theoria Motus Corporum Coelestium (Theory of the Motion of Heavenly Bodies) *contained rigorous methods of determining the orbits of planets and comets from observational data via the method of least squares. It is a landmark in the development of modern mathematical astronomy and statistics.*

The Standard Normal Distribution

All normal distributions share the following features: they are symmetric, bell-shaped curves, and virtually all the data (99.74%) lie within three standard deviations of the mean. Depending on whether the standard deviation is large or small, the bell curve will be either flat and spread out or peaked and narrow, as shown in Figure 4.105.

To find the area under any portion of any bell curve, mathematicians have devised a means of comparing the proportions of any curve with the proportions of a

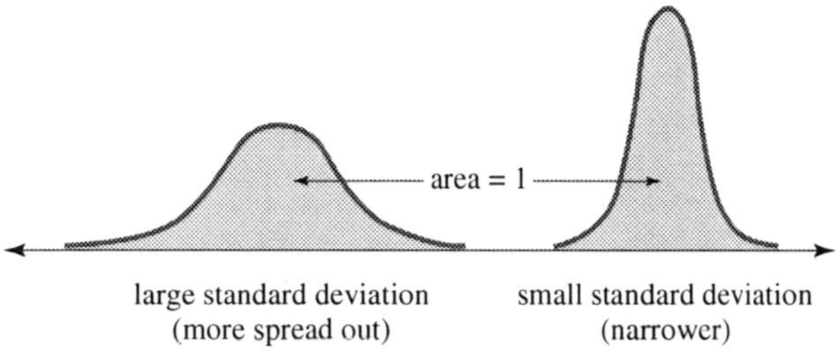

FIGURE 4.105 Large versus small standard deviation.

special curve defined as "standard." To find probabilities involving normally distributed data, we utilize the bell curve associated with the standard normal distribution.

The **standard normal distribution** is the normal distribution whose mean is 0 and standard deviation is 1, as shown in Figure 4.106. The standard normal distribution is also called the ***z*-distribution;** we will always use the letter z to refer to the standard normal. By convention, we will use the letter x to refer to any other normal distribution.

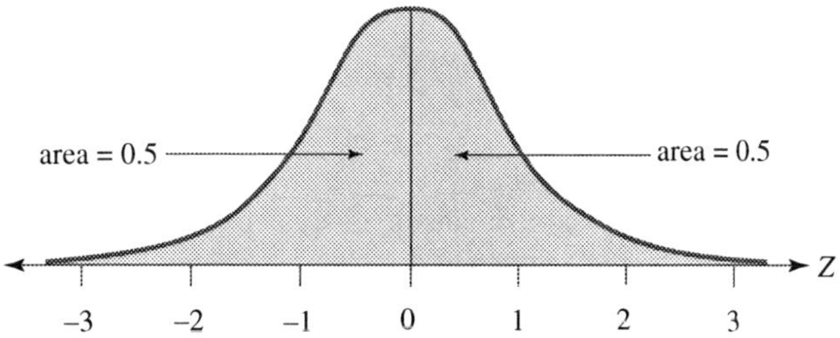

FIGURE 4.106 The standard normal distribution (mean = 0, standard deviation = 1).

Tables have been developed for finding areas under the standard normal curve using the techniques of calculus. Graphing calculators will also give these areas. We will use the table in Appendix F to find $p(0 < z < z^*)$, the probability that z is between 0 and a positive number z^*, as shown in Figure 4.107(a). The table in Appendix F is known as the **body table** because it gives the probability of an interval located in the middle, or body, of the bell curve.

The tapered end of a bell curve is known as a **tail.** To find the probability of a tail—that is, to find $p(z > z^*)$ or $p(z < z^*)$ where z^* is a positive real number—subtract the probability of the corresponding body from 0.5, as shown in Figure 4.107(b).

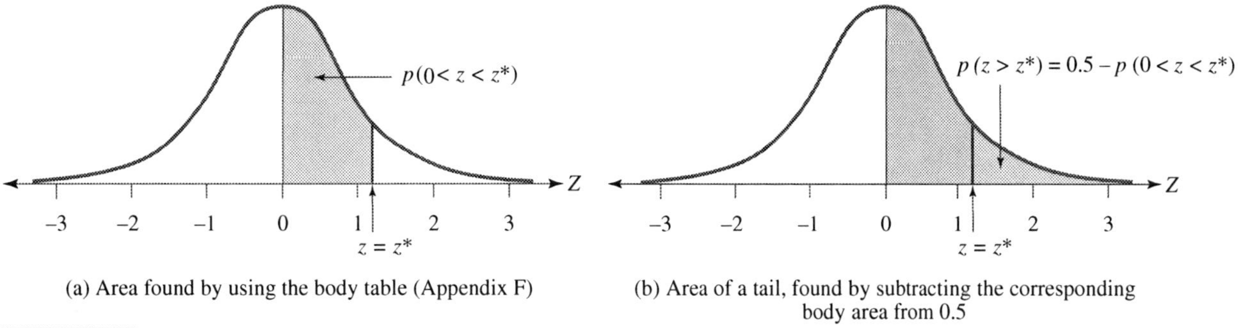

FIGURE 4.107

EXAMPLE 2

FINDING PROBABILITIES OF THE STANDARD NORMAL DISTRIBUTION Find the following probabilities (that is, the areas), where z represents the standard normal distribution.

a. $p(0 < z < 1.25)$ **b.** $p(z > 1.87)$

SOLUTION

a. As a first step, it is always advisable to draw a picture of the z-curve and shade in the desired area. We will use the body table directly, because we are working with a central area (see Figure 4.108).

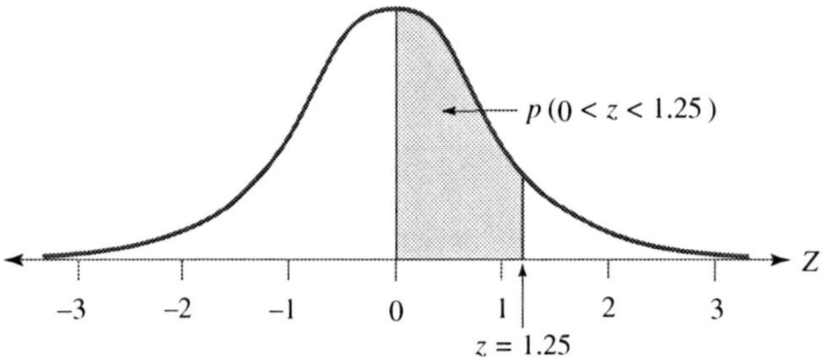

FIGURE 4.108 A central region, or body.

The z-numbers are located along the left edge and the top of the table. Locate the whole number and the first-decimal-place part of the number (1.2) along the left edge; then locate the second-decimal-place part of the number (0.05) along the top. The desired probability (area) is found at the intersection of the row and column of the two parts of the z-number. Thus, $p(0 < z < 1.25) = 0.3944$, as shown in Figure 4.109.

z*	0.00	0.01	0.02	0.03	0.04	0.05	0.06	0.07	0.08	0.09
∘										
∘										
∘										
1.1	0.3643	0.3665	0.3686	0.3708	0.3729	0.3749	0.3770	0.3790	0.3810	0.3830
1.2	0.3849	0.3869	0.3888	0.3907	0.3925	0.3944	0.3962	0.3980	0.3997	0.4015
1.3	0.4032	0.4049	0.4066	0.4082	0.4099	0.4115	0.4131	0.4147	0.4162	0.4177
∘										
∘										

FIGURE 4.109 A portion of the body table.

Hence, we could say that about 39% of the z-distribution lies between $z = 0$ and $z = 1.25$.

b. To find the area of a tail, we subtract the corresponding body area from 0.5, as shown in Figure 4.110. Therefore,

$$p(z > 1.87) = 0.5 - p(0 < z < 1.87)$$
$$= 0.5 - 0.4692$$
$$= 0.0308$$

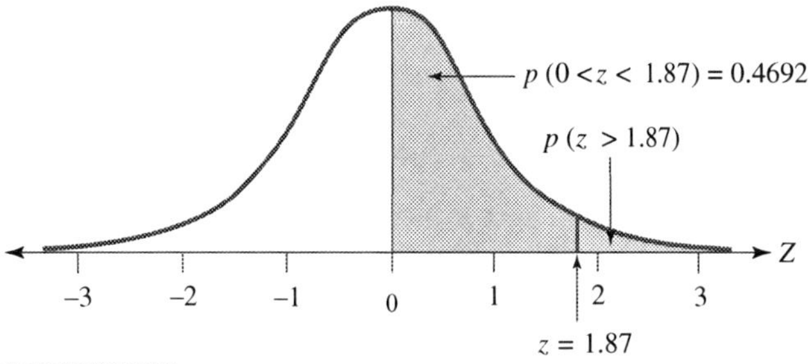

FIGURE 4.110 Finding the area of a tail.

The body table can also be used to find areas other than those given explicitly as $p(0 < z < z^*)$ and $p(z > z^*)$ where z^* is a positive number. By adding or subtracting two areas, we can find probabilities of the type $p(a < z < b)$, where a and b are positive or negative numbers, and probabilities of the type $p(z < c)$, where c is a positive or negative number.

EXAMPLE 3

FINDING PROBABILITIES OF THE STANDARD NORMAL DISTRIBUTION Find the following probabilities (the areas), where z represents the standard normal distribution.

a. $p(0.75 < z < 1.25)$ **b.** $p(-0.75 < z < 1.25)$

SOLUTION

a. Because the required region, shown in Figure 4.111, doesn't begin exactly at $z = 0$, we cannot look up the desired area directly in the body table. Whenever z is between two nonzero numbers, we will take an indirect approach to finding the required area.

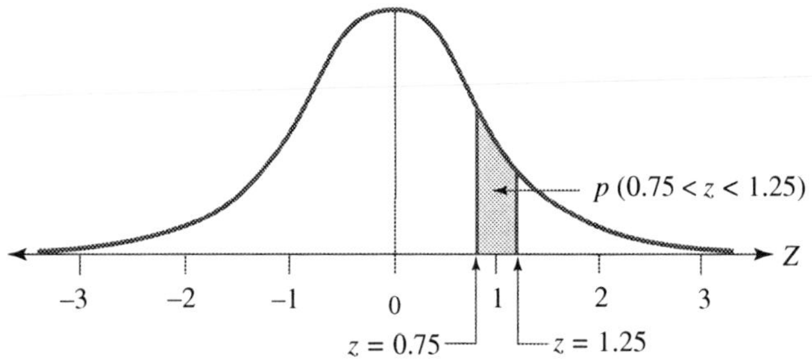

FIGURE 4.111 A strip.

The total area under the curve from $z = 0$ to $z = 1.25$ can be divided into two portions: the area under the curve from 0 to 0.75 and the area under the curve from 0.75 to 1.25.

To find the area of the "strip" between $z = 0.75$ and $z = 1.25$, we *subtract* the area of the smaller body (from $z = 0$ to $z = 0.75$) from that of the larger body (from $z = 0$ to $z = 1.25$), as shown in Figure 4.112.

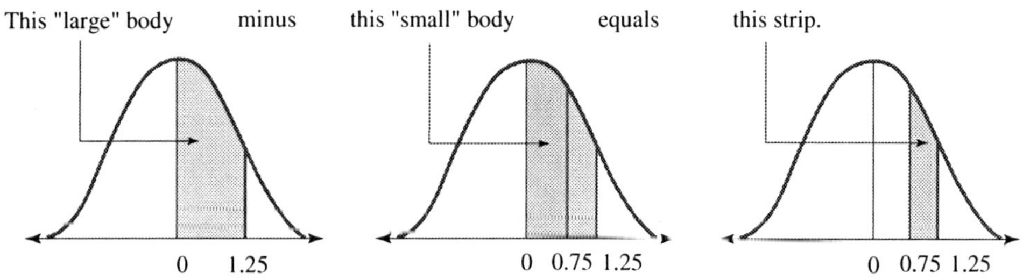

FIGURE 4.112 Area of a strip.

$$\text{area of strip} = \text{area of large body} - \text{area of small body}$$

$$\begin{aligned} p(0.75 < z < 1.25) &= p(0 < z < 1.25) - p(0 < z < 0.75) \\ &= 0.3944 - 0.2734 \\ &= 0.1210 \end{aligned}$$

Therefore, $p(0.75 < z < 1.25) = 0.1210$. Hence, we could say that about 12.1% of the z-distribution lies between $z = 0.75$ and $z = 1.25$.

b. The required region, shown in Figure 4.113, can be divided into two regions: the area from $z = -0.75$ to $z = 0$ and the area from $z = 0$ to $z = 1.25$.

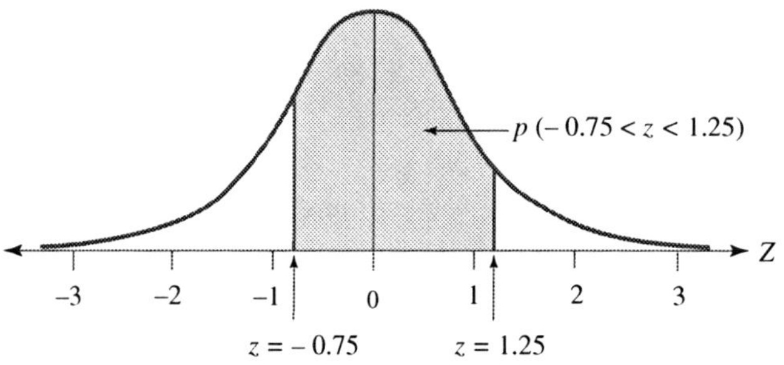

FIGURE 4.113 A central region.

To find the total area of the region between $z = -0.75$ and $z = 1.25$, we *add* the area of the "left" body (from $z = -0.75$ to $z = 0$) to the area of the "right" body (from $z = 0$ to $z = 1.25$), as shown in Figure 4.114.

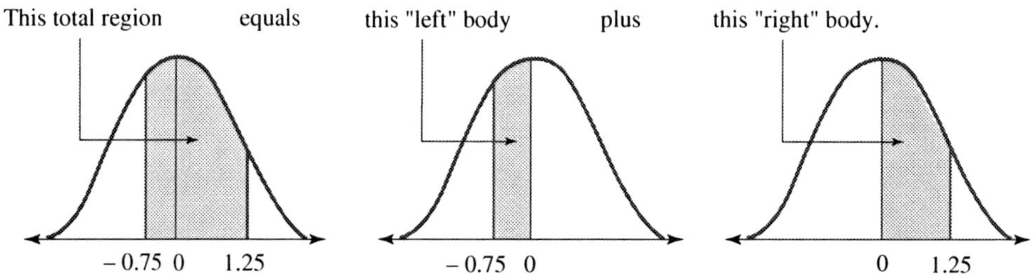

FIGURE 4.114 Left body plus right body.

This example is different from our previous examples in that it contains a negative z-number. A glance at the tables reveals that negative numbers are not included! However, recall that normal distributions are symmetric. Therefore, the area of the body from $z = -0.75$ to $z = 0$ is the same as that from $z = 0$ to $z = 0.75$; that is, $p(-0.75 < z < 0) = p(0 < z < 0.75)$. Therefore,

$$\text{total area of region} = \text{area of left body} + \text{area of right body}$$

$$\begin{aligned} p(-0.75 < z < 1.25) &= p(-0.75 < z < 0) + p(0 < z < 1.25) \\ &= p(0 < z < 0.75) + p(0 < z < 1.25) \\ &= 0.2734 + 0.3944 \\ &= 0.6678 \end{aligned}$$

Therefore, $p(-0.75 < z < 1.25) = 0.6678$. Hence, we could say that about 66.8% of the z-distribution lies between $z = -0.75$ and $z = 1.25$.

EXAMPLE 4

FINDING PROBABILITIES OF THE STANDARD NORMAL DISTRIBUTION Find the following probabilities (the areas), where z represents the standard normal distribution.

a. $p(z < 1.25)$

b. $p(z < -1.25)$

SOLUTION

a. The required region is shown in Figure 4.115. Because 50% of the distribution lies to the left of 0, we can add 0.5 to the area of the body from $z = 0$ to $z = 1.25$:

$$\begin{aligned} p(z < 1.25) &= p(z < 0) + p(0 < z < 1.25) \\ &= 0.5 + 0.3944 \\ &= 0.8944 \end{aligned}$$

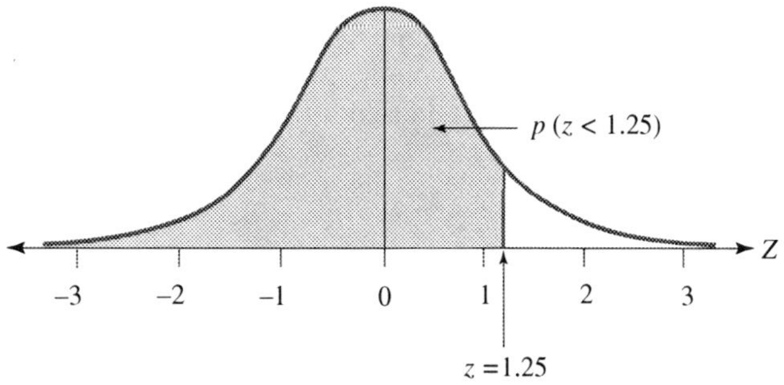

FIGURE 4.115 A body plus 50%.

Therefore, $p(z < 1.25) = 0.8944$. Hence, we could say that about 89.4% of the z-distribution lies to the left of $z = 1.25$.

b. The required region is shown in Figure 4.116. Because a normal distribution is symmetric, the area of the left tail ($z < -1.25$) is the same as the area of the corresponding right tail ($z > 1.25$).

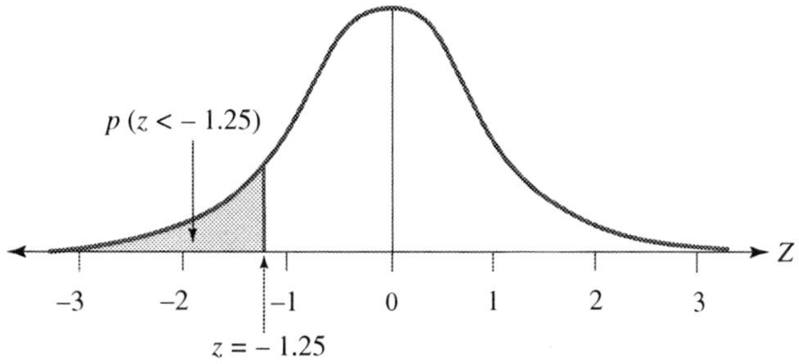

FIGURE 4.116 A left tail.

Therefore,

$$p(z < -1.25) = p(z > 1.25)$$
$$= 0.5 - p(0 < z < 1.25)$$
$$= 0.5 - 0.3944$$
$$= 0.1056$$

Hence, we could say that about 10.6% of the z-distribution lies to the left of $z = -1.25$.

Converting to the Standard Normal

Weather forecasters in the United States usually report temperatures in degrees Fahrenheit. Consequently, if a temperature is given in degrees Celsius, most people convert it to Fahrenheit in order to judge how hot or cold it is. A similar situation arises when we are working with a normal distribution. Suppose we know that a large set of data is normally distributed with a mean value of 68 and a standard deviation of 4. What percent of the data will lie between 65 and 73? We are asked to find $p(65 < x < 73)$. To find this probability, we must first convert the given normal distribution to the standard normal distribution and then look up the approximate z-numbers.

The body table (Appendix F) applies to the standard normal z-distribution. When we are working with any other normal distribution (denoted by X), we must first convert the x-distribution into the standard normal z-distribution. This conversion is done with the help of the following rule.

Given a number x, its corresponding z-number counts the number of standard deviations the number lies from the mean. For example, suppose the mean and standard deviation of a normal distribution are $\mu = 68$ and $\sigma = 4$. The z-number corresponding to $x = 78$ is

$$z = \frac{x - \mu}{\sigma} = \frac{78 - 68}{4} = 2.5$$

This implies that $x = 78$ lies two and one-half standard deviations above the mean, 68. Similarly, for $x = 65$,

$$z = \frac{65 - 68}{4} = -0.75$$

Therefore, $x = 65$ lies three-quarters of a standard deviation below the mean, 68.

CONVERTING A NORMAL DISTRIBUTION INTO THE STANDARD NORMAL Z

Every number x in a given normal distribution has a corresponding number z in the standard normal distribution. The **z-number** that corresponds to the number x is

$$z = \frac{x - \mu}{\sigma}$$

where μ is the mean and σ the standard deviation of the given normal distribution.

EXAMPLE 5

FINDING A PROBABILITY OF A NONSTANDARD NORMAL DISTRIBUTION Suppose a population is normally distributed with a mean of 24.6 and a standard deviation of 1.3. What percent of the data will lie between 25.3 and 26.8?

SOLUTION

We are asked to find $p(25.3 < x < 26.8)$, the area of the region shown in Figure 4.117. Because we need to find the area of the strip between 25.3 and 26.8, we must find the body of each and subtract, as in part (a) of Example 3.

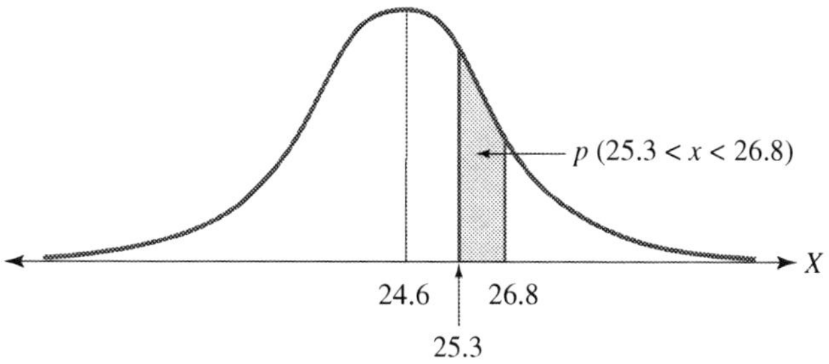

FIGURE 4.117 A strip.

Using the Conversion Formula $z = (x - \mu)/\sigma$ with $\mu = 24.6$ and $\sigma = 1.3$, we first convert $x = 25.3$ and $x = 26.8$ into their corresponding z-numbers.

Converting $x = 25.3$	Converting $x = 26.8$	
$z = \dfrac{x - \mu}{\sigma}$	$z = \dfrac{x - \mu}{\sigma}$	
$= \dfrac{25.3 - 24.6}{1.3}$	$= \dfrac{26.8 - 24.6}{1.3}$	
$= 0.5384615$	$= 1.6923077$	
$= 0.54$	$= 1.69$	rounding off z-numbers to two decimal places

Therefore,

$$\begin{aligned} p(25.3 < x < 26.8) &= p(0.54 < z < 1.69) \\ &= p(0 < z < 1.69) - p(0 < z < 0.54) \\ &= 0.4545 - 0.2054 \quad \text{using the body table} \\ &= 0.2491 \end{aligned}$$

Assuming a normal distribution, approximately 24.9% of the data will lie between 25.3 and 26.8.

EXAMPLE 6

FINDING PROBABILITIES OF A NONSTANDARD NORMAL DISTRIBUTION The heights of a large group of people are assumed to be normally distributed. Their mean height is 68 inches, and the standard deviation is 4 inches. What percentage of these people are the following heights?

a. taller than 73 inches **b.** between 60 and 75 inches

SOLUTION

a. Let x represent the height of a randomly selected person. We need to find $p(x > 73)$, the area of a tail, as shown in Figure 4.118.

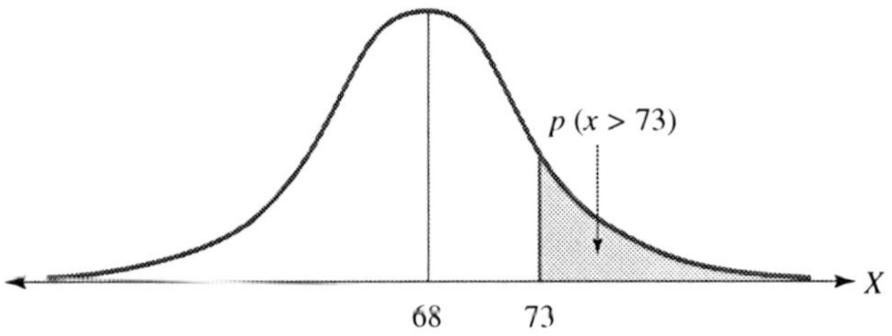

FIGURE 4.118 A right tail.

First, we must convert $x = 73$ to its corresponding z-number. Using the Conversion Formula with $x = 73$, $\mu = 68$, and $\sigma = 4$, we have

$$\begin{aligned} z &= \frac{x - \mu}{\sigma} \\ &= \frac{73 - 68}{4} \\ &= 1.25 \end{aligned}$$

Therefore,

$$\begin{aligned} p(x > 73) &= p(z > 1.25) \\ &= 0.5 - p(0 < z < 1.25) \\ &= 0.5 - 0.3944 \\ &= 0.1056 \end{aligned}$$

Approximately 10.6% of the people will be taller than 73 inches.

b. We need to find $p(60 < x < 75)$, the area of the central region shown in Figure 4.119. Notice that we will be adding the areas of the two bodies.

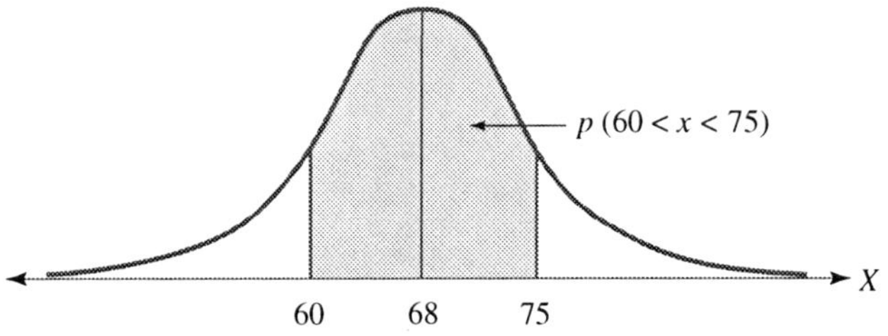

FIGURE 4.119 A central region.

First, we convert $x = 60$ and $x = 75$ to their corresponding z-numbers:

$$\begin{aligned} p(60 < x < 75) &= p\left(\frac{60 - 68}{4} < z < \frac{75 - 68}{4}\right) && \text{using the Conversion Formula} \\ &= p(-2.00 < z < 1.75) \\ &= p(-2.00 < z < 0) + p(0 < z < 1.75) && \text{expressing the area as two bodies} \\ &= p(0 < z < 2.00) + p(0 < z < 1.75) && \text{using symmetry} \\ &= 0.4772 + 0.4599 && \text{using the body table} \\ &= 0.9371 \end{aligned}$$

Approximately 93.7% of the people will be between 60 and 75 inches tall.

All the preceding examples involved finding probabilities that contained only the strict < or > inequalities, never ≤ or ≥ inequalities; the endpoints were never included. What if the endpoints are included? How does $p(a < x < b)$ compare with $p(a \leq x \leq b)$? Because probabilities for continuous data are found by determining *area* under a curve, including the endpoints does not affect the probability! The probability of a single point $p(x = a)$ is 0, because there is no "area" over a single point. (We obtain an area only when we are working with an interval of numbers.) Consequently, if x represents continuous data, then $p(a \leq x \leq b) = p(a < x < b)$; it makes no difference whether the endpoints are included.

EXAMPLE 7

FINDING THE VALUE OF A VARIABLE THAT WILL PRODUCE A SPECIFIC PROBABILITY Tall Dudes is a clothing store that specializes in fashions for tall men. Its informal motto is "Our customers are taller than 80% of the rest." Assuming the heights of men to be normally distributed with a mean of 67 inches and a standard deviation of 5.5 inches, find the heights of Tall Dudes' clientele.

SOLUTION

Let c = the height of the shortest customer at Tall Dudes, and let x represent the height of a randomly selected man. We are given that the heights of all men are normally distributed with $\mu = 67$ and $\sigma = 5.5$.

Assuming Tall Dudes' clientele to be taller than 80% of all men implies that $x < c$ 80% of the time and $x > c$ 20% of the time. Hence, we can say that the probability of selecting someone shorter than the shortest tall dude is $p(x < c) = 0.80$ and that the probability of selecting a tall dude is $p(x > c) = 0.20$, as shown in Figure 4.120.

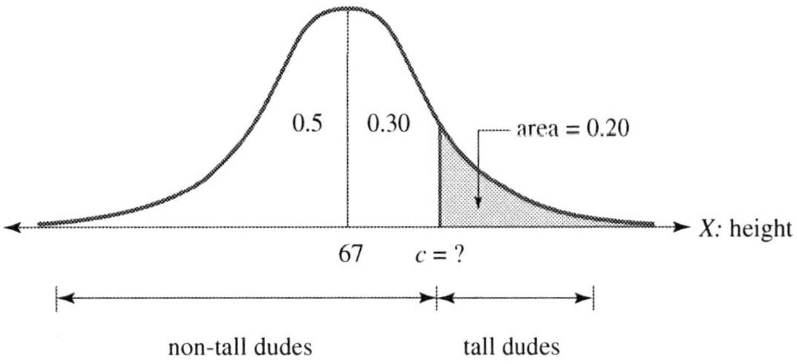

FIGURE 4.120 Finding a 20% right tail.

We are given that the area of the right tail is 0.20 (and that of the right body is 0.30), and we need to find the appropriate cutoff number c. This is exactly the reverse of all the previous examples, in which we were given the cutoff numbers and asked to find the area. Thus, our goal is to find the z-number that corresponds to a body of area 0.30 and convert it into its corresponding x-number.

When we scan through the *interior* of the body table, the number closest to the desired area of 0.30 is 0.2995, which is the area of the body when $z = 0.84$. This means that $p(0 < z < 0.84) = 0.2995$, as shown in Figure 4.121.

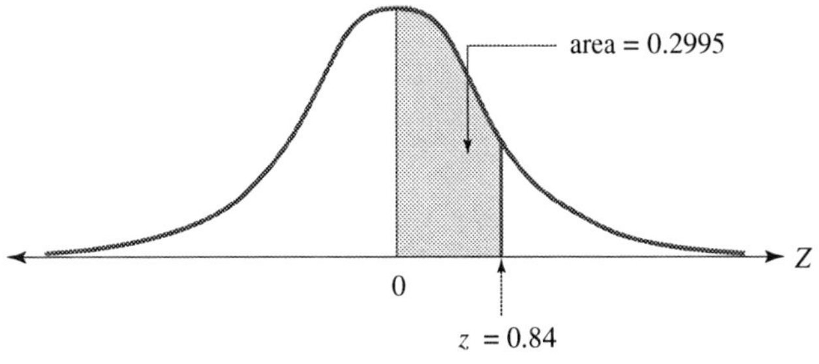

FIGURE 4.121 A body close to 30%.

Therefore, the number c that we are seeking lies 0.84 standard deviations above the mean. All that remains is to convert $z = 0.84$ into its corresponding x-number by substituting $x = c$, $z = 0.84$, $\mu = 67$, and $\sigma = 5.5$ into the Conversion Formula:

$$z = \frac{x - \mu}{\sigma}$$

$$0.84 = \frac{c - 67}{5.5}$$

$$(5.5)(0.84) = c - 67$$

$$4.62 = c - 67$$

$$c = 71.62 \quad (\approx 72 \text{ inches, or 6 feet})$$

Therefore, Tall Dudes caters to men who are at least 71.62 inches tall, or about 6 feet tall.

4.4 Exercises

1. The weights (in ounces) of several bags of corn chips are given in Figure 4.122. Construct a histogram for the data using the groups $15.40 \le x < 15.60$, $15.60 \le x < 15.80, \ldots, 16.60 \le x < 16.80$. Do the data appear to be approximately normally distributed? Explain.

16.08	16.49	15.61	16.66	15.80	15.87
16.02	15.82	16.48	16.08	15.63	16.02
16.00	16.25	15.41	16.22	16.04	15.68
16.45	16.41	16.01	15.82	16.08	15.82
16.29	16.26	16.05	16.25	15.86	

FIGURE 4.122 Weights (in ounces) of bags of corn chips.

2. The weights (in grams) of several bags of chocolate chip cookies are given in Figure 4.123. Construct a histogram for the data using the groups $420 \le x < 430$, $430 \le x < 440, \ldots, 480 \le x < 490$. Do the data appear to be approximately normally distributed? Explain.

451	435	482	449	454	451
479	448	432	423	461	475
467	453	448	459	454	444
475	461	450	466	446	458

FIGURE 4.123 Weights (in grams) of bags of chocolate chip cookies.

3. The time (in minutes) spent waiting in line for several students at the campus bookstore are given in Figure 4.124. Construct a histogram for the data using the groups $0 \le x < 1$, $1 \le x < 2, \ldots, 4 \le x < 5$. Do the data appear to be normally distributed? Explain.

1.75	2	0.5	1.5	0
1	2.5	0.75	2.25	3
1.5	3.5	2.5	1	0.5
1.75	0.25	4.5	0	1.5

FIGURE 4.124 Time (in minutes) spent waiting in line.

4. A die was rolled several times, and the results are given in Figure 4.125. Construct a histogram for the data using the single-values 1, 2, 3, 4, 5, and 6. Do the data appear to be normally distributed? Explain.

5	3	1	1	4	2
6	4	1	2	5	5
4	3	1	6	4	2
3	1	2	5	2	1
4	6	5	1	4	5

FIGURE 4.125 Rolling a die.

5. What percent of the standard normal z-distribution lies between the following values?
   a. $z = 0$ and $z = 1$
   b. $z = -1$ and $z = 0$
   c. $z = -1$ and $z = 1$
   (*Note:* This interval represents one standard deviation of the mean.)
6. What percent of the standard normal z-distribution lies between the following values?
   a. $z = 0$ and $z = 2$
   b. $z = -2$ and $z = 0$
   c. $z = -2$ and $z = 2$
   (*Note:* This interval represents two standard deviations of the mean.)
7. What percent of the standard normal z-distribution lies between the following values?
   a. $z = 0$ and $z = 3$
   b. $z = -3$ and $z = 0$
   c. $z = -3$ and $z = 3$
   (*Note:* This interval represents three standard deviations of the mean.)
8. What percent of the standard normal z-distribution lies between the following values?
   a. $z = 0$ and $z = 1.5$
   b. $z = -1.5$ and $z = 0$
   c. $z = -1.5$ and $z = 1.5$
   (*Note:* This interval represents one and one-half standard deviations of the mean.)
9. A population is normally distributed with mean 24.7 and standard deviation 2.3.
   a. Find the intervals representing one, two, and three standard deviations of the mean.
   b. What percentage of the data lies in each of the intervals in part (a)?
   c. Draw a sketch of the bell curve.
10. A population is normally distributed with mean 18.9 and standard deviation 1.8.
   a. Find the intervals representing one, two, and three standard deviations of the mean.
   b. What percent of the data lies in each of the intervals in part (a)?
   c. Draw a sketch of the bell curve.
11. Find the following probabilities.
   a. $p(0 < z < 1.62)$
   b. $p(1.30 < z < 1.84)$
   c. $p(-0.37 < z < 1.59)$
   d. $p(z < -1.91)$
   e. $p(-1.32 < z < -0.88)$
   f. $p(z < 1.25)$
12. Find the following probabilities.
   a. $p(0 < z < 1.42)$
   b. $p(1.03 < z < 1.66)$
   c. $p(-0.87 < z < 1.71)$
   d. $p(z < -2.06)$
   e. $p(-2.31 < z < -1.18)$
   f. $p(z < 1.52)$
13. Find c such that each of the following is true.
   a. $p(0 < z < c) = 0.1331$
   b. $p(c < z < 0) = 0.4812$
   c. $p(-c < z < c) = 0.4648$
   d. $p(z > c) = 0.6064$
   e. $p(z > c) = 0.0505$
   f. $p(z < c) = 0.1003$
14. Find c such that each of the following is true.
   a. $p(0 < z < c) = 0.3686$
   b. $p(c < z < 0) = 0.4706$
   c. $p(-c < z < c) = 0.2510$
   d. $p(z > c) = 0.7054$
   e. $p(z > c) = 0.0351$
   f. $p(z < c) = 0.2776$
15. A population X is normally distributed with mean 250 and standard deviation 24. For each of the following values of x, find the corresponding z-number. Round off your answers to two decimal places.
   a. $x = 260$ b. $x = 240$
   c. $x = 300$ d. $x = 215$
   e. $x = 321$ f. $x = 197$
16. A population X is normally distributed with mean 72.1 and standard deviation 9.3. For each of the following values of x, find the corresponding z-number. Round off your answers to two decimal places.
   a. $x = 90$ b. $x = 80$
   c. $x = 75$ d. $x = 70$
   e. $x = 60$ f. $x = 50$
17. A population is normally distributed with mean 36.8 and standard deviation 2.5. Find the following probabilities.
   a. $p(36.8 < x < 39.3)$ b. $p(34.2 < x < 38.7)$
   c. $p(x < 40.0)$ d. $p(32.3 < x < 41.3)$
   e. $p(x = 37.9)$ f. $p(x > 37.9)$
18. A population is normally distributed with mean 42.7 and standard deviation 4.7. Find the following probabilities.
   a. $p(42.7 < x < 47.4)$ b. $p(40.9 < x < 44.1)$
   c. $p(x < 50.0)$ d. $p(33.3 < x < 52.1)$
   e. $p(x = 45.3)$ f. $p(x > 45.3)$

19. The mean weight of a box of cereal filled by a machine is 16.0 ounces, with a standard deviation of 0.3 ounce. If the weights of all the boxes filled by the machine are normally distributed, what percent of the boxes will weigh the following amounts?
   a. less than 15.5 ounces
   b. between 15.8 and 16.2 ounces

20. The amount of time required to assemble a component on a factory assembly line is normally distributed with a mean of 3.1 minutes and a standard deviation of 0.6 minute. Find the probability that a randomly selected employee will take the given amount of time to assemble the component.
   a. more than 4.0 minutes
   b. between 2.0 and 2.5 minutes

21. The time it takes an acrylic paint to dry is normally distributed. If the mean is 2 hours 36 minutes with a standard deviation of 24 minutes, find the probability that the drying time will be as follows.
   a. less than 2 hours 15 minutes
   b. between 2 and 3 hours

   HINT: Convert everything to minutes (or to hours).

22. The shrinkage in length of a certain brand of blue jeans is normally distributed with a mean of 1.1 inches and a standard deviation of 0.2 inch. What percent of this brand of jeans will shrink the following amounts?
   a. more than 1.5 inches
   b. between 1.0 and 1.25 inches

23. The mean volume of a carton of milk filled by a machine is 1.0 quart, with a standard deviation of 0.06 quart. If the volumes of all the cartons are normally distributed, what percent of the cartons will contain the following amounts?
   a. at least 0.9 quart
   b. at most 1.05 quarts

24. The amount of time between taking a pain reliever and getting relief is normally distributed with a mean of 23 minutes and a standard deviation of 4 minutes. Find the probability that the time between taking the medication and getting relief is as follows.
   a. at least 30 minutes
   b. at most 20 minutes

25. The results of a statewide exam for assessing the mathematics skills of realtors were normally distributed with a mean score of 72 and a standard deviation of 12. The realtors who scored in the top 10% are to receive a special certificate, while those in the bottom 20% will be required to attend a remedial workshop.
   a. What score does a realtor need in order to receive a certificate?
   b. What score will dictate that the realtor attend the workshop?

26. Professor Harde assumes that exam scores are normally distributed and wants to grade "on the curve." The mean score was 58, with a standard deviation of 16.
   a. If she wants 14% of the students to receive an A, find the minimum score to receive an A.
   b. If she wants 19% of the students to receive a B, find the minimum score to receive a B.

27. The time it takes an employee to package the components of a certain product is normally distributed with $\mu = 8.5$ minutes and $\sigma = 1.5$ minutes. To boost productivity, management has decided to give special training to the 34% of employees who took the greatest amount of time to package the components. Find the amount of time taken to package the components that will indicate that an employee should get special training.

28. The time it takes an employee to package the components of a certain product is normally distributed with $\mu = 8.5$ and $\sigma = 1.5$ minutes. As an incentive, management has decided to give a bonus to the 20% of employees who took the shortest amount of time to package the components. Find the amount of time taken to package the components that will indicate that an employee should get a bonus.

Answer the following questions using complete sentences and your own words.

• CONCEPT QUESTIONS

29. What are the characteristics of a normal distribution?
30. Are all distributions of data normally distributed? Support your answer with an example.
31. Why is the total area under a bell curve equal to 1?
32. Why are there no negative z-numbers in the body table?
33. When converting an x-number to a z-number, what does a negative z-number tell you about the location of the x-number?
34. Is it logical to assume that the heights of all high school students in the United States are normally distributed? Explain.
35. Is it reasonable to assume that the ages of all high school students in the United States are normally distributed? Explain.

• HISTORY QUESTIONS

36. Who is known as "the Prince of Mathematics"? Why?
37. What mathematician was instrumental in the creation of the normal distribution? What application prompted this person to create the normal distribution?

4.5 Polls and Margin of Error

OBJECTIVES

- Find the margin of error in a poll
- Determine the effect of sample size on the margin of error
- Find the level of confidence for a specific sample and margin of error
- Find the minimum sample size to obtain a specified margin of error

One of the most common applications of statistics is the evaluation of the results of surveys and public opinion polls. Most editions of the daily newspaper contain the results of at least one poll. Headlines announce the attitude of the nation toward a myriad of topics ranging from the actions of politicians to controversial current issues, such as abortion, as shown in the newspaper article on the next page. How are these conclusions reached? What do they mean? How valid are they? In this section, we investigate these questions and obtain results concerning the "margin of error" associated with the reporting of "public opinion."

Sampling and Inferential Statistics

The purpose of conducting a survey or poll is to obtain information about a population—for example, adult Americans. Because there are approximately 230 million Americans over the age of eighteen, it would be very difficult, time-consuming, and expensive to contact every one of them. The only realistic alternative is to poll a sample and use the science of inferential statistics to draw conclusions about the population as a whole. Different samples have different characteristics depending on, among other things, the age, sex, education, and locale of the people in the sample. Therefore, it is of the utmost importance that a sample be representative of the population. Obtaining a representative sample is the most difficult aspect of inferential statistics.

Another problem facing pollsters is determining *how many* people should be selected for the sample. Obviously, the larger the sample, the more likely that it will reflect the population. However, larger samples cost more money, so a limited budget will limit the sample size. Conducting surveys can be very costly, even for a small to moderate sample. For example, a survey conducted in 1989 by the Gallup Organization that contacted 1,005 adults and 500 teenagers would have cost $100,000 (the pollsters donated their services for this survey). The results of this poll indicated that Americans thought the "drug crisis" was the nation's top problem (stated by 27% of the adults and 32% of the teenagers).

After a sample has been selected and its data have been analyzed, information about the sample is generalized to the entire population. Because 27% of the 1,005 adults in a poll stated that the drug crisis was the nation's top problem, we would like to conclude that 27% of *all* adults have the same belief. Is this a valid generalization? That is, how confident is the pollster that the feelings of the people in the sample reflect those of the population?

FEATURED IN THE NEWS

MORE AMERICANS "PRO-LIFE" THAN "PRO-CHOICE" FOR THE FIRST TIME

Princeton, NJ—A new Gallup Poll, conducted May 7–10, 2009, finds 51% of Americans calling themselves "pro-life" on the issue of abortion and 42% "pro-choice." This is the first time a majority of U.S. adults have identified themselves as pro-life since Gallup began asking this question in 1995. The new results, obtained from Gallup's annual Values and Beliefs survey, represent a significant shift from a year ago, when 50% were pro-choice and 44% pro-life. Prior to now, the highest percentage identifying as pro-life was 46%, in both August 2001 and May 2002.

The source of the shift in abortion views is clear in the Gallup Values and Belief survey. The percentage of Republicans (including independents who lean Republican) calling themselves "pro-life" rose by 10 points over the past year, from 60% to 70%, while there has been essentially no change in the views of Democrats and Democratic leaners. Similarly, by ideology, all of the increase in pro-life sentiment is seen among self-identified conservatives and moderates; the abortion views of political liberals have not changed.

A year ago, Gallup found more women calling themselves pro-choice than pro-life, by 50% to 43%, while men were more closely divided: 49% pro-choice, 46% pro-life. Now, because of heightened pro-life sentiment among both groups, women as well as men are more likely to be pro-life. Men and women have been evenly divided on the issue in previous years; however, this is the first time in nine years of Gallup Values surveys that significantly more men and women are pro-life than pro-choice.

Results are based on telephone interviews with 1,015 national adults, aged 18 and older, conducted May 7–10, 2009. For results based on the total sample of national adults, one can say with 95% confidence that the maximum margin of sampling error is ± 3 percentage points.

By Lydia Saad
Gallup News Service

Sample Proportion versus Population Proportion

If x members (for example, people, automobiles, households) in a sample of size n have a certain characteristic, then the proportion of the sample, or **sample proportion,** having this characteristic is given by $\frac{x}{n}$. For instance, in a sample of $n = 70$ automobiles, if $x = 14$ cars have a defective fan switch, then the proportion of the sample having a defective switch is $\frac{14}{70} = 0.2$, or 20%. The true proportion of the entire population, or **population proportion,** having the characteristic is represented by the letter P. A sample proportion $\frac{x}{n}$ is an estimate of the population proportion P.

Sample proportions $\frac{x}{n}$ vary from sample to sample; some will be larger than P, and some will be smaller. Of the 1,005 adults in the Gallup Poll sample mentioned on page 295, 27% viewed the drug crisis as the nation's top problem. If a different sample of 1,005 had been chosen, 29% might have had this view. If still another 1,005 had been selected, this view might have been shared by only 25%. We will assume that the sample proportions $\frac{x}{n}$ are normally distributed around the population proportion P. The set of all sample proportions, along with their probabilities of occurring, can be represented by a bell curve like the one in Figure 4.126.

In general, a sample estimate is not 100% accurate; although a sample proportion might be close to the true population proportion, it will have an error term associated with it. The difference between a sample estimate and the true (population) value is called the **error of the estimate.** We can use a bell curve (like the one in Figure 4.126) to predict the probable error of a sample estimate.

Before developing this method of predicting the error term, we need to introduce some special notation. The symbol z_α (read "z alpha") will be used to represent the positive z-number that has a right body of area α. That is, z_α is the number such that $p(0 < z < z_\alpha) = \alpha$, as shown in Figure 4.127.

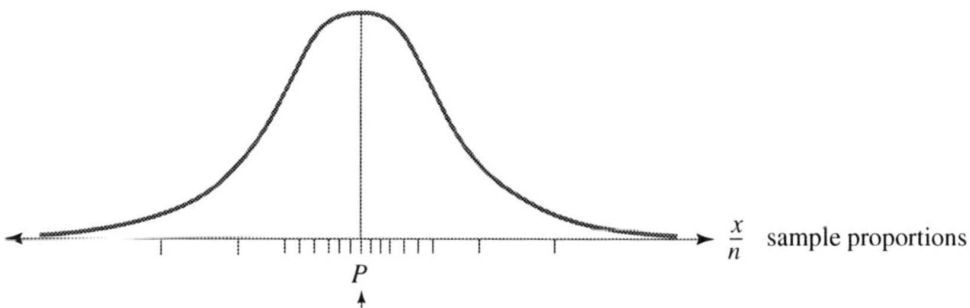

FIGURE 4.126 Sample proportions normally distributed around the true (population) proportion.

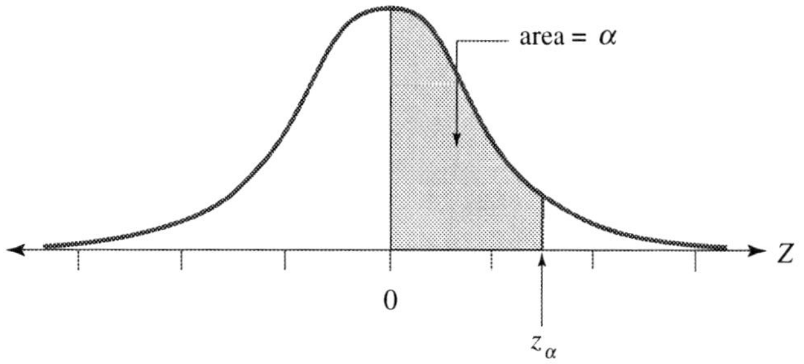

FIGURE 4.127 $p(0 < z < z_\alpha) = \alpha$.

EXAMPLE 1

FINDING A VALUE OF THE STANDARD NORMAL DISTRIBUTION THAT WILL PRODUCE A SPECIFIC PROBABILITY Use the body table in Appendix F to find the following values:

a. $z_{0.3925}$ **b.** $z_{0.475}$ **c.** $z_{0.45}$ **d.** $z_{0.49}$

SOLUTION

a. $z_{0.3925}$ represents the z-number that has a body of area 0.3925. Looking at the interior of the body table, we find 0.3925 and see that it corresponds to the z-number 1.24—that is, $p(0 < z < 1.24) = 0.3925$. Therefore, $z_{0.3925} = 1.24$.

b. In a similar manner, we find that a body of area 0.4750 corresponds to $z = 1.96$. Therefore, $z_{0.475} = 1.96$.

c. Looking through the interior of the table, we cannot find a body of area 0.45. However, we do find a body of 0.4495 (corresponding to $z = 1.64$) and a body of 0.4505 (corresponding to $z = 1.65$). Because the desired body is *exactly halfway* between the two listed bodies, we use a z-number that is exactly halfway between the two listed z-numbers, 1.64 and 1.65. Therefore, $z_{0.45} = 1.645$.

d. We cannot find a body of the desired area, 0.49, in the interior of the table. The closest areas are 0.4898 (corresponding to $z = 2.32$) and 0.4901 (corresponding to $z = 2.33$). Because the desired area (0.49) is *closer* to 0.4901, we use $z = 2.33$. Therefore, $z_{0.49} = 2.33$.

Margin of Error

Sample proportions $\frac{x}{n}$ vary from sample to sample; some will have a small error, and some will have a large error. Knowing that sample estimates have inherent errors, statisticians make predictions concerning the largest possible error associated with

HISTORICAL NOTE

GEORGE H. GALLUP, 1901–1984

© Corbis/Bettmann

To many people, the name *Gallup* is synonymous with opinion polls. George Horace Gallup, the founder of the American Institute of Public Opinion, began his news career while attending the University of Iowa. During his junior year as a student of journalism, Gallup became the editor of his college newspaper, the *Daily Iowan*. After receiving his bachelor's degree in 1923, Gallup remained at the university nine years as an instructor of journalism.

In addition to teaching, Gallup continued his own studies of human nature and public opinion. Interest in how the public reacts to advertisements and perceives various issues of the day led Gallup to combine his study of journalism with the study of psychology. In 1925, he received his master's degree in psychology. Gallup's studies culminated in 1928 with his doctoral thesis, *A New Technique for Objective Methods for Measuring Reader Interest in Newspapers*. Gallup's new technique of polling the public was to utilize a stratified sample, that is, a sample that closely mirrors the composition of the entire population. Gallup contended that a stratified sample of 1,500 people was sufficient to obtain reliable estimates. For his pioneering work in this new field, Gallup was awarded his Ph.D. in journalism in 1928.

Gallup founded the American Institute of Public Opinion in 1935 with the stated purpose "to impartially measure and report public opinion on political and social issues of the day without regard to the rightness or wisdom of the views expressed." His first triumph was his prediction of the winner of the 1936 presidential election between Franklin D. Roosevelt and Alfred Landon. While many, including the prestigious *Literary Digest*, predicted that Landon would win, Gallup correctly predicted Roosevelt as the winner.

Gallup Polls have correctly predicted all presidential elections since, with the exception of the 1948 race between Thomas Dewey and Harry S Truman. Much to his embarrassment, Gallup predicted Dewey as the winner. Truman won the election with 49.9% of the vote, while Gallup had predicted that he would receive only 44.5%. Gallup's explanation was that he had ended his poll too far in advance of election day and had disregarded the votes of those who were undecided. Of his error, Gallup said, "We are continually experimenting and continually learning."

Although some people criticize the use of polls, citing their potential influence and misuse, Gallup considered the public opinion poll to be "one of the most useful instruments of democracy ever devised." Answering the charge that he and his polls influenced elections, Gallup retorted, "One might as well insist that a thermometer makes the weather!" In addition, Gallup confessed that he had not voted in a presidential election since 1928. Above all, Gallup wanted to ensure the impartiality of his polls.

Besides polling people regarding their choices in presidential campaigns, Gallup was the first pollster to ask the public to rate a president's performance and popularity. Today, these "presidential report cards" are so common we may take them for granted. In addition to presidential politics, Gallup also dealt with sociological issues, asking questions such as "What is the most important problem facing the country?"

Polling has become a multimillion-dollar business. In 2009, the Gallup Organization had revenues totaling $271.6 million, and it had over 2,000 employees. Today, Gallup Polls are syndicated in newspapers across the country and around the world.

AP Photo/Byron Rollins

The Gallup Organization predicted that Thomas Dewey would win the 1948 presidential election. Much to Gallup's embarrassment, Harry S. Truman won the election and triumphantly displayed a newspaper containing Gallup's false prediction. With the exception of this election, Gallup Polls have correctly predicted every presidential election since 1936.

a sample estimate. This error is called the **margin of error** of the estimate and is denoted by **MOE.** Because the margin of error is a prediction, we cannot guarantee that it is absolutely correct; that is, the probability that a prediction is correct might be 0.95, or it might be 0.75.

In the field of inferential statistics, the probability that a prediction is correct is referred to as the **level of confidence** of the prediction. For example, we might say that we are 95% confident that the maximum error of an opinion poll is plus or minus 3 percentage points; that is, if 100 samples were analyzed, 95 of them would have proportions that differ from the true population proportion by an amount less than or equal to 0.03, and 5 of the samples would have an error greater than 0.03.

Assuming that sample proportions are normally distributed around the population proportion (as in Figure 4.126), we can use the z-distribution to determine the margin of error associated with a sample proportion. In general, the margin of error depends on the sample size and the level of confidence of the estimate.

MARGIN OF ERROR FORMULA

Given a sample size n, the **margin of error,** denoted by **MOE,** for a poll involving sample proportions is

$$\text{MOE} = \frac{z_{\alpha/2}}{2\sqrt{n}}$$

where α represents the level of confidence of the poll. That is, the probability is α that the sample proportion has an error of at most MOE.

EXAMPLE 2

FINDING THE MARGIN OF ERROR Assuming a 90% level of confidence, find the margin of error associated with each of the following:

a. sample size $n = 275$ **b.** sample size $n = 750$

SOLUTION

a. The margin of error depends on two things: the sample size and the level of confidence. For a 90% level of confidence, $\alpha = 0.90$. Hence, $\frac{\alpha}{2} = 0.45$, and $z_{\alpha/2} = z_{0.45} = 1.645$.

Substituting this value and $n = 275$ into the MOE formula, we have the following:

$$\begin{aligned}\text{MOE} &= \frac{z_{\alpha/2}}{2\sqrt{n}}\\ &= \frac{1.645}{2\sqrt{275}}\\ &= 0.049598616\ldots\\ &\approx 0.050 && \text{rounding off to three decimal places}\\ &= 5.0\%\end{aligned}$$

When we are polling a sample of 275 people, we can say that we are 90% confident that the maximum possible error in the sample proportion will be plus or minus 5.0 percentage points.

b. For a 90% level of confidence, $\alpha = 0.90$, $\frac{\alpha}{2} = 0.45$, and $z_{\alpha/2} = z_{0.45} = 1.645$. Substituting this value and $n = 750$ into the MOE formula, we have the following:

$$\begin{aligned}\text{MOE} &= \frac{z_{\alpha/2}}{2\sqrt{n}}\\ &= \frac{1.645}{2\sqrt{750}}\\ &= 0.030033453\ldots\\ &\approx 0.030 && \text{rounding off to three decimal places}\\ &= 3.0\%\end{aligned}$$

When we are polling a sample of 750 people, we can say that we are 90% confident that the maximum possible error in the sample proportion will be plus or minus 3.0 percentage points.

If we compare the margins of error in parts (a) and (b) of Example 2, we notice that by increasing the sample size (from 275 to 750), the margin of error was reduced (from 5.0 to 3.0 percentage points). Intuitively, this should make sense; a larger sample gives a better estimate (has a smaller margin of error).

EXAMPLE 3

FINDING A SAMPLE PROPORTION AND MARGIN OF ERROR To obtain an estimate of the proportion of all Americans who think the president is doing a good job, a random sample of 500 Americans is surveyed, and 345 respond, "The president is doing a good job."

a. Determine the sample proportion of Americans who think the president is doing a good job.
b. Assuming a 95% level of confidence, find the margin of error associated with the sample proportion.

SOLUTION

a. $n = 500$ and $x = 345$. The sample proportion is

$$\frac{x}{n} = \frac{345}{500} = 0.69$$

Sixty-nine percent of the sample think the president is doing a good job.

b. We must find MOE when $n = 500$ and $\alpha = 0.95$. Because $\alpha = 0.95$, $\frac{\alpha}{2} = 0.475$ and $z_{\alpha/2} = z_{0.475} = 1.96$.

Therefore,

$$\begin{aligned} \text{MOE} &= \frac{z_{\alpha/2}}{2\sqrt{n}} \\ &= \frac{1.96}{2\sqrt{500}} \\ &= 0.043826932\ldots \\ &\approx 0.044 && \text{rounding off to three decimal places} \\ &= 4.4\% \end{aligned}$$

The margin of error associated with the sample proportion is plus or minus 4.4 percentage points. We are 95% confident that 69% (±4.4%) of all Americans think the president is doing a good job. In other words, on the basis of our sample proportion ($\frac{x}{n}$) of 69%, we predict (with 95% certainty) that the true population proportion (P) is somewhere between 64.6% and 73.4%.

Example 2 indicated that increasing the sample size will decrease the margin of error, that is, larger samples give better estimates. If larger samples give better estimates, how large should a sample be? This question can be answered by manipulating the margin of error formula. That is, in its original form, we plug values of $z_{\alpha/2}$ (based upon the level of confidence) and n (the sample size) into the formula, and we calculate the margin of error. However, if we first solve the margin of error formula for n, we will have a new version of the formula for determining how large a sample should be. We proceed as follows:

$$\text{MOE} = \frac{z_{\alpha/2}}{2\sqrt{n}} \quad \text{the margin of error formula}$$

$$\sqrt{n}\cdot(\text{MOE}) = \frac{z_{\alpha/2}}{2} \quad \text{multiplying each side by } \sqrt{n}$$

$$\sqrt{n} = \frac{z_{\alpha/2}}{2(\text{MOE})} \quad \text{dividing each side by MOE}$$

$$n = \left(\frac{z_{\alpha/2}}{2(\text{MOE})}\right)^2 \quad \text{squaring both sides}$$

SAMPLE SIZE FORMULA

The required sample size n, necessary to have a desired margin of error of at most MOE, is given

$$n = \left(\frac{z_{\alpha/2}}{2(\text{MOE})}\right)^2$$

where $z_{\alpha/2}$ is determined by the given level of confidence.

EXAMPLE 4

FINDING A SAMPLE SIZE With a 98% level of confidence, how large should a sample be so that the margin of error is at most 4%?

SOLUTION

For a 98% level of confidence, $\alpha = 0.98$, $\frac{\alpha}{2} = 0.49$, and $z_{\alpha/2} = z_{0.49} = 2.33$. Substituting this value and MOE = 0.04 into the sample size formula, we have

$$\begin{aligned} n &= \left(\frac{z_{\alpha/2}}{2(\text{MOE})}\right)^2 \\ &= \left(\frac{2.33}{2(0.04)}\right)^2 \\ &= 848.265625\ldots \end{aligned}$$

Therefore, we should have roughly 848 and "one-quarter" people ($0.265625 \approx 0.25 = \frac{1}{4}$) in the sample. However, we cannot have part of a person, so it is customary to round this number *up* to the next highest whole number (include the whole person!); thus, the required sample size is $n = 849$ people.

EXAMPLE 5

FINDING THE MARGIN OF ERROR FOR DIFFERENT LEVELS OF CONFIDENCE The article shown on the next page was released by the Gallup Organization in October 2009.

a. The poll states that 44% of the Americans questioned think that the laws covering firearm sales should be made more strict. Assuming a 95% level of confidence (the most commonly used level of confidence), find the margin of error associated with the survey.

b. Assuming a 98% level of confidence, find the margin of error associated with the survey.

FEATURED IN THE NEWS

IN U.S., RECORD-LOW SUPPORT FOR STRICTER GUN LAWS

Princeton, NJ—Gallup finds a new low of 44% of Americans saying the laws covering firearm sales should be made more strict. That is down 5 points in the last year and 34 points from the high of 78% recorded the first time the question was asked, in 1990.

Today, Americans are as likely to say the laws governing gun sales should be kept as they are now (43%) as to say they should be made more strict. Until this year, Gallup had always found a significantly higher percentage advocating stricter laws. At the same time, 12% of Americans believe the laws should be less strict, which is low in an absolute sense but ties the highest Gallup has measured for this response.

These results are based on Gallup's annual Crime Poll, conducted October 1–4, 2009.

The Poll also shows a new low in the percentage of Americans favoring a ban on handgun possession except by the police and other authorized persons, a question that dates back to 1959. Only 28% now favor such a ban. The high point in support for a handgun-possession ban was 60% in the initial measurement in 1959. Since then, less than a majority has been in favor, and support has been below 40% since December 1993.

The trends on the questions about gun-sale laws and a handgun-possession ban indicate that Americans' attitudes have moved toward being more pro-gun rights. But this is not due to a growth in personal gun ownership, which has held steady around 30% this decade, or to an increase in household gun ownership, which has been steady in the low 40% range since 2000.

Results are based on telephone interviews with 1,013 national adults, aged 18 and older, conducted October 1–4, 2009.

By Jeffrey M. Jones
Gallup News Service

SOLUTION

a. We must find MOE when $n = 1{,}013$ and $\alpha = 0.95$. Because $\alpha = 0.95$, $\alpha/2 = 0.475$ and $z_{\alpha/2} = z_{0.475} = 1.96$.

Therefore,

$$\begin{aligned} \text{MOE} &= \frac{z_{\alpha/2}}{2\sqrt{n}} \\ &= \frac{1.96}{2\sqrt{1{,}013}} \\ &= 0.030790827\ldots \\ &\approx 0.031 \qquad \text{rounding off to three decimal places} \end{aligned}$$

The margin of error associated with the survey is plus or minus 3.1%. We are 95% confident that 44% ± 3.1% of all Americans think that the laws covering firearm sales should be made more strict.

b. We must find MOE when $n = 1{,}013$ and $\alpha = 0.98$. Because $\alpha = 0.98$, $\alpha/2 = 0.49$ and $z_{\alpha/2} = z_{0.49} = 2.33$.

Therefore,

$$\begin{aligned} \text{MOE} &= \frac{z_{\alpha/2}}{2\sqrt{n}} \\ &= \frac{2.33}{2\sqrt{1{,}013}} \\ &= 0.036603381\ldots \\ &\approx 0.037 \qquad \text{rounding off to three decimal places} \end{aligned}$$

The margin of error associated with the survey is plus or minus 3.7%. We are 98% confident that 44% ± 3.7% of all Americans think that the laws covering firearm sales should be made more strict.

FEATURED IN THE NEWS

WHO SUPPORTS MARIJUANA LEGALIZATION?

SUPPORT RISING; VARIES MOST BY AGE AND GENDER

Since the late 1960s, Gallup has periodically asked Americans whether the use of marijuana should be made legal in the United States. Although a majority of Americans have consistently opposed the idea of legalizing marijuana, public support has slowly increased over the years. In 1969, just 12% of Americans supported making marijuana legal, but by 1977, roughly one in four endorsed it. Support edged up to 31% in 2000, and now, about a third of Americans say marijuana should be legal.

Support for marijuana legalization varies greatest by gender and age. Overall, younger Americans (aged 18 to 29) are essentially divided, with 47% saying marijuana should be legal and 50% saying it should not be. Support for legalization is much lower among adults aged 30 to 64 (35%) and those aged 65 and older (22%). Men (39%) are somewhat more likely than women (30%) to support the legalization of marijuana in the country.

Americans residing in the western parts of the country are more likely than those living elsewhere to support the legalization of marijuana. These differences perhaps result from the fact that six Western states have, in various ways, already legalized marijuana for medicinal use. Overall, the data show that Westerners are divided about marijuana, with 47% saying it should be legal and 49% saying it should not be. No more than a third of adults living in other parts of the country feel marijuana should be legal.

Support for legalizing marijuana is much lower among Republicans than it is among Democrats or independents. One in five Republicans (21%) say marijuana should be made legal in this country, while 37% of Democrats and 44% of independents share this view.

*Results are based on telephone interviews with 2,034 national adults, aged 18 and older, conducted Aug. 3–5, 2001, Nov. 10–12, 2003, and Oct. 21–23, 2005. For results based on the total sample of national adults, one can say with 95% confidence that the maximum margin of sampling error is ±2 percentage points.

By Joseph Carroll, Gallup Poll Assistant Editor

If we compare the margins of error found in parts (a) and (b) of Example 4, we notice that as the level of confidence went up (from 95% to 98%) the margin of error increased (from 3.1% to 3.7%). Intuitively, if we want to be more confident in our predictions, we should give our prediction more leeway (a larger margin of error).

When the results of the polls are printed in a newspaper, the sample size, level of confidence, margin of error, date of survey, and location of survey may be given as a footnote, as shown in the above article.

EXAMPLE 6

VERIFYING A STATED MARGIN OF ERROR Verify the margin of error stated in the article shown above.

SOLUTION

The footnote to the article states that for a sample size of 2,034 and a 95% level of confidence, the margin of error is ±2%, that is, MOE = 0.02 for $n = 2{,}034$ and $\alpha = 0.95$. Because $\alpha = 0.95$, $\frac{\alpha}{2} = 0.475$ and $z_{\alpha/2} = z_{0.475} = 1.96$. Therefore,

$$\begin{aligned}\text{MOE} &= \frac{z_{\alpha/2}}{2\sqrt{n}} \\ &= \frac{1.96}{2\sqrt{2{,}034}} \\ &= 0.02195127\ldots \\ &\approx 0.02 \qquad \text{rounding off to two decimal places}\end{aligned}$$

Therefore, the stated margin of error of ±2% is correct.

FEATURED IN THE NEWS

TO MANY AMERICANS, UFOS ARE REAL AND HAVE VISITED EARTH IN SOME FORM

Most Americans appear comfortable with and even excited about the thought of the discovery of extraterrestrial life. More than half (56 percent) of the American public think that UFOs are something real and not just in people's imagination. Nearly as many (48 percent) believe that UFOs have visited earth in some form. Males are significantly more likely to believe in the reality of UFOs, as are those under the age of 65. A significant drop is witnessed in the percentage of believers among the 65+ age group.

Two-thirds (67 percent) of adults think there are other forms of intelligent life in the universe. This belief tends to be more prevalent among males, adults ages 64 or younger, and residents of the Northeast as opposed to North Central and South.

In the view of many adults (55 percent), the government does not share enough information with the public in general. An even greater proportion (roughly seven in ten) thinks that the government does not tell us everything it knows about extraterrestrial life and UFOs. The younger the age, the stronger the belief that the government is withholding information about these topics.

This study was conducted by RoperASW. The sample consists of 1,021 male and female adults (in approximately equal number), all 18 years of age and over. The telephone interviews were conducted from August 23 through August 25, 2002, using a Random Digit Dialing (RDD) probability sample of all telephone households in the continental United States. The margin of error for the total sample is ±3 percent.

The Roper Poll
UFOs & Extraterrestrial Life
Americans' Beliefs and Personal Experiences
(Prepared for the SCI FI Channel—September 2002)

News articles do not always mention the level of confidence of a survey. However, if the sample size and margin of error are given, the level of confidence can be determined, as shown in Example 7.

EXAMPLE 7

FINDING THE LEVEL OF CONFIDENCE OF AN OPINION POLL Do you think that UFOs are real? The news article shown above presents the results of a Roper poll pertaining to this question. Find the level of confidence of this poll.

SOLUTION

We are given $n = 1{,}021$ and MOE $= 0.03$. To find α, the level of confidence of the poll, we must first find $z_{\alpha/2}$ and the area of the bodies, as shown in Figure 4.128.

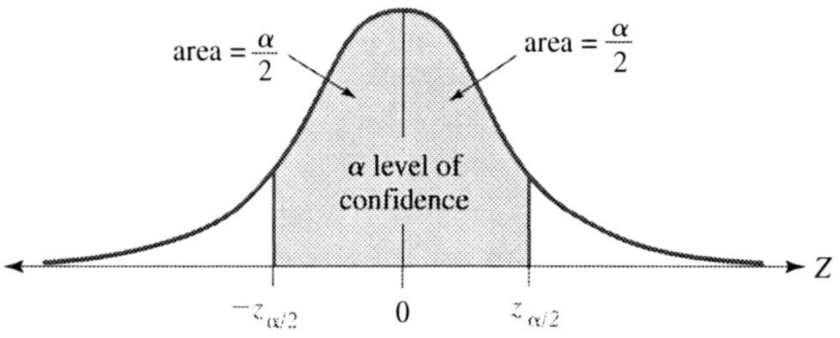

FIGURE 4.128 Level of confidence equals central region.

Substituting the given values into the MOE formula, we have

$$\text{MOE} = \frac{z_{\alpha/2}}{2\sqrt{n}}$$

$$0.03 = \frac{z_{\alpha/2}}{2\sqrt{1{,}021}}$$

TOPIC X RANDOM SAMPLING AND OPINION POLLS
STATISTICS IN THE REAL WORLD

National opinion polls are pervasive in today's world; hardly a day goes by without the release of yet another glimpse at the American psyche. Topics ranging from presidential performance and political controversy to popular culture and alien abduction vie for our attention in the media: "A recent survey indicates that a majority of Americans. . . ." Where do these results come from? Are they accurate? Can they be taken seriously, or are they mere entertainment and speculation?

Although many people are intrigued by the "findings" of opinion polls, others are skeptical or unsure of the fundamental premise of statistical sampling. How can the opinions of a diverse population of nearly 300 million Americans be measured by a survey of a mere 1,000 to 1,500 individuals? Wouldn't a survey of 10,000 people give better, or more accurate, results than those obtained from the typical sample of about a thousand people? The answer is "no, not necessarily." Although intuition might dictate that a survey's reliability is driven by the size of the sample (bigger is better), in reality, the most important factor in obtaining reliable results is the *method* by which the sample was selected. That is, depending on how it was selected, a sample of 1,000 people can yield far better results than a sample of 10,000, 20,000, or even 50,000 people.

The basic premise in statistical sampling is that the views of a small portion of a population *can* accurately represent the views of the entire population *if* the sample is selected properly, that is, if the sample is selected randomly. So what does it mean to say that a sample is selected "randomly"? The answer is simple: A sample is random if every member of the population has an *equal chance* of being selected. For example, suppose that a remote island has a population of 1,234 people and we wish to select a random sample of 30 inhabitants to interview. We could write each islander's name on a slip of paper, put all of the slips in a large box, shake up the box, close our eyes, and select 30 slips. Each person on the island would have an equal chance of being selected; consequently, we would be highly confident (say, 95%) that the views of the sample (plus or minus a margin or error) would accurately represent the views of the entire population.

The key to reliable sampling is the selection of a random sample; to select a random sample, each person in the population must have an equal chance of being selected. Realistically, how can this be accomplished with such a vast population of people in the United States or elsewhere? Years ago, the most accurate polls (especially Gallup polls) were based on data gathered from knocking on doors and conducting face-to-face interviews. However, in today's world, almost every American adult has a telephone, and telephone surveys have replaced the door-to-door surveys of the past. Although a poll might state that the target population is "all Americans aged 18 and over," it really means "all Americans aged 18 and over who have an accessible telephone number." Who will be excluded from this population? Typically, active members of the military forces and people in prisons, institutions, or hospitals are excluded from the sampling frame of today's opinion polls.

How can a polling organization obtain a list of the telephone numbers of all Americans aged 18 and over? Typically, no such list exists. Telephone directories are not that useful because of the large number of unlisted telephone numbers. However, using high-speed computers and a procedure known as random digit dialing, polling organizations are able to create a list of all possible phone numbers and thus are able to select a random sample. Finally, each telephone number in the sample is called, and "an American aged 18 and over" (or whatever group is being targeted) is interviewed. If no one answers the phone or the appropriate person is not at home, the polling organization makes every effort to establish contact at a later time. This ensures that the random sampling process is accurately applied and therefore that the results are true to the stated level of confidence with an acceptable margin of error.

$$z_{\alpha/2} = 0.03(2\sqrt{1{,}021}) \quad \text{multiplying each side by } 2\sqrt{1{,}021}$$

$$z_{\alpha/2} = 1.917185437\ldots$$

$$z_{\alpha/2} \approx 1.92 \quad \text{rounding off to two decimal places}$$

Using the body table in Appendix F, we can find the area under the bell curve between $z = 0$ and $z = 1.92$; that is, $p(0 < z < 1.92) = 0.4726$.

Therefore, $\frac{\alpha}{2} = 0.4726$, and multiplying by 2, we have $\alpha = 0.9452$. Thus, the level of confidence is $\alpha = 0.9452$ (or 95%), as shown in Figure 4.129.

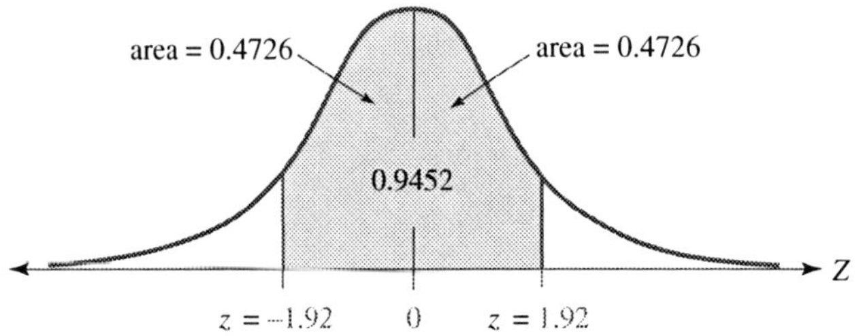

FIGURE 4.129 Finding the level of confidence.

4.5 EXERCISES

In Exercises 1–4, use the body table in Appendix F to find the specified z-number.

1. a. $z_{0.2517}$ b. $z_{0.1217}$ c. $z_{0.4177}$ d. $z_{0.4960}$
2. a. $z_{0.0199}$ b. $z_{0.2422}$ c. $z_{0.4474}$ d. $z_{0.4936}$
3. a. $z_{0.4250}$ b. $z_{0.4000}$ c. $z_{0.3750}$ d. $z_{0.4950}$
4. a. $z_{0.4350}$ b. $z_{0.4100}$ c. $z_{0.2750}$ d. $z_{0.4958}$
5. Find the z-number associated with a 92% level of confidence.
6. Find the z-number associated with a 97% level of confidence.
7. Find the z-number associated with a 75% level of confidence.
8. Find the z-number associated with an 85% level of confidence.

In Exercises 9–22, round off your answers (sample proportions and margins of error) to three decimal places (a tenth of a percent).

9. The Gallup Poll in Example 6 states that one-third (33%) of respondents support general legalization of marijuana. For each of the following levels of confidence, find the margin of error associated with the sample.
   a. a 90% level of confidence
   b. a 98% level of confidence
10. The Roper poll in Example 7 states that 56% of the Americans questioned think that UFOs are real. For each of the following levels of confidence, find the margin of error associated with the sample.
    a. an 80% level of confidence
    b. a 98% level of confidence
11. A survey asked, "How important is it to you to buy products that are made in America?" Of the 600 Americans surveyed, 450 responded, "It is important." For each of the following levels of confidence, find the sample proportion and the margin of error associated with the poll.
    a. a 90% level of confidence
    b. a 95% level of confidence
12. In the survey in Exercise 11, 150 of the 600 Americans surveyed responded, "It is not important." For each of the following levels of confidence, find the sample proportion and the margin of error associated with the poll.
    a. an 85% level of confidence
    b. a 98% level of confidence
13. A survey asked, "Have you ever bought a lottery ticket?" Of the 2,710 Americans surveyed, 2,141 said yes, and 569 said no.*
    a. Determine the sample proportion of Americans who have purchased a lottery ticket.
    b. Determine the sample proportion of Americans who have not purchased a lottery ticket.
    c. With a 90% level of confidence, find the margin of error associated with the sample proportions.
14. A survey asked, "Which leg do you put into your trousers first?" Of the 2,710 Americans surveyed, 1,138 said left, and 1,572 said right.*
    a. Determine the sample proportion of Americans who put their left leg into their trousers first.
    b. Determine the sample proportion of Americans who put their right leg into their trousers first.
    c. With a 90% level of confidence, find the margin of error associated with the sample proportions.

* Data from Poretz and Sinrod, *The First Really Important Survey of American Habits* (Los Angeles: Price Stern Sloan Publishing, 1989).

15. A survey asked, "Do you prefer showering or bathing?" Of the 1,220 American men surveyed, 1,049 preferred showering, and 171 preferred bathing. In contrast, 1,043 of the 1,490 American women surveyed preferred showering, and 447 preferred bathing.*

a. Determine the sample proportion of American men who prefer showering.

b. Determine the sample proportion of American women who prefer showering.

c. With a 95% level of confidence, find the margin of error associated with the sample proportions.

16. A survey asked, "Do you like the way you look in the nude?" Of the 1,220 American men surveyed, 830 said yes, and 390 said no. In contrast, 328 of the 1,490 American women surveyed said yes, and 1,162 said no.*

a. Determine the sample proportion of American men who like the way they look in the nude.

b. Determine the sample proportion of American women who like the way they look in the nude.

c. With a 95% level of confidence, find the margin of error associated with the sample proportions.

Exercises 17–20 are based on a survey (published in April 2000) of 129,593 students in grades 6–12 conducted by USA WEEKEND *magazine.*

17. When asked, "Do you, personally, feel safe from violence in school?" 92,011 said yes, and 37,582 said no.

a. Determine the sample proportion of students who said yes.

b. Determine the sample proportion of students who said no.

c. With a 95% level of confidence, find the margin of error associated with the sample proportions.

18. When asked, "Do kids regularly carry weapons in your school?" 14,255 said yes, and 115,338 said no.

a. Determine the sample proportion of students who said yes.

b. Determine the sample proportion of students who said no.

c. With a 95% level of confidence, find the margin of error associated with the sample proportions.

19. When asked, "Is there a gun in your home?" 58,317 said yes, 60,908 said no, and 10,368 said they did not know.

a. Determine the sample proportion of students who said yes.

b. Determine the sample proportion of students who said no.

c. Determine the sample proportion of students who said they did not know.

d. With a 90% level of confidence, find the margin of error associated with the sample proportions.

20. When asked, "How likely do you think it is that a major violent incident could occur at your school?" 18,143 said very likely, 64,797 said somewhat likely, and 46,653 said not likely at all.

a. Determine the sample proportion of students who said "very likely."

b. Determine the sample proportion of students who said "somewhat likely."

c. Determine the sample proportion of students who said "not likely at all."

d. With a 90% level of confidence, find the margin of error associated with the sample proportions.

21. A survey asked, "Can you imagine a situation in which you might become homeless?" Of the 2,503 Americans surveyed, 902 said yes.*

a. Determine the sample proportion of Americans who can imagine a situation in which they might become homeless.

b. With a 90% level of confidence, find the margin of error associated with the sample proportion.

c. With a 98% level of confidence, find the margin of error associated with the sample proportion.

d. How does your answer to part (c) compare to your answer to part (b)? Why?

22. A survey asked, "Do you think that homeless people are responsible for the situation they are in?" Of the 2,503 Americans surveyed, 1,402 said no.*

a. Determine the sample proportion of Americans who think that homeless people are not responsible for the situation they are in.

b. With an 80% level of confidence, find the margin of error associated with the sample proportion.

c. With a 95% level of confidence, find the margin of error associated with the sample proportion.

d. How does your answer to part (c) compare to your answer to part (b)? Why?

23. A sample consisting of 430 men and 765 women was asked various questions pertaining to international affairs. With a 95% level of confidence, find the margin of error associated with the following samples.

a. the male sample

b. the female sample

c. the combined sample

24. A sample consisting of 942 men and 503 women was asked various questions pertaining to the nation's economy. For a 95% level of confidence, find the margin of error associated with the following samples.

---

* Data from Poretz and Sinrod, *The First Really Important Survey of American Habits* (Los Angeles: Price Stern Sloan Publishing, 1989).

* Data from Mark Clements, "What Americans Say about the Homeless," *Parade Magazine*, Jan. 9, 1994: 4–6.

a. the male sample
b. the female sample
c. the combined sample

25. A poll pertaining to environmental concerns had the following footnote: "Based on a sample of 1,763 adults, the margin of error is plus or minus 2.5 percentage points." Find the level of confidence of the poll.

HINT: See Example 7.

26. A poll pertaining to educational goals had the following footnote: "Based on a sample of 2,014 teenagers, the margin of error is plus or minus 2 percentage points." Find the level of confidence of the poll.

HINT: See Example 7.

27. A recent poll pertaining to educational reforms involved 640 men and 820 women. The margin of error for the combined sample is 2.6%. Find the level of confidence for the entire poll.

HINT: See Example 7.

28. A recent poll pertaining to educational reforms involved 640 men and 820 women. The margin of error is 3.9% for the male sample and 3.4% for the female sample. Find the level of confidence for the male portion of the poll and for the female portion of the poll.

HINT: See Example 7.

In Exercises 29–32, you are planning a survey for which the findings are to have the specified level of confidence. How large should your sample be so that the margin of error is at most the specified amount?

29. 95% level of confidence
a. margin of error = 3%
b. margin of error = 2%
c. margin of error = 1%
d. Comparing your answers to parts (a)–(c), what conclusion can be made regarding the margin of error and the sample size?

30. 96% level of confidence
a. margin of error = 3%
b. margin of error = 2%
c. margin of error = 1%
d. Comparing your answers to parts (a)–(c), what conclusion can be made regarding the margin of error and the sample size?

31. 98% level of confidence
a. margin of error = 3%
b. margin of error = 2%
c. margin of error = 1%
d. Comparing your answers to parts (a)–(c), what conclusion can be made regarding the margin of error and the sample size?

32. 99% level of confidence
a. margin of error = 3%
b. margin of error = 2%
c. margin of error = 1%
d. Comparing your answers to parts (a)–(c), what conclusion can be made regarding the margin of error and the sample size?

Answer the following questions using complete sentences and your own words.

• CONCEPT QUESTIONS

33. What is a sample proportion? How is it calculated?
34. What is a margin of error? How is it calculated?
35. If the sample size is increased in a survey, would you expect the margin of error to increase or decrease? Why?
36. What is a random sample?
37. What is random digit dailing (RDD)?

• HISTORY QUESTIONS

38. Who founded the American Institute of Public Opinion? When? In what two academic fields did this person receive degrees?

WEB PROJECTS

39. What are the nation's current opinions concerning major issues in the headlines? Pick a current issue (such as abortion, same-sex marriage, gun control, war, or the president's approval rating) and find a recent survey regarding the issue. Summarize the results of the survey. Be sure to include the polling organization, date(s) of the survey, sample size, level of confidence, margin of error, and any pertinent information.

40. What is random digit dialing (RDD), and how is it used? Visit the web site of a national polling or news organization, and conduct a search of its FAQs regarding the organization's methods of sampling and the use of RDD. Write a report in which you summarize the use of RDD by the polling or news organization. Some useful links for this web project are listed on the text web site:

www.cengage.com/math/johnson

• PROJECTS

41. The purpose of this project is to conduct an opinion poll. Select a topic that is relevant to you and/or your community. Create a multiple-choice question to gather people's opinions concerning this issue.

For example:

If you could vote today, how would you vote on Proposition X?
i. support ii. oppose iii. undecided

a. Ask fifty people your question, and record their responses. Calculate the sample proportion for each category of response. Use a 95% level of confidence, and calculate the margin of error for your survey.

b. Ask 100 people your question, and record their responses. Calculate the sample proportion for each category of response. Use a 95% level of confidence, and calculate the margin of error for your survey.

c. How does the margin of error in part (b) compare to the margin of error in part (a)? Explain.

4.6 Linear Regression

OBJECTIVES

- Find the line of best fit
- Use linear regression to make a prediction
- Calculate the coefficient of linear correlation

When x and y are variables and m and b are constants, the equation $y = mx + b$ has infinitely many solutions of the form (x, y). A specific ordered pair (x_1, y_1) is a solution of the equation if $y_1 = mx_1 + b$. Because every solution of the given equation lies on a straight line, we say that x and y are *linearly related.*

If we are given two ordered pairs (x_1, y_1) and (x_2, y_2), we should be able to "work backwards" and find the equation of the line passing through them; assuming that x and y are linearly related, we can easily find the equation of the line passing through the points (x_1, y_1) and (x_2, y_2). The process of finding the equation of a line passing through given points is known as **linear regression;** the equation thus found is called the **mathematical model** of the linear relationship. Once the model has been constructed, it can be used to make predictions concerning the values of x and y.

EXAMPLE 1

USING A LINEAR EQUATION TO MAKE A PREDICTION Charlie is planning a family reunion and wants to place an order for custom T-shirts from Prints Alive (the local silk-screen printer) to commemorate the occasion. He has ordered shirts from Prints Alive on two previous occasions; on one occasion, he paid \$164 for twenty-four shirts; on another, he paid \$449 for eighty-four. Assuming a linear relationship between the cost of T-shirts and the number ordered, predict the cost of ordering 100 shirts.

SOLUTION

Letting x = the number of shirts ordered and y = the total cost of the shirts, the given data can be expressed as two ordered pairs: $(x_1, y_1) = (24, 164)$ and $(x_2, y_2) = (84, 449)$. We must find $y = mx + b$, the equation of the line passing through the two points.

First, we find m, the slope:

$$\begin{aligned} m &= \frac{y_2 - y_1}{x_2 - x_1} \\ &= \frac{449 - 164}{84 - 24} \\ &= \frac{285}{60} \\ &= 4.75 \end{aligned}$$

Now we use one of the ordered pairs to find b, the y-intercept. Either point will work; we will use $(x_1, y_1) = (24, 164)$.

The slope-intercept form of a line is $y = mx + b$. Solving for b, we obtain

$$\begin{aligned} b &= y - mx \\ &= 164 - 4.75(24) \\ &= 164 - 114 \\ &= 50 \end{aligned}$$

Therefore, the equation of the line is $y = 4.75x + 50$. We use this linear model to predict the cost of ordering $x = 100$ T-shirts.

$$\begin{aligned} y &= 4.75x + 50 \\ &= 4.75(100) + 50 \\ &= 475 + 50 \\ &= 525 \end{aligned}$$

We predict that it will cost \$525 to order 100 T-shirts.

Linear Trends and Line of Best Fit

Example 1 illustrates the fact that two points determine a unique line. To find the equation of the line, we must find the slope and the y-intercept. If we are given more than two points, the points might not be collinear. In collecting real-world data, this is usually the case. However, after the scatter of points is plotted on an x-y coordinate system, it may appear that the points "almost" fit on a line. If a sample of ordered pairs tend to "go in the same general direction," we say that they exhibit a **linear trend.** See Figure 4.130.

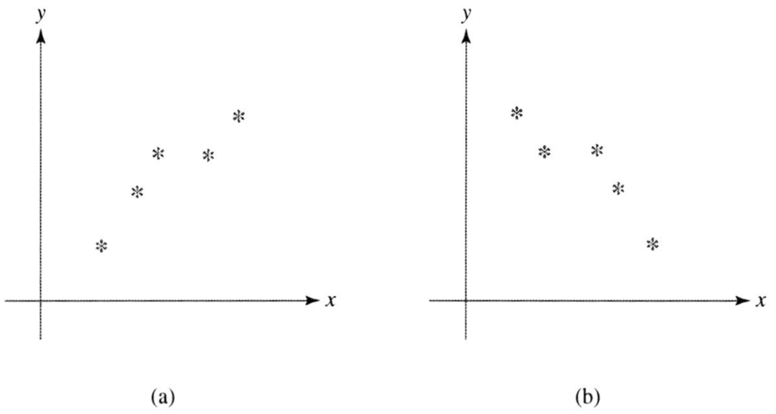

FIGURE 4.130 Ordered pairs that exhibit a linear trend.

When a scatter of points exhibits a linear trend, we construct the line that best approximates the trend. This line is called the **line of best fit** and is denoted by $\hat{y} = mx + b$. The "hat" over the y indicates that the calculated value of y is a prediction based on linear regression. See Figure 4.131.

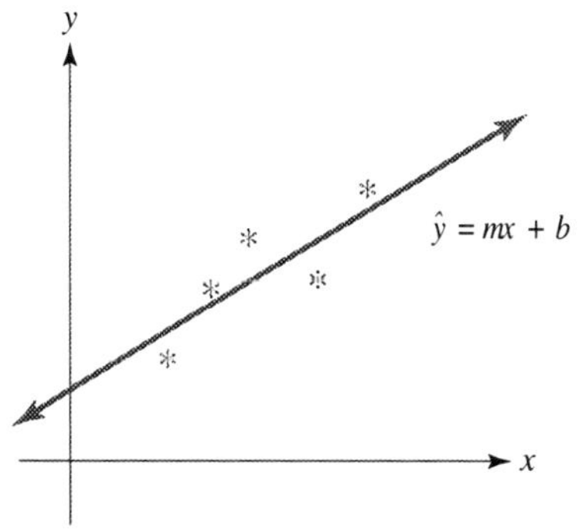

FIGURE 4.131 The line of best fit $\hat{y} = mx + b$.

To calculate the slope and y-intercept of the line of best fit, mathematicians have developed formulas based on the method of least squares. (See the Historical Note on Carl Gauss in Section 4.4.)

Recall that the symbol Σ means "sum." Therefore, Σx represents the sum of the x-coordinates of the points, and Σy represents the sum of the y-coordinates. To find Σxy, multiply the x- and y-coordinates of each point and sum the results.

LINE OF BEST FIT

Given a sample of n ordered pairs $(x_1, y_1), (x_2, y_2), \ldots, (x_n, y_n)$, the **line of best fit** (the line that best represents the data) is denoted by $\hat{y} = mx + b$, where the slope m and y-intercept b are given by

$$m = \frac{n(\Sigma xy) - (\Sigma x)(\Sigma y)}{n(\Sigma x^2) - (\Sigma x)^2} \quad \text{and} \quad b = \bar{y} - m\bar{x}$$

$\bar{x}$ and $\bar{y}$ denote the means of the x- and y-coordinates, respectively.

EXAMPLE 2

FINDING AND GRAPHING THE LINE OF BEST FIT We are given the ordered pairs (5, 14), (9, 17), (12, 16), (14, 18), and (17, 23).

a. Find the equation of the line of best fit.

b. Plot the given data and sketch the graph of the line of best fit on the same coordinate system.

SOLUTION

a. Organize the data in a table and compute the appropriate sums, as shown in Figure 4.132.

(x, y)	x	x^2	y	xy
(5, 14)	5	25	14	$5 \cdot 14 = 70$
(9, 17)	9	81	17	$9 \cdot 17 = 153$
(12, 16)	12	144	16	$12 \cdot 16 = 192$
(14, 18)	14	196	18	$14 \cdot 18 = 252$
(17, 23)	17	289	23	$17 \cdot 23 = 391$
$n = 5$ ordered pairs	$\Sigma x = 57$	$\Sigma x^2 = 735$	$\Sigma y = 88$	$\Sigma xy = 1,058$

FIGURE 4.132 Table of sums.

First, we find the slope:

$$m = \frac{n(\Sigma xy) - (\Sigma x)(\Sigma y)}{n(\Sigma x^2) - (\Sigma x)^2}$$

$$= \frac{5(1,058) - (57)(88)}{5(735) - (57)^2}$$

$$= 0.643192488\ldots$$

(5 × 1058 − 57 × 88) ÷ (5 × 735 − 57 x^2) =

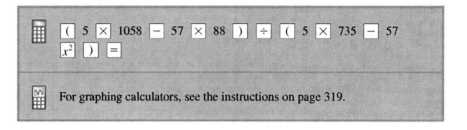

For graphing calculators, see the instructions on page 319.

Once m has been calculated, we store it in the memory of our calculator. We will need it to calculate b, the y-intercept.

$$b = \bar{y} - m\bar{x}$$

$$= \left(\frac{88}{5}\right) - 0.643192488\left(\frac{57}{5}\right)$$

$$= 10.26760564\ldots$$

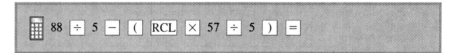

88 ÷ 5 − (RCL × 57 ÷ 5) =

Therefore, the line of best fit, $\hat{y} = mx + b$, is

$$\hat{y} = 0.643192488x + 10.26760564$$

Rounding off to one decimal place, we have

$$\hat{y} = 0.6x + 10.3$$

b. To graph the line, we need to plot two points. One point is the y-intercept $(0, b) = (0, 10.3)$. To find another point, we pick an appropriate value for x—say, $x = 18$—and calculate $\hat{y}$:

$$\hat{y} = 0.6x + 10.3$$

$$= 0.6(18) + 10.3$$

$$= 21.1$$

Therefore, the point $(x, \hat{y}) = (18, 21.1)$ is on the line of best fit.

Plotting (0, 10.3) and (18, 21.1), we construct the line of best fit. It is customary to use asterisks (*) to plot the given ordered pairs, as shown in Figure 4.133.

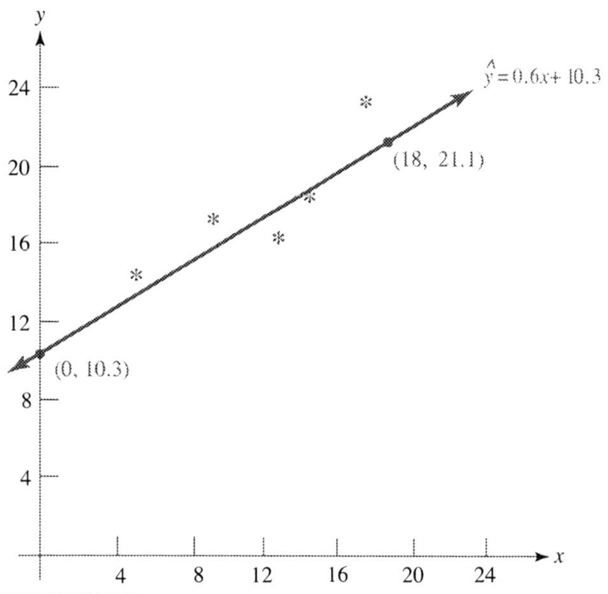

FIGURE 4.133 Scatter diagram and line of best fit.

Coefficient of Linear Correlation

Given a sample of n ordered pairs, we can always find the line of best fit. Does the line accurately portray the data? Will the line give accurate predictions? To answer these questions, we must consider the relative strength of the linear trend exhibited by the given data. If the given points are close to the line of best fit, there is a strong linear relation between x and y; the line will generate good predictions. If the given points are widely scattered about the line of best fit, there is a weak linear relation, and predictions based on it are probably not reliable.

One way to measure the strength of a linear trend is to calculate the **coefficient of linear correlation,** denoted by r. The formula for calculating r is shown in the following box.

COEFFICIENT OF LINEAR CORRELATION

Given a sample of n ordered pairs, $(x_1, y_1), (x_2, y_2), \ldots, (x_n, y_n)$

The **coefficient of linear correlation,** denoted by r, is given by

$$r = \frac{n(\Sigma xy) - (\Sigma x)(\Sigma y)}{\sqrt{n(\Sigma x^2) - (\Sigma x)^2}\sqrt{n(\Sigma y^2) - (\Sigma y)^2}}$$

The calculated value of r is always between -1 and 1, inclusive; that is, $-1 \le r \le 1$. If the given ordered pairs lie perfectly on a line whose slope is *positive,* then the calculated value of r will equal 1 (think 100% perfect with positive slope). In this case, both variables have the same behavior: As one increases (or decreases), so will the other. On the other hand, if the data points fall perfectly on a line whose slope is *negative,* the calculated value of r will equal -1 (think 100% perfect with negative slope). In this case, the variables have opposite behavior: As one increases, the other decreases, and vice versa. See Figure 4.134.

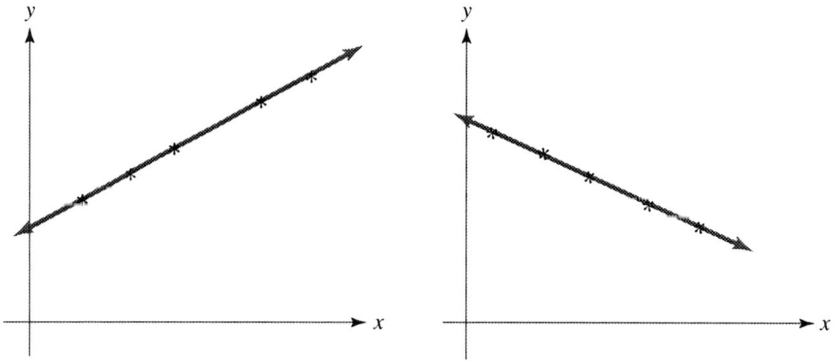

$r = 1$
Perfect, positive
linear relation
(a)

$r = -1$
Perfect, negative
linear relation
(b)

FIGURE 4.134

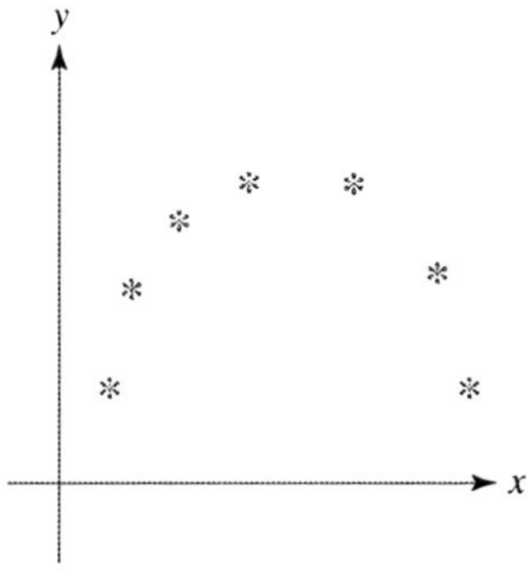

r close to 0

FIGURE 4.135

If the value of r is close to 0, there is little or no *linear* relation between the variables. This does not mean that the variables are not related. It merely means that no *linear* relation exists; the variables might be related in some nonlinear fashion, as shown in Figure 4.135.

In summary, the closer r is to 1 or -1, the stronger is the linear relation between x and y; the line of best fit will generate reliable predictions. The closer r is to 0, the weaker is the linear relation; the line of best fit will generate unreliable predictions. If r is positive, the variables have a direct relationship (as one increases, so does the other); if r is negative, the variables have an inverse relationship (as one increases, the other decreases).

EXAMPLE 3

CALCULATING THE COEFFICIENT OF LINEAR CORRELATION
Calculate the coefficient of linear correlation for the ordered pairs given in Example 2.

SOLUTION

The ordered pairs are (5, 14), (9, 17), (12, 16), (14, 18), and (17, 23). We add a y^2 column to the table in Example 2, as shown in Figure 4.136.

(x, y)	x	x^2	y	y^2	xy
(5, 14)	5	25	14	196	$5 \cdot 14 = 70$
(9, 17)	9	81	17	289	$9 \cdot 17 = 153$
(12, 16)	12	144	16	256	$12 \cdot 16 = 192$
(14, 18)	14	196	18	324	$14 \cdot 18 = 252$
(17, 23)	17	289	23	529	$17 \cdot 23 = 391$
$n = 5$ ordered pairs	$\Sigma x = 57$	$\Sigma x^2 = 735$	$\Sigma y = 88$	$\Sigma y^2 = 1{,}594$	$\Sigma xy = 1{,}058$

FIGURE 4.136 Table of sums.

Now use the formula to calculate r:

$$r = \frac{n(\Sigma xy) - (\Sigma x)(\Sigma y)}{\sqrt{n(\Sigma x^2) - (\Sigma x)^2}\sqrt{n(\Sigma y^2) - (\Sigma y)^2}}$$

$$= \frac{5(1{,}058) - (57)(88)}{\sqrt{5(735) - (57)^2}\sqrt{5(1{,}594) - (88)^2}}$$

$$= 0.883062705\ldots$$

(5 × 1058 − 57 × 88) ÷ ((5 × 735 − 57 x^2) $\sqrt{x}$ × (5 × 1594 − 88 x^2) $\sqrt{x}$) =

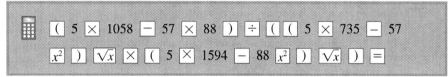

The coefficient of linear correlation is reasonably close to 1, so the line of best fit will generate reasonably reliable predictions. (Notice that the data points in Figure 4.133 are fairly close to the line of best fit.)

EXAMPLE 4

USING LINEAR CORRELATION AND REGRESSION TO MAKE PREDICTIONS Unemployment and personal income are undoubtedly related; we would assume that as the national unemployment rate increases, total personal income would decrease. Figure 4.137 gives the unemployment rate and the total personal income for the United States for various years.

a. Use linear regression to predict the total personal income of the United States if the unemployment rate is 5.0%.
b. Use linear regression to predict the unemployment rate if the total personal income of the United States is $10 billion.
c. Are the predictions in parts (a) and (b) reliable? Why or why not?

Year	Unemployment Rate (percent)	Total Personal Income (billions)
1975	8.5	$1.3
1980	7.1	2.3
1985	7.2	3.4
1990	5.6	4.9
1995	5.6	6.1
2000	4.0	8.4
2006	4.6	11.0

FIGURE 4.137 *Sources:* Bureau of Labor Statistics, U.S. Department of Labor; Bureau of Economic Analysis, U.S. Department of Commerce.

SOLUTION

a. Letting x = unemployment rate and y = total personal income, we have $n = 7$ ordered pairs, as shown in Figure 4.138.

Using a calculator, we find the following sums:

$\Sigma x^2 = 274.38$ $\Sigma y^2 = 271.32$ $\Sigma xy = 197.66$

x (percent)	y (billions)
8.5	\$1.3
7.1	2.3
7.2	3.4
5.6	4.9
5.6	6.1
4.0	8.4
4.6	11.0
$\Sigma x = 42.6$	$\Sigma y = 37.4$

FIGURE 4.138

x = unemployment rate,
y = total personal income.

First we find the slope:

$$m = \frac{n(\Sigma xy) - (\Sigma x)(\Sigma y)}{n(\Sigma x^2) - (\Sigma x)^2}$$
$$= \frac{7(197.66) - (42.6)(37.4)}{7(274.38) - (42.6)^2}$$
$$= -1.979414542\ldots$$

Now we calculate b, the y-intercept:

$$b = \bar{y} - m\bar{x}$$
$$= \left(\frac{37.4}{7}\right) - (-1.979414542)\left(\frac{42.6}{7}\right)$$
$$= 17.3890085\ldots$$

Therefore, the line of best fit, $\hat{y} = mx + b$, is

$$\hat{y} = -1.979414542x + 17.3890085$$

Rounding off to two decimal places (one more than the data), we have

$$\hat{y} = -1.98x + 17.39$$

Now, substituting $x = 5.0$ (5% unemployment) into the equation of the line of best fit, we have

$$\hat{y} = -1.98x + 17.39$$
$$= -1.98(5.0) + 17.39$$
$$= -9.9 + 17.39$$
$$= 7.49$$

If the unemployment rate is 5.0%, we predict that the total personal income of the United States will be approximately \$7.49 billion.

b. To predict the unemployment rate when the total personal income is \$10 billion, we let $y = 10$, substitute into $\hat{y}$, and solve for x:

$$10 = -1.98x + 17.39$$
$$1.98x + 10 = 17.39 \quad \text{adding } 1.98x \text{ to both sides}$$
$$1.98x = 17.39 - 10 \quad \text{subtracting 10 from both sides}$$
$$1.98x = 7.39$$
$$x = \frac{7.39}{1.98} \quad \text{dividing by 1.98}$$
$$= 3.7323232\ldots$$

We predict that the unemployment rate will be approximately 3.7% when the total personal income is \$10 billion.

c. To investigate the reliability of our predictions (the strength of the linear trend), we must calculate the coefficient of linear correlation:

$$r = \frac{n(\Sigma xy) - (\Sigma x)(\Sigma y)}{\sqrt{n(\Sigma x^2) - (\Sigma x)^2}\sqrt{n(\Sigma y^2) - (\Sigma y)^2}}$$
$$r = \frac{7(197.66) - (42.6)(37.4)}{\sqrt{7(274.38) - (42.6)^2}\sqrt{7(271.32) - (37.4)^2}}$$
$$= -0.9105239486\ldots$$

Because r is close to -1, we conclude that our predictions are very reliable; the linear relationship between x and y is high. Furthermore, since r is negative, we know that y (total personal income) decreases as x (unemployment rate) increases.

4.6 EXERCISES

1. A set of $n = 6$ ordered pairs has the following sums:

$$\Sigma x = 64 \quad \Sigma x^2 = 814 \quad \Sigma y = 85$$
$$\Sigma y^2 = 1{,}351 \quad \Sigma xy = 1{,}039$$

   a. Find the line of best fit.
   b. Predict the value of y when $x = 11$.
   c. Predict the value of x when $y = 19$.
   d. Find the coefficient of linear correlation.
   e. Are the predictions in parts (b) and (c) reliable? Why or why not?

2. A set of $n = 8$ ordered pairs has the following sums:

$$\Sigma x = 111 \quad \Sigma x^2 = 1{,}869 \quad \Sigma y = 618$$
$$\Sigma y^2 = 49{,}374 \quad \Sigma xy = 7{,}860$$

   a. Find the line of best fit.
   b. Predict the value of y when $x = 8$.
   c. Predict the value of x when $y = 70$.
   d. Find the coefficient of linear correlation.
   e. Are the predictions in parts (b) and (c) reliable? Why or why not?

3. A set of $n = 5$ ordered pairs has the following sums:

$$\Sigma x = 37 \quad \Sigma x^2 = 299 \quad \Sigma y = 38$$
$$\Sigma y^2 = 310 \quad \Sigma xy = 279$$

   a. Find the line of best fit.
   b. Predict the value of y when $x = 5$.
   c. Predict the value of x when $y = 7$.
   d. Find the coefficient of linear correlation.
   e. Are the predictions in parts (b) and (c) reliable? Why or why not?

4. Given the ordered pairs (4, 40), (6, 37), (8, 34), and (10, 31):
   a. Find and interpret the coefficient of linear correlation.
   b. Find the line of best fit.
   c. Plot the given ordered pairs and sketch the graph of the line of best fit on the same coordinate system.

5. Given the ordered pairs (5, 5), (7, 10), (8, 11), (10, 15), and (13, 16):
   a. Plot the ordered pairs. Do the ordered pairs exhibit a linear trend?
   b. Find the line of best fit.
   c. Predict the value of y when $x = 9$.
   d. Plot the given ordered pairs and sketch the graph of the line of best fit on the same coordinate system.
   e. Find the coefficient of linear correlation.
   f. Is the prediction in part (c) reliable? Why or why not?

6. Given the ordered pairs (5, 20), (6, 15), (10, 14), (12, 15), and (13, 10):
   a. Plot the ordered pairs. Do the ordered pairs exhibit a linear trend?
   b. Find the line of best fit.
   c. Predict the value of y when $x = 8$.
   d. Plot the given ordered pairs and sketch the graph of the line of best fit on the same coordinate system.
   e. Find the coefficient of linear correlation.
   f. Is the prediction in part (c) reliable? Why or why not?

7. Given the ordered pairs (2, 6), (3, 12), (6, 15), (7, 4), (10, 6), and (11, 12):
   a. Plot the ordered pairs. Do the ordered pairs exhibit a linear trend?
   b. Find the line of best fit.
   c. Predict the value of y when $x = 8$.
   d. Plot the given ordered pairs and sketch the graph of the line of best fit on the same coordinate system.
   e. Find the coefficient of linear correlation.
   f. Is the prediction in part (c) reliable? Why or why not?

8. The unemployment rate and the amount of emergency food assistance (food made available to hunger relief organizations such as food banks and soup kitchens) provided by the federal government in the United States in certain years are given in Figure 4.139.

Year	Unemployment Rate	Emergency Food Assistance (in millions)
2000	4.0%	$225
2003	6.0	456
2004	5.5	420
2005	5.1	373
2006	4.6	300
2007	4.6	256

FIGURE 4.139 Unemployment rates and emergency food assistance. *Sources:* Bureau of Labor Statistics (U.S. Department of Labor); Food and Nutrition Service (U.S. Department of Agriculture).

   a. Letting x = the unemployment rate and y = the amount of emergency food assistance, plot the data. Do the data exhibit a linear trend?
   b. Find the line of best fit.

c. Predict the amount of emergency food assistance when the unemployment rate is 5.0%.

d. Predict the unemployment rate when the amount of emergency food assistance is $350 million.

e. Find the coefficient of correlation.

f. Are the predictions in parts (a) and (b) reliable? Why or why not?

9. The average hourly earnings and the average tuition at public four-year institutions of higher education in the United States for 1997–2002 are given in Figure 4.140.

Year	Average Hourly Earnings	Average Tuition at Four-Year Institutions
1997	$12.49	$3,110
1998	13.00	3,229
1999	13.47	3,349
2000	14.00	3,501
2001	14.53	3,735
2002	14.95	4,059

FIGURE 4.140 Average hourly earnings and average tuition, 1997–2002. *Source:* Bureau of Labor Statistics and National Center for Education Statistics.

a. Letting x = the average hourly earnings and y = the average tuition, plot the data. Do the data exhibit a linear trend?

b. Find the line of best fit.

c. Predict the average tuition when the average hourly earning is $14.75.

d. Predict the average hourly earning when the average tuition is $4,000.

e. Find the coefficient of correlation.

f. Are the predictions in parts (a) and (b) reliable? Why or why not?

10. The number of domestic and imported retail car sales (in hundreds of thousands) in the United States for 1999–2007 are given in Figure 4.141.

a. Letting x = the number of domestic car sales and y = the number of imported car sales, plot the data. Do the data exhibit a linear trend?

b. Find the line of best fit.

c. Predict the number of imported car sales when there are 5,600,000 domestic car sales.

d. Predict the number of domestic car sales when there are 2,200,000 imported car sales.

e. Find the coefficient of correlation.

f. Are the predictions in parts (a) and (b) reliable? Why or why not?

Year	Domestic Car Sales	Imported Car Sales
1999	69.8	17.2
2000	68.3	20.2
2001	63.2	21.0
2002	58.8	22.3
2003	55.3	20.8
2004	53.6	21.5
2005	54.8	21.9
2006	54.4	23.4
2007	52.5	23.7

FIGURE 4.141 Domestic and imported retail car sales (hundred thousands), 1999–2007. *Source:* Ward's Commission.

11. The numbers of marriages and divorces (in millions) in the United States are given in Figure 4.142.

Year	1965	1970	1975	1980	1985	1990	2000
Marriages	1.800	2.158	2.152	2.413	2.425	2.448	2.329
Divorces	0.479	0.708	1.036	1.182	1.187	1.175	1.135

FIGURE 4.142 Number of marriages and divorces (millions). *Source:* National Center for Health Statistics.

a. Letting x = the number of marriages and y = the number of divorces in a year, plot the data. Do the data exhibit a linear trend?

b. Find the line of best fit.

c. Predict the number of divorces in a year when there are 2,750,000 marriages.

d. Predict the number of marriages in a year when there are 1,500,000 divorces.

e. Find the coefficient of linear correlation.

f. Are the predictions in parts (c) and (d) reliable? Why or why not?

12. The median home price and average mortgage rate in the United States for 1999–2004 are given in Figure 4.143.

a. Letting x = the median price of a home and y = the average mortgage rate, plot the data. Do the data exhibit a linear trend?

b. Find the line of best fit.

c. Predict the average mortgage rate if the median price of a home is $165,000.

d. Predict the median home price if the average mortgage rate is 7.25%.

e. Find the coefficient of correlation.

f. Are the predictions in parts (a) and (b) reliable? Why or why not?

Year	Median Price (dollars)	Mortgage Rate (percent)
1999	133,300	7.33
2000	139,000	8.03
2001	147,800	7.03
2002	158,100	6.55
2003	180,200	5.74
2004	195,200	5.73

FIGURE 4.143 Median home price and average mortgage rate, 1999–2004. *Source:* National Association of Realtors.

Answer the following questions using complete sentences and your own words.

CONCEPT QUESTIONS

13. What is a line of best fit? How do you find it?
14. How do you measure the strength of a linear trend?
15. What is a positive linear relation? Give an example.
16. What is a negative linear relation? Give an example.

PROJECTS

17. Measure the heights and weights of ten people. Let x = height and y = weight.
    a. Plot the ordered pairs. Do the ordered pairs exhibit a linear trend?
    b. Use the data to find the line of best fit.
    c. Find the coefficient of linear correlation.
    d. Will your line of best fit produce reliable predictions? Why or why not?

TECHNOLOGY AND LINEAR REGRESSION

In Example 2 of this section, we computed the slope and y-intercept of the line of best fit for the five ordered pairs (5, 14), (9, 17), (12, 16), (14, 18), and (17, 23). These calculations can be tedious when done by hand, even with only five data points. In the real world, there are always a large number of data points, and the calculations are always done with the aid of technology.

LINEAR REGRESSION ON A GRAPHING CALCULATOR

Graphing calculators can draw a scatter diagram, compute the slope and y-intercept of the line of best fit, and graph the line.

ON A TI-83/84:

- *Put the calculator into statistics mode* by pressing STAT.
- *Set the calculator up for entering the data* from Example 2 by selecting "Edit" from the "EDIT" menu, and the "List Screen" appears, as shown in Figure 4.144. If data already appear in a list (as they do in Figure 4.144) and you want to clear the list, use the arrow buttons to highlight the name of the list and press CLEAR ENTER.
- Use the arrow buttons and the ENTER button to enter the x-coordinates in list L_1 and the corresponding y-coordinates in list L_2.

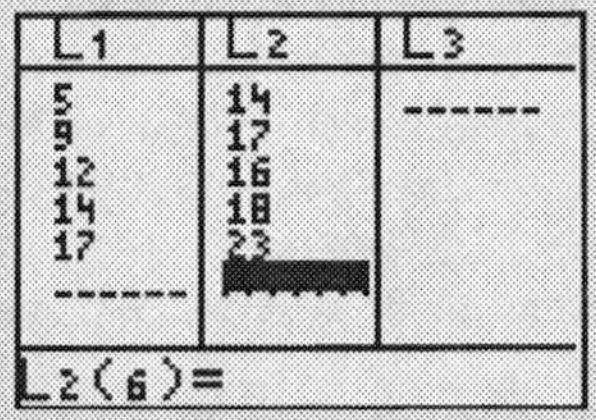

FIGURE 4.144 A TI-83/84's list screen.

ON A CASIO:

- *Put the calculator into statistics mode* by pressing [MENU], highlighting STAT, and pressing [EXE].
- Use the arrow buttons and the [EXE] button to enter the x-coordinates from Example 2 in List 1 and the corresponding y coordinates in List 2. When you are done, your screen should look like Figure 4.145. If data already appears in a list and you want to erase it, use the arrow keys to move to that list, press [F6] and then [F4] (i.e., [DEL-A], which stands for "delete all") and then F1 (i.e., [YES]).

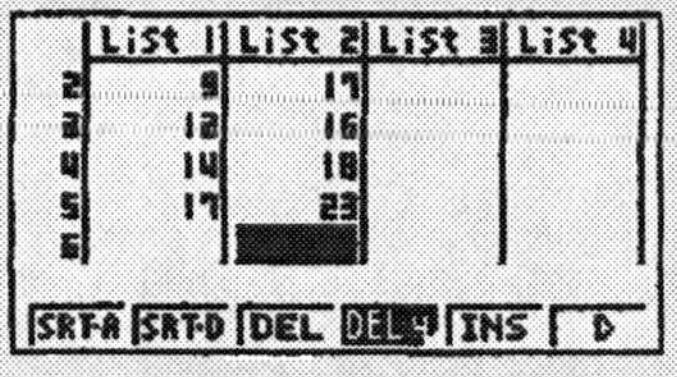

FIGURE 4.145 A Casio's list screen.

Finding the Equation of the Line of Best Fit

Once the data have been entered, it is easy to find the equation. Recall from Example 2 that the equation is $y = 0.643192488x + 10.26760564$.

ON A TI-83/84:

- Press [2nd] [QUIT].
- The TI-83/84 does not display the correlation coefficient unless you tell it to. To do so, press [2nd] [CATALOG], scroll down, and select "DiagnosticOn." When "DiagnosticOn" appears on the screen, press ENTER.
- Press [STAT], scroll to the right, and select the "CALC" menu, and select "LinReg(ax + b)" from the "CALC" menu.
- When "LinReg(ax + b)" appears on the screen, press [ENTER].
- The slope, the y-intercept and the correlation coefficient will appear on the screen, as shown in Figure 4.146. They are not labeled m, b, and r. Their labels are different and are explained in Figure 4.147.

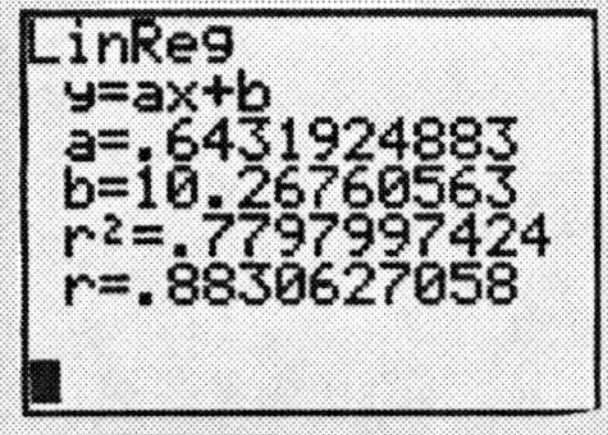

FIGURE 4.146 Finding the equation of the line of best fit on a TI-83/84.

	TI-83/84 Labels
slope *ma*	
y-intercept *b*	b
correlation coefficient *r*	r

FIGURE 4.147 Graphing calculator line of best fit labels.

ON A CASIO:

- Press CALC (i.e., F2) to calculate the equation.
- Press REG (i.e., F3); "REG" stands for "regression."
- Press X (i.e., F1) for linear regression. (See Figure 4.148.)

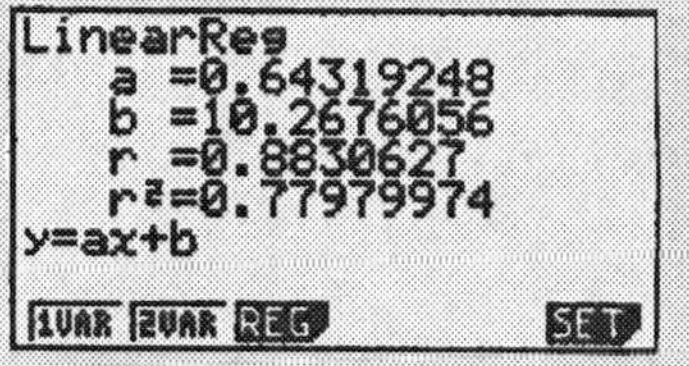

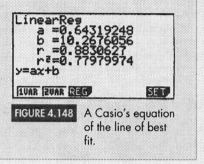

FIGURE 4.148 A Casio's equation of the line of best fit.

Drawing a Scatter Diagram and the Line of Best Fit

Once the equation has been found, the line can be graphed.

ON A TI-83/84:

- Press Y =, and clear any functions that may appear.
- Set the calculator up to draw a scatter diagram by pressing 2nd STAT PLOT and selecting "Plot 1." Turn the plot on and select the scatter icon.
- Tell the calculator to put the data entered in list L_1 on the x-axis by selecting "L_1" for "Xlist," and to put the data entered in list L_2 on the y-axis by selecting "L_2" for "Ylist."
- Automatically set the range of the window and obtain a scatter diagram by pressing ZOOM and selecting option 9: "ZoomStat."
- If you don't want the data points displayed on the line of best fit, press 2nd STAT PLOT and turn off plot 1.
- Quit the statistics mode by pressing 2nd QUIT.
- Enter the equation of the line of best fit by pressing Y = VARS, selecting "Statistics," scrolling to the right to select the "EQ" menu, and selecting "RegEQ" (for regression equation).
- Automatically set the range of the window and obtain a scatter diagram by pressing ZOOM and selecting option 9: "ZoomStat." (See Figure 4.149.)
- Press TRACE to read off the data points as well as points on the line of best fit. Use the up and down arrows to switch between data points and points on the line of best fit. Use the left and right arrows to move left and right on the graph.

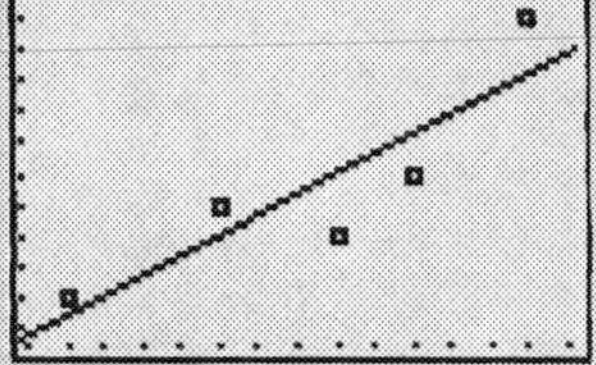

FIGURE 4.149 A TI-83/84's graph of the line of best fit and the scatter diagram.

ON A CASIO:

- Press SHIFT QUIT to return to List 1 and List 2.
- Press GRPH (i.e., F1).
- Press SET (i.e., F6).

- Make the resulting screen, which is labeled "StatGraph1," read as follows:

Graph Type	:Scatter
Xlist	:List1
Ylist	:List2
Frequency	:1

To make the screen read as described above,

- Use the down arrow button to scroll down to "Graph Type."
- Press Scat (i.e., F1).
- In a similar manner, change "Xlist," "Ylist," and "Frequency" if necessary.
- Press SHIFT QUIT to return to List1 and List2.
- Press GRPH (i.e., F1).
- Press GPH1 (i.e., F1) to obtain the scatter diagram.
- Press X (i.e., F1).
- Press DRAW (i.e., F6), and the calculator will display the scatter diagram as well as the line of best fit. (See Figure 4.150.)

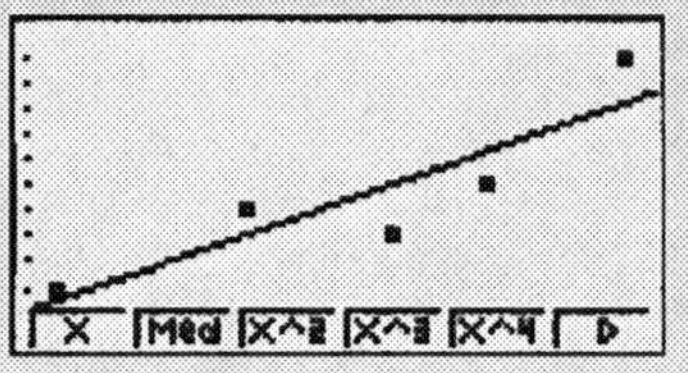

FIGURE 4.150 A Casio's graph of the line of best fit and the scatter diagram.

Linear Regression on Excel

Entering the Data

We will use the data from Example 2. Start by entering the information as shown in Figure 4.151. Then save your spreadsheet.

	A	B	C
1	x-coordinate	y-coordinate	
2	5	14	
3	9	17	
4	12	16	
5	14	18	
6	17	23	
7			

FIGURE 4.151 The given data.

Using the Chart Wizard to Draw the Scatter Diagram

- Press the "Chart Wizard" button at the top of the spreadsheet. (It looks like a bar chart, and it might have a magic wand.)

- Select "XY (Scatter)" and choose the Chart sub-type that displays points without connecting lines.
- Press the "Next" button and a "Chart Source Data" box will appear.
- Click on the triangle to the right of "Data range." This will cause the "Chart Source Data" box to shrink.
- Use your mouse to draw a box around all of the x-coordinates and y-coordinates. This will cause "= Sheet1!$A$2:$B$6" to appear in the "Chart Source Data" box.
- Press the triangle to the right of the "Chart Source Data" box and the box will expand.
- Press the "Next" button.
- Under "Chart Title" type an appropriate title. Our example lacks content, so we will use the title "Example 2."
- After "Category (X)" type an appropriate title for the x-axis. Our example lacks content, so we will use the title "x-axis."
- After "Category (Y)" type an appropriate title for the y-axis.
- Click on the "Gridlines" tab and check the box under "value (X) axis," next to "Major gridlines."
- Click on the "Legend" tab and remove the check from the box next to "show legend."
- Press the "Finish" button and the scatter diagram will appear.
- Save the spreadsheet.

After you have completed this step, your spreadsheet should include the scatter diagram shown in Figure 4.152.

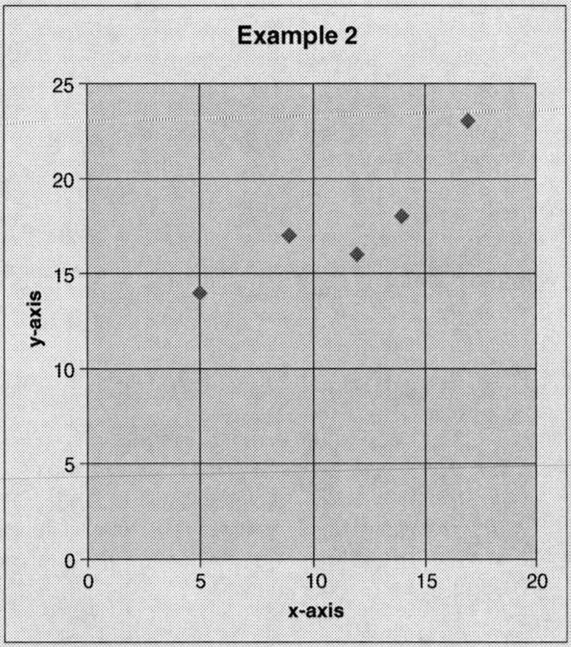

FIGURE 4.152 An Excel-generated scatter diagram.

Finding and Drawing the Line of Best Fit

- Click on the chart until it is surrounded by a thicker border.
- Use your mouse to select "Chart" at the very top of the screen. Then pull your mouse down until "Add Trendline . . ." is highlighted, and let go.
- Press the "Linear" button.

- Click on "Options".
- Select "Display Equation on Chart".
- Select "Display R-Squared value on Chart".
- Press the "OK" button, and the scatter diagram, the line of best fit, the equation of the line of best fit, and r^2 (the square of the correlation coefficient) appear. If the equation is in the way, you can click on it and move it to a better location.
- Use the square root button on your calculator to find r, the correlation coefficient.
- Save the spreadsheet.

After completing this step, your spreadsheet should include the scatter diagram, the line of best fit, and the equation of the line of best fit, as shown in Figure 4.153.

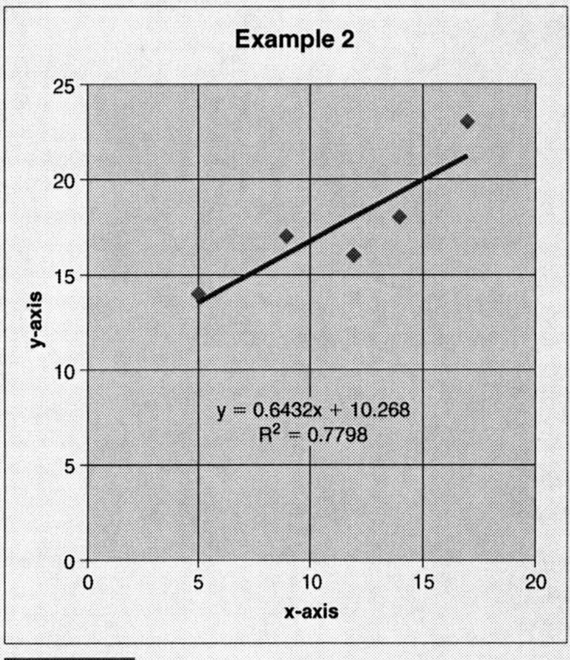

FIGURE 4.153 The scatter diagram with the line of best fit and its equation.

Printing the Scatter Diagram and Line of Best Fit

- Click on the diagram, until it is surrounded by a thicker border.
- Use your mouse to select "File" at the very top of the screen. Then pull your mouse down until "Print" is highlighted, and let go.
- Respond appropriately to the "Print" box that appears. Usually, it is sufficient to press a "Print" button or an "OK" button, but the actual way that you respond to the "Print" box depends on your printer.

EXERCISES

18. Throughout the twentieth century, the record time for the mile run steadily decreased, from 4 minutes 15.4 seconds in 1911 to 3 minutes 43.1 seconds in 1999. Some of the record times are given in Figure 4.154.

a. Convert the given data to ordered pairs (x, y), where x is the number of years since 1900 and y is the time in seconds.

b. Draw a scatter diagram, find the equation of the line of best fit, and find the correlation coefficient.

Year	1911	1923	1933	1942	1945	1954
Time (min:sec)	4:15.4	4:10.4	4:07.6	4:04.6	4:01.4	3:59.4
Year	1964	1967	1975	1980	1985	1999
Time (min:sec)	3:54.1	3:51.1	3:49.4	3:48.8	3:46.3	3:43.1

FIGURE 4.154 Record times for the mile run.

c. Predict the record time for the mile run in the year 2020.

d. Predict when the record time for the mile run will reach 3:30.

e. Predict when the record time for the mile run will reach 0:30.

f. On what assumption are these predictions based? If this assumption is correct, how accurate are these predictions?

19. Heart disease is much less of a problem in France than it is in the United States and other industrialized countries, despite a high consumption of saturated fats in France. One theory is that this lower heart disease rate is due to the fact that the French drink more wine. This so-called French paradox heightened an interest in moderate wine drinking. Figure 4.155 gives data on wine consumption and heart disease for nineteen industrialized countries, including France and the United States, in 1994.

a. Draw a scatter diagram, find the equation of line of best fit, and find the correlation for wine consumption versus annual deaths.

b. Do the data support the theory that wine consumption lowers the heart disease rate?

Country	Wine Consumption (liters per person)	Annual Deaths from Heart Disease per 100,000 People
Australia	2.5	211
Austria	3.9	167
Belgium	2.9	131
Canada	2.4	191
Denmark	2.9	220
Finland	0.8	297
France	9.1	71
Iceland	0.8	211
Ireland	0.6	300
Italy	7.9	107
Netherlands	1.8	266
New Zealand	1.9	266
Norway	0.8	227
Spain	6.5	86
Sweden	1.6	207
Switzerland	5.8	115
United Kingdom	1.3	285
United States	1.2	199
West Germany	2.7	172

FIGURE 4.155 Wine consumption and heart disease. *Source: New York Times*, December 28, 1994.

4 CHAPTER REVIEW

TERMS

bell-shaped curve
body table
categorical data
cells
coefficient of linear correlation
computerized spreadsheet
continuous variable
data point
density
descriptive statistics
deviation from the mean
discrete variable
error of the estimate
extreme value
frequency
frequency distribution
histogram
Inferential statistics
level of confidence
linear regression
linear trend
line of best fit
margin of error (MOE)
mathematical model
mean
measures of central tendency
measures of dispersion
median
mode
normal distribution
outlier
pie chart
population
population proportion
relative frequency
relative frequency density
sample
sample proportion
spreadsheet
standard deviation
standard normal distribution
statistics
summation notation
tail
variance
z-distribution
z-number

REVIEW EXERCISES

1. Find (a) the mean, (b) the median, (c) the mode, and (d) the standard deviation of the following set of raw data:

5 8 10 4 8 10 6 8 7 5

2. To study the composition of families in Winslow, Arizona, forty randomly selected married couples were surveyed to determine the number of children in each family. The following results were obtained:

3 1 0 4 1 3 2 2 0 2 0 2 2 1
4 3 1 1 3 4 2 1 3 0 1 0 2 5
1 2 3 0 0 1 2 3 1 2 0 2

a. Organize the given data by creating a frequency distribution.
b. Find the mean number of children per family.
c. Find the median number of children per family.
d. Find the mode number of children per family.
e. Find the standard deviation of the number of children per family.
f. Construct a histogram using single-valued classes of data.

3. The frequency distribution in Figure 4.156 lists the number of hours per day that a randomly selected sample of teenagers spent watching television. Where possible, determine what percent of the teenagers spent the following number of hours watching television.

x = Hours per Day	Frequency
$0 \le x < 2$	23
$2 \le x < 4$	45
$4 \le x < 6$	53
$6 \le x < 8$	31
$8 \le x \le 10$	17

FIGURE 4.156 Time watching television.

a. less than 4 hours
b. not less than 6 hours
c. at least 2 hours
d. less than 2 hours
e. at least 4 hours but less than 8 hours
f. more than 3.5 hours

4. To study the efficiency of its new oil-changing system, a local service station monitored the amount of time it took to change the oil in customers' cars. The frequency distribution in Figure 4.157 summarizes the findings.

x = Time (in minutes)	Number of Customers
$3 \le x < 6$	18
$6 \le x < 9$	42
$9 \le x < 12$	64
$12 \le x < 15$	35
$15 \le x \le 18$	12

FIGURE 4.157 Time to change oil.

a. Find the mean number of minutes to change the oil in a car.

b. Find the standard deviation of the amount of time to change the oil in a car.

c. Construct a histogram to represent the data.

5. If your scores on the first four exams (in this class) are 74, 65, 85, and 76, what score do you need on the next exam for your overall mean to be at least 80?

6. The mean salary of twelve men is \$37,000, and the mean salary of eight women is \$28,000. Find the mean salary of all twenty people.

7. Timo and Henke golfed five times during their vacation. Their scores are given in Figure 4.158.

Timo	103	99	107	93	92
Henke	101	92	83	96	111

FIGURE 4.158 Golf scores.

a. Find the mean score of each golfer. Who has the lowest mean?

b. Find the standard deviation of each golfer's scores.

c. Who is the more consistent golfer? Why?

8. Suzanne surveyed the prices for a quart of a certain brand of motor oil. The sample data, in dollars per quart, are summarized in Figure 4.159.

Price per Quart	Number of Stores
1.99	2
2.09	3
2.19	7
2.29	10
2.39	14
2.49	4

FIGURE 4.159 Price of motor oil.

a. Find the mean and standard deviation of the prices.

b. What percent of the data lies within one standard deviation of the mean?

c. What percent of the data lies within two standard deviations of the mean?

d. What percent of the data lies within three standard deviations of the mean?

9. Classify the following types of data as discrete, continuous, or neither.

a. weights of motorcycles

b. colors of motorcycles

c. number of motorcycles

d. ethnic background of students

e. number of students

f. amounts of time spent studying

10. What percent of the standard normal z-distribution lies in the following intervals?

a. between $z = 0$ and $z = 1.75$

b. between $z = -1.75$ and $z = 0$

c. between $z = -1.75$ and $z = 1.75$

11. A large group of data is normally distributed with mean 78 and standard deviation 7.

a. Find the intervals that represent one, two, and three standard deviations of the mean.

b. What percent of the data lies in each interval in part (a)?

c. Draw a sketch of the bell curve.

12. The time it takes a latex paint to dry is normally distributed. If the mean is $3\frac{1}{2}$ hours with a standard deviation of 45 minutes, find the probability that the drying time will be as follows.

a. less than 2 hours 15 minutes

b. between 3 and 4 hours

HINT: Convert everything to hours (or to minutes).

13. All incoming freshmen at a major university are given a diagnostic mathematics exam. The scores are normally distributed with a mean of 420 and a standard deviation of 45. If the student scores less than a certain score, he or she will have to take a review course. Find the cutoff score at which 34% of the students would have to take the review course.

14. Find the specified z-number.

a. $z_{0.4441}$ **b.** $z_{0.4500}$ **c.** $z_{0.1950}$ **d.** $z_{0.4975}$

15. A survey asked, "Do you think that the president is doing a good job?" Of the 1,200 Americans surveyed, 800 responded yes. For each of the following levels of confidence, find the sample proportion and the margin of error associated with the poll.

a. a 90% level of confidence

b. a 95% level of confidence

16. A survey asked, "Do you support capital punishment?" Of the 1,000 Americans surveyed, 750 responded no. For each of the following levels of confidence, find the sample proportion and the margin of error associated with the poll.

a. an 80% level of confidence

b. a 98% level of confidence

17. A sample consisting of 580 men and 970 women was asked various questions pertaining to international affairs. For a 95% level of confidence, find the margin of error associated with the following samples.

a. the male sample

b. the female sample

c. the combined sample

18. A poll pertaining to environmental concerns had the following footnote: "Based on a sample of 1,098 adults, the margin of error is plus or minus 2 percentage points." Find the level of confidence of the poll.

19. You are planning a survey with a 94% level of confidence in your findings. How large should your sample be so that the margin of error is at most 2.5%?

20. A set of $n = 5$ ordered pairs has the following sums:

$$\sum x = 66 \quad \sum x^2 = 1{,}094 \quad \sum y = 273$$
$$\sum y^2 = 16{,}911 \quad \sum xy = 4{,}272$$

a. Find the line of best fit.

b. Predict the value of y when $x = 16$.

c. Predict the value of x when $y = 57$.

d. Find the coefficient of linear correlation.

e. Are the predictions in parts (b) and (c) reliable? Why or why not?

21. Given the ordered pairs (5, 38), (10, 30), (20, 33), (21, 25), (24, 18), and (30, 20),

a. Plot the ordered pairs. Do the ordered pairs exhibit a linear trend?

b. Find the line of best fit.

c. Predict the value of y when $x = 15$.

d. Plot the given ordered pairs and sketch the graph of the line of best fit on the same coordinate system.

e. Find the coefficient of linear correlation.

f. Is the prediction in part (c) reliable? Why or why not?

22. The value of agricultural exports and imports in the United States for 1999–2007 are given in Figure 4.160.

a. Letting x = the value of agricultural exports and y = the value of agricultural imports, plot the data. Do the data exhibit a linear trend?

b. Find the line of best fit.

Year	Agricultural Exports (billion dollars)	Agricultural Imports (billion dollars)
1999	48.4	37.7
2000	51.2	39.0
2001	53.7	39.4
2002	53.1	41.9
2003	59.5	47.3
2004	62.4	52.7
2005	62.5	57.7
2006	68.7	64.0
2007	82.2	70.1

FIGURE 4.160 Agricultural exports and imports (billion dollars), 1999–2007. *Source:* U.S. Department of Agriculture.

c. Predict the value of agricultural imports when the value of agricultural exports is 75.0 billion dollars.

d. Predict the value of agricultural exports when the value of agricultural imports is 60.0 billion dollars.

e. Find the coefficient of correlation.

f. Are the predictions in parts (a) and (b) reliable? Why or why not?

Answer the following questions using complete sentences and your own words.

• CONCEPT QUESTIONS

23. What are the three measures of central tendency? Briefly explain the meaning of each.

24. What does standard deviation measure?

25. What is a normal distribution? What is the standard normal distribution?

26. What is a margin of error? How is it related to a sample proportion?

27. How do you measure the strength of a linear trend?

• HISTORY QUESTIONS

28. What role did the following people play in the development of statistics?

- George Gallup
- Carl Friedrich Gauss

APPENDIX A
Using a Scientific Calculator

WHAT KIND OF CALCULATOR DO YOU HAVE?

There are three kinds of calculators: graphing calculators, scientific calculators, and basic calculators. You probably have a:

- *graphing calculator* if your calculator has at least one button labeled "GRAPH" and its screen is about 1.5 or 2 inches tall. Read Appendix B.
- *basic calculator* if you get 20 after typing

  2 [+] 3 [×] 4 [=] or 2 [+] 3 [×] 4 [ENTER]

  A basic calculator is insufficient for this text.
- *scientific calculator* if you get 14 after typing

  2 [+] 3 [×] 4 [=] or 2 [+] 3 [×] 4 [ENTER]

Some scientific calculators work like graphing calculators, except that they don't graph (and therefore they don't have bigger screens). If you get:

- 1.609 . . . after typing

  [ln] 5 [=] or [ln] 5 [ENTER]

  then your scientific calculator works more like a graphing calculator. Read Appendix B.
- 1.609 . . . after typing

  5 [ln]

  then your scientific calculator doesn't work like a graphing calculator. Read this appendix.

USING YOUR SCIENTIFIC CALCULATOR

OBJECTIVE

- Become familiar with a scientific calculator

Read this appendix with your calculator by your side. When a calculation is discussed, do that calculation on your calculator.

Unfortunately, scientific calculators don't all work exactly the same way. Even if you do everything correctly, your answer may have a few more or less decimal places than the one given, or may differ in the last decimal place. Occasionally, the way a calculation is performed on your calculator will differ slightly from that discussed below. If so, experiment a little, or consult your instructor.

The Equals Button

Some buttons perform operations on a pair of numbers. For example, the [+] button is used to add two numbers; it only makes sense to use this button in conjunction with a pair of numbers. *When using such a button, you must finish the typing with the* [=] *button.* That is, to add 3 and 2, type

3 [+] 2 [=]

Some buttons perform operations on a single number. For example, the [x^2] button is used to square a number; it only makes sense to use this button in conjunction with a single number. *When using such a button, you do **not** type the* [=] *button.* That is, to square 5, type

5 [x^2]

The reason for this distinction is that the calculator must be told when you are done entering information and you are ready for it to compute. If you typed

3 [+] 2

the calculator would have no way of knowing whether you were ready for it to compute "3 + 2" or you were in the middle of instructing it to compute "3 + 22." You must follow "3 + 2" with the [=] button to tell the calculator that you are done entering information and are ready for it to compute. On the other hand, when you type

5 [x^2]

the calculator knows that you are done entering information; if you meant to square 53, you would not have pressed the [x^2] button after the "5." There is no need to follow "5 [x^2]" with the [=] button.

The Subtraction Symbol and the Negative Symbol

If "3 − −2" were read aloud, you could say "three subtract negative two" or "three minus minus two," and you would be understood. The expression is understandable even if the distinction between the negative symbol and the subtraction symbol is not made clear. With a calculator, however, this distinction is crucial. The subtraction button is labeled [−]. There is no negative button; instead, there is a [+/−] button or [+○−] button that changes the sign of whatever number is on the display. Typing

5 [+/−]

makes the display read "−5." Typing

5 [+/−] [+/−]

makes the display read "5," because two sign changes undo each other. Typing

[+/−] 5

makes the display read "5," not "−5." *You must press the sign-change button after the number itself.*

EXAMPLE 1

USING THE SUBTRACTION AND NEGATIVE SYMBOLS Calculate 3 − −2, both (a) by hand and (b) with a calculator.

SOLUTION

a. 3 − −2 = 3 + 2 = 5

b. To do this, we must use the subtraction button, which is labeled [−], and the sign-change button, which is labeled [+/−]. Typing

3 [−] 2 [+/−] [=]

makes the display read "5." Remember that you must press the sign-change button *after* the number itself; you press "2" followed by [+/−], not [+/−] followed by "2." Also, you must finish the typing with the [=] button, because the [=] button performs an operation on a pair of numbers.

Order of Operations and Use of Parentheses

Scientific calculators are programmed so that they follow the standard order of operations. That is, they perform calculations in the following order:

1. Parentheses-enclosed work
2. Exponents
3. Multiplication and
Division, from left to right
4. Addition and
Subtraction, from left to right

This order can be remembered by remembering the word *PEMDAS*, which stands for

Parentheses/Exponents/Multiplication/Division/Addition/Subtraction

Frequently, the fact that calculators are programmed to follow the order of operations means that you perform a calculation on your calculator in exactly the same way that it is written.

EXAMPLE 2

PARENTHESES COME FIRST Calculate 2(3 + 4), both (a) by hand and (b) with a calculator.

SOLUTION

a. $2(3 + 4) = 2(7)$ parentheses-enclosed work comes first
$= 14$

b. Type

2 [×] [(] 3 [+] 4 [)] [=]

and the display reads "14." Notice that this is typed on a calculator exactly as it is written, with two important exceptions:

- We must press the [×] button to mean multiplication; we cannot use parentheses to mean multiplication, as is done in part (a).
- We must finish the typing with the [=] button.

EXAMPLE 3

EXPONENTS BEFORE MULTIPLICATION Calculate $2 \cdot 3^2$, both (a) by hand and (b) with a calculator.

SOLUTION

a. $2 \cdot 3^2 = 2 \cdot 9$ exponents come before multiplication
$= 18$

b. Type

2 [×] 3 [x^2] [=]

and the display reads "18." Notice that this is typed on a calculator exactly as it is written, except that we must finish the typing with the [=] button and we must press the [×] button to mean multiplication. The [•] button is a decimal-point button, not a multiplication-dot button.

EXAMPLE 4

MULTIPLICATION BEFORE EXPONENTS Calculate $(2 \cdot 3)^2$, both (a) by hand and (b) with a calculator.

SOLUTION

a. $(2 \cdot 3)^2 = 6^2$ parentheses-enclosed work comes first
$= 36$

b. Type

[(] 2 [×] 3 [)] [x^2]

and the display reads "36." Notice that this is typed on a calculator exactly as it is written, except that we must use the [×] button to mean multiplication.
Also, we do not use the [=] button, because the [x^2] button performs an operation on a single number.

EXAMPLE 5

CUBING Calculate $4 \cdot 2^3$, both (a) by hand and (b) with a calculator.

SOLUTION

a. $4 \cdot 2^3 = 4 \cdot 8$ exponents come before multiplication
$= 32$

b. To do this, we must use the exponent button, which is labeled either [y^x] or [x^y]. Type

4 [×] 2 [y^x] 3 [=] or 4 [×] 2 [x^y] 3 [=]

and the display reads "32."

Sometimes, you don't perform a calculation on your calculator in the same way that it is written, even though calculators are programmed to follow the order of operations.

EXAMPLE 6

A FRACTION BAR ACTS AS PARENTHESES Calculate $\frac{2}{3 \cdot 4}$ with a calculator.

SOLUTION

Wrong It is incorrect to type

2 [÷] 3 [×] 4 [=]

According to the order of operations, multiplication and division are done *from left to right,* so the above typing is algebraically equivalent to

$= \frac{2}{3} \cdot 4$ first dividing then multiplying, since division is on the left and multiplication is on the right

$= \frac{2}{3} \cdot \frac{4}{1} = \frac{2 \cdot 4}{3}$ multiplying

which is not what we want. The difficulty is that the large fraction bar in the expression $\frac{2}{3 \cdot 4}$ groups the "$3 \cdot 4$" together in the denominator; in the above typing, nothing groups the "$3 \cdot 4$" together, and only the 3 ends up in the denominator.

Right The calculator needs parentheses inserted in the following manner:

$$\frac{2}{(3 \cdot 4)}$$

Thus, it is correct to type

2 [÷] [(] 3 [×] 4 [)] [=]

This makes the display read 0.166666667, the correct answer.

Also Right It is correct to type

2 [÷] 3 [÷] 4 [=]

According to the order of operations, multiplication and division are done from left to right, so the above typing is algebraically equivalent to

$$\frac{2}{3} \div 4 \quad \text{doing the left-hand division first}$$
$$= \frac{2}{3} \cdot \frac{1}{4} \quad \text{inverting}$$
$$= \frac{2}{3 \cdot 4} \quad \text{multiplying}$$

which is what we want. *When you're calculating something that involves only multiplication and division and you don't use parentheses, the* [×] *button places a factor in the numerator, and the* [÷] *button places a factor in the denominator.*

EXAMPLE 7

A FRACTION BAR ACTS AS PARENTHESES Calculate $\frac{2}{3/4}$ with a calculator.

SOLUTION

Wrong It is incorrect to type

2 [÷] 3 [÷] 4 [=]

even though that matches the way the problem is written algebraically. As discussed in Example 6, this typing is algebraically equivalent to

$$\frac{2}{3 \cdot 4}$$

which is not what we want.

Right The calculator needs parentheses inserted in the following manner:

$$\frac{2}{(3 \div 4)}$$

Thus, it is correct to type

2 [÷] [(] 3 [÷] 4 [)] [=]

since, according to the order of operations, parentheses-enclosed work is done first. This makes the display read 2.66666667, the correct answer.

EXAMPLE 8

A FRACTION BAR ACTS AS PARENTHESES Calculate $\frac{2+3}{4}$ with a calculator.

SOLUTION

Wrong It is incorrect to type

2 [+] 3 [÷] 4 [=]

According to the order of operations, division is done before addition, so this typing is algebraically equivalent to

$$2 + \frac{3}{4}$$

which is not what we want. The large fraction bar in the expression $\frac{2+3}{4}$ groups the "2 + 3" together in the numerator; in the above typing, nothing groups the "2 + 3" together, and only the 3 ends up in the numerator.

Right The calculator needs parentheses inserted in the following manner:

$$\frac{(2 + 3)}{4}$$

Thus, it is correct to type

[(] 2 [+] 3 [)] [÷] 4 [=]

This makes the display read 1.25, the correct answer.

Also Right It is correct to type

2 [+] 3 [=] [÷] 4 [=]

The first [=] makes the calculator perform all prior calculations before continuing. This too makes the display read "1.25," the correct answer.

The 2nd Button

Many buttons have two labels and two uses. To use the label on a button, you just press that button. To use the label above a button, you press [2nd] and then the button. For example, your calculator might have a button labeled "LN" on the button itself and "e^x" above the button. If so, typing

5 [LN]

makes the display read "1.609 . . ." because ln(5) = 1.609 . . . . Typing

5 [2nd] [e^x]

makes the display read 148.413 . . . because $e^5 = 148.143$ . . . .

Memory

The memory is a place to store a number for later use, without having to write it down. If a number is on your display, you can place it into the memory (or **store** it) by pressing the button labeled [STO] (or [$x \rightarrow$ M] or [M in]), and you can take it out of the memory (or **recall** it) by pressing the button labeled [RCL] (or [RM] or [MR]).

Typing

5 [STO]

makes the calculator store a 5 in its memory. If you do other calculations or just clear your display and later press

[RCL]

then your display will read "5."

Some calculators have more than one memory. If yours does, then pressing the button labeled [STO] won't do anything; pressing [STO] and then "1" will store it in memory number 1; pressing [STO] and then "2" will store it in memory number 2, and so on. Pressing [RCL] and then "1" will recall what has been stored in memory number 1.

EXAMPLE 9

USING THE MEMORY Use the quadratic formula and your calculator's memory to solve

$$2.3x^2 + 4.9x + 1.5 = 0$$

SOLUTION

The quadratic formula says that if $ax^2 + bx + c = 0$, then

$$x = \frac{-b \pm \sqrt{b^2 - 4ac}}{2a}$$

We have $2.3x^2 + 4.9x + 1.5 = 0$, so $a = 2.3$, $b = 4.9$, and $c = 1.5$. This gives

$$x = \frac{-4.9 \pm \sqrt{4.9^2 - 4 \cdot 2.3 \cdot 1.5}}{2 \cdot 2.3}$$

The quickest way to do this calculation is to calculate the radical, store it, and then calculate the two fractions.

Step 1 *Calculate the radical.* To do this, type

4.9 [x^2] [−] 4 [×] 2.3 [×] 1.5 [=] [$\sqrt{x}$] [STO]

This makes the display read "3.195309" and stores the number in the memory. Notice the use of the [=] button; this makes the calculator finish the prior calculation before taking a square root. If the [=] button were not used, the order of operations would require the calculator to take the square root of 1.5.

Step 2 *Calculate the first fraction.* To do this, type

4.9 [+/−] [+] [RCL] [=] [÷] 2 [÷] 2.3 [=]

This makes the display read "−0.3705849." Notice the use of the [=] button.

Step 3 *Calculate the second fraction.* To do this, type

4.9 [+/−] [−] [RCL] [=] [÷] 2 [÷] 2.3 [=]

This makes the display read "−1.7598498."

The solutions to $2.3x^2 + 4.9x + 1.5 = 0$ are $x = -0.3705849$ and $x = -1.7598498$. These are approximate solutions in that they show only the first seven decimal places.

Step 4 *Check your solutions.* These solutions can be checked by seeing whether they satisfy the equation $2.3x^2 + 4.9x + 1.5 = 0$. To check the first solution, type

2.3 [×] .3705849 [+/−] [x^2] [+] 4.9 [×] .3705849 [+/−] [+] 1.5 [=]

and the display will read either "0" or a number very close to 0.

Scientific Notation

Typing

4000000 [×] 8000000 [=]

makes the display read "3.2 13" rather than "32000000000000." This is because the calculator does not have enough room on its display for "32000000000000." When the display shows "3.2 13," read it as "3.2×10^{13}," which is written in scientific notation. Literally, "3.2×10^{13}" means "multiply 3.2 by 10, thirteen times," but as a shortcut, you can interpret it as "move the decimal point in the '3.2' thirteen places to the right."

Typing

.0000005 [×] .0000007 [=]

makes the display read "3.5 −13" rather than "0.00000000000035," because the calculator does not have enough room on its display for "0.00000000000035." Read "3.5 −13" as "3.5×10^{-13}." Literally, this means "divide 3.5 by 10, thirteen times," but as a shortcut, you can interpret it as "move the decimal point in the '3.5' thirteen places to the left."

You can type a number in scientific notation by using the button labeled [EXP] (which stands for *exponent*) or [EE] (which stands for *enter exponent*). For example, typing

5.2 [EXP] 8 or 5.2 [EE] 8

makes the display read "5.2 8," which means "5.2×10^8," and typing

3 [EXP] 17 [+/−] or 3 [EE] 17 [+/−]

makes the display read "3 −17," which means "3×10^{-17}." Notice that the sign-change button is used to make the exponent negative.

Be careful that you don't confuse the [y^x] button with the [EXP] button. The [EXP] button does *not* allow you to type in an exponent; it allows you to type in scientific notation. For example, typing

3 [EXP] 4

makes the display read "3 4," which means "3×10^4," and typing

3 [y^x] 4

makes the display read "81," since $3^4 = 81$.

EXERCISES

Perform the calculations in Exercises 1–32. The correct answer is given in brackets [ ]. In your homework, write down what you type on your calculator to get that answer. Answers are not given in the back of the book.

1. $-3 - -5$ [2]
2. $-6 - 3$ [−9]
3. $4 - -9$ [13]
4. $-6 - -8$ [2]
5. $-3 - (-5 - -8)$ [−6]
6. $-(-4 - 3) - (-6 - -2)$ [11]
7. $-8 \cdot -3 \cdot -2$ [−48]
8. $-9 \cdot -3 -2$ [25]
9. $(-3)(-8) - (-9)(-2)$ [6]
10. $2(3 - 5)$ [−4]
11. $2 \cdot 3 - 5$ [1]
12. $4 \cdot 11^2$ [484]
13. $(4 \cdot 11)^2$ [1,936]
14. $4 \cdot (-11)^2$ [484]
15. $4 \cdot (-3)^3$ [−108]

WARNING: Some calculators will not raise a negative number to a power. If yours has this characteristic, how can you use your calculator on this exercise?

16. $(4 \cdot -3)^3$ [−1,728]
17. $\dfrac{3 + 2}{7}$ [0.7142857]
18. $\dfrac{3 \cdot 2}{7}$ [0.8571429]
19. $\dfrac{3}{2 \cdot 7}$ [0.2142857]
20. $\dfrac{3 \cdot 2}{7 \cdot 5}$ [0.1714286]
21. $\dfrac{3 + 2}{7 \cdot 5}$ [0.1428571]
22. $\dfrac{3 \cdot -2}{7 + 5}$ [−0.5]
23. $\dfrac{3}{7/2}$ [0.8571429]
24. $\dfrac{3/7}{2}$ [0.2142857]
25. 1.8^2 [3.24]
26. $\sqrt{1.8}$ [1.3416408]
27. $47{,}000{,}000^2$ [2.209×10^{15}]
28. $\sqrt{0.0000000000027}$ [1.643168×10^{-6}]
29. $(-3.92)^7$ [−14,223.368737]
30. $(5.72 \times 10^{19})^4$ [1.070494×10^{79}]
31. $(3.76 \times 10^{-12})^{-5}$ [1.330641×10^{57}]
32. $(3.76 \times 10^{-12}) -5$ [−5]
33. Solve $4.2x^2 + 8.3x + 1.1 = 0$ for x. Check your two answers by substituting them back into the equation.
34. Solve $5.7x^2 + 12.3x - 8.1 = 0$ for x. Check your two answers by substituting them back into the equation.
35. Which of the following buttons must be used in conjunction with the [=] button, and why?

[+] [−] [×] [÷] [x^2] [$\sqrt{x}$] [y^x] [$1/x$] [+/−]

APPENDIX B
Using a Graphing Calculator

OBJECTIVE

- Become familiar with a graphing calculator

The following discussion was written specifically for Texas Instruments (or "TI") and Casio graphing calculators, but it frequently applies to other brands as well. Texas Instruments TI-83 and TI-84 models and Casio CFX-9850, FX-9860, and CFX-9950 models are specifically addressed. Read this discussion with your calculator close at hand. When a calculation is discussed, do that calculation on your calculator.

To do any of the calculations discussed in this section on a Casio, start by pressing [MENU]. Then use the arrow buttons to select "RUN" on the main menu, and press [EXE] (which stands for "execute").

The Enter Button

A TI graphing calculator will never perform a calculation until the [ENTER] button is pressed; a Casio will never perform a calculation until the [EXE] button is pressed. A Casio's [EXE] button functions like a TI's [ENTER] button; often, we will refer to either of these buttons as the [ENTER] button.
To add 3 and 2, type

3 [+] 2 [ENTER]

and the display will read 5. To square 4, type

4 [x^2] [ENTER]

and the display will read 16. If the [ENTER] button isn't pressed, the calculation will not be performed.

The 2nd and Alpha Buttons

Most calculator buttons have more than one label and more than one use; you select from these uses with the [2nd] and [ALPHA] buttons. A Casio's [SHIFT] button functions like a TI's [2nd] button; often, we will refer to either of these buttons as the [2nd] button. For example, one button is labeled "x^2" on the button itself, "$\sqrt{\ }$" above the button, and either "I" or "K" above and to the right of the button. If it is used without the [2nd] or [ALPHA] buttons, it will square a number. Typing

4 [x^2] [ENTER]

makes the display read 16, since $4^2 = 16$. If it is used with the [2nd] button, it will take the square root of a number. Typing

[2nd] [$\sqrt{\ }$] 4 [ENTER]

makes the display read 2, since $\sqrt{4} = 2$. If it is used with the [ALPHA] button, it will display the letter I or K.

Notice that to square 4, you press the [x^2] button *after* the 4, but to take the square root of 4, you press the [$\sqrt{\ }$] button *before* the 4. This is because graphing calculators are designed so that the way you type something is as similar as possible to the way it is written algebraically. When you write 4^2, you write the 4 first and then the squared symbol; thus, on your graphing calculator, you press the 4 first and then the [x^2] button. When you write $\sqrt{4}$, you write the square root symbol first and then the 4; thus, on your graphing calculator, you press the [$\sqrt{\ }$] button first and then the 4.

Frequently, the two operations that share a button are operations that "undo" each other. For example, typing

3 [x^2] [ENTER]

makes the display read 9, since $3^2 = 9$, and typing

[2nd] [$\sqrt{\ }$] 9 [ENTER]

makes the display read 3, since $\sqrt{9} = 3$. This is done as a memory device; it is easier to find the various operations on the keyboard if the two operations that share a button also share a relationship.

Two operations that *always* undo each other are called **inverses.** The x^2 and $\sqrt{x}$ operations are not inverses because $(-3)^2 = 9$, but $\sqrt{9} \neq -3$. However, there is an inverse-type relationship between the x^2 and $\sqrt{x}$ operations—they undo each other sometimes. When two operations share a button, they are inverses or they share an inverse-type relation.

Correcting Typing Errors

If you've made a typing error *and you haven't yet pressed* [ENTER], you can correct that error with the [◄] button. For example, if you typed "$5 \times 2 + 7$" and then realized you wanted "$5 \times 3 + 7$," you can replace the incorrect 2 with a 3 by pressing the [◄] button until the 2 is flashing, and then press "3."

If you realize *after* you pressed [ENTER] that you've made a typing error, just press [2nd] [ENTRY] to reproduce the previously entered line. (With a Casio, press [◄].) Then correct the error with the [◄] button, as described above.

The [INS] button allows you to insert a character. For example, if you typed "5×27" and you meant to type "5×217," press the [◄] button until the 7 is flashing, and then insert 1 by typing

[2nd] [INS] 1

The [DEL] button allows you to delete a character. For example, if you typed "5×217" and you meant to type "5×27," press the [◄] button until the 1 is flashing, and then press [DEL].

If you haven't yet pressed [ENTER], the [CLEAR] button erases an entire line. If you have pressed [ENTER], the [CLEAR] button clears everything off of the screen. With a Casio, the [AC] button functions in the same way. "AC" stands "all clear."

The Subtraction Symbol and the Negative Symbol

If you read "$3 - -2$" aloud, you could say, "three subtract negative two" or "three minus minus two," and you would be understood. The expression is understandable even if the distinction between the negative symbol and the subtraction symbol is not made clear. With a calculator, however, this distinction is crucial. The subtraction button is labeled "−," and the negative button is labeled "(−)."

EXAMPLE 1

THE SUBTRACTION AND NEGATIVE SYMBOLS Calculate $3 - -2$, both (a) by hand and (b) with a calculator.

SOLUTION

a. $3 - -2 = 3 + 2 = 5$

b. Type

3 [−] [(−)] 2 [ENTER]

and the display will read 5.

If you had typed

3 [−] [−] 2 [ENTER] or 3 [(−)] [−] 2 [ENTER]

the calculator would have responded with an error message.

The Multiplication Symbol

In algebra, we do not use "x" for multiplication. Instead, we use "x" as a variable, and we use "·" for multiplication. However, Texas Instruments graphing calculators use "×" as the label on the multiplication button and "*" for multiplication on the display screen. (The "variable x" button is labeled "X,T,θ,n".) This is one of the few instances in which you don't type things in the same way that you write them algebraically.

Order of Operations and Use of Parentheses

Texas Instruments and Casio graphing calculators are programmed to follow the order of operations. That is, they perform calculations in the following order:

1. **P**arentheses-enclosed work
2. **E**xponents
3. **M**ultiplication and **D**ivision, from left to right
4. **A**ddition and **S**ubtraction, from left to right

You can remember this order by remembering the word "**PEMDAS**," which stands for

Parentheses/**E**xponents/**M**ultiplication/**D**ivision/**A**ddition/**S**ubtraction

EXAMPLE 2

PARENTHESES COME FIRST Calculate $2(3 + 4)$, both (a) by hand and (b) with a calculator.

SOLUTION

a. $2(3 + 4) = 2(7)$ parentheses-enclosed work comes first

$= 14$

b. Type

2 [×] [(] 3 [+] 4 [)] [ENTER]

and the display will read 14.

In the instructions to Example 2, notice that we wrote "$2(3 + 4)$" rather than "$2 \cdot (3 + 4)$"; in this case, it's not necessary to write the multiplication symbol. Similarly, it's not necessary to type the multiplication symbol. Example 2b could be computed by typing

2 [(] 3 [+] 4 [)] [ENTER]

EXAMPLE 3

EXPONENTS BEFORE MULTIPLICATION Calculate $2 \cdot 3^3$, both (a) by hand and (b) with a calculator.

SOLUTION

a. $2 \cdot 3^3 = 2 \cdot 27$ exponents come before multiplication

$= 54$

b. To do this, we must use the exponent button, which is labeled "∧." Type

2 [×] 3 [∧] 3 [ENTER]

and the display will read 54.

EXAMPLE 4

MULTIPLICATION BEFORE EXPONENTS Calculate $(2 \cdot 3)^3$, both (a) by hand and (b) with a calculator.

SOLUTION

a. $(2 \cdot 3)^3 = 6^3$ parentheses-enclosed work comes first

$= 216$

b. Type

[(] 2 [×] 3 [)] [∧] 3 [ENTER]

and the display will read 216.

In Example 3, the exponent applies only to the 3, because the order of operations dictates that exponents come before multiplication. In Example 4, the exponent applies to the $(2 \cdot 3)$, because the order of operations dictates that parentheses-enclosed work comes before exponents. In each example, the way you type the problem matches the way it is written algebraically, because the calculator is programmed to follow the order of operations. Sometimes, however, the way you type a problem doesn't match the way it is written algebraically.

EXAMPLE 5

A FRACTION BAR ACTS LIKE PARENTHESES Calculate $\frac{2}{3 \cdot 4}$ with a calculator.

SOLUTION

Wrong It is incorrect to type

2 [÷] 3 [×] 4 [ENTER]

even though that matches the way the problem is written algebraically. According to the order of operations, multiplication and division are done *from left to right,* so the above typing is algebraically equivalent to

$= \frac{2}{3} \cdot 4$ first dividing and then multiplying, since division is on the left and multiplication is on the right

$= \frac{2}{3} \cdot \frac{4}{1} = \frac{2 \cdot 4}{3}$

which is not what we want. The difficulty is that the large fraction bar in the expression $\frac{2}{3 \cdot 4}$ groups the "$3 \cdot 4$" together in the denominator; in the above typing, nothing groups the "$3 \cdot 4$" together, and only the 3 ends up in the denominator.

Right The calculator needs parentheses inserted in the following manner:

$$\frac{2}{(3 \cdot 4)}$$

Thus, it is correct to type

2 [÷] [(] 3 [×] 4 [)] [ENTER]

This makes the display read 0.166666667, the correct answer.

Also Right It is correct to type

2 [÷] 3 [÷] 4 [ENTER]

According to the order of operations, multiplication and division are done from left to right, so the above typing is algebraically equivalent to

$$\frac{2}{3} \div 4 \quad \text{doing the left-hand division first}$$

$$= \frac{2}{3} \cdot \frac{1}{4} \quad \text{inverting and multiplying}$$

$$= \frac{2}{3 \cdot 4}$$

which is what we want. *When you're calculating something that involves only multiplication and division and you don't use parentheses, the* [×] *button places a factor in the numerator, and the* [÷] *button places a factor in the denominator.*

EXAMPLE 6

A FRACTION BAR ACTS AS PARENTHESES Calculate $\frac{2}{3/4}$ with a calculator.

SOLUTION

Wrong It is incorrect to type

2 [÷] 3 [÷] 4 [ENTER]

even though that matches the way the problem is written algebraically. As was discussed in Example 5, this typing is algebraically equivalent to

$$\frac{2}{3 \cdot 4}$$

which is not what we want.

Right The calculator needs parentheses inserted in the following manner:

$$\frac{2}{(3/4)}$$

Thus, it is correct to type

2 [÷] [(] 3 [÷] 4 [)] [ENTER]

since, according to the order of operations, parentheses-enclosed work is done first. This makes the display read 2.6666667, the correct answer.

EXAMPLE 7

A FRACTION BAR ACTS AS PARENTHESES Calculate $\frac{2 + 3}{4}$ with a calculator.

SOLUTION

Wrong It is incorrect to type

2 [+] 3 [÷] 4 [ENTER]

even though that matches the way the problem is written algebraically. According to the order of operations, division is done before addition, so this typing is algebraically equivalent to

$$2 + \frac{3}{4}$$

which is not what we want. The large fraction bar in the expression $\frac{2+3}{4}$ groups the "2 + 3" together in the numerator; in the above typing, nothing groups the "2 + 3" together, and only the 3 ends up in the numerator.

Right The calculator needs parentheses inserted in the following manner:

$$\frac{(2+3)}{4}$$

Thus, it is correct to type

[(] 2 [+] 3 [)] [÷] 4 [ENTER]

This makes the display read 1.25, the correct answer.

Also Right It is correct to type

2 [+] 3 [ENTER] [÷] 4 [ENTER]

The first [ENTER] makes the calculator perform all prior calculations before continuing. This too makes the display read 1.25, the correct answer.

Memory

The **memory** is a place to store a number for later use, without having to write it down. Graphing calculators have a memory for each letter of the alphabet; that is, you can store one number in memory A, a second number in memory B, and so on. Pressing

TI-83/84	5 [STO➤] [ALPHA] [A] [ENTER]
Casio	5 [→] [ALPHA] [A] [EXE]

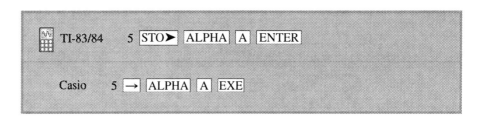

will store 5 in memory A. Similar keystrokes will store in memory B. Pressing [ALPHA] [A] will recall what has been stored in memory A.

EXAMPLE 8

USING THE MEMORY Calculate $\dfrac{3+\dfrac{5+7}{2}}{4}$ (a) by hand and (b) by first calculating $\frac{5+7}{2}$ and storing the result.

SOLUTION

a. $\dfrac{3+\dfrac{5+7}{2}}{4} = \dfrac{3+\dfrac{12}{2}}{4} = \dfrac{3+6}{4} = \dfrac{9}{4} = 2.25$

b. First, calculate $\frac{5+7}{2}$ and store the result in memory A. The large fraction bar in this expression groups the "5 + 7" together in the numerator; in our typing, we must group the "5 + 7" together with parentheses. The calculator needs parentheses inserted in the following manner:

$$\frac{(5+7)}{2}$$

Type

TI-83/84	(5 + 7) ÷ 2 STO➤ ALPHA A ENTER
Casio	(5 + 7) ÷ 2 → ALPHA A EXE

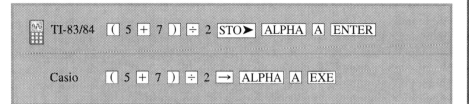

What remains is to compute $\frac{3 + A}{4}$. Again, the large fraction bar groups the "3 + A" together, so the calculator needs parentheses inserted in the following manner:

$$\frac{(3 + A)}{4}$$

Type

(3 + ALPHA A) ÷ 4 ENTER

and the display will read 2.25.

EXAMPLE 9

PARENTHESES WITHIN PARENTHESES Calculate $\dfrac{3 + \dfrac{5 + 7}{2}}{4}$ using one line of instructions and without using the memory.

SOLUTION

There are two large fractions bars, one grouping the "5 + 7" together and one grouping the "3 + $\frac{5 + 7}{2}$" together. In our typing, we must group each of these together with parentheses. The calculator needs parentheses inserted in the following manner:

$$\frac{\left(3 + \dfrac{(5 + 7)}{2}\right)}{4}$$

Type

(3 + (5 + 7) ÷ 2) ÷ 4 ENTER

and the display will read 2.25.

EXAMPLE 10

USING THE MEMORY

a. Use the quadratic formula and your calculator's memory to solve $2.3x^2 + 4.9x + 1.5 = 0$.

b. Check your answers.

SOLUTION

a. According to the Quadratic Formula, if $ax^2 + bx + c = 0$, then

$$x = \frac{-b \pm \sqrt{b^2 - 4ac}}{2a}$$

For our problem, $a = 2.3$, $b = 4.9$ and $c = 1.5$. This gives

$$x = \frac{-4.9 \pm \sqrt{4.9^2 - 4 \cdot 2.3 \cdot 1.5}}{2 \cdot 2.3}$$

One way to do this calculation is to calculate the radical, store it, and then calculate the two fractions.

Step 1 *Calculate the radical.* To do this, type

TI-83/84	2nd √ 4.9 x^2 − 4 × 2.3 × 1.5) STO➤ ALPHA A ENTER
Casio	SHIFT √ (4.9 x^2 − 4 × 2.3 × 1.5) ➤ ALPHA A EXE

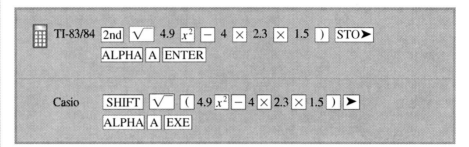

This makes the display read "3.195309062" and stores the number in the memory A. Notice the use of parentheses.

Step 2 *Calculate the first fraction.* The first fraction is

$$\frac{-4.9 \pm \sqrt{4.9^2 - 4 \cdot 2.3 \cdot 1.5}}{2 \cdot 2.3}$$

However, the radical has already been calculated and stored in memory A, so this is equivalent to

$$\frac{-4.9 + \text{A}}{2 \cdot 2.3}$$

Type

TI-83/84	((−) 4.9 + ALPHA A) ÷ (2 × 2.3) STO➤ ALPHA B ENTER
Casio	((−) 4.9 + ALPHA A) ÷ (2 × 2.3) ➤ ALPHA B EXE

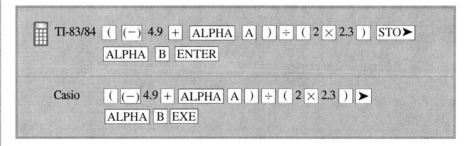

This makes the display read "−0.3705849866" and stores the number in memory B.

Step 3 *Calculate the second fraction.* The second fraction is

$$\frac{-4.9 - \sqrt{4.9^2 - 4 \cdot 2.3 \cdot 1.5}}{2 \cdot 2.3} = \frac{-4.9 - \text{A}}{2 \cdot 2.3}$$

Memories A and B are already in use, so we will store this in memory C. Instead of retyping the line in step 2 with the "+" changed to a "−" and the "B" to a "C," press 2nd ENTRY (← with a Casio) to reproduce that line, and use the ◄ button to make these changes. Press ENTER and the display will read −1.759849796, and that number will be stored in memory C. The solutions to $2.3x^2 + 4.9x + 1.5 = 0$ are $x = -0.370584966$ and $x = -1.759849796$. These are approximate solutions in that they show only the first nine decimal places.

Step 4

Check your solutions. These two solutions are stored in memories B and C; they can be checked by seeing if they satisfy the equation $2.3x^2 + 4.9x + 1.5 = 0$. To check the first solution, type

2.3 [×] [ALPHA] [B] [x^2] [+] 4.9 [×] [ALPHA] [B] [+] 1.5 [ENTER]

and the display should read either "0" or a number very close to 0.

Scientific Notation

Typing

4000000 [×] 8000000 [ENTER]

makes the display read "3.2E13" rather than "32000000000000." This is because the calculator does not have enough room on its display for "32000000000000." When the display shows "3.2E13," read it as "3.2×10^{13}," which is written in scientific notation. Literally, "3.2×10^{13}" means "multiply 3.2 by 10, thirteen times," but as a shortcut, you can interpret it as "move the decimal point in the '3.2' thirteen places to the right."

Typing

.0000005 [×] .0000007 [ENTER]

makes the display read "3.5E-13" rather than "0.00000000000035," because the calculator does not have enough room on its display for "0.00000000000035." Read "3.5E-13" as "3.5×10^{-13}." Literally, this means "divide 3.5 by 10, thirteen times," but as a shortcut, you can interpret it as "move the decimal point in the '3.5' thirteen places to the left."

You can type a number in scientific notation by using the TI button labeled "EE" (which stands for "Enter Exponent") or the Casio button labeled "EXP" (which stands for "Exponent"). For example, typing

5.2 [EE] 8 [ENTER]

makes the display read "520000000." (If the "EE" label is above the button, you will need to use the [2nd] button.) On a Casio, type

5.2 [EXP] 8 [EXE]

Be careful that you don't confuse the [EE] or [EXP] button with the [^] button. The [EE] or [EXP] button does *not* allow you to type in an exponent; it allows you to type in scientific notation. For example, typing

3 [EE] 4 [ENTER]

makes the display read "30000," since $3 \times 10^4 = 30{,}000$. Typing

3 [^] 4 [ENTER]

makes the display read "81," since $3^4 = 81$.

EXERCISES

In Exercises 1–32, use your calculator to perform the given calculation. The correct answer is given in brackets [ ]. In your homework, write down what you type to get that answer. Answers are not given in the back of the book.

1. $-3 - -5$ [2]
2. $-6 - 3$ [−9]
3. $4 - -9$ [13]
4. $-6 - -8$ [2]
5. $-3 - (-5 - -8)$ [−6]
6. $-(-4 - 3) - (-6 - -2)$ [11]
7. $-8 \cdot -3 \cdot -2$ [−48]
8. $-8 \cdot -3 - 2$ [22]
9. $(-3)(-8) - (-9)(-2)$ [6]
10. $2(3 - 5)$ [−4]
11. $2 \cdot 3 - 5$ [1]
12. $4 \cdot 11^2$ [484]

13. $(4 \cdot 11)^2$ [1,936]
14. $4 \cdot (-11)^2$ [484]
15. $4 \cdot (-3)^3$ [−108]
16. $(4 \cdot -3)^3$ [−1,728]
17. $\frac{3 + 2}{7}$ [0.7142857]
18. $\frac{3 \cdot 2}{7}$ [0.8571429]
19. $\frac{3}{2 \cdot 7}$ [0.2142857]
20. $\frac{3 \cdot 2}{7 \cdot 5}$ [0.1714286]
21. $\frac{3 + 5}{7 \cdot 2}$ [0.1428571]
22. $\frac{3 \cdot -2}{7 + 5}$ [−0.5]
23. $\frac{3}{7/2}$ [0.8571429]
24. $\frac{3/7}{2}$ [0.2142857]
25. 1.8^2 [3.24]
26. $\sqrt{1.8}$ [1.3416408]
27. $47{,}000{,}000^2$ [2.209×10^{15}]
28. $\sqrt{0.0000000000027}$ [1.643168×10^{-6}]
29. $(-3.92)^7$ [−14,223.368737]
30. $(5.72 \times 10^{19})^4$ [1.070494×10^{79}]
31. $(3.76 \times 10^{-12})^{-5}$ [1.330641×10^{57}]
32. $(3.76 \times 10^{-12}) - 5$ [−5]

In Exercises 33–36, perform the given calculation (a) by hand; (b) with a calculator, using the memory; and (c) with a calculator, using one line of instruction and without using memory, as shown in Examples 8 and 9. In your homework, for parts (b) and (c), write down what you type. Answers are not given in the back of the book.

33. $\dfrac{\frac{9 - 12}{5} + 7}{2}$
34. $\dfrac{\frac{4 - 11}{6} + 8}{7}$
35. $\dfrac{\frac{7 + 9}{5} + \frac{8 - 14}{3}}{3}$
36. $\dfrac{\frac{4 - 16}{5} + \frac{7 - 22}{2}}{5}$

In Exercises 37–38, use your calculator to solve the given equation for x. Check your two answers, as shown in Example 10. In your homework, write down what you type to get the answers and what you type to check the answers. Answers are not given in the back of the book.

37. $4.2x^2 + 8.3x + 1.1 = 0$
38. $5.7x^2 + 12.3x - 8.1 = 0$
39. Discuss the use of parentheses in step 1 of Example 10a. Why are they necessary? What would happen if they were omitted?
40. Discuss the use of parentheses in step 2 of Example 10a. Why are they necessary? What would happen if they were omitted?
41. **a.** Calculate $\dfrac{\frac{5 + 7.1}{3} + \frac{2 - 7.1}{5}}{7}$ using one line of instruction and without using the memory. In your homework, write down what you type as well as the solution.
    b. Use the [2nd] [ENTRY] feature to calculate $\dfrac{\frac{5 + 7.2}{3} + \frac{2 - 7.2}{5}}{7}$
    c. Use the [2nd] [ENTRY] feature to calculate $\dfrac{\frac{5 + 9.3}{3} + \frac{2 - 4.9}{5}}{7}$
42. **a.** Calculate $\dfrac{3 + \dfrac{5 + \frac{6 - 8.3}{2}}{3}}{9}$ using one line of instruction and without using the memory. In your homework, write down what you type as well as the solution.
    b. Use the [2nd] [ENTRY] feature to calculate $\dfrac{3 + \dfrac{5 - \frac{6 - 8.3}{2}}{3}}{9}$
    c. Use the [2nd] [ENTRY] feature to calculate $\dfrac{3 - \dfrac{5 + \frac{6 - 8.3}{2}}{3}}{9}$
43. **a.** What is the result of typing "8.1 [EE] 4"?
    b. What is the result of typing "8.1 [EE] 12"?
    c. Why do the instructions in part (b) yield an answer in scientific notation, while the instructions in part (a) yield an answer that's not in scientific notation?
    d. By using the [MODE] button, your calculator can be reset so that all answers will appear in scientific notation. Describe how this can be done.

APPENDIX C

Graphing with a Graphing Calculator

GRAPHING WITH A TEXAS INSTRUMENTS GRAPHING CALCULATOR

OBJECTIVE

- Become familiar with graphing operations

The Graphing Buttons

The graphing buttons on a TI graphing calculator are all at the top of the keypad, directly under the screen. The button labels and their uses are listed in Figure A.1.

TI-83/84	Y=	WINDOW	ZOOM	TRACE	GRAPH
use this button to tell the calculator:	what to graph	what part of the graph to draw	to zoom in or out	to give the coordinates of a highlighted point	to draw the graph

FIGURE A.1

Graphing a Line

To graph $y = 2x - 1$ on a TI graphing calculator, follow these steps.

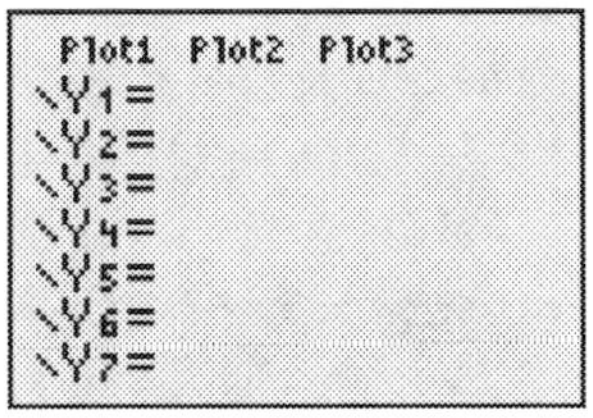

FIGURE A.2
A TI-84's "Y=" screen (other models' screens are similar).

1. *Set the calculator up for instructions on what to graph* by pressing [Y=]. This produces the screen similar to that shown in Figure A.2. If your screen has things written after the equals symbols, use the [▲] and [▼] buttons along with the [CLEAR] button to erase them.

2. *Tell the calculator what to graph* by typing "$2x - 1$" where the screen reads "$Y_1 =$." To type the x symbol, press [X.T.θ.n]. After typing "2x − 1," press the [ENTER] button.

3. *Set the calculator up for instructions on what part of the graph to draw by pressing* [WINDOW].

4. *Tell the calculator what part of the graph to draw* by entering the values shown in Figure A.3. (If necessary, use the [▲] and [▼] buttons to move from line to line).

   - "Xmin" and "Xmax" refer to the left and right boundaries of the graph, respectively.

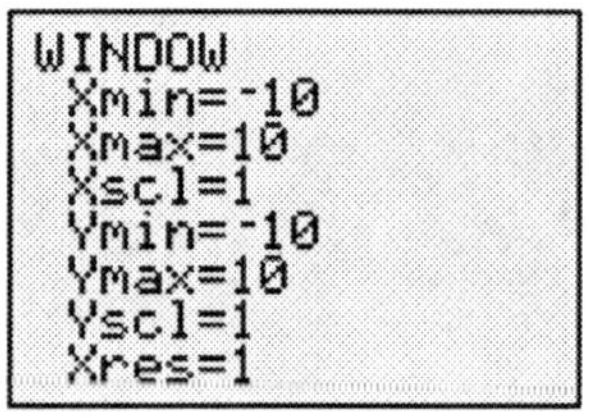

FIGURE A.3 A TI-84's "WINDOW" screen (other models' screens are similar).

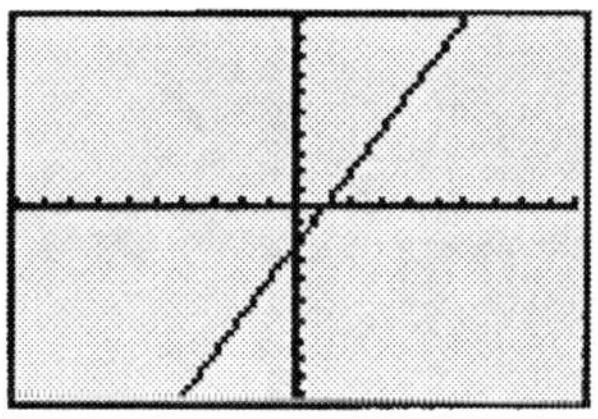

FIGURE A.4 The graph of $y = 2x - 1$.

- "Ymin" and "Ymax" refer to the lower and upper boundaries of the graph, respectively.
- "Xscl" and "Yscl" refer to the scales on the x- and y-axes (i.e., to the location of the tick marks on the axes).

5. *Tell the calculator to draw a graph* by pressing the GRAPH button. This produces the screen shown in Figure A.4.
6. *Discontinue graphing* by pressing 2nd QUIT.

Graphing with a Casio Graphing Calculator

To graph $y = 2x - 1$ on a Casio graphing calculator, follow these steps.

1. *Put the calculator into graphing mode* by pressing Menu, using the arrow buttons to highlight "GRAPH" on the main menu, and then pressing EXE.
2. *Tell the calculator what to graph* by typing "$2x - 1$" where the screen reads "Y1=" (see Figure A.5). (If something is already there, press F2, which now has the label "DEL" directly above it on the screen, and then press F1, which is labeled "YES.") To type the x symbol, press x,θ,T. After typing "$2x - 1$," press the EXE button.

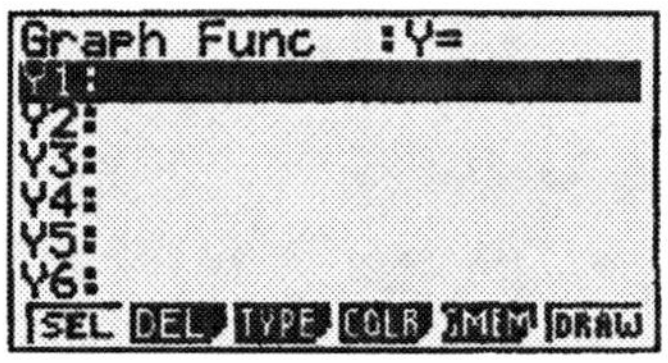

FIGURE A.5 A Casio's "Y=" screen.

3. *Tell the calculator to draw a graph* by pressing the F6 button, which now has the label "DRAW" directly above it on the screen.
4. *Set the calculator up for instructions on what part of the graph to draw* by pressing SHIFT and then the F3 button, which now has the label "V-WIN" directly above it on the screen. ("V-WIN" is short for "viewing window.")
5. *Tell the calculator what part of the graph to draw* by entering the values shown in Figure A.6. Press EXE after each entry. Then press EXIT to return to the "Y=" screen.

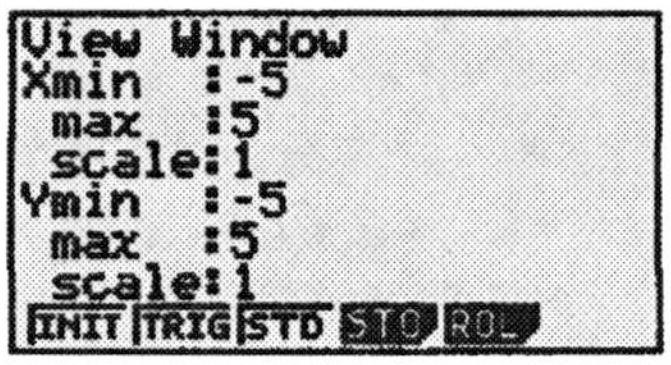

FIGURE A.6 A Casio's view window.

- "Xmin" and "Xmax" refer to the left and right boundaries of the graph, respectively.
- "Ymin" and "Ymax" refer to the lower and upper boundaries of the graph, respectively.
- "Xscale" and "Yscale" refer to scales on the x- and y-axes, respectively (i.e., to the location of the tick marks on the axes).

6. *Tell the calculator to draw a graph* by pressing the F6 button, which now has the label "DRAW" directly above it on the screen (see Figure A.7).

7. *Discontinue graphing* by pressing MENU.

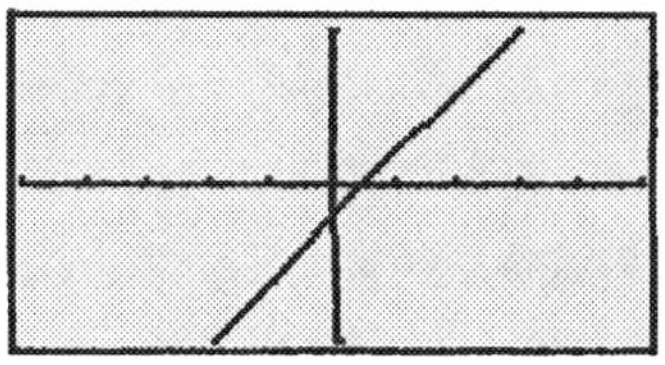

FIGURE A.7

EXERCISES

In the following exercises, you will explore some of your calculator's graphing capabilities. Answers are not given in the back of the book.

1. *Exploring the "Zoom Standard" command.* Use your calculator to graph $y = 2x - 1$, as discussed in this section. When that graph is on the screen, select the TI "Zoom Standard" command from the "Zoom menu" or the Casio "STD" command from the "V-WIN" menu by doing the following:

TI-83/84	• press ZOOM • select option 6 "Standard" or "ZStandard" by either: • using the down arrow to scroll down to that option and pressing ENTER, or • typing the number "6"
Casio	• press SHIFT • press V-WIN (i.e., F3) • press STD (i.e., F3)

   a. What is the result?
   b. How else could you accomplish the same thing without using any zoom commands?

2. *Exploring the "WINDOW" or "V-WIN" screen.* Use your calculator to graph $y = 2x - 1$ as discussed in this section. When that graph is on the screen, use the "WINDOW" or "V-WIN" screen described in this section to reset the following:
   - Xmin to 1
   - Xmax to 20
   - Ymin to 1
   - Ymax to 20

   a. Why are there no axes shown?
   b. Why does the graph start exactly in the lower left corner of the screen?

3. *Exploring the "WINDOW" or "V-WIN" screen.* Use your calculator to graph $y = 2x - 1$, as discussed in this section. When that graph is on the screen, use the "WINDOW" or "V-WIN" screen described in this section to reset the following:
   - Xmin to -1
   - Xmax to -20
   - Ymin to -1
   - Ymax to -20

   Why did the TI calculator respond with an error message? What odd thing did the Casio calculator do? (Casio hint: Use the TRACE button to investigate.)

4. *Exploring the "WINDOW" or "V-WIN" screen.* Use your calculator to graph $y = 2x - 1$, as discussed in this section. When that graph is on the screen, use the "WINDOW" or "V-WIN" screen described in this section to reset the following:
   - Xmin to -5
   - Xmax to 5
   - Ymin to 10
   - Ymax to 25

   a. Why was only one axis shown?
   b. Why was no graph shown?

5. *Exploring the "TRACE" command.* Use your calculator to graph $y = 2x - 1$, as discussed in this section. When that graph is on the screen, press TRACE (F1 on a Casio). This causes two things to happen:
   - A mark appears at a point on the line. This mark can be moved with the left and right arrow buttons.
   - The corresponding ordered pair is printed out at the bottom of the screen.

   a. Use the "TRACE" feature to locate the line's x-intercept, the point at which the line hits the x-axis. (You may need to approximate it.)
   b. Use algebra, rather than the graphing calculator, to find the x-intercept.
   c. Use the "TRACE" feature to locate another ordered pair on the line.
   d. Use substitution to check that the ordered pair found in part (c) is in fact a point on the line.

6. *Exploring the "ZOOM BOX" command.* Use your calculator to graph $y = 2x - 1$ as discussed in this section. When that graph is on your screen, press ZOOM (F2 on a Casio) and select option 1, "ZBOX," in the manner

described in Exercise 1. This seems to have the same result as [TRACE] except that the mark does not have to be a point on the line. Use the four arrow buttons to move to a point of your choice (that may be on or off the line). Press [ENTER]. Use the arrow buttons to move to a different point so that the resulting box encloses a part of the line. Press [ENTER] again. What is the result of using the "Zoom Box" command?

7. *Exploring the "ZOOM IN" command.* Use your calculator to graph $y = 2x - 1$, as discussed in this section. When that graph is on your screen, press [ZOOM] ([F2] on a Casio) and select "Zoom In" in the manner described in Exercise 1. (Press [IN] or [F3] on a Casio.) This causes a mark to appear on the screen. Use the four arrow buttons to move the mark to a point of your choice, either on or near the line. Press [ENTER].

a. What is the result of the "Zoom In" command?

b. How could you accomplish the same thing without using any zoom commands?

c. The "Zoom Out" command is listed right next to the "Zoom In" command? What does it do?

8. *Zooming in on x-intercepts.* Use the "Zoom In" command described in Exercise 7, and the "Trace" command described in Exercise 5, to approximate the location of the x-intercept of $y = 2x - 1$ as accurately as possible. You might need to use these commands more than once.

a. Describe the procedure you used to generate this answer.

b. According to the calculator, what is the x-intercept?

c. Is this answer the same as that of Exercise 5(b)? Why or why not?

9. *Calculating x-intercepts.* The TI-83, TI-84, and Casio will calculate the x-intercept (also called a "root" or "zero") without using the "Zoom In" and "Trace" commands. First, use your calculator to graph $y = 2x - 1$ as discussed in this section. When that graph is on the screen, do the following.

On a TI-83/84:

- Press [2nd] [CALC] and select option 2, "root" or "zero."
- The calculator responds by asking, "Lower Bound?" or "Left Bound?" Use the left and right arrow buttons to move the mark to a point slightly to the left of the x-intercept, and press [ENTER].
- When the calculator asks, "Upper Bound?" or "Right Bound?", move the mark to a point slightly to the right of the x-intercept, and press [ENTER].
- When the calculator asks, "Guess?", move the mark to a point close to the x-intercept, and press [ENTER]. The calculator will then display the location of the x-intercept.

On a Casio:

- Press [G-Solv] (i.e., [F5]).
- Press [ROOT]. Wait and watch.

a. According to the calculator, what is the x-intercept of the line $y = 2x - 1$?

b. Is this answer the same as that of Exercise 5(b)? Why or why not?

APPENDIX D

Finding Points of Intersection with a Graphing Calculator

The graphs of $y = x + 3$ and $y = -x + 9$ intersect on the standard viewing screen. To find the point of intersection with a TI or Casio graphing calculator, first enter the two equations on the "Y=" screen (one as Y_1 and one as Y_2), erase any other equations, and erase the "Y= " screen by pressing [2nd] [QUIT]. Then follow the following instructions.

On a TI-83/84

- Graph the two equations on the standard viewing screen.
- Press [2nd] [CALC] and select option 5, "intersect."
- When the calculator responds with "First curve?" and a mark on the first equation's graph, press [ENTER].
- When the calculator responds with "Second curve?" and a mark on the second equation's graph, press [ENTER].
- When the calculator responds with "Guess?", use the left and right arrows to place the mark near the point of intersection, and press [ENTER].
- Check your answer by substituting the ordered pair into each of the two equations.

On a Casio

- Graph the two equations on the standard viewing screen.
- Press [G-Solv] (i.e., [F5]).
- Press [ISCT] (i.e., [F5]). Watch and wait.
- After a pause, the calculator will display the location of the x-intercept.

Figure A.8 shows the results of computing the point of intersection of $y = x + 3$ and $y = -x + 9$.

The information on the screen indicates that the point of intersection is (3, 6). To check this, substitute 3 for x into each of the two equations; you should get 6.

$$y = x + 3 = 3 + 3 = 6. \checkmark$$
$$y = -x + 9 = -3 + 9 = 6. \checkmark$$

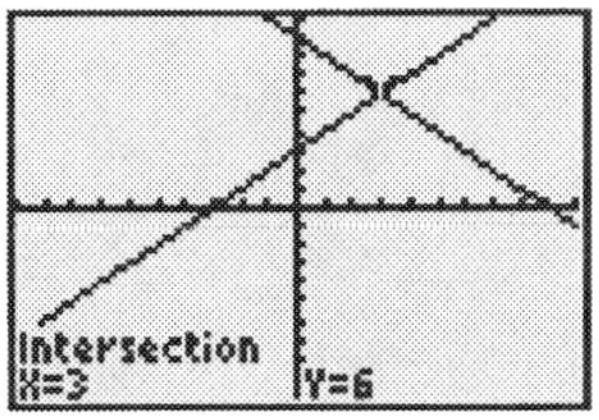

FIGURE A.8 Finding the point of intersection.

EXERCISES

In Exercises 1–6, do the following.

a. Use the graphing calculator to find the point of intersection of the given equations.

b. Check your solutions by substituting the ordered pair into each of the two equations. Answers are not given in the back of the book.

1. $y = 3x + 2$ and $y = 5x + 5$

2. $y = 2x - 6$ and $y = 3x + 4$

3. $y = 8x - 14$ and $g(x) = 11x + 23$

HINT: You will have to change Xmin, Xmax, Ymin, and Ymax to find the point.

4. $y = -7x + 12$ and $y = -12x - 71$

HINT: You will have to change Xmin, Xmax, Ymin, and Ymax to find the point.

5. $y = x^2 - 2x + 3$ and $y = -x^2 - 3x + 12$
(Find two answers.)

6. $f(x) = 8x^2 - 3x - 7$ and $y = 2x + 4$
(Find two answers.)

APPENDIX E
Dimensional Analysis

Most people know how to convert 6 feet into yards (it's $6 \div 3 = 2$ yards). Few people know how to convert 50 miles per hour into feet per minute. The former problem is so commonplace that people just remember how to do it; the latter problem is not so common, and people don't know how to do it. Dimensional analysis is an easy way of converting a quantity from one set of units to another; it can be applied to either of these two problems.

Dimensional analysis involves using a standard conversion (such as 1 yard = 3 feet) to create a fraction, including units (such as "feet" and "yards") in that fraction, and canceling units in the same way that variables are canceled. That is, in the fraction 2 feet/3 feet, we can cancel feet with feet and obtain 2/3, just as we can cancel x with x in the fraction $2x/3x$ and obtain 2/3.

EXAMPLE 1 USING DIMENSIONAL ANALYSIS Convert 6 feet into yards.

SOLUTION Start with the standard conversion:

$$1 \text{ yard} = 3 \text{ feet} \quad \text{a standard conversion}$$

Create a fraction by dividing each side by 3 feet:

$$\frac{1 \text{ yard}}{3 \text{ feet}} = \frac{\cancel{3} \cancel{\text{feet}}}{\cancel{3} \cancel{\text{feet}}}$$

$$\frac{1 \text{ yard}}{3 \text{ feet}} = 1 \quad \text{a fractional version of a standard conversion}$$

To convert 6 feet into yards, multiply 6 feet by the fraction 1 yard/3 feet. This is valid because that fraction is equal to 1, and multiplying something by 1 doesn't change its value.

$$\begin{aligned} 6 \text{ feet} &= 6 \text{ feet} \cdot 1 \\ &= \cancel{6} \cancel{\text{feet}} \cdot \frac{1 \text{ yard}}{\cancel{3} \cancel{\text{feet}}} && \text{substituting } \frac{1 \text{ yard}}{3 \text{ feet}} \text{ for } 1 \\ &= 2 \text{ yards} && \text{canceling} \end{aligned}$$

It is crucial to include units in this work. If at the beginning of this example, we had divided by 1 yard instead of 3 feet, we would have obtained

$$1 \text{ yard} = 3 \text{ feet} \quad \text{a standard conversion}$$

$$\frac{\cancel{1} \cancel{\text{yard}}}{\cancel{1} \cancel{\text{yard}}} = \frac{3 \text{ feet}}{1 \text{ yard}}$$

$$1 = \frac{3 \text{ feet}}{1 \text{ yard}} \quad \text{a fractional version of a standard conversion}$$

Multiplying 6 feet by 3 feet/1 yard would not allow us to cancel feet with feet and would not leave an answer in yards.

$$6 \text{ feet} = 6 \text{ feet} \cdot 1$$
$$= 6 \text{ feet} \cdot \frac{3 \text{ feet}}{1 \text{ yard}} \quad \text{substituting } \frac{3 \text{ feet}}{1 \text{ yard}} \text{ for } 1$$

It's important to include units in dimensional analysis, because it's only by looking at the units that we can tell that multiplying by 1 yard/3 feet is productive and multiplying by 3 feet/1 yard isn't.

EXAMPLE 2

USING DIMENSIONAL ANALYSIS Convert 50 miles/hour to feet/minute.

SOLUTION

The appropriate standard conversions are

$$1 \text{ mile} = 5{,}280 \text{ feet}$$
$$1 \text{ hour} = 60 \text{ minutes} \quad \text{two standard conversions}$$

This problem has two parts, one for each of these two standard conversions.

Part 1 *Use the standard conversion "1 mile = 5,280 feet" to convert miles/hour to feet/hour.* The fraction 50 miles/hour has miles in the numerator, and we are to convert it to a fraction that has feet in the numerator. To replace miles with feet, first rewrite the standard conversion as a fraction that has miles in the denominator, and then multiply by that fraction. This will allow miles to cancel.

$$1 \text{ mile} = 5{,}280 \text{ feet} \quad \text{a standard conversion}$$
$$\frac{\cancel{1 \text{ mile}}}{\cancel{1 \text{ mile}}} = \frac{5280 \text{ feet}}{1 \text{ mile}} \quad \text{placing miles in the denominator}$$
$$1 = \frac{5280 \text{ feet}}{1 \text{ mile}} \quad \text{a fractional version of a standard conversion}$$

Now multiply 50 miles/hour by this fraction and cancel:

$$\frac{50 \text{ miles}}{\text{hour}} = \frac{50 \text{ miles}}{\text{hour}} \cdot 1$$
$$= \frac{50 \cancel{\text{miles}}}{\text{hour}} \cdot \frac{5280 \text{ feet}}{1 \cancel{\text{mile}}} \quad \text{substituting } \frac{5280 \text{ feet}}{1 \text{ mile}} \text{ for } 1$$
$$= \frac{50 \cdot 5280 \text{ feet}}{\text{hour}} \quad \text{canceling miles}$$

Part 2 *Use the standard conversion "1 hour = 60 minutes" to convert feet/hour to feet/minute.* To replace hours with minutes, first rewrite the standard conversion as a fraction that has hours in the numerator, and then multiply by that fraction. This will allow hours to cancel.

$$1 \text{ hour} = 60 \text{ minutes} \quad \text{a standard conversion}$$
$$\frac{1 \text{ hour}}{60 \text{ minutes}} = \frac{\cancel{60 \text{ minutes}}}{\cancel{60 \text{ minutes}}} \quad \text{placing hours in the numerator}$$
$$\frac{1 \text{ hour}}{60 \text{ minutes}} = 1 \quad \text{a fractional version of a standard conversion}$$

Continuing where we left off, multiply by this fraction and cancel:

$$\frac{50 \text{ miles}}{\text{hour}} = \frac{50 \cdot 5280 \text{ feet}}{\text{hour}} \quad \text{from part 1}$$

$$= \frac{50 \cdot 5280 \text{ feet}}{\text{hour}} \cdot 1$$

$$= \frac{50 \cdot 5280 \text{ feet}}{\cancel{\text{hour}}} \cdot \frac{1 \cancel{\text{hour}}}{60 \text{ minutes}} \quad \text{substituting } \frac{1 \text{ hour}}{60 \text{ minutes}} \text{ for } 1$$

$$= \frac{4400 \text{ feet}}{1 \text{ minute}} \quad \text{since } 50 \cdot 5280 / 60 = 4400$$

Thus, 50 miles/hour is equivalent to 4,400 feet/minute.

EXAMPLE **3**

USING DIMENSIONAL ANALYSIS How many feet will a car travel in half a minute if that car's rate is 50 miles/hour?

SOLUTION

This seems to be a standard algebra problem that uses the formula "distance = rate · time"; we're given the rate (50 miles/hour) and the time (1/2 minute), and we are to find the distance. However, the units are not consistent.

$$\text{distance} = \text{rate} \cdot \text{time}$$

$$= \frac{50 \text{ miles}}{\text{hour}} \cdot \frac{1}{2} \text{ minute}$$

None of these units cancels, and we are not left with an answer in feet. However, if the car's rate were in feet/minute rather than miles/hour, the units would cancel, and we would be left with an answer in feet:

$$\text{distance} = \text{rate} \cdot \text{time}$$

$$= \frac{50 \text{ miles}}{\text{hour}} \cdot \frac{1}{2} \text{ minute}$$

$$= \frac{4400 \text{ feet}}{1 \cancel{\text{minute}}} \cdot \frac{1}{2} \cancel{\text{minute}} \quad \text{from Example 2}$$

$$= 2{,}200 \text{ feet}$$

The car would travel 2,200 feet in half a minute.

EXERCISES

In Exercises 1–6, use dimensional analysis to convert the given quantity.

1. **a.** 12 feet into yards
   b. 12 yards into feet

2. **a.** 24 feet into inches
   b. 24 inches into feet

3. **a.** 10 miles into feet
   b. 10 feet into miles (Round off to the nearest ten thousandth of a mile.)

4. **a.** 2 hours into minutes
   b. 2 minutes into hours (Round off to the nearest thousandth of an hour.)

5. 2 miles into inches

HINT: Convert first to feet, then to inches.

6. 3 hours into seconds

HINT: Convert first to minutes, then to seconds.

In Exercises 7–12, use the following information. The metric system is based on three different units:

- the gram (1 gram = 0.0022046 pound)
- the meter (1 meter = 39.37 inches)
- the liter (1 liter = 61.025 cubic inches)

Other units are formed by adding the following prefixes to these basic units:

- kilo-, which means one thousand (for example, 1 kilometer = 1,000 meters)
- centi-, which means one hundredth (for example, 1 centimeter = 1/100 meter)
- milli-, which means one thousandth (for example, 1 millimeter = 1/1,000 meter)

7. Use dimensional analysis to convert
   a. 50 kilometers to meters
   b. 50 meters to kilometers

8. Use dimensional analysis to convert
   a. 30 milligrams to grams
   b. 30 grams to milligrams

9. Use dimensional analysis to convert
   a. 2 centiliters to liters
   b. 2 liters to centiliters

10. Use dimensional analysis to convert
   a. 1 yard to meters (Round off to the nearest hundredth.)
   b. 20 yards to centimeters (Round off to the nearest centimeter.)

11. Use dimensional analysis to convert
   a. 1 pound to grams (Round off to the nearest gram.)
   b. 8 pounds to kilograms (Round off to the nearest tenth of a kilogram.)

12. Use dimensional analysis to convert
   a. 1 cubic inch to liters (Round off to the nearest thousandth.)
   b. 386 cubic inches to liters (Round off to the nearest tenth.)

13. **a.** Use dimensional analysis to convert 60 miles per hour into feet per second.
   b. Leadfoot Larry speeds through an intersection at 60 miles per hour, and he is ticketed for speeding and reckless driving. He pleads guilty to speeding but not guilty to reckless driving, telling the judge that at that speed, he would have plenty of time to react to any cross traffic. The intersection is 80 feet wide. How long would it take Larry to cross the intersection? (Round off to the nearest tenth of a second.)

14. In the United States, a typical freeway speed limit is 65 miles per hour.
   a. Use dimensional analysis to convert this to kilometers per hour. (Round off to the nearest whole number.)
   b. At this speed, how many miles can be traveled in 10 minutes? (Round off to the nearest whole number.)

15. In Germany, a typical autobahn speed limit is 130 kilometers per hour.
   a. Use dimensional analysis to convert this into miles per hour. (Round off to the nearest whole number.)
   b. How many more miles will a car traveling at 130 kilometers per hour go in 1 hour than a car traveling at 65 miles per hour? (Round off to the nearest whole number.)

16. Light travels 6×10^{12} miles per year.
   a. Convert this to miles per hour. (Round off to the nearest whole number.)
   b. How far does light travel in 1 second? (Round off to the nearest whole number.)

17. In December 2009, Massachusetts' Capital Crossing Bank offered a money market account with a 5.75% interest rate. This means that Capital Crossing Bank will pay interest at a rate of 5.75% per year.
   a. Use dimensional analysis to convert this to a percent per day.
   b. If you deposited $10,000 on September 1, how much interest would your account earn by October 1? (There are 30 days in September.) (Round off to the nearest cent.)
   c. If you deposited $10,000 on October 1, how much interest would your account earn by November 1? (There are 31 days in October.) (Round off to the nearest cent.)

18. In May 2010, Utah's Ally Bank offered a money market account with a 1.29% interest rate. This means that Republic Bank will pay interest at a rate of 1.29% per year.
   a. Use dimensional analysis to convert this to a percent per day.
   b. If you deposited $10,000 on September 1, how much interest would your account earn by October 1? (There are 30 days in September.) (Round off to the nearest cent.)
   c. If you deposited $10,000 on October 1, how much interest would your account earn by November 1? (There are 31 days in October.) (Round off to the nearest cent.)

19. You cannot determine whether a person is overweight by merely determining his or her weight; if a short person and a tall person weigh the same, the short person could be overweight and the tall person could be underweight. Body mass index (BMI) is becoming a standard way of determining if a person is overweight, since it takes both weight and height into consideration. BMI is defined as (weight in kilograms)/(height in meters)^2. According to the World Health Organization, a person is overweight if his or her BMI is 25 or greater. In October 1996, Katherine Flegal, a statistician for the National Center for Health Statistics, said that according to this standard, one out of every two Americans is overweight. (*Source: San Francisco Chronicle*, 16 October 1996, page A6.)

a. Lenny is 6 feet tall. Convert his height to meters. (Round off to the nearest hundredth.)

b. Lenny weighs 169 pounds. Convert his weight to kilograms. (Round off to the nearest tenth.)

c. Determine Lenny's BMI. (Round off to the nearest whole number.) Is he overweight?

d. Fred weighs the same as Lenny, but he is 5′5″ tall. Determine Fred's BMI. (Round off to the nearest whole number.) Is Fred overweight?

e. Why does BMI use the metric system (kilograms and meters) rather than the English system (feet, inches, and pounds)?

20. According to the U.S. Census Bureau, in 2009, California, the most populous state, had 36,961,664 people residing in 158,706 square miles of area, while Rhode Island, the smallest state, had 1,053,209 people residing in 1,212 square miles of area. Determine which state is more crowded by computing the number of square feet per person in each state.

21. A wading pool is 4 feet wide, 6 feet long, and 11 inches deep. There are 7.48 gallons of water per cubic foot. How many gallons of water does it take to fill the pool?

22. John ate a 2,000 calorie lunch and immediately felt guilty. Jogging for one minute consumes 0.061 calorie per pound of body weight. John weighs 205 pounds. How long will he have to jog to burn off all of the calories from lunch?

APPENDIX F

Body Table for the Standard Normal Distribution

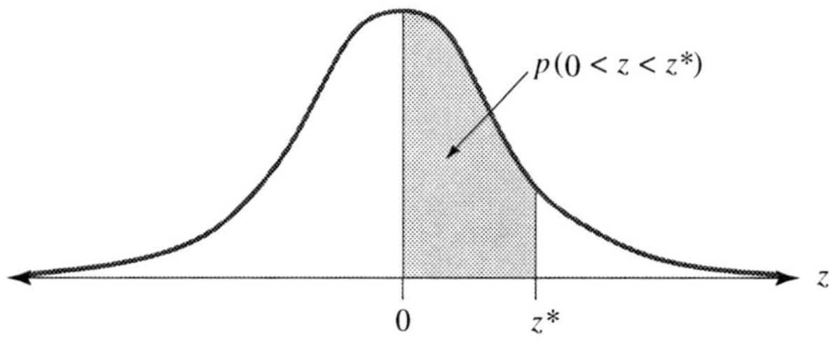

z*	0.00	0.01	0.02	0.03	0.04	0.05	0.06	0.07	0.08	0.09
0.0	0.0000	0.0040	0.0080	0.0120	0.0160	0.0199	0.0239	0.0279	0.0319	0.0359
0.1	0.0398	0.0438	0.0478	0.0517	0.0557	0.0596	0.0636	0.0675	0.0714	0.0753
0.2	0.0793	0.0832	0.0871	0.0910	0.0948	0.0987	0.1026	0.1064	0.1103	0.1141
0.3	0.1179	0.1217	0.1255	0.1293	0.1331	0.1368	0.1406	0.1443	0.1480	0.1517
0.4	0.1554	0.1591	0.1628	0.1664	0.1700	0.1736	0.1772	0.1808	0.1844	0.1879
0.5	0.1915	0.1950	0.1985	0.2019	0.2054	0.2088	0.2123	0.2157	0.2190	0.2224
0.6	0.2257	0.2291	0.2324	0.2357	0.2389	0.2422	0.2454	0.2486	0.2517	0.2549
0.7	0.2580	0.2611	0.2642	0.2673	0.2704	0.2734	0.2764	0.2794	0.2823	0.2852
0.8	0.2881	0.2910	0.2939	0.2967	0.2995	0.3023	0.3051	0.3078	0.3106	0.3133
0.9	0.3159	0.3186	0.3212	0.3238	0.3264	0.3289	0.3315	0.3340	0.3365	0.3389
1.0	0.3413	0.3438	0.3461	0.3485	0.3508	0.3531	0.3554	0.3577	0.3599	0.3621
1.1	0.3643	0.3665	0.3686	0.3708	0.3729	0.3749	0.3770	0.3790	0.3810	0.3830
1.2	0.3849	0.3869	0.3888	0.3907	0.3925	0.3944	0.3962	0.3980	0.3997	0.4015
1.3	0.4032	0.4049	0.4066	0.4082	0.4099	0.4115	0.4131	0.4147	0.4162	0.4177
1.4	0.4192	0.4207	0.4222	0.4236	0.4251	0.4265	0.4279	0.4292	0.4306	0.4319
1.5	0.4332	0.4345	0.4357	0.4370	0.4382	0.4394	0.4406	0.4418	0.4429	0.4441
1.6	0.4452	0.4463	0.4474	0.4484	0.4495	0.4505	0.4515	0.4525	0.4535	0.4545
1.7	0.4554	0.4564	0.4573	0.4582	0.4591	0.4599	0.4608	0.4616	0.4625	0.4633
1.8	0.4641	0.4649	0.4656	0.4664	0.4671	0.4678	0.4686	0.4692	0.4699	0.4706
1.9	0.4713	0.4719	0.4726	0.4732	0.4738	0.4744	0.4750	0.4756	0.4761	0.4767
2.0	0.4772	0.4778	0.4783	0.4788	0.4793	0.4798	0.4803	0.4808	0.4812	0.4817
2.1	0.4821	0.4826	0.4830	0.4834	0.4838	0.4842	0.4846	0.4850	0.4854	0.4857
2.2	0.4861	0.4864	0.4868	0.4871	0.4875	0.4878	0.4881	0.4884	0.4887	0.4890
2.3	0.4893	0.4896	0.4898	0.4901	0.4904	0.4906	0.4909	0.4911	0.4913	0.4916
2.4	0.4918	0.4920	0.4922	0.4925	0.4927	0.4929	0.4931	0.4932	0.4934	0.4936
2.5	0.4938	0.4940	0.4941	0.4943	0.4945	0.4946	0.4948	0.4949	0.4951	0.4952
2.6	0.4953	0.4955	0.4956	0.4957	0.4959	0.4960	0.4961	0.4962	0.4963	0.4964
2.7	0.4965	0.4966	0.4967	0.4968	0.4969	0.4970	0.4971	0.4972	0.4973	0.4974
2.8	0.4974	0.4975	0.4976	0.4977	0.4977	0.4978	0.4979	0.4979	0.4980	0.4981
2.9	0.4981	0.4982	0.4982	0.4983	0.4984	0.4984	0.4985	0.4985	0.4986	0.4986
3.0	0.4987	0.4987	0.4987	0.4988	0.4988	0.4989	0.4989	0.4989	0.4990	0.4990

APPENDIX G

Selected Answers to Odd Exercises

CHAPTER 1 Logic

1.1 Deductive vs. Inductive Reasoning

1. **a.** Valid **b.** Invalid

3. **a.** Invalid **b.** Valid

5. Invalid

7. Valid

9. Valid

11. Valid

13. Valid

15. Invalid

17. Invalid

19. Valid

21. **a.** Inductive
b. Deductive

23. 23

25. 20

27. 25

29. 13

31. 5

33. F, S

35. T, W

37. r

39. f

41. f

43. Answers may vary.

47. $3 \pm \sqrt{2}$

49.

3	5	6	4	1	8	2	7	9
2	1	7	6	5	9	3	4	8
9	8	4	3	7	2	1	5	6
6	2	1	8	9	5	4	3	7
8	3	5	7	4	6	9	2	1
7	4	9	2	3	1	6	8	5
5	9	2	1	8	4	7	6	3
1	6	3	5	2	7	8	9	4
4	7	8	9	6	3	5	1	2

51.

9	5	6	3	4	1	2	8	7
8	2	3	6	5	7	1	4	9
4	7	1	9	8	2	5	3	6
6	8	7	1	3	4	9	5	2
5	9	4	2	6	8	7	1	3
3	1	2	7	9	5	4	6	8
7	6	5	8	1	9	3	2	4
2	4	8	5	7	3	6	9	1
1	3	9	4	2	6	8	7	5

53.

7	6	8	5	4	2	9	1	3
1	5	2	7	3	9	6	4	8
9	3	4	6	1	8	2	5	7
8	7	1	9	6	3	5	2	4
5	2	6	1	7	4	8	3	9
4	9	3	8	2	5	7	6	1
6	8	9	4	5	1	3	7	2
3	4	7	2	9	6	1	8	5
2	1	5	3	8	7	4	9	6

1.2 Symbolic Logic

1. **a.** Statement
b. Statement
c. Not a statement. (question)
d. Not a statement. (opinion)

3. (a) and (d) are negations; (b) and (c) are negations

5. **a.** Her dress is red.
b. No computer is priced under $100. (All computers are $100 or more.)
c. Some dogs are not four-legged animals.
d. Some sleeping bags are waterproof.

7. **a.** $p \wedge q$ **b.** $\sim p \rightarrow \sim q$ **c.** $\sim(p \vee q)$
d. $p \wedge \sim q$ **e.** $p \rightarrow q$ **f.** $\sim q \rightarrow \sim p$

9. **a.** $(p \vee q) \rightarrow r$ **b.** $p \wedge q \wedge r$ **c.** $r \wedge \sim(p \vee q)$
d. $q \rightarrow r$ **e.** $\sim r \rightarrow \sim(p \vee q)$ **f.** $r \rightarrow (q \vee p)$

11. p: A shape is a square.
q: A shape is a rectangle.
$p \rightarrow q$

13. p: A shape is a square.
q: A shape is a triangle.
$p \rightarrow \sim q$

15. p: A number is a whole number.
q: A number is even.
r: A number is odd.
$p \rightarrow (q \vee r)$

17. p: A number is a whole number.
q: A number is greater than 3.
r: A number is less than 4.
$p \rightarrow \sim(q \wedge r)$

19. p: A person is an orthodontist.
q: A person is a dentist.
$p \rightarrow q$

21. p: A person knows Morse code.
q: A person operates a telegraph.
$q \rightarrow p$

23. p: The animal is a monkey.
q: The animal is an ape.
$p \rightarrow \sim q$

25. p: The animal is a monkey.
q: The animal is an ape.
$q \rightarrow \sim p$

27. p: I sleep soundly.
q: I drink coffee.
r: I eat chocolate.
$(q \vee r) \rightarrow \sim p$

29. p: Your check is accepted.
q: You have a driver's license.
r: You have a credit card.
$\sim(q \vee r) \rightarrow \sim p$

31. p: You do drink.
q: You do drive.
r: You are fined.
s: You go to jail.
$(p \wedge q) \rightarrow (r \vee s)$

33. p: You get a refund.
q: You get a store credit.
r: The product is defective.
$r \rightarrow (p \vee q)$

35. **a.** I am an environmentalist and I recycle my aluminum cans.
b. If I am an environmentalist, then I recycle my aluminum cans.
c. If I do not recycle my aluminum cans, then I am not an environmentalist.
d. I recycle my aluminum cans or I am not an environmentalist.

37. **a.** If I recycle my aluminum cans or newspapers, then I am an environmentalist.
b. If I am not an environmentalist, then I do not recycle my aluminum cans or newspapers.
c. I recycle my aluminum cans and newspapers or I am not an environmentalist.
d. If I recycle my newspapers and do not recycle my aluminum cans, then I am not an environmentalist.

39. Statement #1: Cold weather is required in order to have snow.

41. Statement #2: A month with 30 days would also indicate that the month is not February.

55. a. **57.** b.

1.3 Truth Tables

1.

p	q	$\sim q$	$p \vee \sim q$
T	T	F	T
T	F	T	T
F	T	F	F
F	F	T	T

3.

p	$\sim p$	$p \vee \sim p$
T	F	T
F	T	T

5.

p	q	$\sim q$	$p \rightarrow \sim q$
T	T	F	F
T	F	T	T
F	T	F	T
F	F	T	T

7.

p	q	$\sim q$	$\sim p$	$\sim q \rightarrow \sim p$
T	T	F	F	T
T	F	T	F	F
F	T	F	T	T
F	F	T	T	T

9.

p	q	$p \vee q$	$\sim p$	$(p \vee q) \rightarrow \sim p$
T	T	T	F	F
T	F	T	F	F
F	T	T	T	T
F	F	F	T	T

11.

p	q	$p \vee q$	$p \wedge q$	$(p \vee q) \rightarrow (p \wedge q)$
T	T	T	T	T
T	F	T	F	F
F	T	T	F	F
F	F	F	F	T

13.

p	q	r	$q \vee r$	$\sim(q \vee r)$	$p \wedge \sim(q \vee r)$
T	T	T	T	F	F
T	T	F	T	F	F
T	F	T	T	F	F
T	F	F	F	T	T
F	T	T	T	F	F
F	T	F	T	F	F
F	F	T	T	F	F
F	F	F	F	T	F

15.

p	q	r	$\sim q$	$\sim q \wedge r$	$p \vee (\sim q \wedge r)$
T	T	T	F	F	T
T	T	F	F	F	T
T	F	T	T	T	T
T	F	F	T	F	T
F	T	T	F	F	F
F	T	F	F	F	F
F	F	T	T	T	T
F	F	F	T	F	F

17.

p	q	r	$\sim r$	$\sim r \vee p$	$q \wedge p$	$(\sim r \vee p) \rightarrow (q \wedge p)$
T	T	T	F	T	T	T
T	T	F	T	T	T	T
T	F	T	F	T	F	F
T	F	F	T	T	F	F
F	T	T	F	F	F	T
F	T	F	T	T	F	F
F	F	T	F	F	F	T
F	F	F	T	T	F	F

19.

p	q	r	$\sim r$	$p \vee r$	$q \wedge (\sim r)$	$(p \vee r) \rightarrow (q \wedge \sim r)$
T	T	T	F	T	F	F
T	T	F	T	T	T	T
T	F	T	F	T	F	F
T	F	F	T	T	F	F
F	T	T	F	T	F	F
F	T	F	T	F	T	T
F	F	T	F	T	F	F
F	F	F	T	F	F	T

21. p: It is raining.
q: The streets are wet.
$p \rightarrow q$

p	q	$p \rightarrow q$
T	T	T
T	F	F
F	T	T
F	F	T

23. p: It rains.
q: The water supply is rationed.
$\sim p \rightarrow q$

p	q	$\sim p$	$\sim p \rightarrow q$
T	T	F	T
T	F	F	T
F	T	T	T
F	F	T	F

25. p: A shape is a square.
q: A shape is a rectangle.
$p \rightarrow q$

p	q	$p \rightarrow q$
T	T	T
T	F	F
F	T	T
F	F	T

27. p: It is a square.
q: It is a triangle.
$p \rightarrow \sim q$

p	q	$\sim q$	$p \rightarrow \sim q$
T	T	F	F
T	F	T	T
F	T	F	T
F	F	T	T

29. p: The animal is a monkey.
q: The animal is an ape.
$p \rightarrow \sim q$

p	q	$\sim q$	$p \rightarrow \sim q$
T	T	F	F
T	F	T	T
F	T	F	T
F	F	T	T

31. p: The animal is a monkey.
q: The animal is an ape.
$q \rightarrow \sim p$

p	q	$\sim p$	$q \rightarrow \sim p$
T	T	F	F
T	F	F	T
F	T	T	T
F	F	T	T

33. p: You have a driver's license.
q: You have a credit card.
r: Your check is approved.
$(p \vee q) \rightarrow r$

p	q	r	$p \vee q$	$(p \vee q) \rightarrow r$
T	T	T	T	T
T	T	F	T	F
T	F	T	T	T
T	F	F	T	F
F	T	T	T	T
F	T	F	T	F
F	F	T	F	T
F	F	F	F	T

35. p: Leaded gas is used.
q: The catalytic converter is damaged.
r: The air is polluted.
$p \rightarrow (q \wedge r)$

p	q	r	$q \wedge r$	$p \rightarrow (q \wedge r)$
T	T	T	T	T
T	T	F	F	F
T	F	T	F	F
T	F	F	F	F
F	T	T	T	T
F	T	F	F	T
F	F	T	F	T
F	F	F	F	T

37. p: I have a college degree.
q: I have a job.
r: I own a house.
$p \wedge \sim(q \vee r)$

p	q	r	$q \vee r$	$\sim(q \vee r)$	$p \wedge \sim(q \vee r)$
T	T	T	T	F	F
T	T	F	T	F	F
T	F	T	T	F	F
T	F	F	F	T	T
F	T	T	T	F	F
F	T	F	T	F	F
F	F	T	T	F	F
F	F	F	F	T	F

39. p: Proposition A passes.
q: Proposition B passes.
r: Jobs are lost.
s: New taxes are imposed.
$(p \wedge \sim q) \rightarrow (r \vee s)$

p	q	r	s	$\sim q$	$p \wedge \sim q$	$r \vee s$	$(p \wedge \sim q) \rightarrow (r \vee s)$
T	T	T	T	F	F	T	T
T	T	T	F	F	F	T	T
T	T	F	T	F	F	T	T
T	T	F	F	F	F	F	T
T	F	T	T	T	T	T	T
T	F	T	F	T	T	T	T
T	F	F	T	T	T	T	T
T	F	F	F	T	T	F	F
F	T	T	T	F	F	T	T
F	T	T	F	F	F	T	T
F	T	F	T	F	F	T	T
F	T	F	F	F	F	F	T
F	F	T	T	T	F	T	T
F	F	T	F	T	F	T	T
F	F	F	T	T	F	T	T
F	F	F	F	T	F	F	T

41. The statements are equivalent.
43. The statements are equivalent.
45. The statements are not equivalent.
47. The statements are equivalent.
49. The statements are not equivalent.
51. i. and iv. are equivalent.
ii. and iii. are equivalent.
53. i. and iv. are equivalent.
ii. and iii. are equivalent.
55. They are equivalent.
57. p: I have a college degree.
q: I am employed.
$p \wedge \sim q$ Negation: $\sim(p \wedge \sim q) \equiv \sim p \vee \sim(\sim q) \equiv \sim p \vee q$
I do not have a college degree or I am employed.
59. p: The television set is broken.
q: There is a power outage.
$p \vee q$ Negation: $\sim(p \vee q) \equiv \sim p \wedge \sim q$
The television set is not broken and there is not a power outage.
61. p: The building contains asbestos.
q: The original contractor is responsible.
$p \rightarrow q$ Negation: $\sim(p \rightarrow q) \equiv p \wedge \sim q$
The building contains asbestos and the original contractor is not responsible.
63. p: The lyrics are censored.
q: The First Amendment has been violated.
$p \rightarrow q$ Negation: $\sim(p \rightarrow q) \equiv p \wedge \sim q$
The lyrics are censored and the First Amendment has not been violated.

65. p: It is rainy weather.
q: I am washing my car.
$p \rightarrow \sim q$ Negation: $\sim(p \rightarrow \sim q) \equiv p \wedge \sim(\sim q) \equiv p \wedge q$
It is rainy weather and I am washing my car.

67. p: The person is talking.
q: The person is listening.
$q \rightarrow \sim p$ Negation: $\sim(q \rightarrow \sim p) \equiv q \wedge \sim(\sim p) \equiv q \wedge p$
The person is listening and talking.

1.4 More on Conditionals

1. a. If she is a police officer, then she carries a gun.
b. If she carries a gun, then she is a police officer.
c. If she is not a police officer, then she does not carry a gun.
d. If she does not carry a gun, then she is not a police officer.
e. Parts (a) and (d) are equivalent; parts (b) and (c) are equivalent. The contrapositive statement is always equivalent to the original.

3. a. If I watch television, then I do not do my homework.
b. If I do not do my homework, then I watch television.
c. If I do not watch television, then I do my homework.
d. If I do my homework, then I do not watch television.
e. Parts (a) and (d) are equivalent; parts (b) and (c) are equivalent. The contrapositive statement is always equivalent to the original.

5. a. If you do not pass this mathematics course, then you do not fulfill a graduation requirement.
b. If you fulfill a graduation requirement, then you pass this mathematics course.
c. If you do not fulfill a graduation requirement, then you do not pass this mathematics course.

7. a. If the electricity is turned on, then the television set does work.
b. If the television set does not work, then the electricity is turned off.
c. If the television set does work, then the electricity is turned on.

9. a. If you eat meat, then you are not a vegetarian.
b. If you are a vegetarian, then you do not eat meat.
c. If you are not a vegetarian, then you do eat meat.

11. a. The person not being a dentist is sufficient to not being an orthodontist.
b. Not being an orthodontist is necessary for not being a dentist.

13. a. Not knowing Morse code is sufficient to not operating a telegraph.
b. Not being able to operate a telegraph is necessary to not knowing Morse code.

15. a. *Premise*: I take public transportation. *Conclusion*: Public transportation is convenient.
b. If I take public transportation, then it is convenient.
c. The statement is false when I take public transportation and it is not convenient.

17. a. *Premise*: I buy foreign products. *Conclusion*: Domestic products are not available.
b. If I buy foreign products, then domestic products are not available.
c. The statement is false when I buy foreign products and domestic products are available.

19. a. *Premise*: You may become a U. S. senator. *Conclusion*: You are at least 30 years old and have been a citizen for nine years.
b. If you become a U. S. senator, then you are at least 30 years old and have been a citizen for nine years.
c. The statement is false when you become a U. S. senator and you are not at least 30 years old or have not been a citizen for nine years, or both.

21. If you obtain a refund, then you have a receipt and if you have a receipt, then you obtain a refund.

23. If the quadratic equation $ax^2 + bx + c = 0$ has two distinct real solutions, then $b^2 - 4ac > 0$ and if $b^2 - 4ac > 0$, then the quadratic equation $ax^2 + bx + c = 0$ has two distinct real solutions.

25. If a polygon is a triangle, then the polygon has three sides and if the polygon has three sides, then the polygon is a triangle.

27. If $a^2 + b^2 = c^2$, then the triangle has a 90° angle, and if the triangle has a 90° angle, than $a^2 + b^2 = c^2$.

29. p: I can have surgery.
q: I have health insurance.
$\sim q \rightarrow \sim p$ and $p \rightarrow q$
The statements are equivalent.

31. p: You earn less than $12,000 per year.
q: You are eligible for assistance.
$p \rightarrow q$ and $\sim q \rightarrow \sim p$
The statements are equivalent.

33. p: I watch television.
q: The program is educational.
$p \rightarrow q$ and $\sim q \rightarrow \sim p$
The statements are equivalent.

35. p: The automobile is American-made.
q: The automobile hardware is metric.
$p \rightarrow \sim q$ and $q \rightarrow \sim p$
The statements are equivalent.

37. If I do not walk to work, then it is raining.

39. If it is not cold, then it is not snowing.

41. If you are a vegetarian, then you do not eat meat.

43. If the person does not own guns, then the person is not a policeman.

45. If the person is eligible to vote, then the person is not a convicted felon.

47. ii and iii are equivalent and i and iv are equivalent.

49. i and iii are equivalent and ii and iv are equivalent.

51. i and iii are equivalent and ii and iv are equivalent.

59. b. **61.** c.

1.5 Analyzing Arguments

1. $p \rightarrow q$, p, $\therefore q$

3. $p \rightarrow q$, $\sim q$, $\therefore \sim p$

5. $p \rightarrow q$, $\sim p$, $\therefore \sim q$

7. $q \rightarrow p$, $r \wedge p$, $\therefore r \wedge q$

9. $p \rightarrow \sim q$, $r \wedge q$, $\therefore r \wedge \sim p$

11. The argument is valid.

13. The argument is valid.

15. The argument is invalid if you don't exercise regularly and you are healthy.

17. The argument is invalid if (1) the person isn't Nikola Tesla, operates a telegraph, and knows Morse code; *or* (2) knows Morse code, doesn't operate a telegraph, and is Nikola Tesla; *or* (3) knows Morse code, doesn't operate a telegraph, and isn't Nikola Tesla.

19. The argument is valid.

21. p: The Democrats have a majority.
q: Smith is appointed.
r: Student loans are funded.

$$\begin{array}{l} p \rightarrow (q \wedge r) \\ q \vee \sim r \\ \hline \sim p \end{array}$$

$\{[p \rightarrow (q \wedge r)] \wedge [q \vee \sim r]\} \rightarrow \sim p$
The argument is invalid when the Democrats have a majority and Smith is appointed and student loans are funded.

23. p: You argue with a police officer.
q: You get a ticket.
r: You break the speed limit.

$$\begin{array}{l} p \rightarrow q \\ \sim r \rightarrow \sim q \\ \hline r \rightarrow p \end{array}$$

$\{(p \rightarrow q) \wedge (\sim r \rightarrow \sim q)\} \rightarrow (r \rightarrow p)$
The argument is invalid when (1) you do not argue with a police officer, you do get a ticket and you do break the speed limit, *or* (2) you do not argue with a police officer, you do not get a ticket and you do break the speed limit.

25. Rewriting the argument:
If it is a pesticide, then it is harmful to the environment.
If it is a fertilizer, then it is not a pesticide.

If it is a fertilizer, then it is not harmful to the environment.
p: It is a pesticide.
q: It is harmful to the environment.
r: It is a fertilizer.

$$\begin{array}{l} p \rightarrow q \\ r \rightarrow \sim p \\ \hline r \rightarrow \sim q \end{array}$$

$\{(p \rightarrow q) \wedge (r \rightarrow \sim p)\} \rightarrow (r \rightarrow \sim q)$
The argument is invalid if it is not a pesticide and if it is harmful to the environment and it is a fertilizer.

27. Rewriting the argument:
If you are a poet, then you are a loner.
If you are a loner, then you are a taxi driver.

If you are a poet, then you are a taxi driver.
p: You are a poet.
q: You are a loner.
r: You are a taxi driver.

$$\begin{array}{l} p \rightarrow q \\ q \rightarrow r \\ \hline p \rightarrow r \end{array}$$

$\{(p \rightarrow q) \wedge (q \rightarrow r)\} \rightarrow (p \rightarrow r)$
The argument is valid.

29. Rewriting the argument:
If you are a professor, then you are not a millionaire.
If you are a millionaire, then you are literate.

If you are a professor, then you are literate.
p: You are a professor.
q: You are a millionaire.
r: You are illiterate.

$$\begin{array}{l} p \rightarrow \sim q \\ q \rightarrow \sim r \\ \hline p \rightarrow \sim r \end{array}$$

$\{(p \rightarrow \sim q) \wedge (q \rightarrow \sim r)\} \rightarrow (p \rightarrow \sim r)$
The argument is invalid if you are a professor who is not a millionaire and is illiterate.

31. Rewriting the argument:
If you are a lawyer, then you study logic.
If you study logic, then you are a scholar.
You are not a scholar.

You are not a lawyer.
p: You are a lawyer.
q: You study logic
r: You are a scholar.

$$\begin{array}{l} p \rightarrow q \\ q \rightarrow r \\ \sim r \\ \hline \sim p \end{array}$$

$\{(p \rightarrow q) \wedge (q \rightarrow r) \wedge \sim r\} \rightarrow \sim p$
The argument is valid.

33. Rewriting the argument:
If you are drinking espresso, then you are not sleeping.
If you are on a diet, then you are not eating dessert.
If you are not eating dessert then you are drinking.

If you are sleeping, then you are not on a diet.
p: You are drinking espresso.
q: You are sleeping.
r: You are eating dessert.
t: You are on a diet.

$$\begin{array}{l} p \rightarrow \sim q \\ t \rightarrow \sim r \\ \sim r \rightarrow p \\ \hline q \rightarrow \sim t \end{array}$$

$\{(p \rightarrow \sim q) \wedge (t \rightarrow \sim r) \wedge (\sim r \rightarrow p)\} \rightarrow (q \rightarrow \sim t)$
The argument is valid.

35. p: The defendant is innocent.
q: The defendant goes to jail.

$$\begin{array}{l} p \rightarrow \sim q \\ q \\ \hline \sim p \end{array}$$

$\{(p \rightarrow \sim q) \wedge q\} \rightarrow \sim p$
The argument is valid.

37. p: You are in a hurry.
q: You eat at Lulu's Diner.
r: You eat good food.

$$\begin{array}{l} \sim p \rightarrow q \\ p \rightarrow \sim r \\ q \\ \hline r \end{array}$$

$\{(\sim p \rightarrow q) \wedge (p \rightarrow \sim r) \wedge q\} \rightarrow r$
The argument is invalid when (1) you are in a hurry and you eat at Lulu's Diner and you do not eat good food, *or* (2) you are not in a hurry and you eat at Lulu's Diner and you do not eat good food.

39. p: You listen to rock and roll.
q: You go to heaven.
r: You are a moral person.

$$\begin{array}{l} p \rightarrow \sim q \\ r \rightarrow q \\ \hline p \rightarrow \sim r \end{array}$$

$\{(p \rightarrow \sim q) \wedge (r \rightarrow q)\} \rightarrow (p \rightarrow \sim r)$
The argument is valid.

41. p: The water is cold.
q: You go to swimming.
r: You have goggles.

$$\begin{array}{l} \sim p \rightarrow q \\ q \rightarrow r \\ \underline{\sim r} \\ p \end{array}$$

$\{(\sim p \rightarrow q) \wedge (q \rightarrow r) \wedge \sim r\} \rightarrow p$
The argument is valid.

43. p: It is medicine.
q: It is nasty.

$$\begin{array}{l} p \rightarrow q \\ \underline{p} \\ q \end{array}$$

$\{(p \rightarrow q) \wedge p\} \rightarrow q$
The argument is valid.

45. p: It is intelligible.
q: It puzzles me.
r: It is logic.

$$\begin{array}{l} p \rightarrow \sim q \\ \underline{r \rightarrow q} \\ r \rightarrow \sim p \end{array}$$

$\{(p \rightarrow \sim q) \wedge (r \rightarrow q)\} \rightarrow (r \rightarrow \sim p)$
The argument is valid.

47. p: A person is a Frenchman.
q: A person likes plum pudding.
r: A person is an Englishman.

$$\begin{array}{l} p \rightarrow \sim q \\ \underline{r \rightarrow q} \\ r \rightarrow \sim p \end{array}$$

$\{(p \rightarrow \sim q) \wedge (r \rightarrow q)\} \rightarrow (r \rightarrow \sim p)$
The argument is valid.

49. p: An animal is a wasp.
q: An animal is friendly.
r: An animal is a puppy.

$$\begin{array}{l} p \rightarrow \sim q \\ \underline{r \rightarrow q} \\ r \rightarrow \sim p \end{array}$$

$\{(p \rightarrow \sim q) \wedge (r \rightarrow q)\} \rightarrow (r \rightarrow \sim p)$
The argument is valid.

Chapter 1 Review

1. a. Inductive **b.** Deductive
3. 9 **5.** Invalid
7. Valid **9.** Valid
11. a. Statement. It is either true or false.
b. Statement. It is either true or false.
c. Not a statement. It is a question.
d. Not a statement. It is an opinion.
13. a. His car is new.
b. No building is earthquake proof.
c. Some children do not eat candy.
d. Sometimes I cry in a movie theater.
15. a. $q \rightarrow p$ **b.** $\sim r \wedge \sim q \wedge p$
c. $r \rightarrow \sim p$ **d.** $p \wedge \sim(q \vee r)$
e. $p \rightarrow (\sim r \vee \sim q)$ **f.** $(r \wedge q) \rightarrow \sim p$
17. a. If the movie is critically acclaimed or a box office hit, then the movie is available on videotape.
b. If the movie is critically acclaimed and not a box office hit, then the movie is not available on videotape.
c. The movie is not critically acclaimed or a box office hit and it is available on DVD.
d. If the movie is not available on video tape, then the movie is not critically acclaimed and it is not a box office hit.

19.

p	q	$\sim q$	$p \wedge \sim q$
T	T	F	F
T	F	T	T
F	T	F	F
F	F	T	F

21.

p	q	$p \wedge q$	$\sim q$	$(p \wedge q) \rightarrow \sim q$
T	T	T	F	F
T	F	F	T	T
F	T	F	F	T
F	F	F	T	T

23.

p	q	r	$\sim p$	$q \vee r$	$\sim p \rightarrow (q \vee r)$
T	T	T	F	T	T
T	T	F	F	T	T
T	F	T	F	T	T
T	F	F	F	F	T
F	T	T	T	T	T
F	T	F	T	T	T
F	F	T	T	T	T
F	F	F	T	F	F

25.

p	q	r	$\sim r$	$p \vee r$	$q \wedge \sim r$	$(p \vee r) \rightarrow (q \wedge \sim r)$
T	T	T	F	T	F	F
T	T	F	T	T	T	T
T	F	T	F	T	F	F
T	F	F	T	T	F	F
F	T	T	F	T	F	F
F	T	F	T	F	T	T
F	F	T	F	T	F	F
F	F	F	T	F	F	T

27. The statements are not equivalent.
29. The statements are equivalent.
31. Jesse did not have a party or somebody came.
33. I am not the winner and you are not blind.
35. His application is ignored and the selection procedure has not been violated.
37. If a person is sleeping, then the person is not drinking espresso.

39. **a.** If you are an avid jogger, then you are healthy.
b. If you are healthy, then you are an avid jogger.
c. If you are not an avid jogger, then you are not healthy.
d. If you are not healthy, then you are not an avid jogger.
e. You are an avid jogger if and only if you are healthy.

41. **a.** Being lost is sufficient for not having a map.
b. Not having a map is necessary for being lost.

43. **a.** *Premise*: The economy improves.
Conclusion: Unemployment goes down.
b. If the economy improves, then unemployment goes down.

45. **a.** *Premise*: It is a computer. *Conclusion*: It is repairable.
b. If it is a computer, then it is repairable.

47. **a.** *Premise*: The fourth Thursday in November.
Conclusion: The U. S. Post Office is closed.
b. If it is the fourth Thursday in November, then the U. S. Post Office is closed.

49. p: You are allergic to dairy products.
q: You can eat cheese.
$p \rightarrow \sim q$ and $\sim q \rightarrow p$
The statements are not equivalent.

51. i and iii are equivalent; ii and iv are equivalent.

53. p: You pay attention.
q: You learn the new method.

$$\begin{array}{l} \sim p \rightarrow \sim q \\ \underline{q \quad\quad} \\ p \end{array}$$

$[(\sim p \rightarrow \sim q) \wedge q] \rightarrow p$
The argument is valid.

55. p: The Republicans have a majority.
q: Farnsworth is appointed.
r: No new taxes are imposed.

$$\begin{array}{l} p \rightarrow (q \wedge r) \\ \underline{\sim r \quad\quad} \\ \sim p \vee \sim q \end{array}$$

$\{[p \rightarrow (q \wedge r)] \wedge \sim r\} \rightarrow (\sim p \vee \sim q)$
The argument is valid.

57. p: You are practicing.
q: You are making mistakes.
r: You receive an award.

$$\begin{array}{l} p \rightarrow \sim q \\ \sim r \rightarrow q \\ \underline{\sim r \quad\quad} \\ \sim p \end{array}$$

$\{[p \rightarrow \sim q] \wedge [\sim r \rightarrow q] \wedge \sim r\} \rightarrow \sim p$
The argument is valid.

59. p: I will go to the concert.
q: You buy me a ticket.

$$\begin{array}{l} p \rightarrow q \\ \underline{q \quad\quad} \\ p \end{array}$$

$[(p \rightarrow q) \wedge q] \rightarrow p$
The argument is invalid.

61. p: Our oil supply is cut off.
q: Our economy collapses.
r: We go to war.

$$\begin{array}{l} p \rightarrow q \\ \underline{r \rightarrow \sim q} \\ \sim p \rightarrow \sim r \end{array}$$

$\{[p \rightarrow q] \wedge [r \rightarrow \sim q]\} \rightarrow (\sim p \rightarrow \sim r)$
The argument is invalid.

63. p: A person is a professor.
q: A person is educated.
r: An animal is a monkey.

$$\begin{array}{l} p \rightarrow q \\ \underline{r \rightarrow \sim p} \\ r \rightarrow \sim q \end{array}$$

$\{[p \rightarrow q] \wedge [r \rightarrow \sim p]\} \rightarrow (r \rightarrow \sim q)$
The argument is invalid.

65. p: You are investing in the stock market.
q: The invested money is to be guaranteed.
r: You retire at an early age.

$$\begin{array}{l} q \rightarrow \sim p \\ \underline{\sim q \rightarrow \sim r} \\ r \rightarrow \sim p \end{array}$$

$\{(q \rightarrow \sim p) \wedge (\sim q \rightarrow \sim r)\} \rightarrow (r \rightarrow \sim p)$
The argument is valid.

67. The argument is valid.

CHAPTER 2 Sets and Counting

2.1 Sets and Set Operations

1. **a.** well-defined **b.** not well-defined
c. well-defined **d.** not well-defined

3. Proper: { }, {Lennon}, {McCartney}; Improper: {Lennon, McCartney}

5. Proper: { }, {yes}, {no}, {undecided}, {yes, no}, {yes, undecided}, {no, undecided}
Improper: {yes, no, undecided}

7. **a.** {4, 5} **b.** {1, 2, 3, 4, 5, 6, 7, 8}
c. {0, 6, 7, 8, 9} **d.** {0, 1, 2, 3, 9}

9. **a.** { } **b.** {0, 1, 2, 3, 4, 5, 6, 7, 8, 9}
c. {0, 2, 4, 6, 8} **d.** {1, 3, 5, 7, 9}

11. {Friday}

13. {Monday, Tuesday, Wednesday, Thursday}

15. {Friday, Saturday, Sunday}

17.

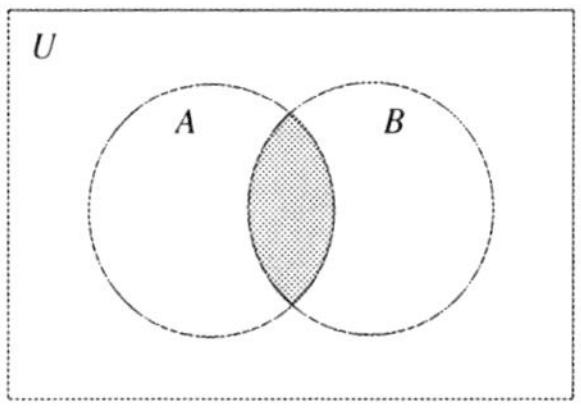

19.

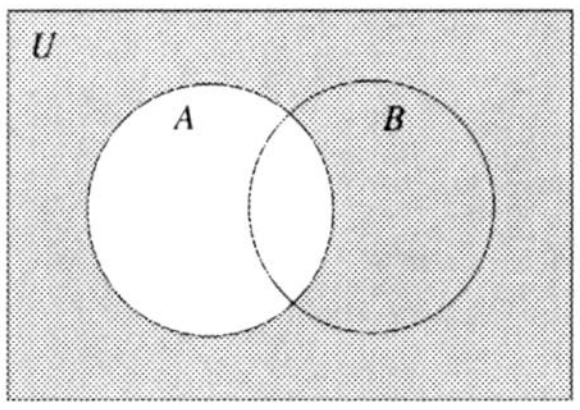

21.
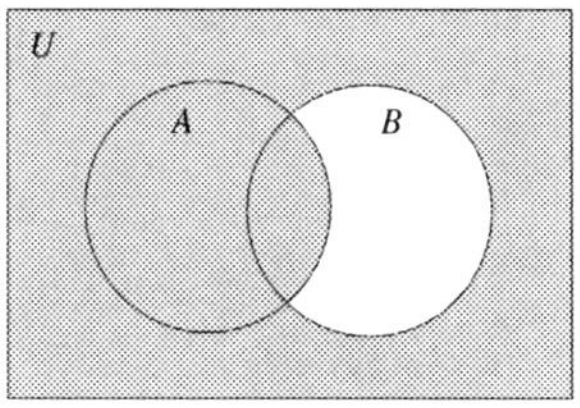

23.
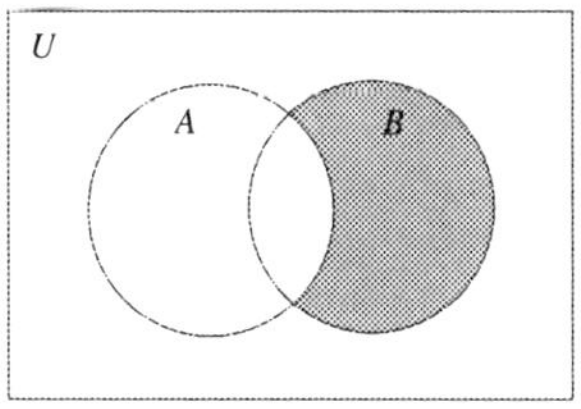

25.
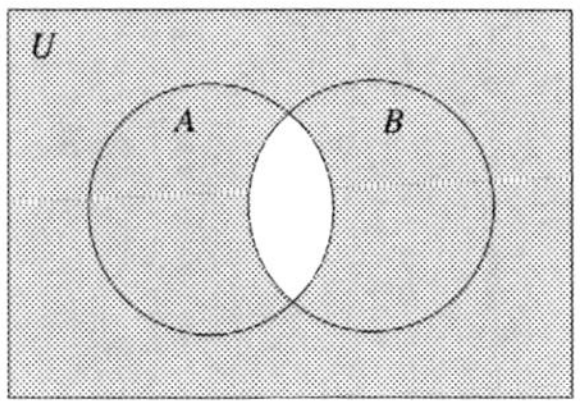

27. a. $n(A \cap B) = 21$
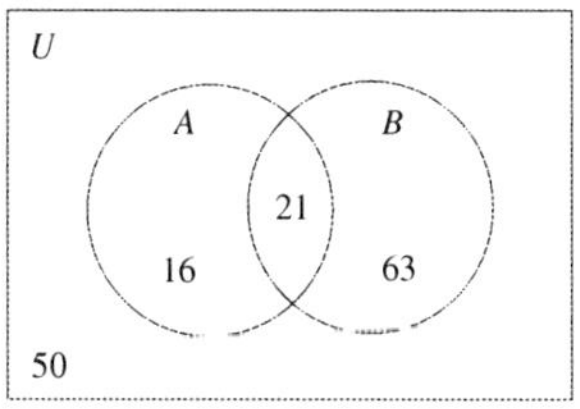

b. $n(A \cap B) = 0$
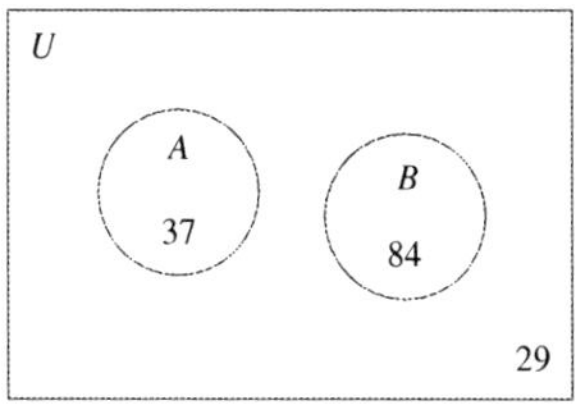

29. a.
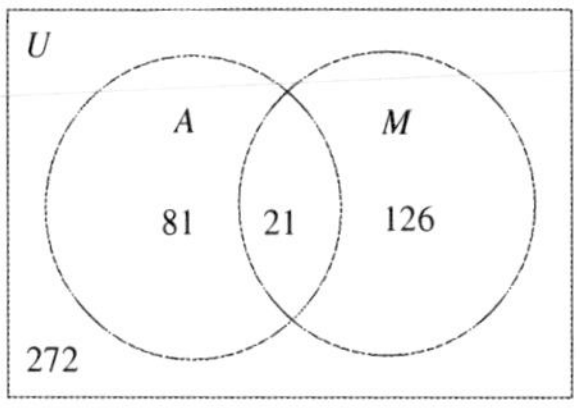

b. 45.6%

31. a.
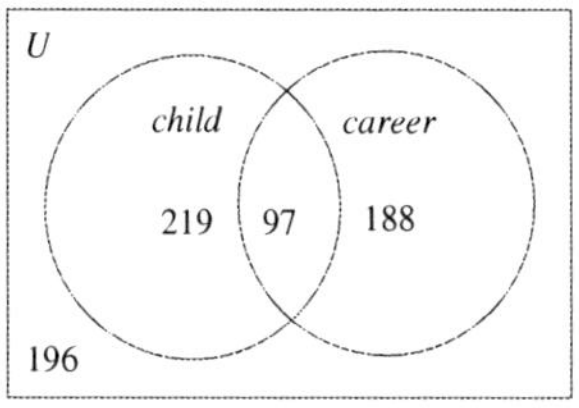

b. 13.857%

33. 42 **35.** 43 **37.** 8 **39.** 5 **41.** 16
43. 32 **45.** 6 **47.** 8 **49.** 0
51. a. $\{1, 2, 3\}$ **b.** $\{1, 2, 3, 4, 5, 6\}$
c. $E \subseteq F$ **d.** $E \subseteq F$
53. a. $\{\ \}, \{a\}$ 2 subsets **b.** $\{\ \}, \{a\}, \{b\}, \{a, b\}$ 4 subsets
c. $\{\ \}, \{a\}, \{b\}, \{c\}, \{a, b\}, \{a, c\}, \{b, c\}, \{a, b, c\}$ 8 subsets
d. $\{\ \}, \{a\}, \{b\}, \{c\}, \{d\}, \{a, b\}, \{a, c\}, \{a, d\}, \{b, c\}, \{b, d\}, \{c, d\}, \{a, b, c\}, \{a, b, d\}, \{a, c, d\}, \{b, c, d\}, \{a, b, c, d\}$ 16 subsets
e. Yes! The number of subsets of $A = 2^{n(A)}$ **f.** 64
65. d. conforms
67. c. **69.** e

2.2 Applications of Venn Diagrams

1. a. 143 **b.** 16 **c.** 49 **d.** 57
3. a. 408 **b.** 1343 **c.** 664 **d.** 149
5. a. 106 **b.** 448 **c.** 265 **d.** 159
7. a. 51.351% **b.** 24.324% **c.** 16.216%
9. a. $x + y - z$ **b.** $x - z$
c. $y - z$ **d.** $w - x - y + z$
11. a. 43.8% **b.** 10.8%
13. a. 66.471% **b.** 27.974%
15. a. 0% **b.** 54.1%
17. a. 44.0% **b.** 12.8%
19. a. 20% **b.** 58.8%
21. 16 **23.** $\{0, 4, 5\}$ **25.** $\{1, 2, 3, 6, 7, 8, 9\}$

27.
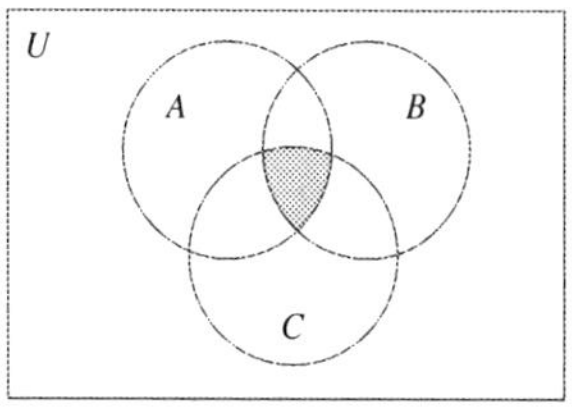

29.
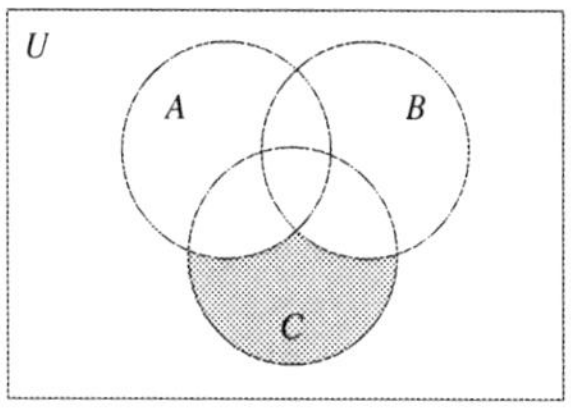

31.
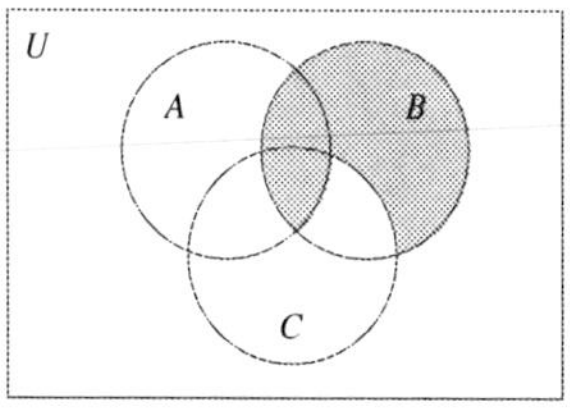

35. a. 84% **b.** 16% **37. a.** 85% **b.** 15%
39. a. 75% **b.** 25% **41.** Type B or Type O
43. Type O
47. e **49.** b **51.** b

2.3 Introduction to Combinatorics

1. a. 8
b. {HHH, HHT, HTH, HTT, THH, THT, TTH, TTT}
3. a. 12
b. {Mega-BW-BN, Mega-BW-NW, Mega-WW-BN, Mega-WW-NW, Mega-GW-BN, Mega-GW-NW, BB-BW-BN, BB-BW-NW, BB-WW-BN, BB-WW-NW, BB-GW-BN, BB-GW-NW}

5. 24 **7.** 720 **9.** 2,646 **11.** 216
13. $10^9 = 1{,}000{,}000{,}000$ **15.** $10^{10} = 10{,}000{,}000{,}000$
17. a. 128 **b.** 800 **c.** The phone company needed more.
19. a. $1{,}179{,}360{,}000 = 1.17936 \times 10^9$
b. $67{,}600{,}000{,}000 = 6.76 \times 10^{10}$
c. $42{,}000{,}000{,}000 = 4.2 \times 10^{10}$
21. 540,000 **23.** 24 **25.** 3,628,800
27. $2{,}432{,}902{,}008{,}176{,}640{,}000 = 2.432902008 \times 10^{18}$
29. 17,280
31. a. 30 **b.** 360
33. 56 **35.** 70 **37.** 3,321
39. $10{,}461{,}394{,}944{,}000 = 1.046139494 \times 10^{13}$
41. 120 **43.** 35 **45.** 1
51. b **53.** d

2.4 Permutations and Combinations

1. a. 210 **b.** 35
3. a. 120 **b.** 1
5. a. 14 **b.** 14
7. a. 970,200 **b.** 161,700
9. a. $x!$ **b.** x
11. a. $x \cdot (x-1) = x^2 - x$ **b.** $\dfrac{x \cdot (x-1)}{2} = \dfrac{x^2 - x}{2}$
13. a. 6
b. $\{a, b\}, \{a, c\}, \{b, c\}, \{b, a\}, \{c, a\}, \{c, b\}$
15. a. 6
b. $\{a, b\}, \{a, c\}, \{a, d\}, \{b, c\}, \{b, d\}, \{c, d\}$
17. a. 39,916,800 **b.** 1
19. 360 **21.** 78 **23.** 1,716
25. a. 540 **b.** 1,365 **c.** 630
27. 2,598,960
29. a. 4,512 **b.** 58,656
31. 123,552 **33.** 22,957,480
35. 376,992
37. It is easier to win a 5/36 lottery since there are fewer possible tickets.
39. a. 1 **b.** 2 **c.** 4
d. 8 **e.** 16
f. Yes. The sum of each row is twice the previous sum, or, sum of entries in n^{th} row is 2^{n-1}.
g. 32
h. 32. Our prediction was correct.
i. 2^{n-1}
41. a. fifth row **b.** $(n+1)^{\text{st}}$ row
c. No. **d.** Yes.
e. $(r+1)^{\text{st}}$ number in the $(n+1)^{\text{st}}$ row
43. 120 **45.** 3,360
47. 630 **49.** 831,600
51. a. 24 **b.** 12
53. a. 120 **b.** 60
57. c **59.** c **61.** e

2.5 Infinite Sets

1. equivalent; Match each state to its capital.
3. not equivalent **5.** equivalent; $3n \leftrightarrow 4n$
7. not equivalent **9.** equivalent; $2n - 1 \leftrightarrow 2n + 123$
11. a. For n starting at 1, match the term described by n in N to the term described by $2n - 1$ in O; $n \leftrightarrow 2n - 1$
b. $n = 918$ **c.** $n = \dfrac{x+1}{2}$
d. 1,563 **e.** $n \to 2n - 1$
13. a. For n starting at 1, match the term described by n in N to the term described by $3n$ in T; $n \leftrightarrow 3n$
b. $n = 312$ **c.** $n = \frac{1}{3}x$
d. 2,808 **e.** $n \to 3n$
15. a. -344 **b.** $248 \to 248$
c. $n = 755$ **d.** $n(A) = \aleph_0$
17. Match any real number $x \in [0, 1]$ to $3x \in [0, 3]$.
19. From the center of the circle, match the corresponding points that lie on the same radial line.
21. From the inside circle, match the corresponding points that lie on the same radial line.
23. First, the semicircle is equivalent to $[0, 1]$. Draw a vertical line to match points. To match the semicircle and line, draw a line passing through $(\frac{1}{2}, 0)$ (not horizontal).

Chapter 2 Review

1. a. Well defined. **b.** Not well defined
c. Not well defined. **d.** Well defined.
3. a. $A \cup B$ = {Maria, Nobuku, Leroy, Mickey, Kelly, Rachel, Deanna}
b. $A \cap B$ = {Leroy, Mickey}
5. a. 18
b.

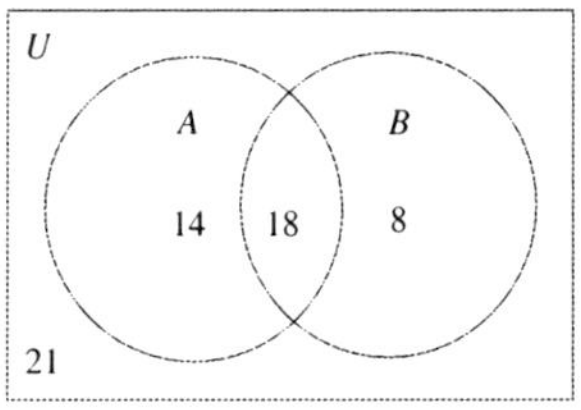

7.

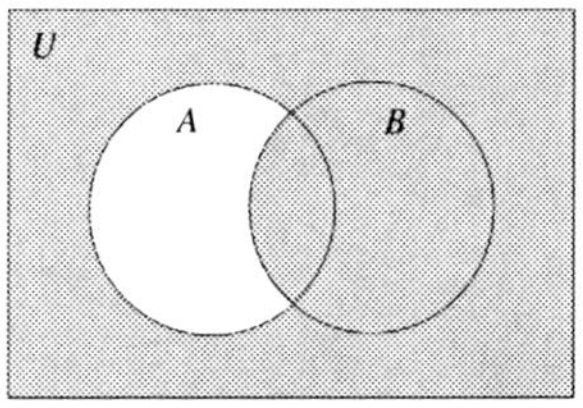

9.

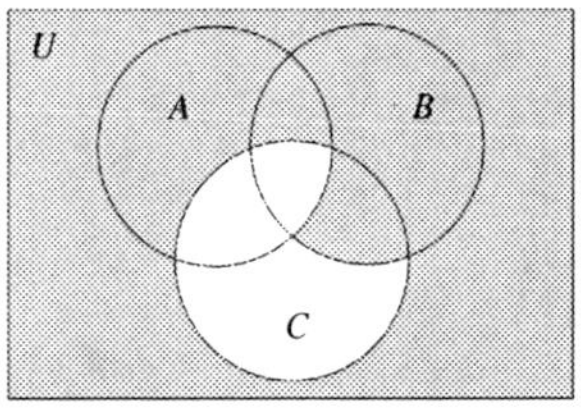

11. a. 31.154% **b.** 9.038% **c.** 17.692%
13. 27.875%
15. a. 85 **b.** 0 **c.** 15
17. a. 12
b. {MMA-S-LC, MMA-S-L, MMA-J-LC, MMA-J-L, MMA-C-LC, MMA-C-L, NPG-S-LC, NPG-S-L, NPG-J-LC, NPG-J-L, NPG-C-LC, NPG-C-L}
19. 1,080,000
21. a. 165 **b.** 990
23. a. 660 **b.** 1,540 **c.** 880
25. 720 **27.** 177,100,560
29. a. 5,040 **b.** 2,520 **c.** 1,260
31. Order is important in permutations, whereas it is not important in combinations.

33. a. 3 **b.** 3 **c.** 1 **d.** 1 **e.** 8
f. The number of subsets that set S has equals $2^{n(S)}$.
$2^{n(S)} = 2^3 = 8$

35. They are equivalent. $2n + 1 \leftrightarrow 2n$

37. a. Match each number in N to its square in S: $n \leftrightarrow n^2$
b. 29 **c.** $\sqrt{x}$ **d.** 20,736 **e.** $n \rightarrow n^2$

39. Match any real number $x \in [0, 1]$ to $\pi x \in [0, \pi]$.

41. $\{0\}$ is a set with one element (namely 0).
$\varnothing$ is a set with no elements.

43. If order of selection matters, we use permutations. If not, we use combinations.

45. e. **47.** b.

CHAPTER 3 Probability

3.1 History of Probability

1. Answers will vary.
3. a. Answers will vary. **b.** Answers will vary.
5. a. Answers will vary. **b.** Answers will vary.
7. a. Answers will vary. **b.** Answers will vary.
9. a. win \$350 **b.** lose \$10
11. a. win \$85 **b.** win \$85 **c.** lose \$5
13. a. lose \$20 **b.** win \$160 **c.** lose \$20
15. a. lose \$10 **b.** win \$50
17. a. win \$50 **b.** lose \$25
19. a. lose \$50 **b.** win \$50
21. a. lose \$45 **b.** win \$855 **c.** win \$675
23. a. win \$495 **b.** win \$525
c. lose \$45 **d.** loss of \$15
25. \$3 **27.** \$500 **29. a.** 13 **b.** $\frac{1}{4}$
31. a. 12 **b.** $\frac{3}{13}$
33. a. 4 **b.** $\frac{1}{13}$

3.2 Basic Terms of Probability

1. Choose 1 jellybean and look at its color.
3. 34% **5.** 51% **7.** 71% **9.** 0% **11.** 12:23
13. 18:17 **15.** Drawing or selecting a card from a deck
17. a. $\frac{1}{2}$ **b.** 1:1
19. a. $\frac{1}{13}$ **b.** 1:12
21. a. $\frac{1}{52}$ **b.** 1:51
23. a. $\frac{3}{13}$ **b.** 3:10
25. a. $\frac{10}{13}$ **b.** 10:3
27. a. $\frac{3}{13}$ **b.** 3:10
29. a. $\frac{1}{38}$ **b.** 1:37
31. a. $\frac{3}{38}$ **b.** 3:35
33. a. $\frac{5}{38}$ **b.** 5:33
35. a. $\frac{6}{19}$ **b.** 6:13
37. a. $\frac{9}{19}$ **b.** 9:10
39. a. 50.7% **b.** 17.5%
41. a. 4% **b.** 32% **c.** 64%
43. $\frac{6}{29}$ **45.** 1 : 4
47. $\frac{3}{5}$ **49.** $a : (b - a)$
51. a. $\frac{9}{19}$ **b.** 9 : 10
53. a. House odds. sportsbook.com determines their own odds based on their own set of rules.
b. New York Yankees: $p(\text{win}) = \frac{5}{7}$
New York Mets: $p(\text{win}) = \frac{7}{8}$
c. The New York Mets are more likely to win because $\frac{7}{8} > \frac{5}{7}$
55. a. Intentional self-harm has the highest odds (1 : 9,085), so it is the most likely cause of death in one year.
b. Once again, intentional self-harm has the highest odds (1 : 117), so it is the most likely cause of death in a lifetime.
57. a. $\frac{1}{20{,}332}$. **b.** $\frac{1}{502{,}555}$.
59. a. $\frac{1}{20{,}332}$, car transportation accident (intentional self-harm).
b. $\frac{1}{502{,}555}$, airplane and space accidents (war).
61. a. $S = \{\text{bb, gb, bg, gg}\}$ **b.** $E = \{\text{gb, bg}\}$
c. $F = \{\text{gb, bg, gg}\}$ **d.** $G = \{\text{gg}\}$
e. $\frac{1}{2}$ **f.** $\frac{3}{4}$ **g.** $\frac{1}{4}$
h. 1 : 1 **i.** 3 : 1 **j.** 1 : 3
63. a. $S = \{\text{ggg, ggb, gbg, bgg, gbb, bgb, bbg, bbb}\}$
b. $E = \{\text{ggb, gbg, bgg}\}$ **c.** $F = \{\text{ggb, gbg, bgg, ggg}\}$
d. $G = \{\text{ggg}\}$
e. $\frac{3}{8}$ **f.** $\frac{1}{2}$ **g.** $\frac{1}{8}$
h. 3 : 5 **i.** 1 : 1 **j.** 1 : 7
65. $S = \{\text{bb, bg, gb, gg}\}$ **a.** $\frac{1}{4}$ **b.** $\frac{1}{2}$
c. $\frac{1}{4}$
d. same sex: $\frac{2}{4}$, different sex: $\frac{2}{4}$. They are equally likely.
67. Same sex: $\frac{1}{4}$, Different sex: $\frac{3}{4}$ Different sex is more likely. $\frac{3}{4} > \frac{1}{4}$
69. a.

$$S = \left\{\begin{matrix} (1,1), (1,2), (1,3), (1,4), (1,5), (1,6), \\ (2,1), (2,2), (2,3), (2,4), (2,5), (2,6), \\ (3,1), (3,2), (3,3), (3,4), (3,5), (3,6), \\ (4,1), (4,2), (4,3), (4,4), (4,5), (4,6), \\ (5,1), (5,2), (5,3), (5,4), (5,5), (5,6), \\ (6,1), (6,2), (6,3), (6,4), (6,5), (6,6) \end{matrix}\right\}$$

b. $E = \{(1, 6), (2, 5), (3, 4), (4, 3), (5, 2), (6, 1)\}$
c. $F = \{(5, 6), (6, 5)\}$
d. $G = \{(1, 1), (2, 2), (3, 3), (4, 4), (5, 5), (6, 6)\}$
e. $\frac{1}{6}$ **f.** $\frac{1}{18}$ **g.** $\frac{1}{6}$
h. 1: 5 **i.** 1 : 17 **j.** 1: 5
71. a. $\frac{1}{4}$ **b.** $\frac{1}{4}$ **c.** $\frac{2}{4} = \frac{1}{2}$
73. a. $\frac{1}{4}$ **b.** $\frac{2}{4} = \frac{1}{2}$ **c.** $\frac{1}{4}$
75. a. $\frac{0}{4} = 0$ **b.** $\frac{2}{4} = \frac{1}{2}$ **c.** $\frac{4}{4} = 1$
77. a. $\frac{2}{4} = \frac{1}{2}$ **b.** $\frac{0}{4} = 0$ **c.** $\frac{2}{4} = \frac{1}{2}$

3.3 Basic Rules of Probability

1. E and F are not mutually exclusive. There are many women doctors.
3. E and F are mutually exclusive. One cannot be both single and married.
5. E and F are not mutually exclusive. A brown haired person may have some gray.
7. E and F are mutually exclusive. One can't wear both boots and sandals (excluding one on each foot!)
9. E and F are mutually exclusive. Four is not odd.
11. a. $\frac{1}{26}$ **b.** $\frac{7}{13}$ **c.** $\frac{25}{26}$
13. a. $\frac{1}{52}$ **b.** $\frac{4}{13}$ **c.** $\frac{51}{52}$
15. a. $\frac{2}{13}$ **b.** $\frac{5}{13}$ **c.** 0
d. $\frac{7}{13}$

17. a. $\frac{9}{13}$ **b.** $\frac{8}{13}$ **c.** $\frac{4}{13}$
d. 1
19. $\frac{12}{13}$ **21.** $\frac{10}{13}$ **23.** $\frac{11}{13}$
25. $\frac{9}{13}$ **27.** 9 : 5
29. $o(E) = 2:5, o(E') = 5:2$
31. $b:a$ **33.** 12 : 1
35. 10 : 3 **37.** 10 : 3
39. a. $\frac{71}{175}$ **b.** $\frac{104}{175}$
41. a. $\frac{151}{700}$ **b.** $\frac{97}{140}$
43. a. $\frac{8}{25}$ **b.** $\frac{17}{25}$
45. a. $\frac{9}{25}$ **b.** $\frac{16}{25}$
47. a. $\frac{1}{6}$ **b.** $\frac{1}{9}$ **c.** $\frac{1}{18}$
49. a. $\frac{2}{9}$ **b.** $\frac{7}{18}$
51. a. $\frac{1}{6}$ **b.** $\frac{3}{4}$
53. a. $\frac{1}{6}$ **b.** $\frac{1}{2}$
55. a. 0.65 **b.** 0.30 **c.** 0.35
57. Relative frequencies; data was collected from client's records and the probabilities were calculated from this data.
59. a. 76.7% **b.** 23% **c.** 23.3%
61. Relative frequencies; they were calculated from a poll.
63. a. $\frac{17}{25}$ **b.** $\frac{2}{25}$
65. a. $\frac{212}{1,451}$ **b.** $\frac{514}{1,451}$
67. a. $\frac{7}{58}$ **b.** $\frac{31}{174}$
69. a. $\frac{3}{4}$ **b.** $\frac{1}{2}$ **c.** 1
71. a. $\frac{1}{4}$ **b.** $\frac{1}{2}$ **c.** 0
73. a. 0.35 **b.** 0.25 **c.** 0.15
d. 0.6
75. a. 0.20 **b.** 0.95 **c.** 0.80
83. a. $\frac{6}{11}$ **b.** $\frac{6}{11}$
85. a. $\frac{4}{21}$ **b.** $\frac{4}{21}$
87. a. $\frac{92}{105}$ **b.** $\frac{92}{105}$
89. a. $\frac{23}{21}$ **b.** $\frac{23}{21}$
91. Don't press " → Frac"

3.4 Combinatorics and Probability

1. over 70% **3.** 23 people
5. a. $\frac{1}{13,983,816}$ **b.** $\frac{1}{22,957,480}$
c. $\frac{1}{18,009,460}$ **d.** 64% more likely
e. Answers will vary.
7. a. $\frac{1}{575,757}$ **b.** $\frac{3}{10,000}$
9. a. $\frac{1}{324,632}$ **b.** $\frac{5}{10,000}$
11. $\frac{1}{175,711,536}$ **13.** $\frac{1}{2,718,576}$
15. 4/26 easiest; 6/54 hardest
17.

Outcome	Probability
8 winning spots	0.0000043457
7 winning spots	0.0001604552
6 winning spots	0.0023667137
5 winning spots	0.0183025856
4 winning spots	0.0815037015
Fewer than 4 winning spots	0.8976621984

19. a. 1,000
b. There are 1,000 straight play numbers. **c.** $\frac{1}{1,000}$
21. a. 0.000495 **b.** 0.001981
c. 0.000015 **d.** 0.0019654015
23. ≈0.05 **25.** ≈0.48
27. ≈0.64 **29.** ≈0.36
31. a. ≈0.09 **b.** ≈0.42 **c.** ≈0.49
33. $_6C_6 = \frac{6!}{6!(6-6)!} = \frac{6!}{6!0!} = 1$

3.5 Expected Value

1. a. −$0.0526315789
3. a. −$0.0526315789
5. a. −$0.0526315789
7. a. −$0.0526315789
9. a. −$0.0526315789
11. $549.00 **13.** 1.75 books
15. $10.05
17. Since $-\$1.67 < 0$, don't play.
19. No, since the expected value of 10 games, $50, is less than $100, don't play. Accept $100.
21. $\frac{35 + 37\cdot(-1)}{38} = \frac{35\cdot 1 + 37(-1)}{38} = \frac{35\cdot 1}{38} + \frac{37(-1)}{38}$
$= 35\cdot\frac{1}{38} + (-1)\frac{37}{38} = -\frac{2}{38} \approx -0.053$
23. Decision theory indicates that the bank's savings account is the better investment.
25. The speculative investment is the better choice if $1.1p - 0.6 > 0.045$, that is, if $p > 645/1100 = 0.586...$ ≈ 0.59.
27. a. 0 **b.** $\frac{1}{16}$ **c.** $\frac{3}{8}$
d. If you can eliminate one or more answers, guessing is a winning strategy.
29. −$0.28; You should expect to lose about 28 cents for every dollar you bet, if you play a long time.
31. a. $p(\text{first prize}) = \frac{1}{14,950}$
$p(\text{second prize}) = \frac{3}{500}$
$p(\text{third prize}) \approx 0.09 = \frac{9}{100}$
b. $p(\text{losing}) \approx 0.901 \approx \frac{9}{10}$
c. $EV = 0.00147 \approx \$0.001$
33. a. $p(\text{losing}) = 0.99$
b. $EV \approx \$0.46$
35. To make a profit, the price should be more than $1,200.
37. They should drill in the back yard.
39. a. $12.32 You should buy a ticket.
b. −$1.34 You should not buy a ticket.
c. −$5.89 You should not buy a ticket.
41. a. −$0.473 You should not buy a ticket.
b. −$2.74 You should not buy a ticket.
c. −$3.49 You should not buy a ticket.
43. $7.785 million or $7.8 million
45. a. 13,983,816 **b.** $13,983,816
c. Since we have 100 people buying, we only need 97.109833 days.
47. You can bet and lose 6 times.
The net winnings are $1.

3.6 Conditional Probability

1. a. $p(H|Q)$; Conditional since the sample space is limited to well-qualified candidates.
b. $p(H \cap Q)$; Not conditional.

3. a. $p(S|D)$; Conditional since the sample space is limited to users that are dropped a lot.
b. $p(S \cap D)$; Not conditional.
c. $p(D|S)$; Conditional since the sample space is limited to users that switch carriers.

5. a. $\frac{3}{4}$ **b.** $\frac{1}{2}$ **7. a.** $\frac{3}{8}$ **b.** $\frac{1}{8}$

9. a. ≈ 0.2333 About 23.3% of those surveyed said no.
b. ≈ 0.5333 About 53.3% of those surveyed were women.
c. ≈ 0.1406 About 14.1% of the women said no.
d. ≈ 0.3214 About 32.1% of those who said no were women.
e. ≈ 0.08 About 8% of the respondents said no and were women.
f. ≈ 0.08 About 8% of the respondents were women and said no.

11. a. ≈ 0.13. In one year, 13% of those who die of a transportation accident die of a pedestrian transportation accident.
b. ≈ 0.13. In a lifetime, 13% of those who die of a transportation accident die of a pedestrian transportation accident.
c. 0. In a lifetime, none of those who die of a non-transportation accident die of a pedestrian transportation accident.

13. a. ≈ 0.00054. In a lifetime, approximately 0.05% of those who die of a non transportation accident die from an earthquake.
b. ≈ 0.00053. In one year, approximately 0.05% of those who die of a non transportation accident die from an earthquake.
c. ≈ 0.00031. In one year, approximately 0.03% of those who die from an external cause die from an earthquake.

15. a. $\frac{1}{4}$ **b.** $\frac{4}{17}$ **c.** $\frac{1}{17}$
d.

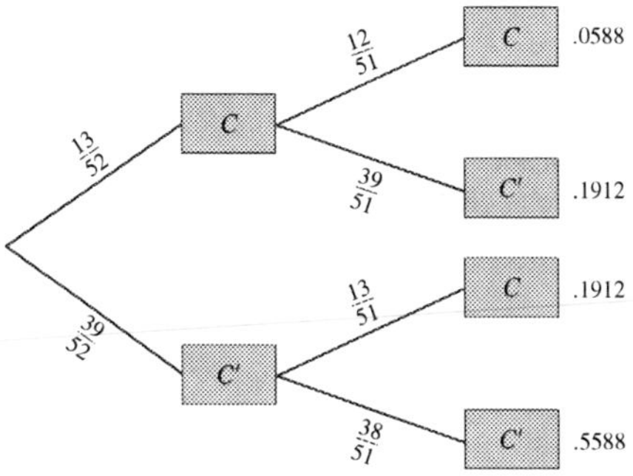

17. a. $\frac{13}{52} = \frac{1}{4}$ **b.** $\frac{13}{51}$ **c.** $\frac{13}{52} \cdot \frac{13}{51} = \frac{13}{204}$
d.

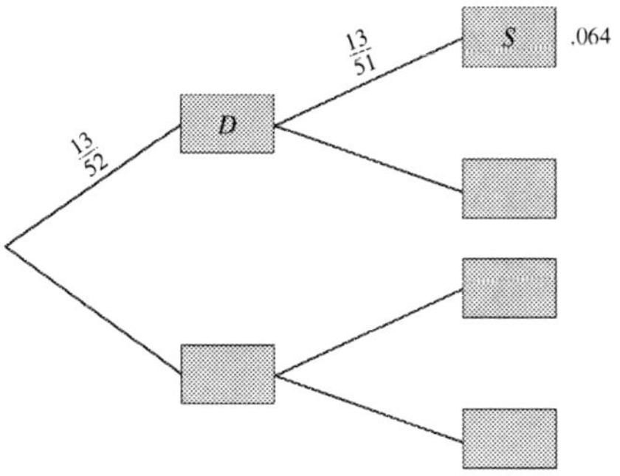

19. a. $p(B|A)$ **b.** $p(A')$ **c.** $p(C|A')$

21. a. $\frac{1}{6}$ **b.** $\frac{1}{3}$ **c.** 0 **d.** 1

23. a. $\frac{5}{36}$ **b.** $\frac{5}{18}$ **c.** 0 **d.** 1

25. a. $\frac{1}{12}$ **b.** $\frac{3}{10}$ **c.** 1

27. E_2 is most likely; E_3 is least likely

29. ≈ 0.46
About 46% of those who were happy with the service made a purchase.

31. 0.0005 **33.** 0.0020 **35.** 0.14 **37.** 0.20

39. 0.07 **41.** 0.6% **43.** 1.3%

45. a. 0.489 **b.** 0.560
c. For those voting for Obama, a higher percentage was women.

47. a. 0.406 **b.** 0.502
c. Those over 45 years were more likely to vote for McCain as compared to those under 45 years.

49. a. $p(\text{male}) = 0.803$
$p(\text{female}) = 0.197$
b. $p(\text{injection drug use} \mid \text{male}) = 0.217$
$p(\text{injection drug use} \mid \text{female}) = 0.404$
c. $p(\text{injection drug use} \cap \text{male}) = 0.174$
$p(\text{injection drug use} \cap \text{female}) = 0.080$
d. $p(\text{heterosexual contact} \mid \text{male}) = 0.079$
$p(\text{heterosexual contact} \mid \text{female}) = 0.565$
e. $p(\text{heterosexual contact} \cap \text{male}) = 0.063$
$p(\text{heterosexual contact} \cap \text{female}) = 0.111$
f. Part (b) is the percentage of males that were exposed by injection drug use. Part (c) is the percentage of the total that are both male and exposed by injection drug use. Similarly for parts (d) and (e).

51. a. 32% **b.** 16% **c.** 35%
d. 18% **e.** 34%
f. Part (c) is the percentage of adult women who are obese. Part (d) is the percentage of adults who are both female and obese.

53. a. 12.6 % **b.** 6.7% **c.** 0% **d.** 5.8%
e. Answers will vary. Possible answer: When it is a mixed race case, black defendants are more likely to have the death penalty imposed than white defendants.
f. Answers will vary. Possible answer:
$p(\text{death not imposed} \mid \text{victim white and defendant white}) \approx 87.4\%$
$p(\text{death not imposed} \mid \text{victim white and defendant black}) \approx 93.3\%$
$p(\text{death not imposed} \mid \text{victim black and defendant white}) = 100\%$
$p(\text{death not imposed} \mid \text{victim black and defendant black}) \approx 94.2\%$
Also see answers to Exercise 54(c) and 54(d).
g. Answers will vary.

55. $\frac{2}{3}$ **57.** 92% **59.** 39% **61.** 0.05

63. a. 86% **b.** 34% **c.** 66%
d. $N' \mid W; p(N \mid W) = \frac{45}{320} = 0.14 = 1 - 0.86$

65. The complement is $A' \mid B$. **67.** 0.14

3.7 Independence; Trees in Genetics

1. a. E and F are dependent. Knowing F affects E's probability.
b. E and F are not mutually exclusive. There are many women doctors.

3. a. E and F are dependent. Knowing F affects E's probability.
b. E and F are mutually exclusive. One cannot be both single and married.
5. a. E and F are independent. Knowing F does not affect E's probability.
b. E and F are not mutually exclusive. A brown haired person may have some gray.
7. a. E and F are dependent. Knowing F affects E's probability.
b. E and F are mutually exclusive. One can't wear both boots and sandals (excluding one on each foot!)
9. a. E and F are dependent.
b. E and F are mutually exclusive.
c. Knowing that you got an odd number changes the probability of getting a 4. You cannot get both a 4 and an odd number.
11. E and F are not independent.
13. a. $\frac{1}{6}$ **b.** 0
c. No. If you roll a 5, the probability of rolling an even number is zero.
d. Yes. 5 is not an even number.
e. Mutually exclusive events are always dependent.
15. a. $\frac{1}{13}$ **b.** $\frac{1}{13}$
c. Yes. Being dealt a red card doesn't change the probability of being dealt a jack.
d. No. There are red jacks.
e. Independent events are always not mutually exclusive. Exercises 14 and 15 show that events not mutually exclusive can be dependent or independent.
17. The events are dependent. Being happy with the service increases the probability of a purchase.
19. E and F are dependent. Knowing that the chip is made in Japan increases the probability that it is defective.
21. a. Since $0.616 \neq 0.673$, they are not independent.
b. Since $0.518 \neq 0$, they are not mutually exclusive.
c. Being a vegetarian increases your chances of being healthy.
23. a. Since $0.638 \neq 0.285$, they are not independent.
b. No; $p(A \cap B) \neq 0$; the circles overlap with 89 in both.
c. Living in Bishop decreases the chances of supporting proposition 3.
25. They are dependent. HAL users were more likely to quit.
27. They are independent. Smell So Good users quit at the same rate as all deodorant users.
29. a. 0.000001
b. 4 backup systems
31. $0.010101 \approx 1.0\%$
33. a. 0.47 **b.** 0.999996 **c.** 0.53 **d.** 0.000004
e. The results from (a) and (c) would be informative because a positive test doesn't necessarily mean you are ill.
f. (c) is a false positive because the test was positive, but he was healthy. (d) is a false negative because the test was negative, but he was ill.
g. Answers will vary.
h. Answers will vary.
35. Answers will vary.
37. a. Therefore, the probability of both cousins being a carrier would be 1/24. And the probability of their child having cystic fibrosis (Type A) would be $1/96 \approx 1\%$.
b. And the probability of their child having cystic fibrosis would be $1/144 \approx 0.7\%$.
39. $p(\text{both test positive}) = (0.85)(0.85) = 0.7225$
$p(\text{don't both test positive}) = 1 - 0.7225 = 0.2775$
41. $\frac{1}{3}$
43. If the first child is albino, Mrs. Jones must be a carrier. Thus, the child has $\frac{1}{2}$ chance of being albino.
45. $\frac{1}{2}$ chance of chestnut hair, $\frac{1}{2}$ chance of shiny dark brown
47. The child has a $\frac{1}{8}$ chance of the following hair colors: dark red, light brown, auburn or medium brown. The child has a $\frac{1}{4}$ chance of the following: reddish brown, chestnut.
49. The child has a $\frac{1}{8}$ chance of the following hair colors: reddish brown, chestnut, shiny dark brown, shiny black. The child has a $\frac{1}{16}$ chance of the following hair colors: dark red, light brown, auburn, medium brown, glossy dark brown, dark brown, glossy black, black.
51. a. $\frac{1}{6}$ **b.** $\frac{5}{6}$ **c.** 0.48
d. 0.52 **e.** \$0.04
53. He won because the expected value was positive. He lost because the expected value was negative. $\$0.04 > -\$0.017$
55. a. $p(\text{die}) = (0.25)(0.12)(0.9) = 0.027$
$p(\text{live}) = 1 - 0.027 = 0.973 = 97.3\%$
b. $p(\text{die}) = (0.25)(0.9) = 0.225$
$p(\text{live}) = 1 - 0.225 = 0.775$
c. $p(\text{good health returned}) = 1 - 0.02 = 0.98$
d. Yes.

Chapter 3 Review

1. Experiment is pick one card from a deck of 52 cards. Sample space S = {possible outcomes} = {jack of hearts, ace of spades, . . .}; n(S) = 52.
3. $p = \frac{1}{4}$, odds 1 : 3 If you deal a card many times, you should expect to be dealt a club approximately 1/4 of the time, and you should expect to be dealt a club approximately one time for every 3 times you are dealt something else.
5. $p(Q \cup \text{club}) = \frac{4}{13}$ odds 4 : 9 If you deal a card many times, you should expect to be dealt a queen or a club approximately 4/13 of the time, and you should expect to be dealt a queen of clubs approximately four times for every 9 times you are dealt something else.
7. a. Flip three coins and observe the results.
b. S = {HHH, HHT, HTH, HTT, TTT, TTH, THT, THH}
9. F = {HTT, TTH, THT, TTT}
11. $p(F) = \frac{1}{2}$, odds 1 : 1. According to the Law of Large Numbers, if the experiment is repeated a large number of times, we should expect to get two or more tails about half the time. Also, we should expect to get two or more tails about 1 time for every time we get at least two heads.
13. $p = \frac{1}{6}$. According to the Law of Large Numbers, if the experiment is repeated a large number of times, we should expect to roll a 7 about one time out of every six rolls.
15. $p = \frac{7}{18}$. According to the Law of Large Numbers, if the experiment is repeated a large number of times, we should expect to roll a 7, an 11, or doubles about seven times out of every eighteen rolls.
17. $p(3 \text{ or } 5 \text{ or } 7 \text{ or } 9 \text{ or } 10 \text{ or } 11 \text{ or } 12) = \frac{11}{18}$. According to the Law of Large Numbers, if the experiment is repeated a large number of times, we should expect to roll a number that is either odd or greater than 8 about eleven times out of every eighteen rolls.
19. 0.013 **21.** 0.151
23. $\frac{12}{51}$ **25.** $\frac{1}{216}$
27. $\frac{2}{27}$ **29.** $\frac{1}{6}$

31. $\frac{1}{2}$. According to the Law of Large Numbers, if the experiment is repeated a large number of times, we should expect the offspring to be long-stemmed about half the time.

33. $\frac{1}{4}$ **35.** $\frac{1}{4}$ **37.** $\frac{1}{2}$

39. $\frac{1}{2}$ **41.** $\frac{1}{2}$

43. $p(\text{2 tens and 3 jacks}) \approx 9.23 \times 10^{-6}$

45. a.

Winning Spot	Probability
9	${}_{20}C_9 \cdot {}_{60}C_0/{}_{80}C_9 = 0.000000724$
8	${}_{20}C_8 \cdot {}_{60}C_1/{}_{80}C_9 = 0.000032592$
7	${}_{20}C_7 \cdot {}_{60}C_2/{}_{80}C_9 = 0.000591678$
6	${}_{20}C_6 \cdot {}_{60}C_3/{}_{80}C_9 = 0.005719558$
5	${}_{20}C_5 \cdot {}_{60}C_4/{}_{80}C_9 = 0.032601481$
4 or less	0.961053966

b. −\$0.24; You would expect to lose \$0.24.

47. \$10.93

49. $p(\text{O'Neill}\mid\text{rural}) \approx 0.40$; $p(\text{Bell}\mid\text{rural}) \approx 0.55$

51. $p(\text{urban}\mid\text{Bell}) \approx 0.50$; $p(\text{rural}\mid\text{Bell}) \approx 0.50$

53. The urban residents prefer O'Neill. The rural residents prefer Bell.

55. Since the governors election depends only on who gets the most votes, O'Neill is ahead with 49.6% to Bell's $\frac{184 + 181}{800} \approx 45.6\%$.

57. Independent. Not mutually exclusive.

59. Dependent. Not mutually exclusive.

61. 14% **63.** 8% **65.** 43%

67. 0.78% **69.** 1.817% **71.** 43%

73. No, since $p(\text{AK}) = 0.39 \neq p(\text{AK}\mid\text{defective}) = 0.43$, they are not independent. Being made in Arkansas increases the probability of being defective.

75. Answers will vary.

77. Answers will vary.

79. The smallest chance of an event occurring is 0% (an impossible event). A certain event has the largest probability 100%.

81. Answers will vary.

83. Answers will vary.

85. Answers will vary.

CHAPTER 4 Statistics

4.1 Population, Sample, and Data

1. a.

Number of Visits	Frequency
1	9
2	8
3	2
4	5
5	6
	30

b. Library Habits - Number of Student Visits

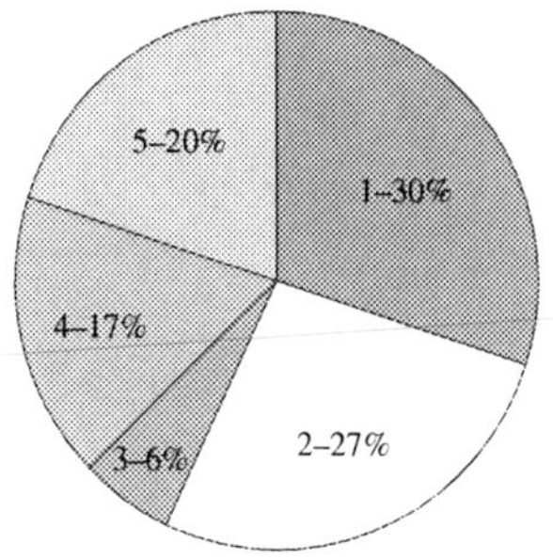

c.

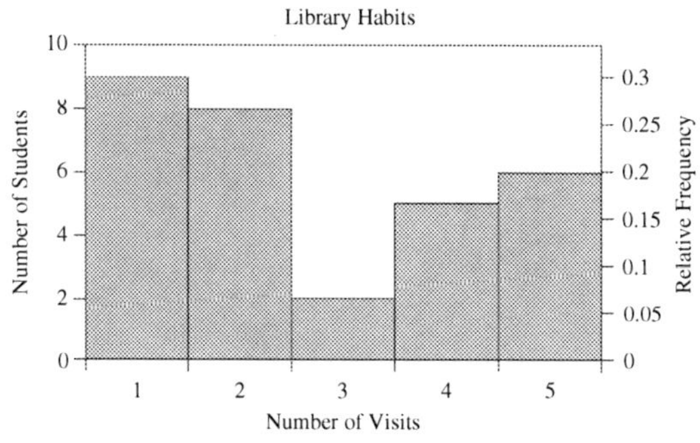

3. a.

Number of Children	Frequency
0	8
1	13
2	9
3	6
4	3
5	1
	40

b. Families in Manistee, Michigan

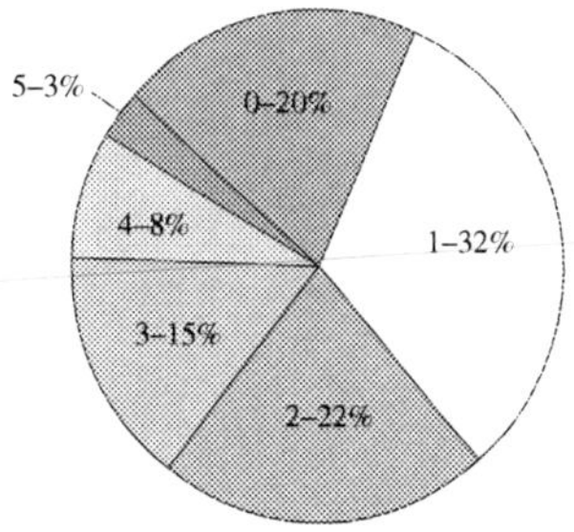

c.

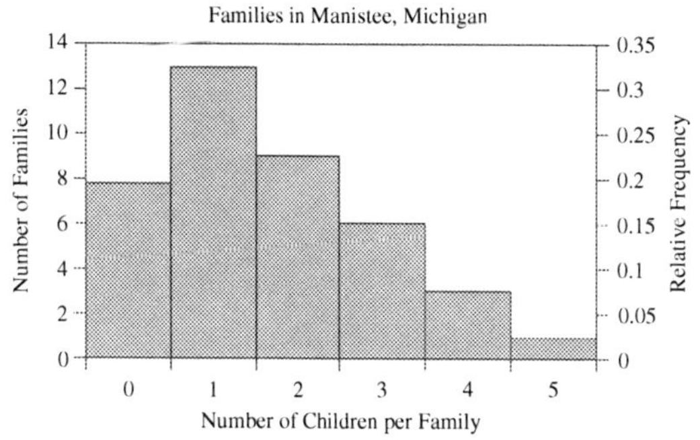

5. a.

x = Speed	Frequency	Relative Frequency
$51 \le x < 56$	2	0.05
$56 \le x < 61$	4	0.1
$61 \le x < 66$	7	0.175
$66 \le x < 71$	10	0.25
$71 \le x < 76$	9	0.225
$76 \le x < 81$	8	0.2
	40	

b.

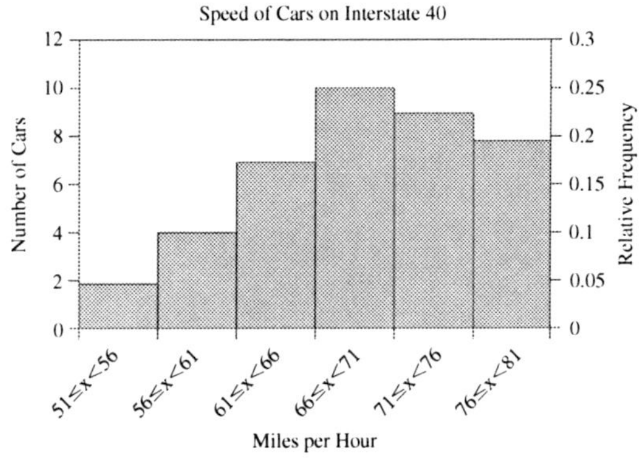

7.

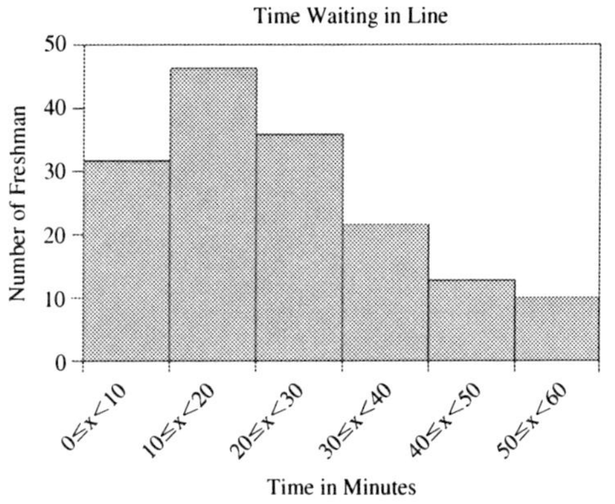

9.

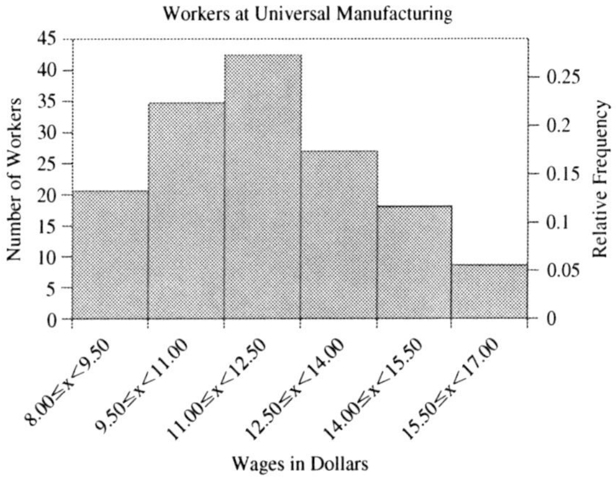

11.

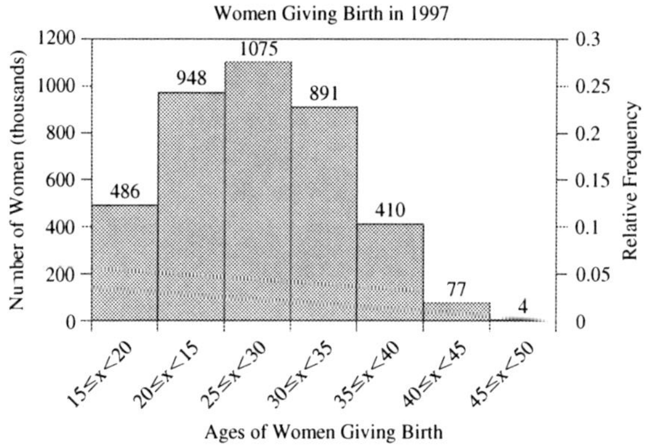

13.

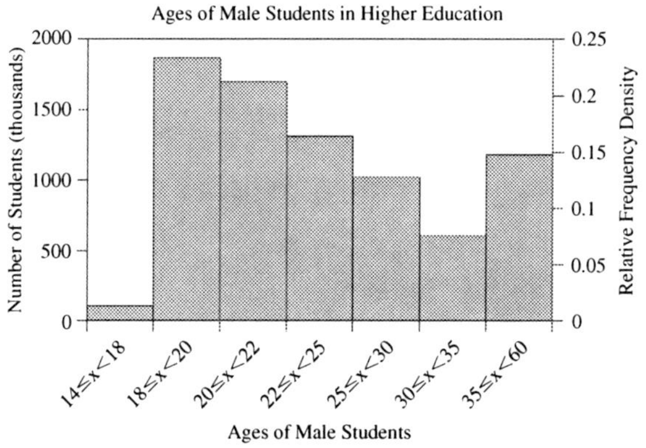

15. a. 28.5% **b.** 32%
c. Not possible **d.** 97.5%
e. 50% **f.** 52.5%

17.

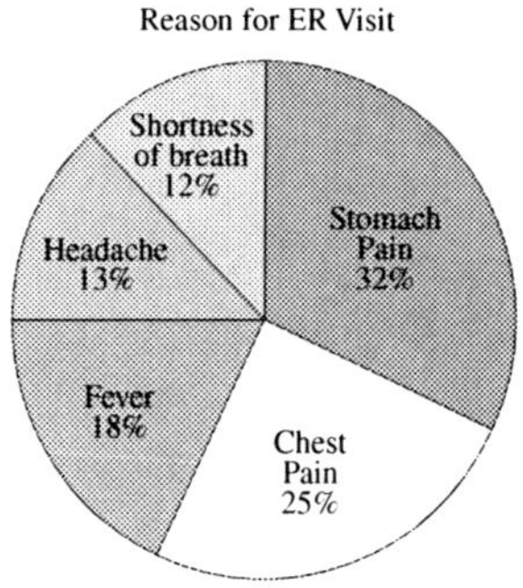

19. a.

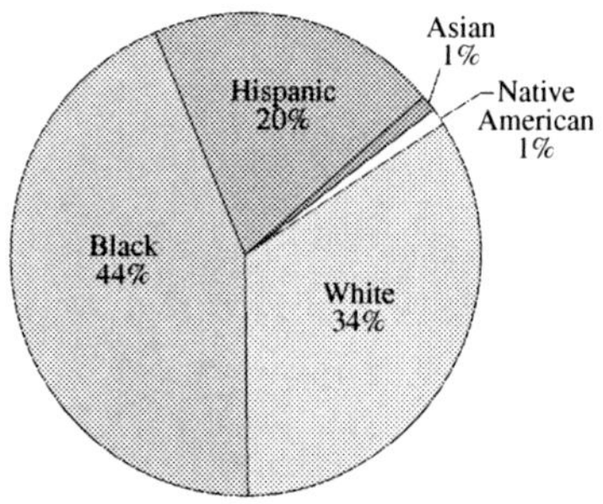

b.

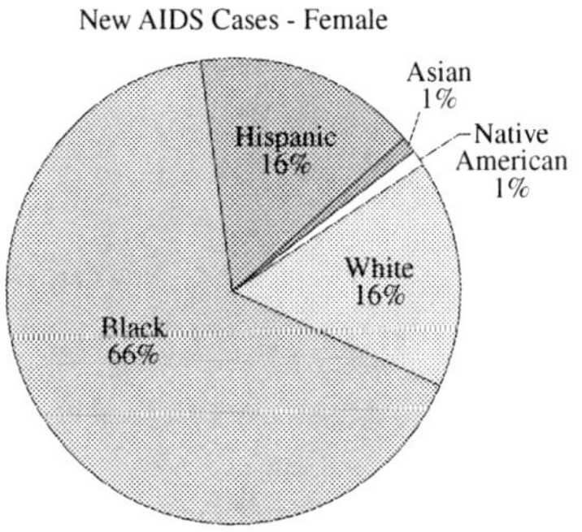

c. Females have a higher percentage of blacks with new AIDS cases than males do. Males have a higher percentage of whites with new AIDS cases than females do. Both Asian and Native American males and females are less than 1%.

d.

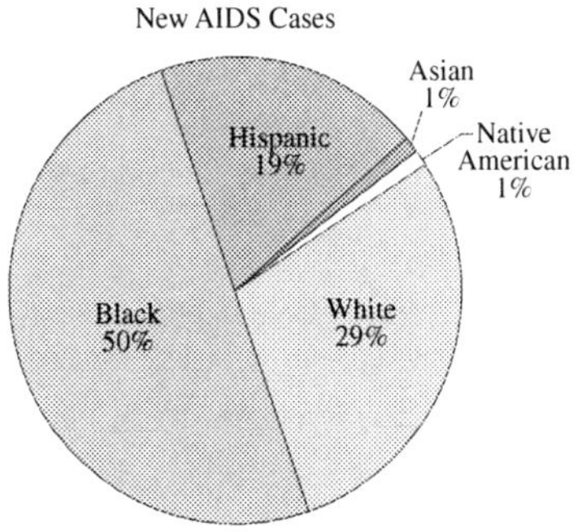

21. a.

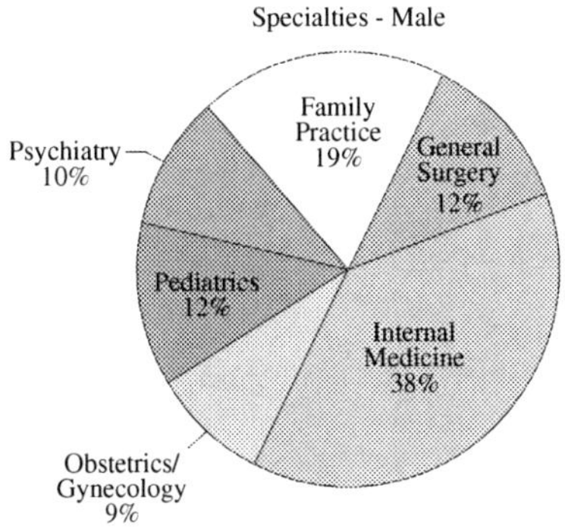

b.

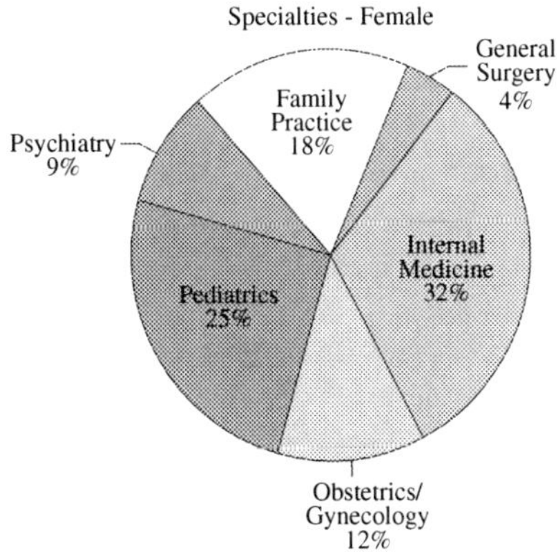

c. A higher percentage of males than females are general surgeons. A higher percentage of females than males are pediatricians.

d.

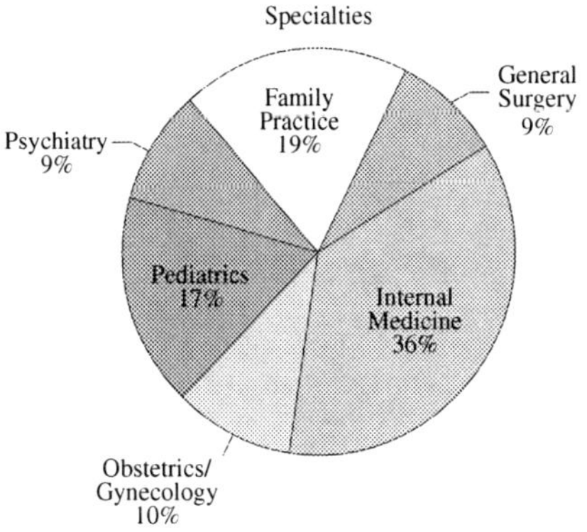

29. a.

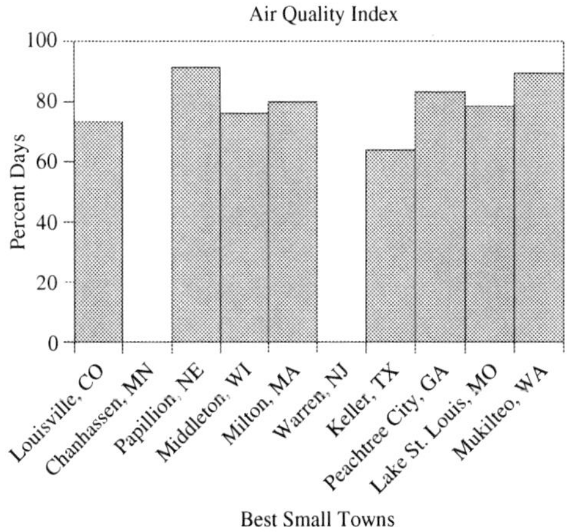

b. The air quality index (percent days AQI ranked good) is well over 50%, specifically from 65% to 92%.

31. a.

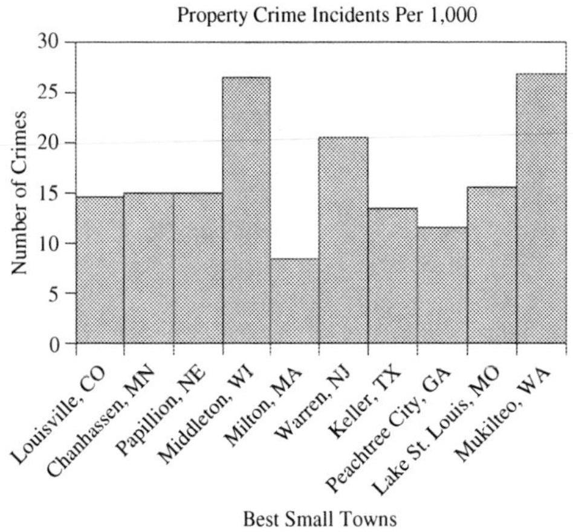

b. These small towns, considered the best places to live, have a very low rate of property crime incidents (less than 3%).

33. a. For comparison, two sets of histograms have been constructed.

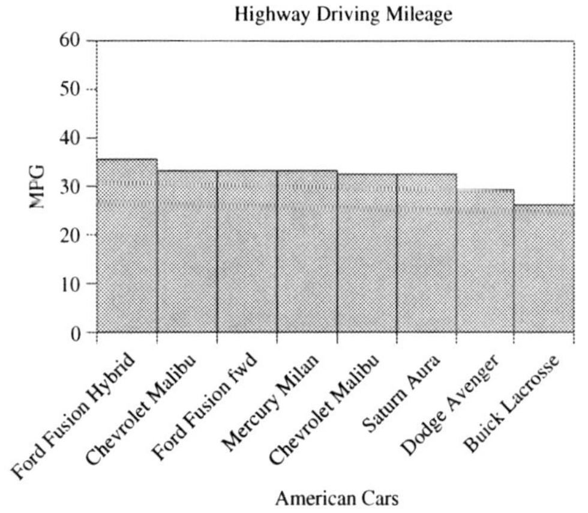

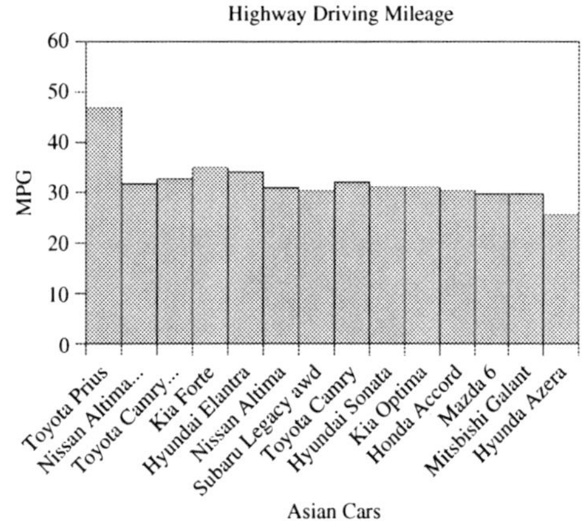

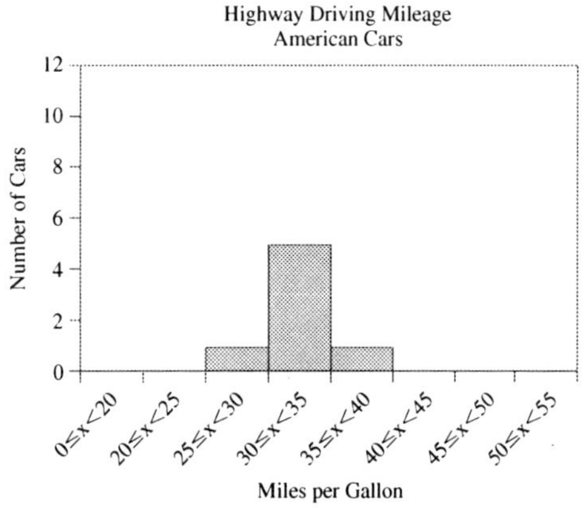

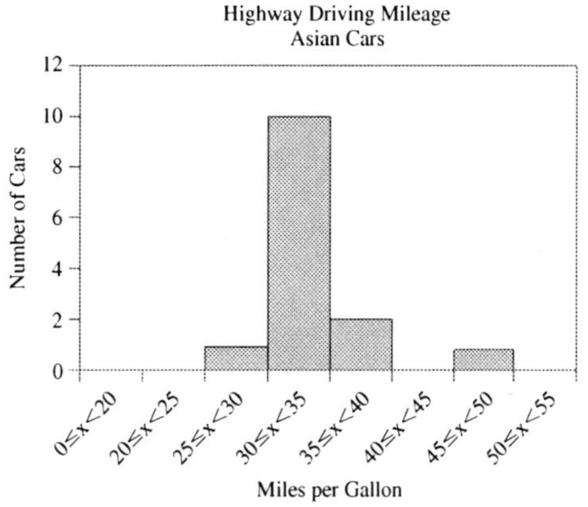

b. In general, the majority of both American and Asian cars have average highway driving mileage between 30 and 35 miles per gallon.

4.2 Measures of Central Tendency

1. Mean = 12.5
Median = 11.5; The mode is 9.

3. Mean = 1.45
Median = 1.5; The mode is 1.7.

5. a. Mean = 11
Median = 10.5
The mode is 9.

b. Mean = 25.5
Median = 10.5
The mode is 9.

c. The mean is affected by the change. The median and mode stay the same.

7. a. Mean = 7
Median = 7
There is no mode.

b. Mean = 107
Median = 107
There is no mode.

c. The data in part (b) is 100 more than the data in part (a).

d. The mean and median are 100 more. Neither data set had a mode.

9. Mean = 3 : 14.
Median = 3 : 14.
The mode is 3 : 12.

11. Mean = 9.85
Median = 9.
The mode is 9 and 15.

13. Mean = 34.64
Median = 34.
Mode: 25, 33, 34, 37, 45, 46 (all have frequency = 2)

15. Mean = 7.6
Median = 8
The mode is 9 and 6.

17. Mean = 16.028 oz

19. a. 42 **b.** 92

c. Impossible, it would require a score of 142.

21. Average = \$51,600

23. Mean speed = 51.4 mph

25. 39 years old

27. a. mean = \$40,750

b. The median does not change: \$42,000.

29. a. Envionmental Protection Agency; \$10,780

b. Department of the Army; \$2,761

c. mean = \$9,627

d. mean = \$4,442

31. a. Mean = 35.7 years old **b.** Mean = 36.5 years old

33. a. Mean = 23.0 years old **b.** Mean = 28.5 years old

4.3 Measures of Dispersion

1. a. Variance: $s^2 = 16.4$
Standard deviation: $s \approx 4.0$

b. Variance: $s^2 = 16.4$
Standard deviation: $s \approx 4.0$

3. a. Variance: $s^2 = 0$

b. Standard deviation: $s = 0$

5. a. Mean = 22
Standard deviation: $s \approx 7.5$

b. Mean = 1,100
Standard deviation: $s \approx 374.2$

c. The data in (b) are 50 times the data in (a).

d. The mean and standard deviation are 50 times larger in (b).

7. a. Joey's mean = 168
Dee Dee's mean = 167
Joey's mean is higher.

b. Joey: $s \approx 30.9$
Dee Dee: $s \approx 18.2$
Dee Dee's standard deviation is smaller.

c. Dee Dee is more consistent than Joey because his standard deviation is lower.

9. $s \approx 6.74$ ounces ≈ 0.421 lb $= 0:07$

11. $s \approx 14.039$

13. a. $\bar{x} = 67.5$ **b.** 80% **c.** 90%
$s \approx 7.1$

15. a. Mean $= \bar{x} = 3.175$ **b.** 58% **c.** 100%
$s \approx 1.807$

17. a. Mean $= 8$ **b.** 71% **c.** 94% **d.** 100%
$s \approx 1.4$

19. $s \approx 0.285$ **21.** $s \approx 6.80$

29. a. $\bar{x} = 80$

b. Sample standard deviation would be more appropriate because this does not include all the cities.

c. $s \approx 8.62$

d. The air quality index averages 80.0% with 62.5% of all indices falling within one standard deviation of the mean air quality index.

31. a. $\bar{x} = 17.3$

b. Sample standard deviation would be more appropriate because this does not include all the towns identified as best places to live.

c. $s \approx 5.96$

d. The property crime incidents averages 17.3 per 1,000 with 70% of all property crime incidents falling within one standard deviation of the mean property crime incidents.

33. a. American: $\bar{x} = \frac{261}{8} = 32.625$; $s \approx 2.83$
Asian: $\bar{x} = \frac{463}{14} \approx 33.07$; $s \approx 4.94$
Sample standard deviation was used as this does not include every car that could fall into this category.

b. Asian cars have the highest mpg for city driving on the average, but it varies over 50% more than the average mpg varies for American cars.

4.4 The Normal Distribution

1. Yes, it looks like a bell curve.

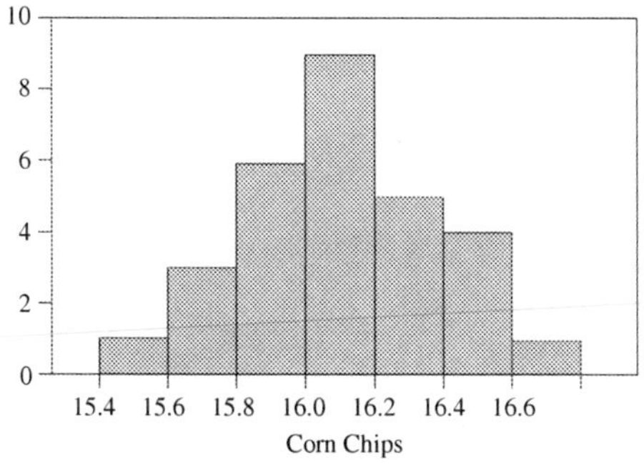

3. No, it is right-tailed.

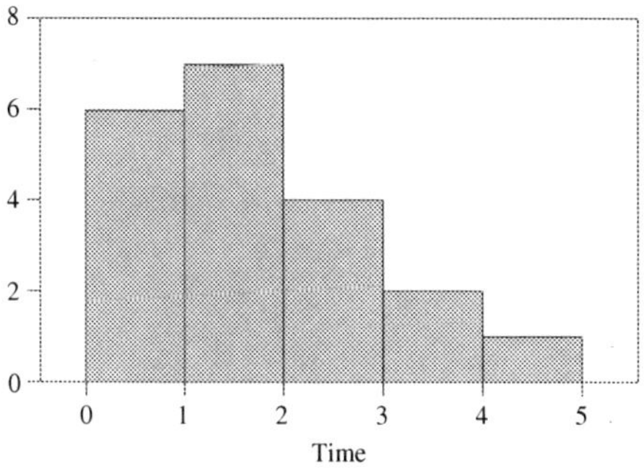

5. a. 34.13% **b.** 34.13% **c.** 68.26%

7. a. 49.87% **b.** 49.87% **c.** 99.74%

9. a. One standard deviation: [22.4, 27]
Two standard deviations: [20.1, 29.3]
Three standard deviations: [17.8, 31.6]

b. 68.26% lies in [22.4, 27]
95.44% lies in [20.1, 29.3]
99.74% lies in [17.8, 31.6]

c.

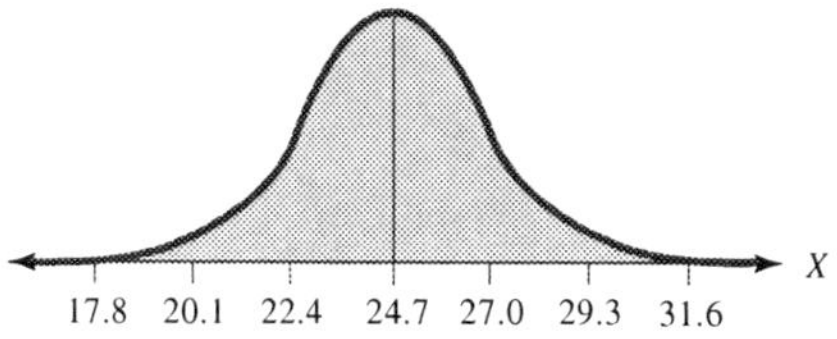

11. a. 0.4474 **b.** 0.0639 **c.** 0.5884
d. 0.0281 **e.** 0.0960 **f.** 0.8944

13. a. 0.34 **b.** −2.08 **c.** 0.62
d. −0.27 **e.** 1.64 **f.** −1.28

15. a. 0.42 **b.** −0.42 **c.** 2.08
d. −1.46 **e.** 2.96 **f.** −2.21

17. a. 0.3413 **b.** 0.6272 **c.** 0.8997
d. 0.9282 **e.** 0 **f.** 0.3300

19. a. 4.75% **b.** 49.72%

21. a. 0.1908 **b.** 0.7745

23. a. 95.25% **b.** 79.67%

25. a. About 87. **b.** 62 or less.

27. 9.1 minutes or more

4.5 Polls and Margin of Error

1. a. $z = 0.68$ **b.** $z = 0.31$ **c.** $z = 1.39$ **d.** $z = 2.65$

3. a. $z = 1.44$ **b.** $z = 1.28$ **c.** $z = 1.15$
d. $z = 2.575$ since it is halfway between 0.4949 and 0.4951

5. 1.75 **7.** 1.15

9. a. 1.8% **b.** 2.6%

11. a. 75.0% ± 3.4% **b.** 75.0% ± 4.0%

13. a. 79.0% **b.** 21.0% **c.** ± 1.6%

15. a. 86.0% **b.** 70.0%
c. For men: ± 2.8%
For women: ± 2.5%

17. a. 71% **b.** 29% **c.** ± 0.3%

19. a. 45.0% **b.** 47.0% **c.** 8.0% **d.** ± 0.2%

21. a. 36.0% **b.** ± 1.6% **c.** ± 2.3%
d. The answer in (c) is larger than the answer in (b). In order to guarantee 98% accuracy, we must increase the error.

23. a. ± 4.7% **b.** ± 3.5% **c.** ± 2.8%

25. about 96.4% **27.** about 95.3%

29. a. 1,068 **b.** 2,401 **c.** 9,604
d. Smaller margin of errors require a larger sample size, and larger margin of errors may have a smaller sample size.

31. a. 1,509 **b.** 3,394 **c.** 13,573
d. Smaller margin of errors require a larger sample size, and larger margin of errors may have a smaller sample size.

4.6 Linear Regression

1. a. $y = 1.0x + 3.5$ **b.** 14.5 **c.** 15.5
d. 0.9529485 **e.** Yes, they are reliable since r is close to 1.

3. a. $y = -0.087x + 8.246$ **b.** 7.811 **c.** 14.322
d. −0.095
e. No, they are not reliable since r is close to 0.

5. a. Yes, the ordered pairs seem to exhibit a linear trend.

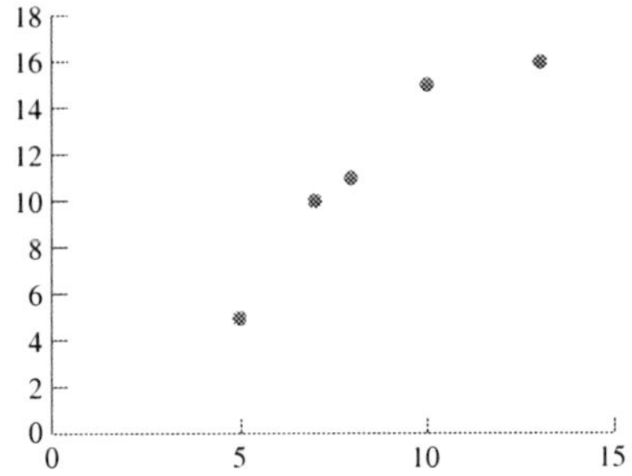

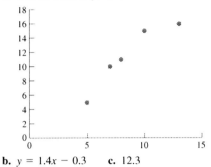

b. $y = 1.4x - 0.3$ **c.** 12.3

d.

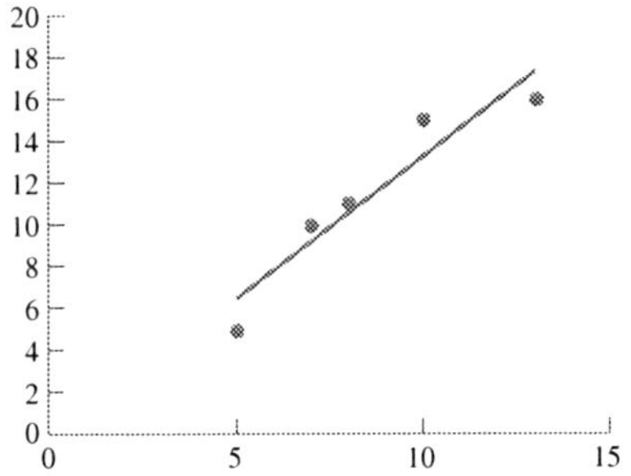

e. 0.9479459

f. Yes, the prediction is reliable since r is close to 1.

7. a. No, the ordered pairs do not seem to exhibit a linear trend.

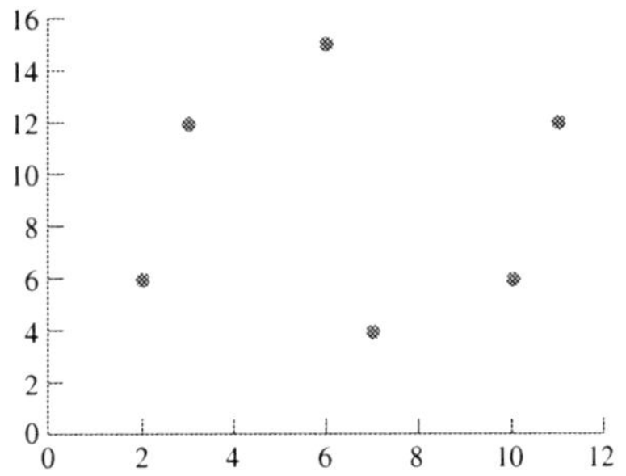

b. $y = 0.008x + 9.1$ **c.** 9.164

d.

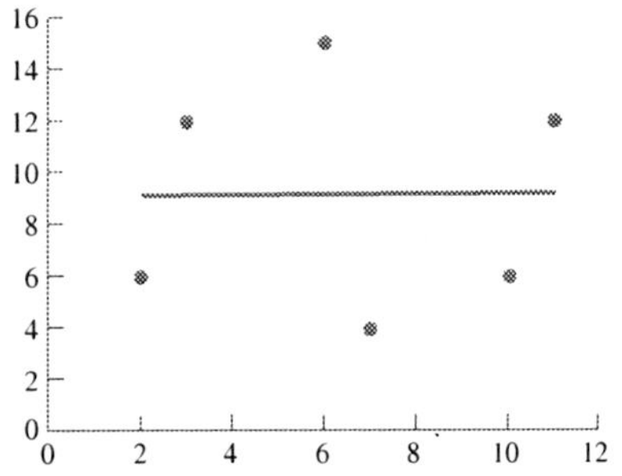

e. 0.0062782255

f. The prediction is not reliable since r is close to 0.

9. a. Yes, the data exhibit a linear trend.

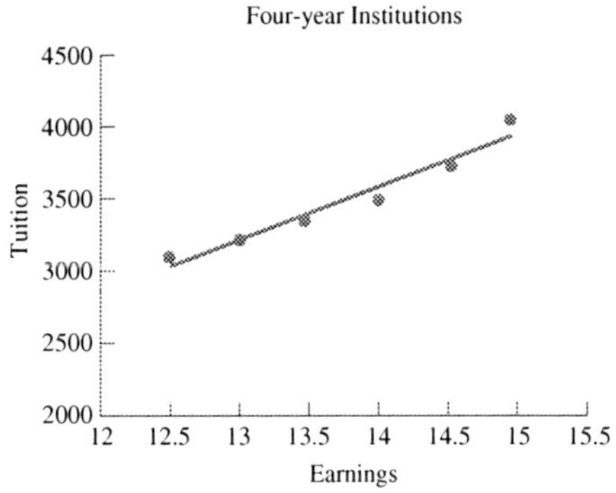

b. $y = 366.72x - 1{,}541.52$ **c.** $3,867.55

d. $15.11 **e.** 0.9732507334

f. Yes, r is close to 1.

11. a. Yes, the data exhibit a linear trend.

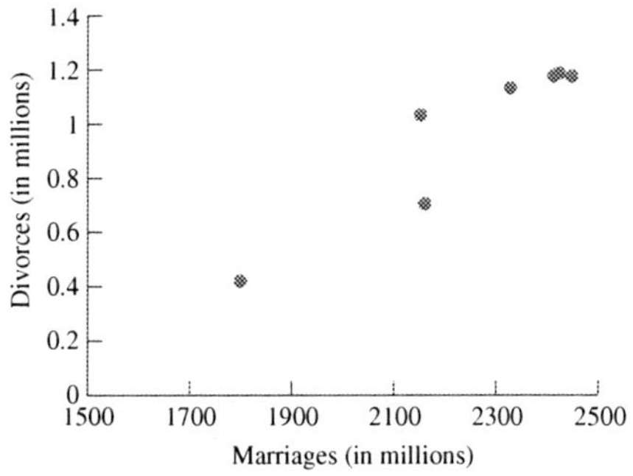

b. "y-hat" $= 1.1301x - 1.5526$ **c.** 1.555 million

d. 2.699 million **e.** 0.93344427

f. Yes, they are reliable since r is close to 1.

19. a. $y = -24.0x + 269$; $r = -0.86$

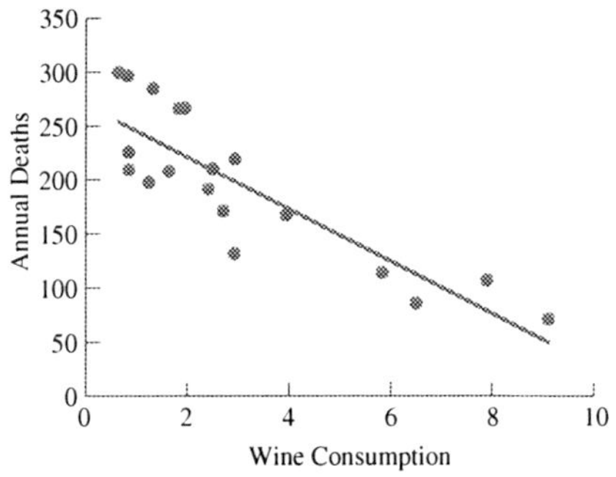

b. Yes
(2.5, 211), (3.9, 167), (2.9, 131), (2.4, 191), (2.9, 220), (0.8, 297), (9.1, 71), (0.8, 211), (0.6, 300), (7.9, 107), (1.8, 266), (1.9, 266), (0.8, 227), (6.5, 86), (1.6, 207), (5.8, 115), (1.3, 285), (1.2, 199), (2.7, 172)

Chapter 4 Review

1. a. 7.1 **b.** 7.5 **c.** 8 **d.** 2.1

3. a. 40% **b.** 28% **c.** 86% **d.** 14%
e. 50% **f.** Cannot determine.

5. 100

7. a. Timo: mean = 98.8
Henke: mean = 96.6
Henke has a lower mean.

b. Timo: $s \approx 6.4$
Henke: $s \approx 10.4$
c. Timo is more consistent because his standard deviation is lower.

9. a. Continuous **b.** Neither **c.** Discrete
d. Neither **e.** Discrete **f.** Continuous

11. a. One standard deviation = [71, 85]
Two standard deviations = [64, 92]
Three standard deviations = [57, 99]
b. 68.26% of the data lies in [71, 85]
95.44% of the data lies in [64, 92]
99.74% of the data lies in [57, 99]
c.

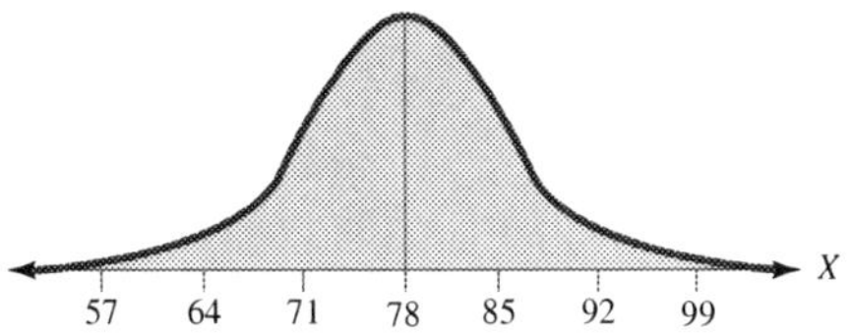

13. Cutoff at 401.

15. a. 66.7% ± 2.4%
b. 66.7% ± 2.8%

17. a. about ± 4.1%
b. about ± 3.1%
c. about ± 2.5%

19. 1,414

21. a. Yes, the ordered pairs seem to exhibit a linear trend.

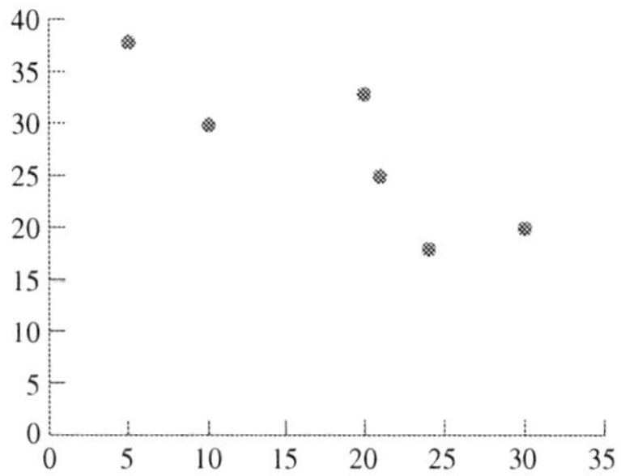

b. $y = -0.70455x + 40.25$ **c.** 29.682
d.

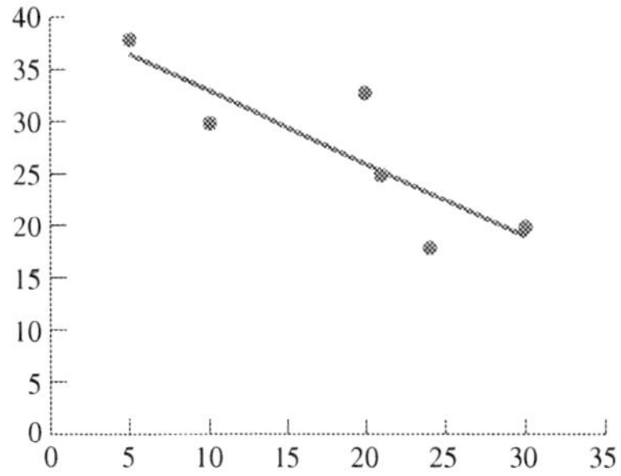

e. −0.839839286
f. Yes, the prediction is reliable since r is close to −1.

CHAPTER 5 Finance

5.1 Simple Interest

1. a. 60 days **b.** 61 days

3. a. 100 days **b.** 101 days

5. \$480

7. \$25.24

9. a. \$28.39 **b.** \$28.87

11. \$4,376.48

13. \$13,750.47

15. \$1,699.25

17. \$6,162.51

19. \$17,042.18

21. \$6,692.61

23. \$1,058.19

25. \$1,255.76

27. 1,013 days

29. \$233,027.39

31. \$714.31

33. \$222.86

35. a. \$112,102.05 **b.** \$2,102.05

37. Average daily balance ≈ \$140.27
Finance charge ≈ \$2.08

39. Average daily balance ≈ \$152.84
Finance charge ≈ \$2.73

41. a. \$8,125.00 **b.** \$146,250
c. \$8,125.00 **d.** \$67.71
e. \$19,500.08 **f.** \$136,500
g. \$156,000.08

43. a. \$38,940 **b.** \$311,520
c. \$38,940 **d.** \$356.95
e. \$95,013.60 **f.** \$288,156
g. \$383,169.60

45. a. Average daily balance: \$989.68
Finance charge: \$17.65
New Balance: \$997.65
b. Average daily balance: \$988.36
Finance charge: \$15.92
New Balance: \$993.57
c. Average daily balance: \$983.25
Finance charge: \$17.54
New Balance: \$991.11
d. The minimum \$20 payment is reducing your debt by about \$3 each month. It will take you roughly 28 years to pay off your debt. This means you will have paid the credit card company roughly \$6,720 on a \$1,000 debt. (572% of your debt was total interest.)

47. a. Average daily balance: \$979.35
Finance charge: \$17.47
New Balance: \$977.47
b. Average daily balance: \$958.90
Finance charge: \$15.45
New Balance: \$952.92
c. Average daily balance: \$932.27
Finance charge: \$16.63
New Balance: \$929.55
d. The change in the minimum required payment policy has had some impact, because you're reducing your debt by approximately \$23 per month.

5.2 Compound Interest

1. a. 0.03 **b.** 0.01
c. 0.000328767 **d.** 0.0046153846
e. 0.005

3. a. 0.00775 **b.** 0.002583333
c. 0.000084932 **d.** 0.001192307
e. 0.001291667

5. a. 0.02425 **b.** 0.008083333
c. 0.000265753 **d.** 0.003730769
e. 0.004041667

7. a. 34 quarters **b.** 102 months
c. 3,102.5 days

9. a. 120 quarters **b.** 360 months
c. 10,950 days

11. a. \$7,189.67
b. After 15 years, the investment is worth \$7,189.67.
13. a. \$9,185.46
b. After 8.5 years, the investment is worth \$9,185.46.
15. a. \$3,951.74
b. After 17 years, the investment is worth \$3,951.74.
17. a. 8.30%
b. The given compound rate is equivalent to 8.30% simple interest.
19. a. 4.34%
b. The given compound rate is equivalent to 4.34% simple interest.
21. a. 10.38%
The given compound rate is equivalent to 10.38% simple interest.
b. 10.47%
The given compound rate is equivalent to 10.47% simple interest.
c. 10.52%
The given compound rate is equivalent to 10.52% simple interest.
23. a. \$583.49
b. One would have to invest \$583.49 now to have the future value in the given time.
25. a. \$470.54
b. One would have to invest \$470.54 now to have the future value in the given time.
27. a.

Month	$FV = P(1 + rt)$
1	\$10,083.33
2	\$10,167.36
3	\$10,252.09
4	\$10,337.52
5	\$10,423.67
6	\$10,510.53

b. \$10,510.53
29. a.

Year	$FV = P(1 + rt)$
1	\$15,900
2	\$16,854
3	\$17,865.24

b. \$17,865.24
31. \$4,032,299.13
33. He would have to invest at least \$9,058.33.
35. a. \$9,664.97
b. The difference is the interest: \$51.77
c. \$5,045.85
37. a. \$19,741.51
b. \$10,535.83
39. a. \$8,108.87
b. The difference is the interest: \$690.65
When Marlene retires, her account will not have exactly \$100,000 in it, so we will not compute the monthly interest on this amount.
41. FNB: 9.55%
CS: 9.45%
First National Bank has the better offer.
43. a. $r \approx 3.45\%$; Does not verify.
b. $r \approx 3.45\%$; Does not verify.
c. $r \approx 3.50\%$
The annual yield for the 5-year certificate verifies if you pro-rate the interest over a 360-day year, and then pay interest for 365 days.
45. $r \approx 9.74\%$; Does not verify.
47. a. 5.12% **b.** \$1,025.26
c. \$25.26 **d.** 2.53%
e. Part (d) is for 6 months; part (a) is for 1 year.
f. You earn interest on interest; that is what compounding means.
49. a. \$387.10 **b.** \$173.96 **c.** \$40.71
51. $r = (1 + i)^n - 1$
53. 7.45%
55. a. 5.70% **b.** 5.74%
c. 5.77% **d.** 5.79%
e. 5.78% **f.** 5.78%
57. compound interest; $i = 5.00\%$ because $£1{,}000 \cdot (1 + 0.05)^{100} \approx £131{,}000$
59. 4.57%

5.3 Annuities

1. \$1,478.56
3. \$5,422.51
5. a. \$226.32 **b.** \$226.32
c. \$225 **d.** \$1.32
7. a. \$455.46 **b.** \$455.46
c. \$450 **d.** \$5.46
9. a. \$283,037.86 **b.** \$54,600.00 **c.** \$228,437.86
11. a. \$251,035.03 **b.** \$69,600.00 **c.** \$181,435.03
13. a. \$1,396.14
15. a. \$222.40
17. a. \$18,680.03
19. \$51.84 **21.** \$33.93 **23.** \$33.63
25. a. \$537,986.93
b.

	Beginning Balance	Interest	Withdrawal	Ending Balance
1	\$537,986.93	\$2,734.77	\$650	\$540,071.70
2	\$540,071.70	\$2,745.36	\$650	\$542,167.06
3	\$542,167.06	\$2,756.02	\$650	\$544,273.08
4	\$544,273.08	\$2,766.72	\$650	\$546,389.80
5	\$546,389.80	\$2,777.48	\$650	\$548,517.28

27. a. \$175,384.62
b. \$111.84
29. a. \$2,411.03 **b.** \$4,890.75 **c.** \$7,565.55
31. a. \$1,484,909.47
b. Shannon: \$91,000
Parents: \$11,000
Total contribution is \$102,000.
c. \$1,382,909.47 **d.** \$979,650.96 **e.** \$643,631.10
33. \$11,954.38 per year
35. \$45.82

37. a. \$595.83 **b.** \$5,367.01

c.

Period	Starting Balance	Interest	Deposit	Ending Balance
1	0	0	\$5,367.01	\$5,367.01
2	\$5,367.01	\$224.74	\$5,367.01	\$10,958.76
3	\$10,958.76	\$458.90	\$5,367.01	\$16,784.67
4	\$16,784.67	\$702.86	\$5,367.01	\$22,854.54
5	\$22,854.54	\$957.03	\$5,367.01	\$29,178.58
6	\$29,178.58	\$1,221.85	\$5,367.01	\$35,767.44
7	\$35,767.44	\$1,497.76	\$5,367.01	\$42,632.21
8	\$42,632.21	\$1,785.22	\$5,367.01	\$49,784.44
9	\$49,784.44	\$2,084.72	\$5,367.01	\$57,236.17
10	\$57,236.17	\$2,396.76	\$5,367.01	\$64,999.94

39. $P = pymt \cdot \dfrac{1-(1+i)^{-n}}{i}$

41. \$1,396.14

43. \$222.40

5.4 Amortized Loans

1. a. \$125.62 **b.** \$1,029.76

3. a. \$193.91 **b.** \$1,634.60

5. a. \$1,303.32 **b.** \$314,195.20

7. a. \$378.35 **b.** \$3,658.36

c.

Payment Number	Principal Portion	Interest Portion	Total Payment	Balance
0				\$14,502.44
1	\$239.37	\$138.98	\$378.35	\$14,263.07
2	\$241.66	\$136.69	\$378.35	\$14,021.41

9. a. \$1,602.91 **b.** \$407,047.60

c.

Payment Number	Principal Portion	Interest Portion	Total Payment	Balance
0				\$170,000.00
1	\$62.28	\$1,540.63	\$1,602.91	\$169,937.72
2	\$62.85	\$1,540.06	\$1,602.91	\$169,874.87

d. \$4,218.18 per month (assuming only home loan payment)

11. a. Dealer: \$404.72
Bank: \$360.29
b. Dealer: \$4,597.24
Bank: \$2,464.60
c. Choose the bank loan since the interest is less.

13. a. *pymt*: \$354.29
Interest: \$1,754.44
b. *pymt*: \$278.33
Interest: \$2,359.84
c. *pymt*: \$233.04
Interest: \$2,982.40

15. a. *pymt*: \$599.55
Interest: \$115,838
b. *pymt*: \$665.30
Interest: \$139,508
c. *pymt*: \$733.76
Interest: \$164,153.60
d. *pymt*: \$804.62
Interest: \$189,663.20
e. *pymt*: \$877.57
Interest: \$215,925.20
f. *pymt*: \$952.32
Interest: \$242,835.20

17. a. *pymt*: \$877.57
Interest: \$215,925.20
b. *pymt*: \$404.89
Interest: \$215,814.20

19. a. 30 year: \$914.74
15 year: \$1,066.97
Both verify, but not exactly. Their computations are actually more accurate than ours, since they are using more accurate round-off rules than we do.
b. 30 year: Total payments: \$329,306.40
15 year: Total payments: \$192,054.60
Savings: \$137,251.80

21. a. \$19,053.71

b.

Payment Number	Principal Portion	Interest Portion	Total Payment	Balance Due
0				\$75,000.00
1	\$18,569.33	\$484.38	\$19,053.71	\$56,430.67
2	\$18,689.26	\$364.45	\$19,053.71	\$37,741.41
3	\$18,809.96	\$243.75	\$19,053.71	\$18,931.45
4	\$18,931.45	\$122.27	\$19,053.72	\$0

23. a. \$23,693.67

b.

Payment Number	Principal Portion	Interest Portion	Total Payment	Balance Due
0				\$93,000.00
1	\$22,986.48	\$707.19	\$23,693.67	\$70,013.52
2	\$23,161.28	\$532.39	\$23,693.67	\$46,852.24
3	\$23,337.40	\$356.27	\$23,693.67	\$23,514.84
4	\$23,514.85	\$178.81	\$23,693.66	\$0

25. **a.** $29.91
b. $797.30
c.

Payment Number	Principal Portion	Interest Portion	Total Payment	Balance
0				$6,243.00
1	$767.39	$29.91	$797.30	$5,475.61
2	$771.06	$26.24	$797.30	$4,704.55
3	$774.76	$22.54	$797.30	$3,929.79
4	$778.47	$18.83	$797.30	$3,151.32
5	$782.20	$15.10	$797.30	$2,369.12
6	$785.95	$11.35	$797.30	$1,583.17
7	$789.71	$7.59	$797.30	$793.46
8	$793.46	$3.80	$797.26	$0

d. $135.36

27. **a.** $89.25 **b.** $1,905.92
c.

Payment Number	Principal Portion	Interest Portion	Total Payment	Balance
0				$12,982.00
1	$1,816.67	$89.25	$1,905.92	$11,165.33
2	$1,829.16	$76.76	$1,905.92	$9,336.17
3	$1,841.73	$64.19	$1,905.92	$7,494.44
4	$1,854.40	$51.52	$1,905.92	$5,640.04
5	$1,867.14	$38.78	$1,905.92	$3,772.90
6	$1,879.98	$25.94	$1,905.92	$1,892.92
7	$1,892.92	$13.01	$1,905.93	$0

d. $359.45

29. $3,591.73
31. $160,234.64
33. **a.** $1,735.74 **b.** $151,437.74
c. $1,204.91 **d.** $472,016.40
e. $343,404.24
f. Yes, they should refinance. Why do you think so?
35. **a.** $1,718.80 **b.** $172,157.40
c. $240,354.60 **d.** $573,981.98
e. $1,200,543.86
37. **a.** $18,950 **b.** $151,600
c. $18,950 **d.** $1,501.28
e. $189.50
39. **a.** $271.61 **b.** $1,962.39
c. $1,501.28
41. Receiving $142,000 was not enough to pay off the loan. They did not make a profit but will have to pay $66,454.72.
43. **a.** $882.36
b. The loan will be paid off sometime between the 279th and 280th payment, or 80 payments early, which is approximately 6.67 years.

5.5 Annual Percentage Rate on a Graphing Calculator

1. 14.6% **3.** 11.2%
5. **a.** $126.50
b. $APR \approx 14.35\%$
Verifies; this is within the tolerance.
7. **a.** $228.77
b. $APR \approx 16.50\%$
Doesn't verify; the advertised APR is incorrect.
9. Either one could be less expensive, depending on the A.P.R.
11. Really Friendly S and L will have lower payments but higher fees and/or more points.
13. $8,109.53
15. **a.** $819.38 **b.** $801.40
c. $8,009.75 **d.** $9,496.12
e. The RTC loan has a lower monthly payment but has higher fees.

5.6 Payout Annuities

1. **a.** $143,465.15 **b.** $288,000
3. **a.** $179,055.64 **b.** $390,000
5. **a.** $96.26
b. Pay in = $34,653.60
Receive = $288,000
Therefore, Suzanne receives $253,346.40 more than she paid.
7. **a.** $138.30 **b.** $331.39
9. **a.** $185,464.46 **b.** $14,000
c. $14,560 **d.** $29,495.89
11. **a.** $36,488.32 **b.** $576,352.60
c. $454.10 **d.** $1,079.79
13. **a.** $12,565.57 **b.** 10.47130674%
c. $138,977.90 **d.** $61.48
15. **a.** $17,502.53 **b.** 9.2721727%
c. $245,972.94 **d.** $372.99
17. $490,907.37

Chapter 5 Review

1. $8,730.15 **3.** $20,121.60
5. $4,067.71 **7.** $29,909.33
9. $23,687.72 **11.** $5,469.15
13. $6,943.26 **15.** 7.25%
17. $109,474.35 **19.** $275,327.05
21. **a.** $525.05 **b.** $6,503
23. $362,702.26
25. Average daily balance: $3,443.70
$I \approx \$57.03$
27. **a.** $10,014.44 **b.** $1,692.71 **29.** $255,048.53
31. $37,738.26
33. **a.** $102,887.14
b.

Month	Beginning	Interest Earned	With-drawal	End
1	$102,887.14	$493.00	$1,000	$102,380.14
2	$102,380.14	$490.57	$1,000	$101,870.71
3	$101,870.71	$488.13	$1,000	$101,358.84
4	$101,358.84	$485.68	$1,000	$100,844.52
5	$100,844.52	$483.21	$1,000	$100,327.73

35. a. \$1,246.53 **b.** \$268,270.80

c.

Payment	Principal Portion	Interest Portion	Total Payment	Balance
0				\$180,480
1	\$137.33	\$1,109.20	\$1,246.53	\$180,342.67
2	\$138.17	\$1,108.36	\$1,246.53	\$180,204.50

37. a. \$268.14 **b.** \$6,085.50

c.

Payment	Principal Portion	Interest Portion	Total Payment	Balance
0				\$41,519.00
1	\$5,817.36	\$268.14	\$6,085.50	\$35,701.64
2	\$5,854.93	\$230.57	\$6,085.50	\$29,846.71
3	\$5,892.74	\$192.76	\$6,085.50	\$23,953.97
4	\$5,930.80	\$154.70	\$6,085.50	\$18,023.17
5	\$5,969.10	\$116.40	\$6,085.50	\$12,054.07
6	\$6,007.65	\$77.85	\$6,085.50	\$6,046.42
7	\$6,046.42	\$39.05	\$6,085.47	\$0.00

d. \$1,079.50

39. a. \$174.50
b. $APR \approx 17.44\%$
Does not verify; the advertised APR is incorrect.

41. a. \$1,217.96
b. \$7,299.98

43. a. \$58,409.10
b. \$881,764.23
c. \$556.55

45. Answers will vary.

47. An account that earns compound interest is earning interest on the account balance, which is the original principal and all previous interest. An annuity, earns compound interest on each periodic payment.

49. This is a law that requires lenders to disclose in writing the annual percentage rate of interest on a loan and other particular terms.

51. The first credit card, Western Union's, offered deferred payments on a customer's account and the assurance of prompt and courteous service. After World War II, credit cards provided a convenient means of paying restaurant, hotel, and airline bills.

CHAPTER 6 Voting and Apportionment

6.1 Voting Systems

1. a. 2,000 **b.** Cruz **c.** Yes

3. a. 10,351 **b.** Edelstein **c.** No

5. a. 30 **b.** Park **c.** 47%
d. Beach wins. **e.** 53% **f.** Beach wins.
g. 63 pts **h.** Beach wins. **i.** 2 pts.

7. a. 65 **b.** Coastline wins. **c.** 48%
d. Coastline wins. **e.** 63% **f.** Coastline wins.
g. 150 pts. **h.** Coastline wins. **i.** 2 pts.

9. a. 140 **b.** Shattuck wins. **c.** 40%
d. Nirgiotis wins. **e.** 51% **f.** Shattuck wins.
g. 284 pts. **h.** Nirgiotis wins. **i.** 2 pts.

11. a. 1,342 **b.** Jones wins. **c.** 44%
d. Jones wins. **e.** 50.4% **f.** Jones wins.
g. 3,960 pts. **h.** Jones wins. **i.** 3 pts.

13. a. 31,754 **b.** Darter wins. **c.** 52%
d. Darter wins **e.** 52% **f.** Darter wins.
g. 138,797 pts. **h.** Darter wins. **i.** 4 pts.

15. 720 **17.** 15

19. a. 75 **b.** 25

21. a. 6 **b.** 0

23. a. A has the majority of the first-place votes and therefore should win.
b. B wins.
c. Yes, A received 7 votes for first choice, which is a majority of the 13 votes, but A did not win.

25. a. C wins by getting a majority of first-choice votes.
b. Using Figure 6.38, B wins the instant runoff.
c. Yes. C won originally and votes were changed in favor of C, but B won the instant runoff.

6.2 Methods of Apportionment

1. a. See table. **b.** Standard divisor: 50.00
c. See table.

State	A	B	C	Total
Population (thousands)	900	700	400	2,000
Std q (d = 50.00)	18.00	14.00	8.00	—
Lower q	18	14	8	40
Upper q	19.00	15.00	9.00	43

3. a. See table. **b.** Standard divisor: 2.64
c. See table.

State	NY	PA	NJ	Total
Population (millions)	18.977	12.281	8.414	39.672
Std q (d = 2.64)	7.19	4.65	3.19	—
Lower q (d = 2.64)	7	4	3	14
Upper q (d = 2.64)	8	5	4	17

5.

State	A	B	C	Total
Seats Hamilton	18	14	8	40

7.

State	NY	PA	NJ	Total
Seats Hamilton	7	5	3	15

9.

School	A	B	C	D	Total
Calculators Hamilton	69	53	43	35	200

11.

Region	N	S	E	W	Total
Seats Hamilton	3	8	3	10	24

13.

Country	D	F	I	N	S	Total
Seats Hamilton	5	4	0	4	7	20

15.

Country	C	E	G	H	N	P	Total
Seats Hamilton	3	4	9	4	3	2	25

17.

	State	A	B	C	Total
a.	Seats Jefferson	18	14	8	40
b.	Seats Adams	18	14	8	40
c.	Seats Webster	18	14	8	40

19.

	State	NY	PA	NJ	Total
a.	Seats Jefferson	7	5	3	15
b.	Seats Adams	7	5	3	15
c.	Seats Webster	7	5	3	15

21.

	School	A	B	C	D	Total
a.	Seats Jefferson	69	53	43	35	200
b.	Seats Adams	69	53	43	35	200
c.	Seats Webster	69	53	43	35	200

23.

	Region	N	S	E	W	Total
a.	Seats Jefferson	3	8	3	10	24
b.	Seats Adams	3	7	4	10	24
c.	Seats Webster	3	8	3	10	24

25.

	Country	D	F	I	N	S	Total
a.	Seats Jefferson	4	4	0	4	8	20
b.	Seats Adams	4	4	1	4	7	20
c.	Seats Webster	5	4	0	4	7	20

27.

	Country	C	E	G	H	N	P	Total
a.	Seats Jefferson	2	4	10	4	3	2	25
b.	Seats Adams	3	4	8	4	4	2	25
c.	Seats Webster	3	4	9	4	3	2	25

29.

State	A	B	C	Total
Seats Hill-Huntington	18	14	8	40

31.

State	NY	PA	NJ	Total
Seats Hill-Huntington	7	5	3	15

33.

School	A	B	C	D	Total
Calculators Hill-Huntington	69	53	43	35	200

35.

Region	N	S	E	W	Total
Seats Hill-Huntington	3	8	3	10	24

37.

Country	D	F	I	N	S	Total
Seats Hill-Huntington	4	4	1	4	7	20

39.

Country	C	E	G	H	N	P	Total
Seats Hill-Huntington	3	4	9	4	3	2	25

41. The satellite campus should receive the new instructor.

43. Banach school should receive the new instructor.

45. The hypothetical Delaware has one more and Virginia has one fewer than the actual. See Table below.

		45.	46.	47.	48.
State	Seats Actual	Seats Hamilton	Seats Adams	Seats Webster	Seats Hill-Huntington
VA	19	18	18	18	18
MA	14	14	14	14	14
PA	13	13	12	13	12
N.C.	10	10	10	10	10
N.Y.	10	10	10	10	10
MD	8	8	8	8	8
CT	7	7	7	7	7
S.C.	6	6	6	6	6
N.J.	5	5	5	5	5
N.H.	4	4	4	4	4
VT	2	2	3	2	3
GA	2	2	2	2	2
KY	2	2	2	2	2
RI	2	2	2	2	2
DE	1	2	2	2	2
Total	105	105	105	105	105

47. Virginia has one fewer and Delaware has one more than the actual. See table in answer 45.

6.3 Flaws of Apportionment

1. **a.** See table. **b.** Standard divisor: 0.77 **c.** See table.

d.

State	A	B	C	Total	Comments
Population (millions)	3.5	4.2	16.8	24.5	Part (a)
Std q (d = 0.77)	4.55	5.45	21.82	—	Part (c)
Lower q (d = 0.77)	4	5	21	30	Part (c)
Upper q (d = 0.77)	5	6	22	33	Part (c)
Modified lower q (d = 0.73)	4	5	23	32	d. Jefferson

e. Yes. C has 23 seats, which is neither upper nor lower quota.

3. **a.** See table. **b.** Standard divisor: 0.31 **c.** See table.

d.

State	A	B	C	Total	Comments
Population (millions)	3.5	4.2	16.8	24.5	Part (a)
Std q (d = 0.31)	11.29	13.55	54.19	—	Part (c)
Lower q (d = 0.31)	11	13	54	78	Part (c)
Upper q (d = 0.31)	12	14	55	81	Part (c)
Modified upper q (d = 0.317)	12	14	53	79	d. Adams

e. Yes. C has 53 seats, which is neither upper nor lower quota.

5. **a.** See table. **b.** Standard divisor: 0.21 **c.** See table. **d.** (Note: For answer, delete line 6, "Rounded q . . .".)

State	A	B	C	D	Total	Comments
Population (millions)	1.2	3.4	17.5	19.4	41.5	Part (a)
Std q (d = 0.21)	5.71	16.19	83.33	92.38	—	Part (c)
Lower q (d = 0.21)	5	16	83	92	196	Part (c)
Upper q (d = 0.21)	6	17	84	93	200	Part (c)
Rounded q (d = 0.21)	6	16	83	92	197	
Modified rounded q (d = 0.20715)	6	16	84	94	200	d. Webster

e. Yes. D has 94 seats, which is neither upper nor lower quota.

7. **a.** See table. **b.** Standard divisor: 0.21 **c.** See table.
d. (Note: For answer, delete line 6 and line 7, "Geometric mean . . ." and "Rounded q . . .")

State	A	B	C	D	Total	Comments
Population (millions)	1.2	3.4	17.5	19.4	41.5	Part (a)
Std q (d = 0.21)	5.71	16.19	83.33	92.38	—	Part (c)
Lower q (d = 0.21)	5	16	83	92	196	Part (c)
Upper q (d = 0.21)	6	17	84	93	200	Part (c)
Geometric mean	5.48	16.49	83.50	92.50	—	
Rounded q (d = 0.21)	6	16	83	92	197	
Modified rounded q (d_m = 0.2085)	6	16	84	94	200	d. H–H

e. Yes. D has 94 seats, which is neither upper nor lower quota.

9. a. See table. **b.** Standard divisor: 106.67 **c.** See table. **d.** New standard divisor: 105.79

	State	A	B	C	Total
a.	Population (thousands)	690	5700	6410	12800
c.	Seats Hamilton (d = 106.67)	7	53	60	120
d.	Seats Hamilton (d = 105.79)	6	54	61	121

e. Yes. State A has lost a seat at the expense of the two larger states even though its population didn't change.

11. a. See table. **b.** Standard divisor: 45.45 **c.** See table.

	State	A	B	C	Total
a.	Population (thousands)	1056	1844	2100	5000
c.	Seats Hamilton (d = 45.45)	23	41	46	110

d. Add 53 new seats.
e. New standard divisor: 45.64

	State	A	B	C	D	Total
a.	Population (thousands)	1056	1844	2100	2440	7440
e.	Seats Hamilton (d = 45.64)	23	40	46	54	163

f. Yes, apportionment changes. State B is altered.

13. a.

	Campus	A	B	C	Total
a.	2005 Specialists Hamilton	1	3	7	11
b.	2006 Specialists Hamilton	2	3	6	11
c.	% increase	5.1	10.6	6.25	
	Change in seats	+1	0	−1	

d. Yes. C lost a seat to A, but C grew at a faster rate.

Chapter 6 Review

1. a. 74 **b.** Beethoven wins. **c.** 34% **d.** Vivaldi wins. **e.** 59% **f.** Beethoven wins.
g. Beethoven received 197 points. **h.** There is a tie between Beethoven and Vivaldi. **i.** They each received 2 points.
3. See Table.

	State	A	B	C	Total
3.	Seats Hamilton	15	29	32	76
4.	Seats Jefferson	15	29	32	76
5.	Seats Adams	15	29	32	76
6.	Seats Webster	15	29	32	76
7.	Seats Hill-Huntington	15	29	32	76

5. See table above. **7.** See table above.

	Country	A	B	C	P	U	Total
8.	Seats Hamilton	13	3	6	2	1	25
9.	Seats Jefferson	14	3	5	2	1	25
10.	Seats Adams	13	3	5	2	2	25
11.	Seats Webster	14	3	5	2	1	25
12.	Seats Hill-Huntington	14	3	5	2	1	25

9. See table above. **11.** See table above. **13.** Napier school should get the new instructor.

15. a. See table. **b.** Standard divisor: 0.28 **c.** See table.

d. See the apportionment shown in the table. (Note: For answer, delete top part only of line 6, "Modified std....")

State	A	B	C	Total	Comments
Population (millions)	1.6	3.5	15.3	20.4	Part (a)
Std q ($d = 0.28$)	5.71	12.5	54.64	—	Part (c)
Lower q ($d = 0.28$)	5	12	54	71	Part (c)
Upper q ($d = 0.28$)	6	13	55	74	Part (c)
Modified std q **(d_m = 0.29)** Upper q	5.52 **6**	12.07 **13**	52.76 **53**	— **72**	Use $d_m > d$ **Adams**

e. Yes. C has 53 seats, which is neither an upper nor a lower quota.

17. a. See table. **b.** Standard divisor: 0.21 **c.** See table.

d. See the apportionment shown in the table. (Note: For answer, delete line 6 and line 7, "Geometric mean..." and "Rounded q...". Also, delete top part <u>only</u> of line 8, "Modified std q...)

State	A	B	C	D	Total	Comments
Population (millions)	1.100	3.500	17.600	19.400	41.6	Part (a)
Std q ($d = 0.21$)	5.24	16.67	83.81	92.38	—	Part (c)
Lower q ($d = 0.21$)	5	16	83	92	196	Part (c)
Upper q ($d = 0.21$)	6	17	84	93	200	Part (c)
Geometric mean ($d = 0.21$)	5.48	16.49	83.50	92.50	—	
Rounded q ($d = 0.21$)	5	17	84	92	198	Use $d_m < d$
Modified std q **(d_m = 0.208)** Rounded q	5.29 **5**	16.83 **17**	84.62 **85**	93.27 **94**	 **201**	Part (d) **H–H**

e. Yes. Neither C nor D has a number of seats that is either an upper or lower quota.

19. a. See table. **b.** Standard divisor: 45.45 **c.** See table.

	State	A	B	C	Total
a.	Population (thousands)	1057	1942	2001	5000
c.	Seats Hamilton ($d = 45.45$)	23	43	44	110

d. Add 53 seats.

e. New standard divisor: 45.71

	State	A	B	C	D	Total
a.	Population (thousands)	1057	1942	2001	2450	7450
e.	Seats Hamilton ($d = 45.71$)	23	42	44	54	163

f. Yes. The apportionment changed.

CHAPTER 7 Number Systems and Number Theory

7.1 Place Systems

1. $891 = 8 \cdot 10^2 + 9 \cdot 10^1 + 1 \cdot 10^0$ or $8 \cdot 10^2 + 9 \cdot 10 + 1$

3. $3{,}258 = 3 \cdot 10^3 + 2 \cdot 10^2 + 5 \cdot 10^1 + 8 \cdot 10^0$ or $3 \cdot 10^3 + 2 \cdot 10^2 + 5 \cdot 10 + 8$

5. $372_8 = 3 \cdot 8^2 + 7 \cdot 8^1 + 2 \cdot 8^0$ or $3 \cdot 8^2 + 7 \cdot 8 + 2$

7. $3592_{16} = 3 \cdot 16^3 + 5 \cdot 16^2 + 9 \cdot 16^1 + 2 \cdot 16^0$ or $3 \cdot 16^3 + 5 \cdot 16^2 + 9 \cdot 16 + 2$

9. $\text{ABCDE0}_{16} = 10 \cdot 16^5 + 11 \cdot 16^4 + 12 \cdot 16^3 + 13 \cdot 16^2 + 14 \cdot 16^1 + 0 \cdot 16^0$ or $10 \cdot 16^5 + 11 \cdot 16^4 + 12 \cdot 16^3 + 13 \cdot 16^2 + 14 \cdot 16$

11. $1011001_2 = 1 \cdot 2^6 + 0 \cdot 2^5 + 1 \cdot 2^4 + 1 \cdot 2^3 + 0 \cdot 2^2 + 0 \cdot 2^1 + 1 \cdot 2^0$ or $2^6 + 2^4 + 2^3 + 1$

13. 1324_2 does not exist because base two only uses numerals 0 and 1.

15. $5,32,85_{60}$ does not exist because base 60 only uses numerals 0 through 59.
17. $4312_5 = 4 \cdot 5^3 + 3 \cdot 5^2 + 1 \cdot 5^1 + 2 \cdot 5^0$ or $4 \cdot 5^3 + 3 \cdot 5^2 + 1 \cdot 5 + 2$
19. $123_4 = 1 \cdot 4^2 + 2 \cdot 4^1 + 3 \cdot 4^0$ or $1 \cdot 4^2 + 2 \cdot 4 + 3$
21. 250 **23.** 13,714 **25.** 11,259,360
27. 89 **29.** $5,32,85_{60}$ does not exist.
31. 582 **33.** 27 **35.** 965
37. 219 **39.** 11,982,839 **41.** 22,424
43. 3,732 **45.** 59,714 **47.** 136
49. 100101010_2 **51.** 1010101100000_2
53. 12_8 **55.** 266_8 **57.** 1010011101000010_2
59. 11101010110000_2 **61.** 14_{16}
63. BA_{16} **65. a.** 1010111010_2 **b.** 1272_8
67. 22_{16} **69.** 160_{16} **71.** 5270_8
73. 100035_8 **75.** 1,332 **77.** 119,665
79. $6,4_{60}$ **81.** $36,4,5_{60}$ **83.** 1162_7
85. 221_5 **87. a.** The base is 3. **b.** The base is 5.
89. a. The base is 8. **b.** The base is 9.
91. a. 0, 1, 2, 3, 4 **b.** 5^0 5^1 5^2 5^3
c. 0_5 1_5 2_5 3_5 4_5
10_5 11_5 12_5 13_5 14_5
20_5 21_5 22_5 23_5 24_5
30_5 31_5 32_5 33_5 34_5
40_5 41_5 42_5 43_5 44_5
93. a. 0, 1, 2, 3, 4, 5, 6, 7, 8, 9, A **b.** 11^0 11^1 11^2 11^3
c. 0_{11} 1_{11} 2_{11} 3_{11} 4_{11} 5_{11} 6_{11} 7_{11} 8_{11} 9_{11} A_{11}
10_{11} 11_{11} 12_{11} 13_{11} 14_{11} 15_{11} 16_{11} 17_{11} 18_{11} 19_{11} $1A_{11}$
20_{11} 21_{11} 22_{11}
95. 16-bit color has $2^{16} = 65,536$ shades, and 24-bit color has $2^{24} = 16,777,216$ shades.

37.

	1_{14}	2_{14}	3_{14}	4_{14}	5_{14}	6_{14}	7_{14}	8_{14}	9_{14}	A_{14}	B_{14}	C_{14}	D_{14}	10_{14}	11_{14}
1_{14}	1_{14}	2_{14}	3_{14}	4_{14}	5_{14}	6_{14}	7_{14}	8_{14}	9_{14}	A_{14}	B_{14}	C_{14}	D_{14}	10_{14}	11_{14}
2_{14}	2_{14}	4_{14}	6_{14}	8_{14}	A_{14}	C_{14}	10_{14}	12_{14}	14_{14}	16_{14}	18_{14}	$1A_{14}$	$1C_{14}$	20_{14}	22_{14}
3_{14}	3_{14}	6_{14}	9_{14}	C_{14}	11_{14}	14_{14}	17_{14}	$1A_{14}$	$1D_{14}$	22_{14}	25_{14}	28_{14}	$2B_{14}$	30_{14}	33_{14}
4_{14}	4_{14}	8_{14}	C_{14}	12_{14}	16_{14}	$1A_{14}$	20_{14}	24_{14}	28_{14}	$2C_{14}$	32_{14}	36_{14}	$3A_{14}$	40_{14}	44_{14}
5_{14}	5_{14}	A_{14}	11_{14}	16_{14}	$1B_{14}$	22_{14}	27_{14}	$2C_{14}$	33_{14}	38_{14}	$3D_{14}$	44_{14}	49_{14}	50_{14}	55_{14}
6_{14}	6_{14}	C_{14}	14_{14}	$1A_{14}$	22_{14}	28_{14}	30_{14}	36_{14}	$3C_{14}$	44_{14}	$4A_{14}$	52_{14}	58_{14}	60_{14}	66_{14}
7_{14}	7_{14}	10_{14}	17_{14}	20_{14}	27_{14}	30_{14}	37_{14}	40_{14}	47_{14}	50_{14}	57_{14}	60_{14}	67_{14}	70_{14}	77_{14}
8_{14}	8_{14}	12_{14}	$1A_{14}$	24_{14}	$2C_{14}$	36_{14}	40_{14}	48_{14}	52_{14}	$5A_{14}$	64_{14}	$6C_{14}$	76_{14}	80_{14}	88_{14}
9_{14}	9_{14}	14_{14}	$1D_{14}$	28_{14}	33_{14}	$3C_{14}$	47_{14}	52_{14}	$5B_{14}$	66_{14}	71_{14}	$7A_{14}$	85_{14}	90_{14}	99_{14}
A_{14}	A_{14}	16_{14}	22_{14}	$2C_{14}$	38_{14}	44_{14}	50_{14}	$5A_{14}$	66_{14}	72_{14}	$7C_{14}$	88_{14}	94_{14}	$A0_{14}$	AA_{14}
B_{14}	B_{14}	18_{14}	25_{14}	32_{14}	$3D_{14}$	$4A_{14}$	57_{14}	64_{14}	71_{14}	$7C_{14}$	89_{14}	96_{14}	$A3_{14}$	$B0_{14}$	BB_{14}
C_{14}	C_{14}	$1A_{14}$	28_{14}	36_{14}	44_{14}	52_{14}	60_{14}	$6C_{14}$	$7A_{14}$	88_{14}	96_{14}	$A4_{14}$	$B2_{14}$	$C0_{14}$	CC_{14}
D_{14}	D_{14}	$1C_{14}$	$2B_{14}$	$3A_{14}$	49_{14}	58_{14}	67_{14}	76_{14}	85_{14}	94_{14}	$A3_{14}$	$B2_{14}$	$C1_{14}$	$D0_{14}$	DD_{14}
10_{14}	10_{14}	20_{14}	30_{14}	40_{14}	50_{14}	60_{14}	70_{14}	80_{14}	90_{14}	$A0_{14}$	$B0_{14}$	$C0_{14}$	$D0_{14}$	100_{14}	110_{14}
11_{14}	11_{14}	22_{14}	33_{14}	44_{14}	55_{14}	66_{14}	77_{14}	88_{14}	99_{14}	AA_{14}	BB_{14}	CC_{14}	DD_{14}	110_{14}	121_{14}

39. 133_5 Rem 11_5
41. $B9_{14}$ Rem 48_{14}

7.2 Addition and Subtraction in Different Bases

1. 12_8 **3.** 11_8 **5.** 10_{16}
7. 14_{16} **9.** 10_2 **11.** 110_2
13. a. 105_8 **15. a.** 751_8 **17. a.** 124_{16}
19. a. 1889_{16} **21. a.** 1011_2 **23. a.** 11010011_2
25. 4_8 **27.** 142_{16} **29.** 0011_2 or 11_2
31. a. 7_8 **33. a.** AE_{16} **35. a.** 00101_2 or 101_2
37. a. 43_5

7.3 Multiplication and Division in Different Bases

1. 74_8 **3.** $3C5_{16}$
5. 1111_2 **7.** 2122_4
9. a. 151_8 **11. a.** 2442_{16}
13. a. 100011110_2 **15. a.** 1313_5
17. 23_8 **19. a.** $6F_{16}$
21. a. 4_8 **23. a.** 10_8 Rem 3_8
25. a. 1_8 Rem 125_8 **27. a.** $22C_{16}$ Rem 11_{16}
29. a. 39_{16} Rem $1B3_{16}$ **31. a.** 10001_2 Rem 1_2
33. a. 1001_2 Rem 11_2
35.

	1_5	2_5	3_5	4_5	10_5	11_5
1_5	1_5	2_5	3_5	4_5	10_5	11_5
2_5	2_5	4_5	11_5	13_5	20_5	22_5
3_5	3_5	11_5	14_5	22_5	30_5	33_5
4_5	4_5	13_5	22_5	31_5	40_5	44_5
10_5	10_5	20_5	30_5	40_5	100_5	110_5
11_5	11_5	22_5	33_5	44_5	110_5	121_5

7.4 Prime Numbers and Perfect Numbers

1. a. $2 \cdot 3 \cdot 7$ **b.** composite
3. a. 23 **b.** prime
5. a. $2 \cdot 3^3$ **b.** composite
7. composite **9.** composite
11. prime
13. The prime numbers are 29, 31, 37, 41, 43, 47.
15. Since $17 = 17 \cdot 1$, it is prime.
17. Since $65 = 5(13)$, it is composite.
19. Since $\frac{15}{3} = 5$, it is composite.
21. Since $\frac{511}{7} = 73$, it is composite.
23. The proper factors are 1, 2, 3, 6, 7, 14, 21 .
25. The proper factors are 1, 2, 3, 6, 9, 18, 27 .
27. Since $1 + 2 + 3 + 6 + 7 + 14 + 21 = 54 > 42$, it is abundant.
29. Since $1 + 2 + 3 + 6 + 9 + 18 + 27 = 66 > 54$, it is abundant.
31. The proper factor is 1 .
Since $1 < 61$, it is deficient.
33. The proper factors are 1, 2, 31 .
Since $1 + 2 + 31 = 34 < 62$, it is deficient.
35. a. 13 **b.** 4,096
c. 8,191 **d.** 33,550,336
37. a. 19 **b.** 262,144
c. 524,287 **d.** $1.374386913 \times 10^{11}$
39. a. $(2^{127-1})(2^{127} - 1)$ **b.** 1.45×10^{76}
41. The private key is 5, 7.
43. The private key is 11, 13.
45. a. $6 = 110_2$
$28 = 11100_2$
$496 = 111110000_2$
$8128 = 1111111000000_2$
b. The nth perfect number is a sequence of m ones, where m is the nth prime, followed by $2(n - 1)$ zeros, $n \neq 1$. (When $n = 1$, there is 1 zero.)

7.5 Fibonacci Numbers and the Golden Ratio

1. After six years, there are 13 total, and 5 newly born.
3. 21, 34, 55, 89, 144, 233, 377, 610, 987, 1597
5. 8 great-great-great grandparents
7. Answers will vary.
9.

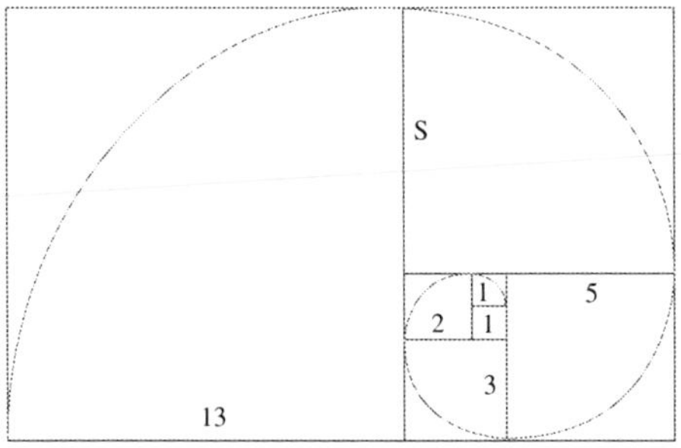

11. a. $n = 30$, 832,040
$n = 40$, 102,334,155
b. Possible answer: It allows you to find a Fibonacci number without listing all the preceding numbers.
c. Possible answer: It is an unnecessarily lengthy formula when dealing with the first numbers, say the first 24 or so numbers.
13. a. $\dfrac{a+b}{a} = \dfrac{a}{b}$
b. $ab \cdot \dfrac{a+b}{a} = ab \cdot \dfrac{a}{b}$
$b(a + b) = a^2$
c. $ab + b^2 = a^2$
d. $\dfrac{ab + b^2}{b^2} = \dfrac{a^2}{b^2}$
$\dfrac{ab}{b^2} + \dfrac{b^2}{b^2} = \dfrac{a^2}{b^2}$
$\dfrac{ab}{b^2} + 1 = \dfrac{a^2}{b^2}$
e. $\dfrac{a}{b} + 1 = \left(\dfrac{a}{b}\right)^2$
f. Let $\dfrac{a}{b} = x$.
$x + 1 = x^2$
g. $x^2 - x - 1 = 0$
$x = \dfrac{1 + \sqrt{5}}{2}$ is the golden ratio.
h. x is a ratio of lengths.
15. Answers may vary. **17.** Answers may vary.
19. 1.61803398896 is an accurate approximation to the golden ratio up to nine decimal places.
21. Answers may vary.
23. Answers may vary. Possible answer: the heads in the bottom right exhibit a golden ratio.
25. Answers will vary.

Chapter 7 Review

1. $5372 = 5 \cdot 10^3 + 3 \cdot 10^2 + 7 \cdot 10^1 + 2 \cdot 10^0$ or $5 \cdot 10^3 + 3 \cdot 10^2 + 7 \cdot 10 + 2$
3. $325_8 = 3 \cdot 8^2 + 2 \cdot 8^1 + 5 \cdot 8^0$ or $3 \cdot 8^2 + 2 \cdot 8 + 5$
5. 390_8; does not exist because base eight only uses numerals 0 through 7.
7. $905_{16} = 9 \cdot 16^2 + 0 \cdot 16^1 + 5 \cdot 16^0$ or $9 \cdot 16^2 + 5$
9. $ABC_{16} = 10 \cdot 16^2 + 11 \cdot 16^1 + 12 \cdot 16^0$ or $10 \cdot 16^2 + 11 \cdot 16 + 12$
11. $11011001_2 = 1 \cdot 2^7 + 1 \cdot 2^6 + 0 \cdot 2^5 + 1 \cdot 2^4 + 1 \cdot 2^3 + 0 \cdot 2^2 + 0 \cdot 2^1 + 1 \cdot 2^0$ or $2^7 + 2^6 + 2^4 + 2^3 + 1$
13. 101112_2 does not exist because base two only uses numerals 0 and 1.
15. $39{,}22{,}54_{60} = 39 \cdot 60^2 + 22 \cdot 60^1 + 54 \cdot 60^0$ or $39 \cdot 60^2 + 22 \cdot 60 + 54$
17. $25{,}44{,}34_{60} = 25 \cdot 60^2 + 44 \cdot 60^1 + 34 \cdot 60^0$ or $25 \cdot 60^2 + 44 \cdot 60 + 34$
19. 332 **21.** 53 **23.** 48,282 **25.** 56
27. 940 **29.** 1002_8 **31.** $20EE_{16}$ **33.** $10{,}54{,}47_{60}$
35. 10111011100011_2 **37.** 335051_7 **39.** The base is 6.
41. The base is 9.
43. a. 134_8 **45. a.** 1011_2
47. a. 144_{16} **49. a.** $1{,}18{,}18{,}17_{60}$
51. a. $6\ 7_8$ **53. a.** 0011_2 or 11_2
55. a. $98D_{16}$ **57. a.** $11{,}49{,}11_{60}$
59. a. 2024_8 **61. a.** 11110_2
63. a. 6616_{16} **65. a.** 3044_5
67. a. 23_8 Rem 3_8 **69. a.** 10_2 Rem 11_2
71. a. $5F_{16}$ Rem 3_{16} **73. a.** 11011_2
75. a. $3 \cdot 11$ **b.** composite
77. a. 41 **b.** prime
79. The prime numbers are 53 and 59.
81. $n = 2, 2^n + 1 = 2^2 + 1 = 4 + 1 = 5$; prime
$n = 3, 2^n + 1 = 2^3 + 1 = 8 + 1 = 9$; composite, $9 = 3 \cdot 3$

83. $2^n - 1 = 2^3 - 1 = 8 - 1 = 7$; prime because $7 = 1 \cdot 7$.
85. The proper factors are 1, 2, 5, 7, 10, 14, 35.
87. The proper factors are 1, 2, 4, 5, 10.
Since $1 + 2 + 4 + 5 + 10 = 22 > 20$, it is abundant.
89. The proper factors are 1 and 7.
Since $1 + 7 = 8 < 49$, it is deficient.
91. **a.** $(2^{17-1})(2^{17} - 1)$ **b.** 8.59×10^9
93. 1, 1, 2, 3, 5, 8, 13, 21, 34, 55, 89, 144, 233, 377, 610, 987, 1597, 2584, 4181, 6765
95. 5 great-great grandparents
97. After 4 years, there are 5 total, and 2 newly born.
99. **a.** $n = 25$: 75,025
$n = 26$: 121,393
b. 196,418
101. The rectangle's length is about 1.6 times its width.
103. Answers may vary.
105. Answers will vary.
107. Answers may vary. Possible answer: 0 represents none in that place value.
109. Answers will vary.
111. Answers may vary. Possible description: 321−74
Step 1: Set 321 on the first (ones) and second (tens) and third (hundreds) rods.
Step 2: To subtract 74, first consider subtracting the 7 in the tens place. Subtract 1 bead from the hundreds rod and add 7's complement (3) to the tens rod.
Step 3: Now subtract the 4 in the ones place. Once again, subtract 1 bead from the tens rod and add 4's complement (6) to the ones rod.
Step 4: Read the result, 247.
113. His first book was on Hindu numerals. It introduced Europe to the simpler calculation techniques of the Hindu system, including the multiplication algorithm and the division algorithm that we use today. His second book was on algebra. It discusses linear and quadratic equations.
115. Answers may vary. Possible answer: They believed that numbers had mystical properties.
117. Answers may vary. Possible answer: He was trying to answer questions in nature.

CHAPTER 8 Geometry

8.1 Perimeter and Area

1. 16.1 square centimeters **3.** 12.5 square inches
5. 21.7 square feet **7.** 176 square meters
9. **a.** 33.2 square inches **b.** 20.4 inches
11. **a.** 30 square meters **b.** 30 meters
13. **a.** 27.7 square meters **b.** 24 meters
15. **a.** 80 square feet **b.** 44 feet
17. **a.** 3927.0 square yards **b.** 257.1 yards
19. **a.** 54 square yards **b.** 30 yards
21. **a.** 49.8 square feet **b.** 28.9 feet
23. **a.** 1963.5 square feet **b.** 863.9 square feet
c. 3,572.6 square feet
25. **a.** 3,136 square inches **b.** 21.8 square feet
27. 8.0 square feet **29.** 3.9 miles
31. 8 feet
33. **a.** 5,256.6 square yards **b.** 325.7 yards
35. You need 7 cans.
37. The large pizza is the best deal.
39. 0.25 A **41.** 640 A
43. 525.03 square miles
45. **a.** 36 million miles **b.** 0.39 AU
c. The distance between Mercury and the earth is 0.39 times the distance between the earth and sun.
47. **a.** 27,200,000 AU **b.** 431 ly **c.** 132 pc
49. **c.** **51.** **c.** **53.** **e.** **55.** **b.**

8.2 Volume and Surface Area

1. **a.** 38.22 cubic meters **b.** 72.94 square meters
3. **a.** 785.40 cubic inches **b.** 471.24 square inches
5. **a.** 2.81 cubic inches **b.** 9.62 square inches
7. 16.76 cubic feet **9.** 21.33 cubic feet
11. 36 cubic feet **13.** 47.12 cubic feet
15. **a.** 136.5 cubic inches **b.** 151 square inches
17. 17,802.36 cubic feet
19. **a.** Hardball: 12.31 cubic inches
Softball: 29.18 cubic inches
Volume of softball is 137% more than that of hardball.
b. Hardball: 25.78 square inches
Softball: 45.83 square inches
Surface area of softball is 78% more than that of hardball.
21. about 49 moons **23.** about 147 Plutos
25. You did not get an honest deal. You should have paid $118.75.
27. 36.82 cubic inches **29.** $175.20
31. 91,445,760 cubic feet **33.** 301.59 cubic feet
35. **e.** **37.** **a.** **39.** **c.** **41.** **b.**

8.3 Egyptian Geometry

1. This is a right triangle. **3.** This is not a right triangle.
5. The regular pyramid (part **b**) holds more.
7. 36; 18; 9; 36; 18; 9; 63; 9; 3; 63; 189; 189
9. square of side 21 **11.** $\pi = \frac{49}{16} = 3.0625$
13. **a.** 28.4444 square palms **b.** 28.2743 square palms
c. $0.00623 \approx 0.6\%$
15. **a.** 113.7778 cubic palms **b.** 113.0973 cubic palms
c. $0.00602 \approx 0.6\%$
17. 11 cubits
19. 1 khar is larger. **21.** 1.5 setats
23. **a.** 2 khet **b.** 200 cubits
25. **a.** 960 khar **b.** 954.2588 khar
c. $0.00602 \approx 0.6\%$

8.4 The Greeks

1. $x = 5$
$y = 60$
3. $x = 64.6$
$y = 2.5$
5. 36 feet
9. 5.3 feet, (rounded down so the object will fit)
11. 7.5 inches
13.

1.	$AD = CD$	Given
2.	$AB = CB$	Given
3.	$DB = DB$	Anything is equal to itself.
4.	$\Delta DBA \cong \Delta DBC$	SSS
5.	$\angle DBA = \angle DBC$	Corresponding parts of congruent triangles are equal.

15. **1.** $AD = BD$ Given
2. $\angle ADC = \angle BDC$ Given
3. $DC = DC$ Anything is equal to itself.
4. $\Delta ACD \cong \Delta BCD$ SAS
5. $AC = BC$ Corresponding parts of congruent triangles are equal.

17. **1.** $AE = CE$ Given
2. $AB = CB$ Given
3. $BE = BE$ Anything is equal to itself.
4. $\Delta ABE \cong \Delta CBE$ SSS
5. $\angle ABE = \angle CBE$ Corresponding parts of congruent triangles are equal.
6. $BD = BD$ Anything is equal to itself.
7. $\Delta ADB \cong \Delta CDB$ SAS
8. $\angle ADB = \angle CDB$ Corresponding parts of congruent triangles are equal.

19. **a.** $\pi \approx 4\sqrt{2 - \sqrt{2}} = 3.061467459$
b. $\pi \approx 8(\sqrt{2} - 1) = 3.313708499$

21. **a.** $\pi \approx 8\sqrt{2 - \sqrt{2 + \sqrt{2}}} = 3.121445152$

b. $\pi \approx 8\left(\dfrac{2\sqrt{2 - \sqrt{2}}}{2 + \sqrt{2 + \sqrt{2}}}\right) = 3.182597878$

23. $\dfrac{V_{\text{sphere}}}{V_{\text{cylinder}}} = \dfrac{\frac{4}{3}\pi r^3}{\pi r^2(2r)} = \dfrac{\frac{4}{3}}{2} = \dfrac{2}{3}$

8.5 Right Triangle Trigonometry

1. $x = 3$, $y = 3\sqrt{3}$, $\theta = 60°$
3. $x = 14$, $y = 7\sqrt{3}$, $\theta = 30°$
5. $y = 3\sqrt{2}$, $x = 3$, $\theta = 45°$
7. $x = \frac{7\sqrt{2}}{2}$, $y = \frac{7\sqrt{2}}{2}$, $\theta = 45°$
9. $B = 53°$, $c = 19.9$, $b = 15.9$
11. $B = 35.7°$, $c = 9.6$, $a = 7.8$
13. $A = 40.1°$, $b = 0.70$, $a = 0.59$
15. $A = 80.85°$, $c = 1{,}565.9$, $b = 249$
17. $B = 36.875°$, $a = 43.52$, $b = 32.64$
19. $A = 85.00°$, $c = 11.497$, $a = 11.453$
21. $b = 17.4$, $A = 40.7°$, $B = 49.3°$
23. $c = 9.2$, $A = 40.6°$, $B = 49.4°$
25. $a = 0.439$, $A = 74.4°$, $B = 15.6°$
27. **a.** 64.0 feet **b.** 92.3 feet
29. 150 feet
31. 18.4°
33. **a.** 176.1 feet **b.** 26.7 feet
35. 26.2 feet
37. 167 feet
39. 630 feet
41. 1,454 feet
43. 24.6 feet
45. 0.5 pc
47. **a.** 185,546 AU **b.** 1.11 seconds
49. **e.**
51. **d.**

8.6 Linear Perspective

1. **a.** one-point perspective **b.** above
c. approximately (17, 14) **d.** $y = 14$

3. **a.** two-point perspective **b.** below
c. approximately (1, 2) and (18, 2) **d.** $y = 2$

5. Answers will vary.
7. Anwers will vary.
9. Answers will vary.
11. Answers may vary.
a. One-point perspective; there is one vanishing point and the wall faces are parallel to the surface of the painting.
b. Between the two figures at the center of the last arch (Plato and Aristotle). Raphael might have chosen that point to put the focus on Plato and Aristotle.
c. In the front, a man is leaning against a small table. Two-point perspective was used.
d. The arch in the very front has its own vanishing point, directly below Plato and Aristotle. Raphael might have done this to make the front arch seem separate from the rest of the painting, thus giving the work more three-dimensionality
e. Yes, the tiles in the front.
f. Yes, the tiles.
g. Albertian grid is used on the tiles.
h. Yes, along the row of people, which includes Plato and Aristotle.

13. Answers may vary.
a. Both one-point and two-point perspective. The work has two parts. The left side includes the big arch and uses one-point perspective. The right side includes the two buildings, and uses two-point perspective.
On the floor at the very bottom of the painting, in the middle (horizontally). It is between the two parts, but not on either part. This unites the two separate parts.
c. No
d. The left side's one vanishing point is on the floor at the very bottom of the painting, in the middle (horizontally). The right part has two vanishing points, one of which is off the work. Their visible vanishing point is the same as the large arch's vanishing point. All vanishing points are on the same horizon.
e. Yes, the tiles on the ceiling.
f. The tiles on the ceiling are somewhat like pavement
g. On the tiles in the ceiling.
h. Yes, it is the floor.

15. Answers may vary.
a. One-point perspective; there is one vanishing point
b. The vanishing point appears to be at the small grove of trees directly above the walking people. Pissaro might have chosen that point to emphasize the distance between the travelers and the vanishing point, thus giving the work more three-dimensionality.
c. No
d. No
e. Yes, the houses get smaller and smaller.
f. No
g. No
h. Yes, horizon is along the tops of trees and buildings in the distance.

17. Answers may vary.
a. One-point perspective; there is one vanishing point and the walls are parallel to the surface of the painting.
b. The vanishing point is on the column that separates the painting's two halves. This serves to emphasize this separation.

c. No
d. No, There are two separate parts, the left side of the column and the right side. They share the same vanishing point.
e. Yes, the tiles in the ceiling and on the floor.
f. Yes, both sets of tiles.
g. Albertian grid is used on both sets of tiles.
h. No

19. Answers may vary.
a. One-point perspective; there is one vanishing point.
b. The vanishing point is at the left end of the line of boys. Homer might have chosen that point to give the sense of movement of the 'whip.'
c. No
d. No
e. Yes, the boys get smaller
f. No
g. No
h. Yes, the horizon is at the edge of the level ground.

21. Eakins drew center lines on the horizontal and vertical. Directly below that intersection, he placed the vanishing point. An Albertian grid was used to draw the boat. Eakins was so accurate that scholars have been able to determine the precise length of the boat and to pinpoint the exact time of day to 7:20 p.m. http://www.philamuseum.org/micro_sites/exhibitions/eakins/1872/main_frameset.html

23. Answers may vary.
a. One-point perspective; there is one vanishing point.
b. The vanishing point appears to be off center at the height of the people's heads. Sargent might have chosen that point to give a sense of the length of the street that the woman walked.
c. No
d. No, there is a central vanishing point.
e. Yes, on the street pavement and the windows.
f. Yes, on the street.
g. Albertian grid is used on the street pavements.
h. No

25. Answers may vary.
a. Three-point perspective portrays three dimensions.
b. Near the left roof, near the right tower, and below the etching.
c. No
d. See (b).
e. All of the windows, and the marching people
f. No
g. No.
h. No.

8.7 Conic Sections and Analytic Geometry

1. Center (0, 0)
$r = 1$

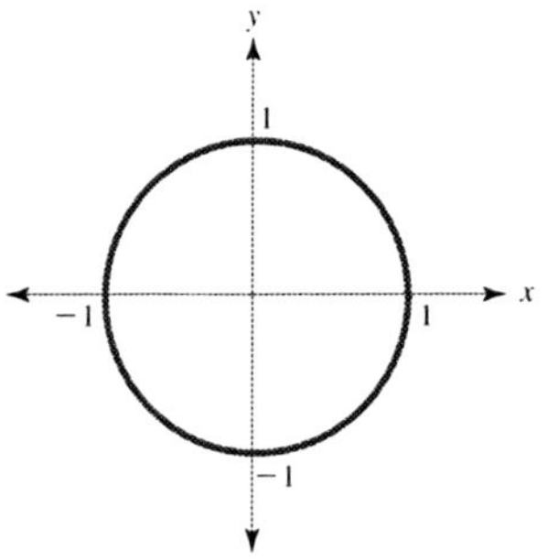

3. Center (2, 0)
$r = 3$

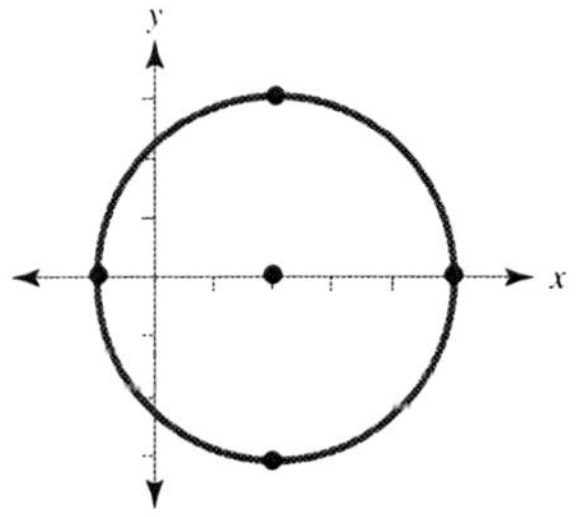

5. Center $(5, -2)$
$r = 4$

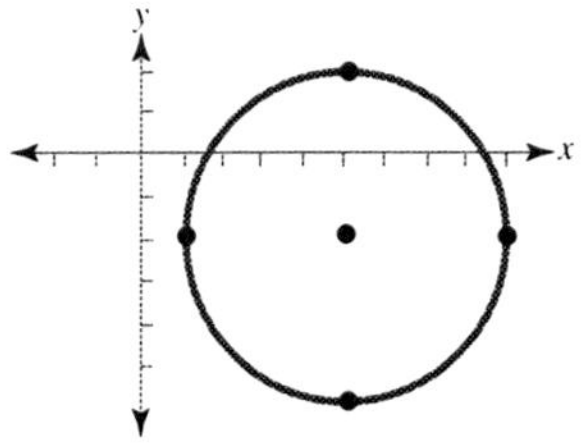

7. Center (5, 5)
$r = 5$

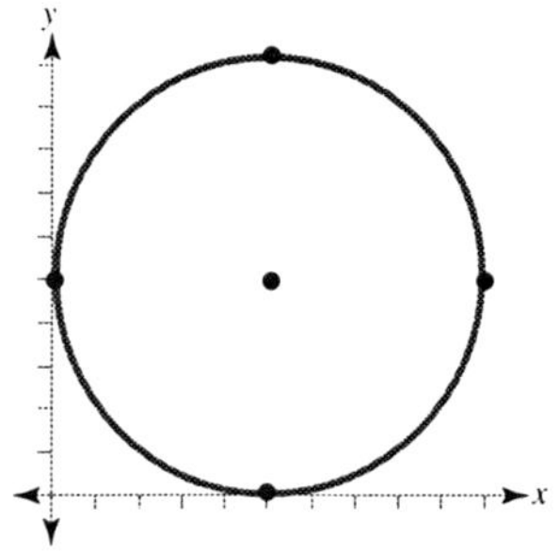

9. Focus $(0, \frac{1}{4})$

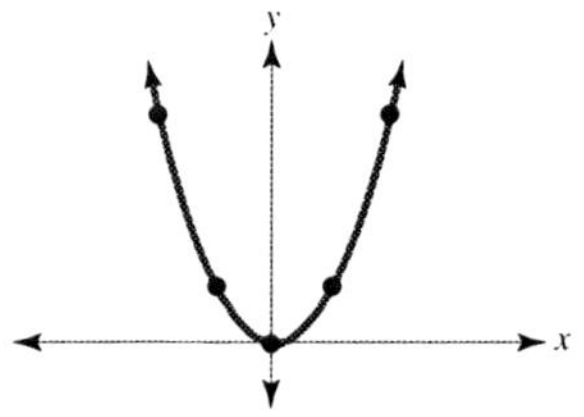

11. Focus $(0, \frac{1}{2})$

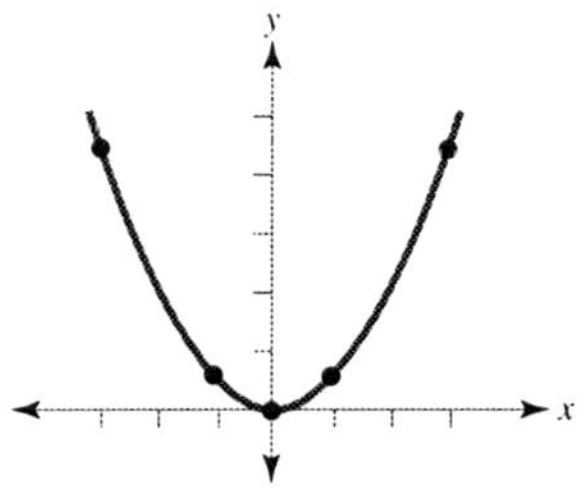

13. The water container should be placed 2.9 ft above the bottom of the dish.

15. The light bulb should be located $\frac{9}{16}$ inch above the bottom of the reflector.

17. Foci $(-\sqrt{5}, 0)$ and $(\sqrt{5}, 0)$

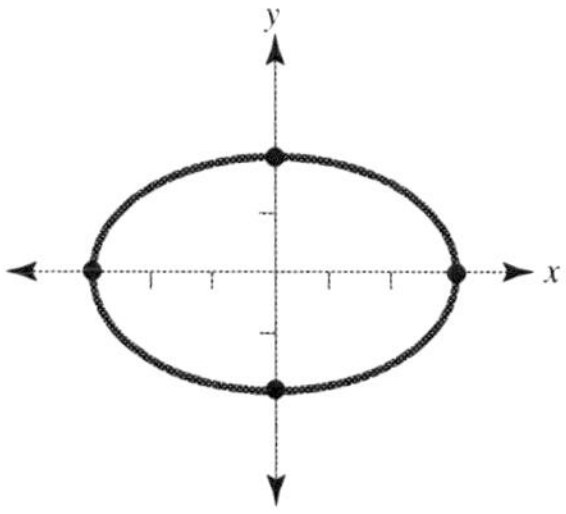

19. Foci $(0, -\sqrt{21})$ and $(0, \sqrt{21})$

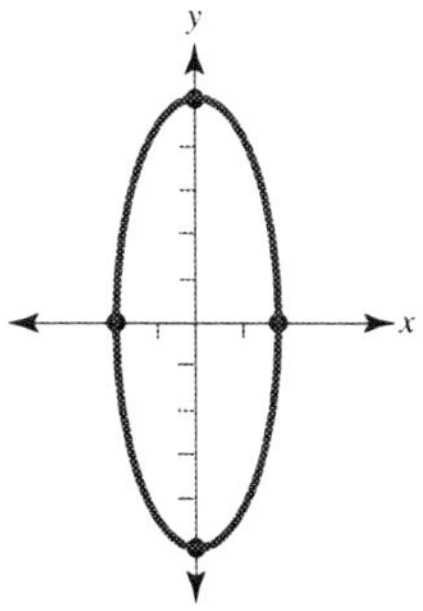

21. Foci $(0, -\sqrt{7})$ and $(0, \sqrt{7})$

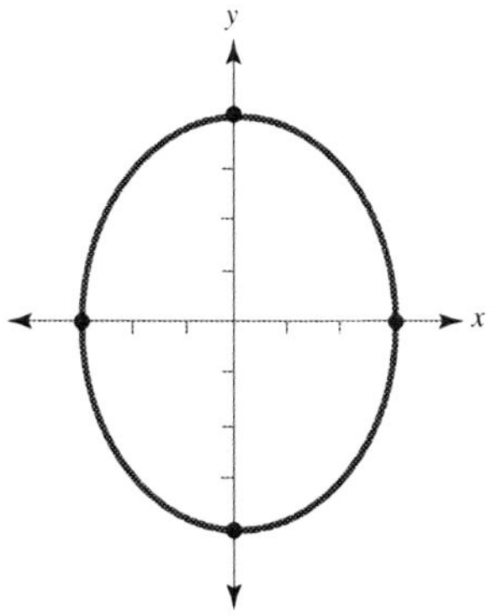

23. Foci $(1, 2 - \sqrt{3})$ and $(1, 2 + \sqrt{3})$

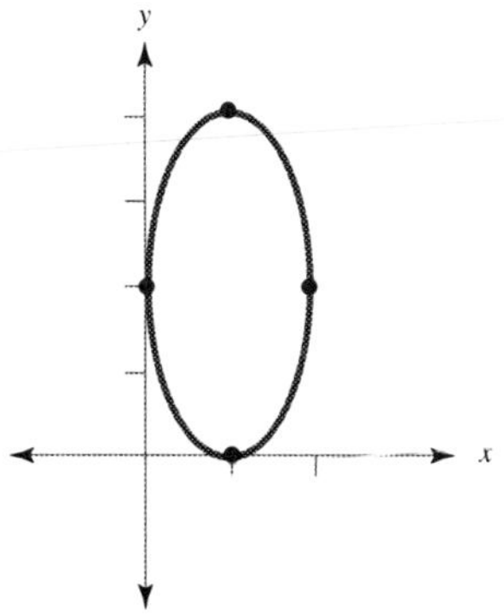

25. They should stand at the foci, which are located 9 ft from the center, in the long direction.

27. $\dfrac{x^2}{92.955^2} + \dfrac{y^2}{92.942^2} = 1$

29. a. Foci $(-\sqrt{2}, 0)$ and $(\sqrt{2}, 0)$

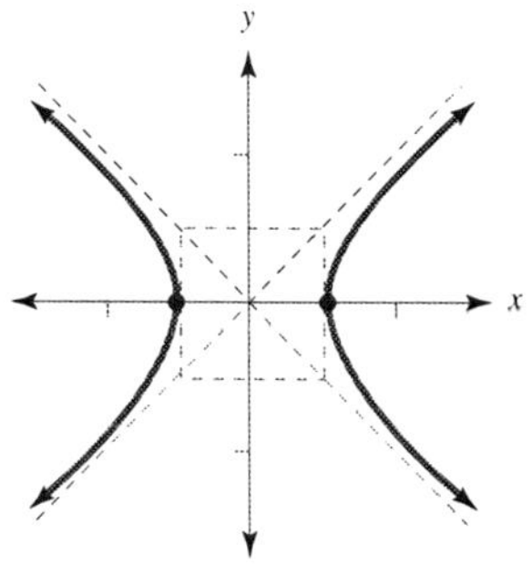

b. Foci $(0, -\sqrt{2})$ and $(0, \sqrt{2})$

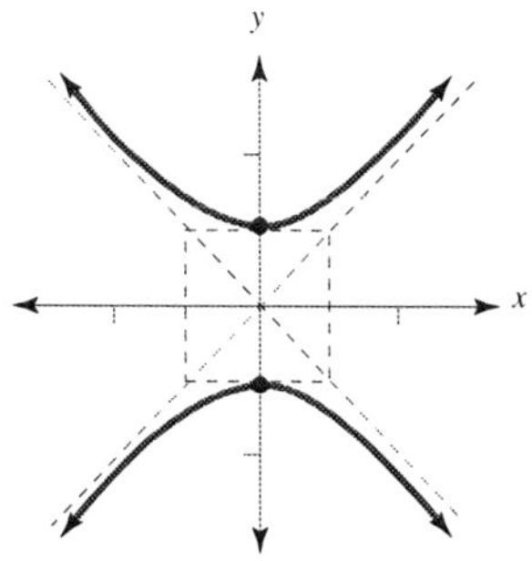

31. Foci $(-\sqrt{13}, 0)$ and $(\sqrt{13}, 0)$

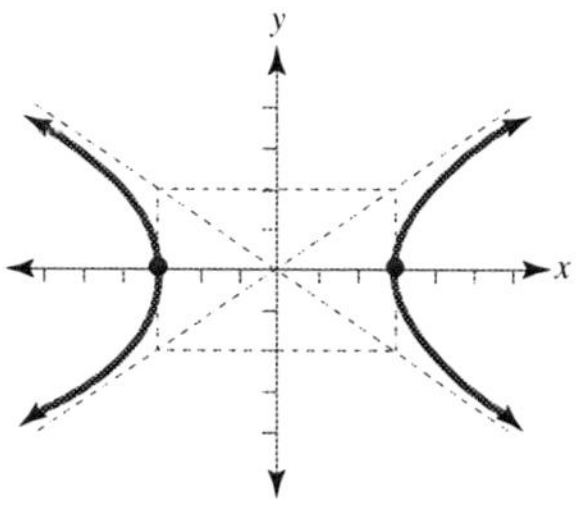

33. Foci $(0, -\sqrt{29})$ and $(0, \sqrt{29})$

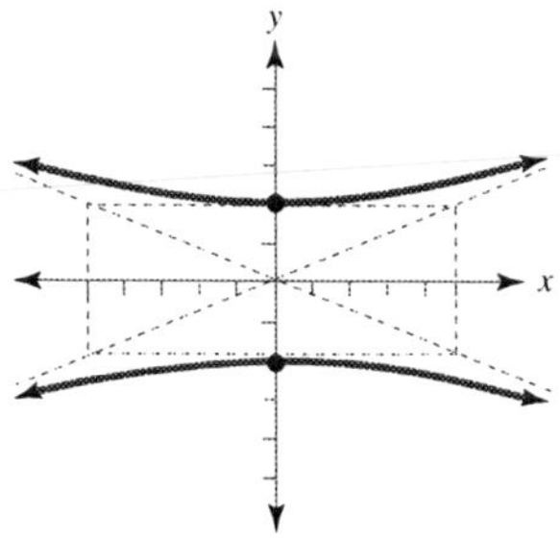

35. Foci $(3 - \sqrt{5}, 0)$ and $(3 + \sqrt{5}, 0)$

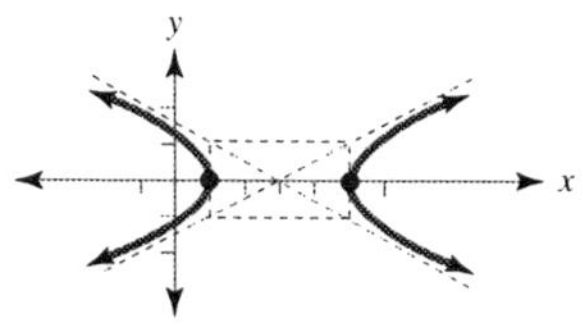

8.8 Non-Euclidean Geometry

1. One
3. Zero or one
5. Zero or one
7. None
9. Zero, one, two, or three
11. Two
13. Infinitely many
15. Zero or one
17. Zero or one

8.9 Fractal Geometry

1. a.

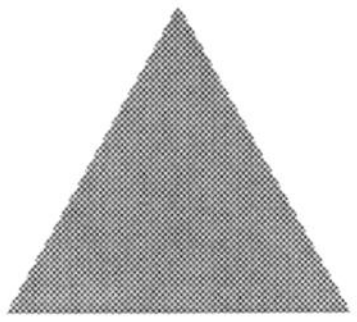

b.

3.

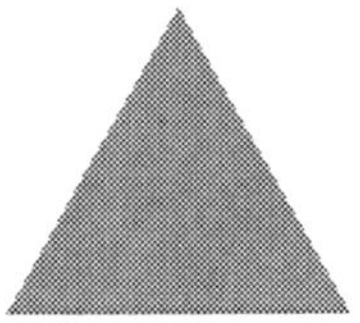

step 1

step 2

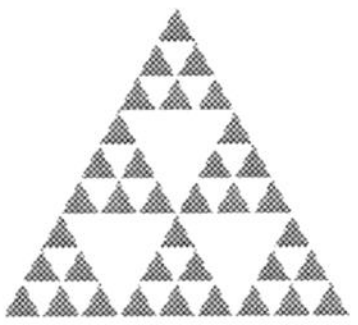

step 3

5. $d = 2$
7. $d = \frac{1}{3}$
9. $d \approx 1.6$ The dimension of the Sierpinski gasket is larger than the dimension of a circle, but smaller than the dimension of a square. It's not a two-dimensional as a regular triangle is.
11. $d \approx 1.6$ The dimension of the Mitsubishi gasket is larger than the dimension of a circle, but smaller than the dimension of a square. It's not a two-dimensional as a regular triangle is.
13. $d = 1.5$ The dimension of the square snowflake is larger than the dimension of a circle, but smaller than the dimension of a square. It's not two-dimensional.
15. $d = 3$, the same dimension that we observed before.
17. $d = 2$. Also, using the formula $s^d = n$: solve $3^d = 9$. Again, $d = 2$.

8.10 The Perimeter and Area of a Fractal

1. a. In step 1, there is only one triangle. If each side is 1 ft. in length, then the perimeter is 3 ft. In step 2, the original triangle is modified by removing a center triangle. This results in three smaller triangles. Each side of the original triangle now has two triangles with sides of length $\frac{1}{2}$ ft. Thus, the perimeter of one of these smaller triangles is $\frac{1}{2} + \frac{1}{2} + \frac{1}{2} = 3 \cdot \frac{1}{2} = \frac{3}{2}$ ft. Since there are three such triangles, the total perimeter is $3 \cdot \frac{3}{2} = \frac{9}{2}$ ft.

b.

Step	Number of Triangles	Length of Each Side (feet)	Perimeter of One Triangle (feet)	Total Perimeter of All Triangles (feet)
1	1	1	3	3
2	3	$\frac{1}{2}$	$3 \cdot \frac{1}{2} = \frac{3}{2}$	$3 \cdot \frac{3}{2} = \frac{9}{2}$
3	9	$\frac{1}{4}$	$3 \cdot \frac{1}{4} = \frac{3}{4}$	$9 \cdot \frac{3}{4} = \frac{27}{4}$
4	27	$\frac{1}{8}$	$3 \cdot \frac{1}{8} = \frac{3}{8}$	$27 \cdot \frac{3}{8} = \frac{81}{8}$
5	81	$\frac{1}{16}$	$3 \cdot \frac{1}{16} = \frac{3}{16}$	$81 \cdot \frac{3}{16} = \frac{243}{16}$
6	243	$\frac{1}{32}$	$3 \cdot \frac{1}{32} = \frac{3}{32}$	$243 \cdot \frac{3}{32} = \frac{729}{32}$

c. 3. Each triangle is divided into four smaller triangles, one of which is deleted.
d. $\frac{1}{2}$. When the middle triangle is taken out, it creates two triangles along each side.
e. $\frac{1}{2}$. Since each side is half as long, the perimeter is half as much. The perimeter of one triangle is decreasing.
f. $\frac{3}{2}$. Three times the number of triangles, but each one has half the perimeter. The total perimeter of all triangles is increasing.
g. $3(\frac{3}{2})^{n-1}$
h. The perimeter is infinite. The numbers get larger and larger without bound.

3. a. In step 1, there is only one square. If each side is 1 ft. in length, then the perimeter is 4 ft. In step 2, the original square is divided into nine smaller squares and modified by removing the center square. The removed square contributes to the perimeter. It has sides of length $\frac{1}{3}$ ft. Thus, the perimeter of this smaller square is $4 \cdot \frac{1}{3} = \frac{4}{3}$ ft. The total perimeter is the sum of the previous perimeter and the new contribution: $4 + \frac{4}{3}$ ft.

b.

Step	Number of New Squares	Length of Each Side (feet)	Perimeter of One New Square (feet)	Total Perimeter of All New Squares (feet)	Total Perimeter of All Squares (feet)
1	1	1	$4 \cdot 1 = 4$	$1 \cdot 4 = 4$	4
2	1	$^1/_3$	$4 \cdot {}^1/_3 = {}^4/_3$	$1 \cdot {}^4/_3 = {}^4/_3$	$4 + {}^4/_3$
3	8	$^1/_9$	$4 \cdot {}^4/_9 = {}^4/_9$	$8 \cdot {}^4/_9 = {}^{32}/_9$	$4 + {}^4/_3 + {}^{32}/_9$
4	64	$^1/_{27}$	$4 \cdot {}^1/_{27} = {}^4/_{27}$	$64 \cdot {}^4/_{27} = {}^{256}/_{27}$	$4 + {}^4/_3 + {}^{32}/_9 + {}^{256}/_{27}$
5	512	$^1/_{81}$	$4 \cdot {}^1/_{81} = {}^4/_{81}$	$512 \cdot {}^4/_{81} = {}^{2,048}/_{81}$	$4 + {}^4/_3 + {}^{32}/_9 + {}^{256}/_{27} + {}^{2048}/_{81}$
6	4,096	$^1/_{243}$	$4 \cdot {}^1/_{243} = {}^4/_{243}$	$4{,}096 \cdot {}^4/_{243} = {}^{16,384}/_{243}$	$4 + {}^4/_3 + {}^{32}/_9 + {}^{256}/_{27} + {}^{2048}/_{81} + {}^{16384}/_{243}$

c. 8. Each step creates nine new squares, one of which is deleted.

d. $^1/_3$. Each original side is divided into three new squares.

e. $^1/_3$. Since each side is one third as long, the perimeter is one third as long.

f. $^8/_3$. Eight new squares each of which has perimeter one third as much as before. The total perimeter of all new squares is increasing.

g. $\dfrac{4 \cdot 8^{n-2}}{3^{n-1}}$ or $\dfrac{4}{3} \cdot \left(\dfrac{8}{3}\right)^{n-2}$, valid for $n > 1$

h. The total perimeter is infinite. The numbers increase without bound.

i. The total perimeter is infinite. It increases without bound.

5. a. $3 \cdot 2^{n-1}$ ft. (If the original triangle has sides of length 1 ft.)

b. Infinite ∞

c. $\left(\dfrac{2}{3}\right)^{n-1} \cdot \dfrac{\sqrt{3}}{4}$ sq. ft. (If the original triangle has sides of length 1 ft.)

d. 0

7. Inductive: noticing the pattern that the number of sides is increasing by a factor of 4, that the length of each side is decreasing by a factor of $\frac{1}{3}$, and that the perimeter is increasing by a factor of $\frac{4}{3}$.
Deductive: applying the general rule that the perimeter of a square is four times the length of a side, applying the general rule that the perimeter of a shape is the sum of all lengths of sides.

21. square with side 16

23. 6.5 ft (rounded down so the object will fit)

25. The first square has area c^2.
Rearranging the area creates two squares: one with area a^2 and another with area b^2. Thus, $a^2 + b^2 = c^2$.

27.

1.	$\angle CBA = \angle DAB$	Given.
2.	$BC = AD$	Given.
3.	$AB = AB$	Anything is equal to itself.
4.	$\Delta CAB \cong \Delta DBA$	SAS
5.	$\angle CAB = \angle DBA$	Corresponding parts of congruent triangles are equal.
6.	$\angle CAD = \angle DBC$	Since $\angle CAD = \angle CAB - \angle DAB$ $= \angle DBA - \angle CBA$ $= \angle DBC$

29. $\pi \approx 6[(\sqrt{6} - \sqrt{2})/2] = 3.105828541$

31. a. Foci $(-\sqrt{13}, 0)$ and $(\sqrt{13}, 0)$

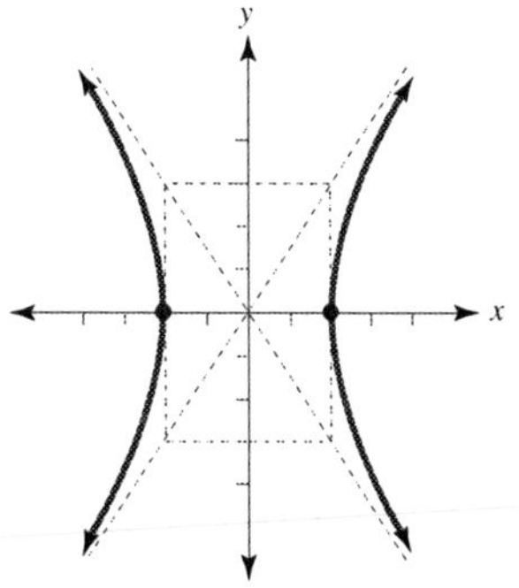

b. Foci $(0, -\sqrt{13})$ and $(0, \sqrt{13})$

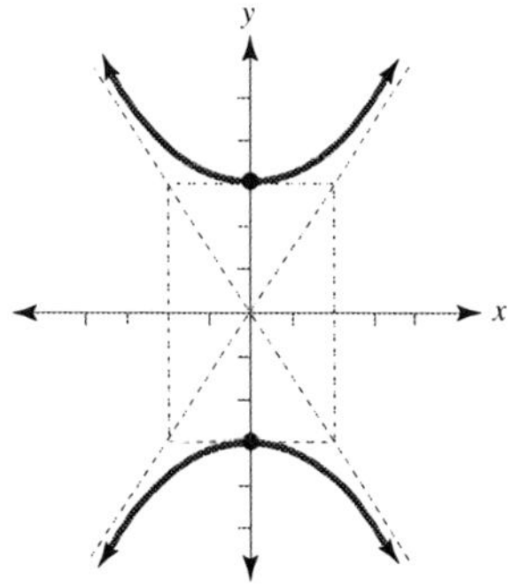

Chapter 8 Review

1. Area = $(x^2 - 2x)$ square feet
Perimeter $= 4x - 4$ feet

3. $A = 73.5$ square inches
$P = 41$ inches

5. $A = 69$ square yards
$P \approx 37.4$ yards

7. a. 1,598,000 AU **b.** 25.3 ly **c.** 7.8 pc

9. $V = 72.8$ cubic centimeters
$SA = 153.9$ square centimeters

11. $V = 1{,}056$ cubic inches
$A = 664$ square inches

13. He will need 9 bags.

15. a. 224 cubic inches **b.** 200 square inches

17. a. 450.7 cubic feet **b.** 268.0 square feet

19. a. 4.2140 cubic cubits **b.** 4.1888 cubic cubits
c. $0.00602 \approx 0.6\%$

33. Foci $(0, -\sqrt{5})$ and $(0, \sqrt{5})$

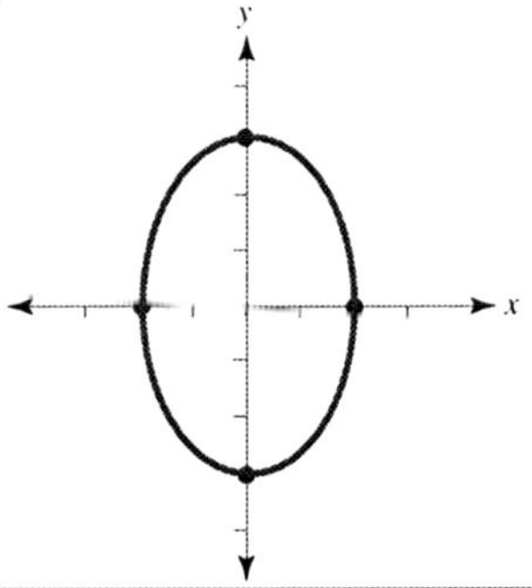

35. $B = 64.6°$
$b = 50.7$ feet
$a = 24.1$ feet

37. 213.2 feet

39. 0.33 pc

41. a. 175,238 AU **b.** 1.18 seconds

43. a. One **b.** More than one **c.** None

45. a. Step 1

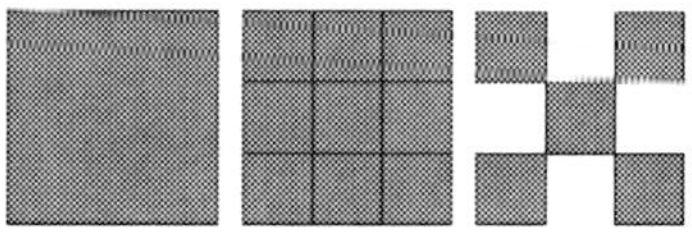

b.

Step	Number of Squares	Length of Side	Perimeter of Square	Total Perimeter
1	1	1	4	4
2	5	$\frac{1}{3}$	$\frac{4}{3}$	$\frac{20}{3}$
3	25	$\frac{1}{9}$	$\frac{4}{9}$	$\frac{100}{9}$
4	125	$\frac{1}{27}$	$\frac{4}{27}$	$\frac{500}{27}$
5	625	$\frac{1}{81}$	$\frac{4}{81}$	$\frac{2500}{81}$
n	5^{n-1}	$\frac{1}{3^{n-1}}$	$\frac{4}{3^{n-1}}$	$4 \cdot \frac{5^{n-1}}{3^{n-1}} = 4\left(\frac{5}{3}\right)^{n-1}$

c. $4\left(\frac{5}{3}\right)^{n-1}$ **d.** The perimeter is infinite.

e.

Step	Number of Squares	Length of Side	Area of Square	Total Area
1	1	1	1	1
2	5	$\frac{1}{3}$	$\frac{1}{9}$	$\frac{5}{9}$
3	25	$\frac{1}{9}$	$\frac{1}{81}$	$\frac{25}{81}$
4	125	$\frac{1}{27}$	$\frac{1}{729}$	$\frac{125}{729}$
5	625	$\frac{1}{81}$	$\frac{1}{6561}$	$\frac{625}{6561}$
n	5^{n-1}	$\frac{1}{3^{n-1}}$	$\left(\frac{1}{3^{n-1}}\right)^2$	$\left(\frac{5}{9}\right)^{n-1}$

f. 0 **g.** $d = 1.465$.
h. If we focus in on any smaller square it will look like the entire box fractal.
i. For any square that is not removed, the recursive rule is to remove the same four squares in the middle of the edges.

47. a. one-point perspective
b. below
c. approximately $(1, 2)$
d. $y = 2$

49. Answers will vary. Possible drawing:

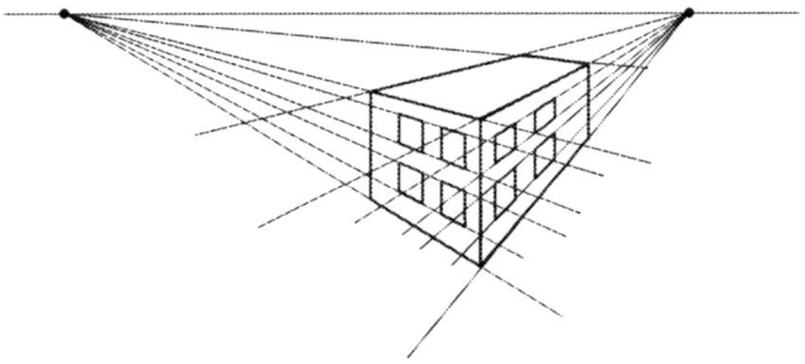

51. Answers may vary.
a. Both. The floor tiles use two-point perspective while the rest uses one-point perspective.
b. The vanishing point is on the girl playing. Vermeer might have chosen that point to draw the viewer's eye to the girl at the piano.
c. No
d. Yes, the floor's two vanishing points are off of the painting.
e. Yes, on the tiles in the floor and on the windows.
f. Yes, there are tiles on the floor.
g. Albertian grid is used on the tiles in the floor. See diagonal lines on painting.
h. Yes, horizon follows the line of the windows and piano.

CHAPTER 9 Graph Theory

9.1 A Walk Through Konigsberg

1. Yes. There are four vertices. If you eliminate one bridge, then two of the vertices have an odd number of edges, and the rest have an even number of edges. Start at one of the odd vertices and end at the other.

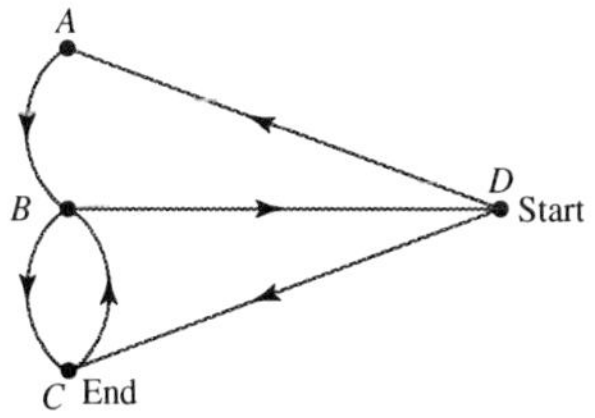

3. No; the closure of any one bridge is sufficient to create a bridge walk.

5. Yes. Start at one odd vertex and end at the other.

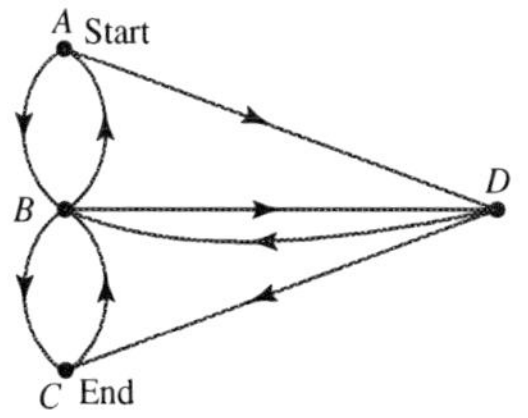

ABDBCBADC

7. 4 vertices, 4 edges, 0 loops

9. 5 vertices, 5 edges (one of which is a loop), 1 loop

11. **a.** 7 vertices, 6 edges, 0 loops
b. A family member is not its own child.

13.

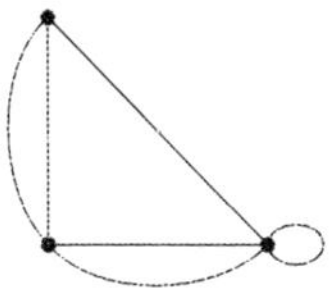

15. **a.** They are all graphs with 4 edges, 4 vertices, and 0 loops, and the edges in both graphs connect the same points.
b.

17.

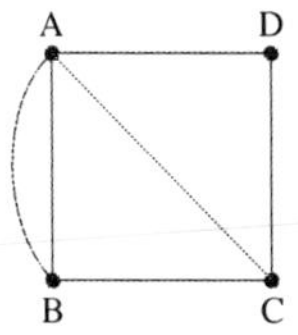

9.2 Graphs and Euler Trials

1. **a.** AB and AC
b. A and B
c. A-3, B-1, C-2, D-2
d. Yes; every pair of vertices is connected by a trail.

3. **a.** AB and BD
b. A and B
c. A-3, B-4, C-2, D-3
d. Yes; every pair of vertices is connected by a trail.

5. **a.** AB and AC **b.** A and B
c. A-4, B-3, C-3, D-2
d. Yes; every pair of vertices is connected by a trail.

7. **a.**

b.

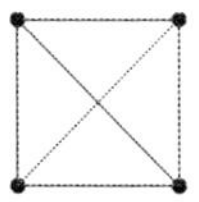

c.

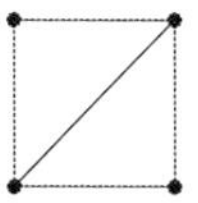

9. **a.**

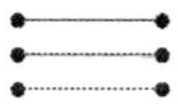

b.

11. **a.**

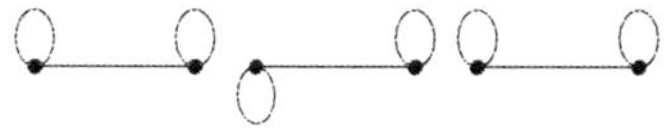

b.

c.

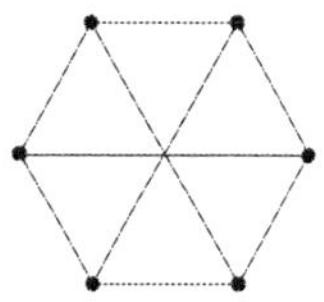

d.

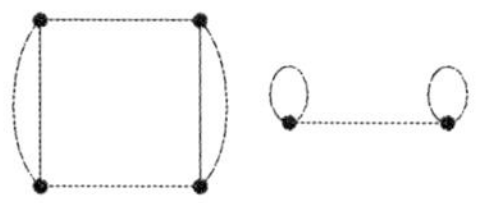

13. **a.** Euler trail (2 odd)
b. BACDA

15. **a.** There is an Euler trail because there are exactly two odd vertices.
b. ACABDBD (Label the vertices AB from L to R in the top row and CD from L to R in the bottom row.)

17. **a.** There is an Euler trail because there are exactly two odd vertices.
b. Start at B; end at C. BADCABC

19. Answers will vary. Possible answer:

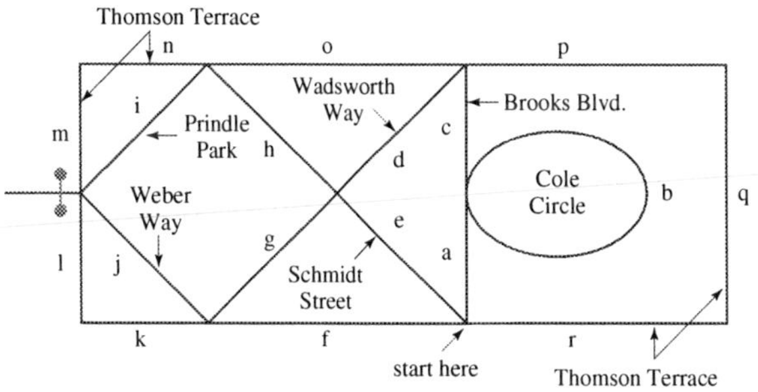

21. **a.**

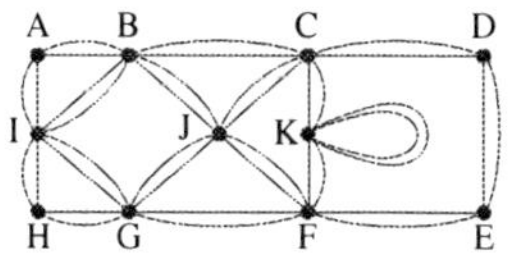

Start in the upper left-hand corner, go clockwise one block, and detour to take Prindle, Weber, Wadsworth, Schmidt, Brooks Road, Brooks Circle, Brooks Road, and Wadsworth and Schmidt back to where you detoured. Continue along the outer border to where you started, and repeat for the other side.

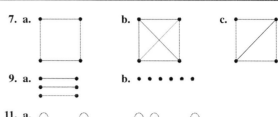

b.

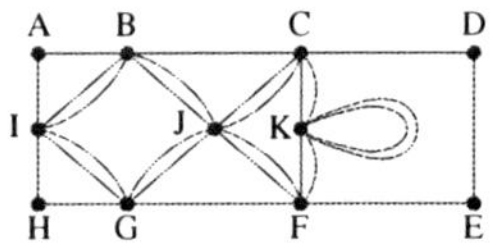

Start in the upper left-hand corner and go one block clockwise. Make a detour around the inside. (See part (a).) When you come back, continue along the border clockwise to your return.

c.

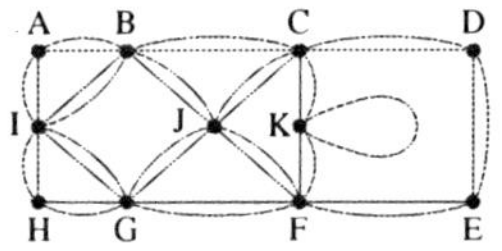

Everything is the same in part (a) except that when you repeat the inside circuit, you skip Cole Circle the second time.

23.

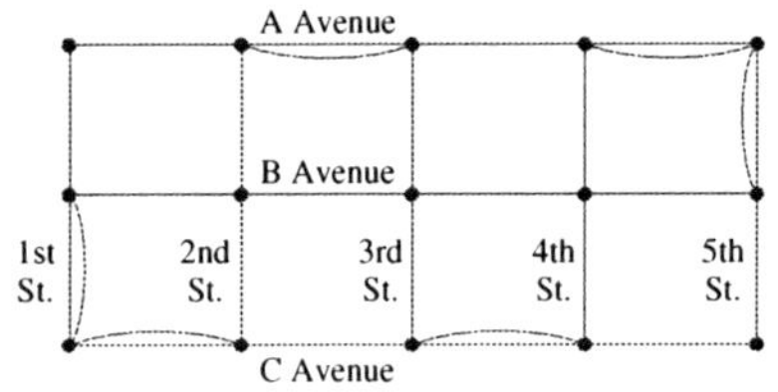

Start at A on 1st. Go over to 5th, down to C, over to 1st, up to B, over to 5th, back up to A, back to 4th, down to C, over to 3rd, up to A, over to 2nd, down to C, over to 1st, up to B and return back to A.

25.

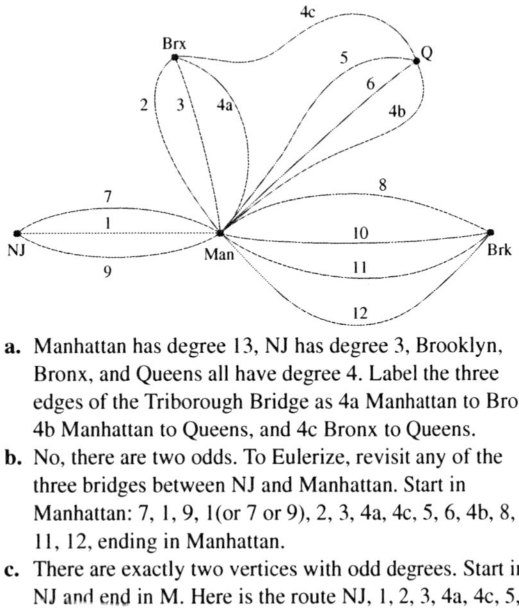

a. Manhattan has degree 13, NJ has degree 3, Brooklyn, Bronx, and Queens all have degree 4. Label the three edges of the Triborough Bridge as 4a Manhattan to Bronx, 4b Manhattan to Queens, and 4c Bronx to Queens.

b. No, there are two odds. To Eulerize, revisit any of the three bridges between NJ and Manhattan. Start in Manhattan: 7, 1, 9, 1(or 7 or 9), 2, 3, 4a, 4c, 5, 6, 4b, 8, 10, 11, 12, ending in Manhattan.

c. There are exactly two vertices with odd degrees. Start in NJ and end in M. Here is the route NJ, 1, 2, 3, 4a, 4c, 5, 6, 4b, 8, 10, 11, 12, 9, 7, M.

27. a.

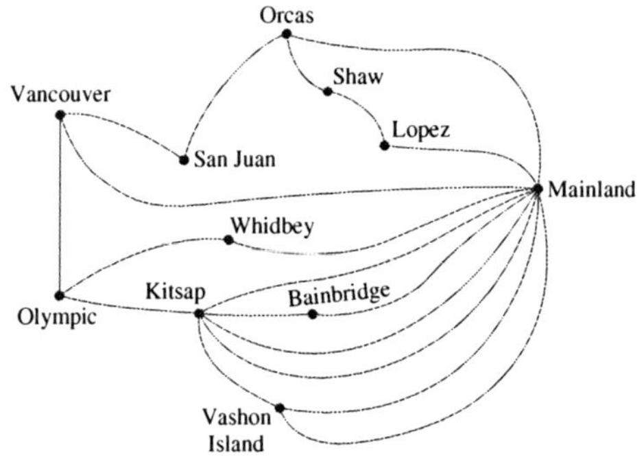

b. No, there are four odd vertices. To Eulerize, revisit a ferry from Vancouver to Orcas, and a ferry from Vashon Island to Olympic Peninsula. Now all the vertices could be considered even, and you can start and end at the same point.

c. No, there are four odd vertices. To Eulerize, revisit a ferry connecting two of the odd vertices. The other pair of odd vertices are the starting and stopping points.

29. a.

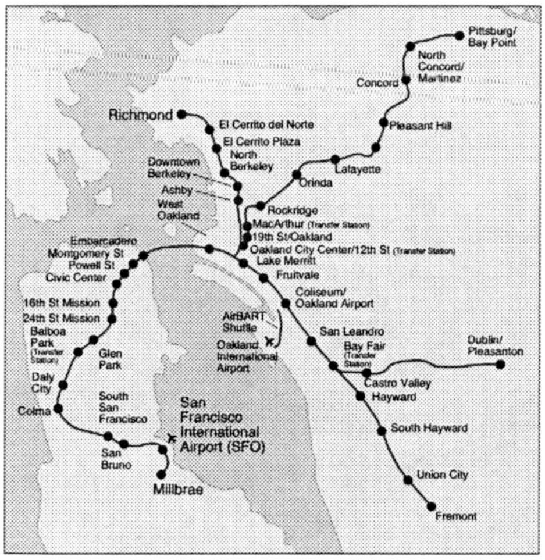

b. No, because there are only two stations with odd-degree: Pittsburg and Dublin, both with 1. Millbrae has 2 (the red and the yellow); Daly City has 6 (2 reds, 2 yellows, 1 blue, and 1 green). To Eulerize, take the yellow from Pittsburg to Millbrae, then the red from Millbrae to Richmond, the orange from Richmond to Fremont, the green from Fremont to Daly City, the blue from Daly City to Dublin, the blue from Dublin to West Oakland and the yellow back to Pittsburg.

c. Yes, it is possible to start and end at different points since there are only two odd-degrees stations. Therefore, follow the route described in part (b) except end at Dublin. Do not proceed from Dublin back to Pittsburg.

d. They can patrol all stations efficiently.

e. Part (b) is more useful because they can start and end their day at the same place.

31. a. Figure 9.15:
Number of edges in the graph: 11
Sum of the degrees of all of the vertices: 22
Figure 9.17:
Number of edges in the graph: 7
Sum of the degrees of all of the vertices: 14
Figure 9.20:
Number of edges in the graph: 18
Sum of the degrees of all of the vertices: 36

b. Sum of degrees of vertices $= 2 \times$ number of edges or

$$\text{number of edges} = \frac{\text{sum of degrees of vertices}}{2}$$

c. 2

d. 1 to each

e. An edge connects two vertices.

9.3 Hamilton Circuits

1. a. 6

b. $A \rightarrow B \rightarrow D \rightarrow P \rightarrow A$: \$927

c. $A \rightarrow B \rightarrow P \rightarrow D \rightarrow A$: \$848

d. $A \rightarrow B \rightarrow P \rightarrow D \rightarrow A$: \$848

e. $A \rightarrow B \rightarrow P \rightarrow D \rightarrow A$: \$848

3. **a.** 24
b. A → B → Po → Px → D → A: $685
c. A → B → Po → Px → D → A: $685
d. A → B → Px → Po → D → A: $918
e. There are too many possibilities.
5. F → I → K → B → C → L → F: 123 min
7. F → K → I → C → B → L → F: 108 min
9. F → I → L → B → C → R → F: 84 min
11. F → I → L → B → C → R → F: 84 min
13. **a.** 5 mm **b.** 7 mm
c. 10 mm **d.** 22 mm
15. (0, 0) → (2, 2) → (3, 1) → (4, 5) → (6, 4) → (7, 5) → (0, 0): 28 mm
Or
(0, 0) → (3, 1) → (2, 2) → (4, 5) → (6, 4) → (7, 5) → (0, 0): 28 mm
17. (0, 0) → (2, 2) → (3, 1) → (4, 5) → (6, 4) → (7, 5) → (0, 0): 28 mm
19. (0, 0) → (5, 1) → (4, 3) → (3, 4) → (1, 8) → (2, 7) → (0, 0): 28 mm
21. (0, 0) → (1, 8) → (2, 7) → (3, 4) → (4, 3) → (5, 1) → (0, 0): 26 mm
23. In some cases answers may differ because of a tie.
ATL → WASH → BOS → SFO → PORT → PHX → DEN → ATL: $748 The cheapest route is $748.
25. In some cases, answers may differ because of a tie.
ATL → WASH → BOS → SFO → PORT → PHX → DEN → ATL: $748
Or
ATL → DEN → PHX → PORT → SFO → BOS → WASH → ATL: $748
27. In some cases, answers may differ because of a tie.
ATL → DEN → SFO → PORT → PHX → BOS → WASH → ATL: $1,096
29. Exercises 23 and 25 both yielded $748.
31. In some cases, answers may differ because of a tie.
NYC → SEA → CHI → MIA → HOU → LAX → NYC: $1,132

9.4 Networks

1. Yes; it is a connected graph that has no circuits.
3. No; it is not connected.
5. No; it has a circuit.
7. No; it has a circuit.
9. Yes; it is a connected graph that has no circuits.
11. No; it is not connected.
19. 65

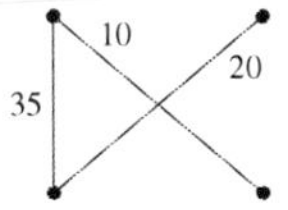

21. 115

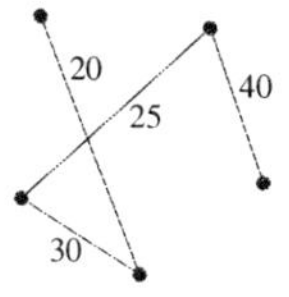

23. 531

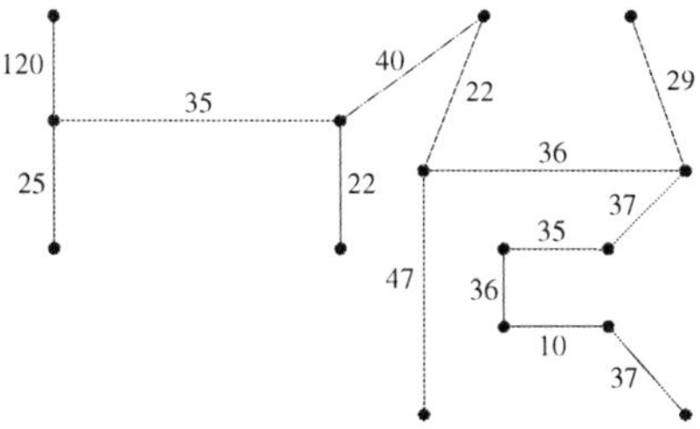

25. N–B–C–K–F: 2808 miles
27. N → Ph → A → D → S → Po: 3,442 miles
29. S → Po → K → C → A → Ph → N → B: 3,954 miles
31. 131

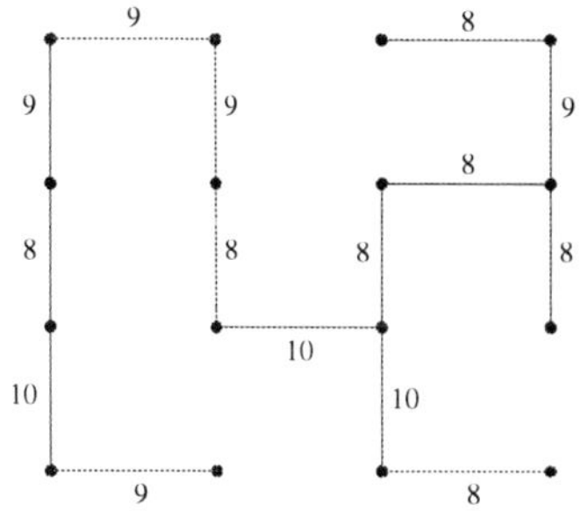

33. B because the sum of the distances is the shortest and it is given that one of them is a Steiner point.
43. **a.** 100 tan 40°
b. 100 tan 20°
c. 100(tan 40° tan 20°)
45. **a.** 50°
b. 70°
c. 110°
47. **a.** 100
b. 100 tan 25°
c. 100 − 100 tan 25°
49. **a.** 45°
b. 65°
c. 115°
51. Since the three points form a triangle with angles that are less than 120°, find a Steiner point inside the triangle. Let y = perpendicular edge length and let h = length of each other edge.
$y + 2h = 50(\tan 40° + \sqrt{3})$
53. Since the three points form a triangle with an angle that is 120° or more (Durham, 130°), then the shortest network consists of the sum of the two shortest sides of the triangle. Let h = length of one short side.
$2h = \frac{560}{\sin 65°}$
55. Therefore, **c**, $500\sqrt{3} + 600$, has the shortest network.
57. Use two Steiner points to form triangles with the shorter sides: $600\sqrt{3} + 700$
59. Therefore, **c**, $500\sqrt{3} + 500$, has the shortest network.
61. Answers will vary. Possible answer: Financial considerations

9.5 Scheduling

1.

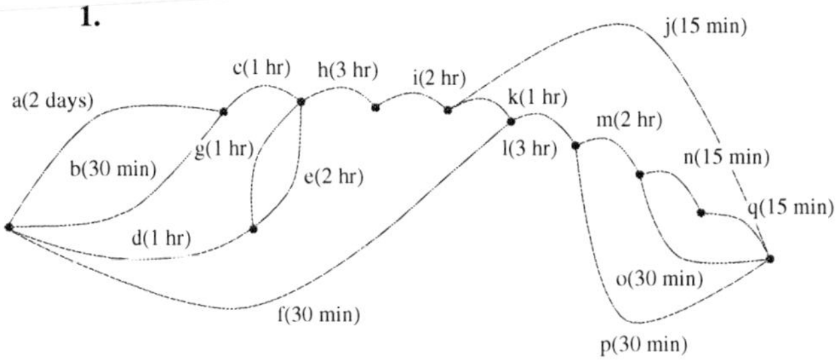

The critical path is a, c, h, i, k, l, m, n, q.

3. 4 workers

5.

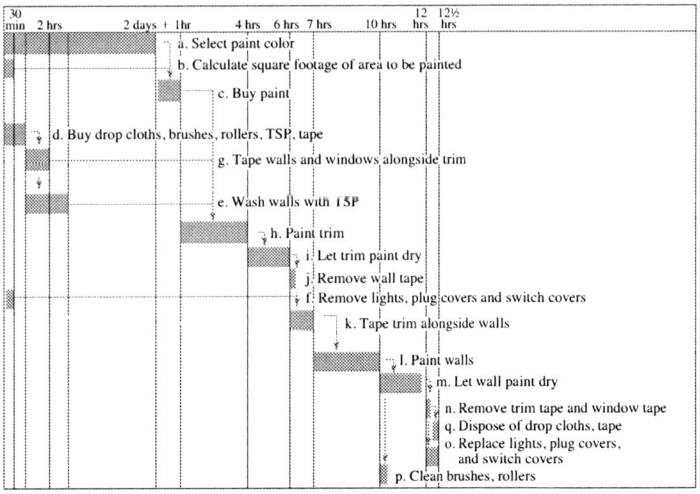

7. 2 days 12 hours. **9.** $8\frac{3}{4}$ hours **11.** $12\frac{1}{4}$ hours

13. Start: $4\frac{1}{2}$ hours
Finish: $5\frac{1}{2}$ hours

15.

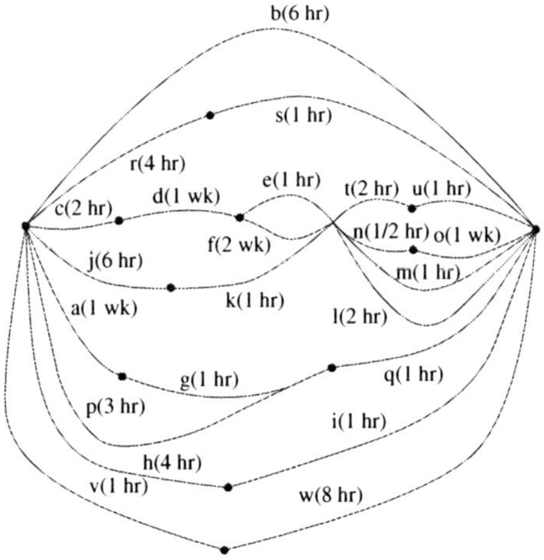

The critical path is c, d, f, n, o.

17. 3 workers

19.

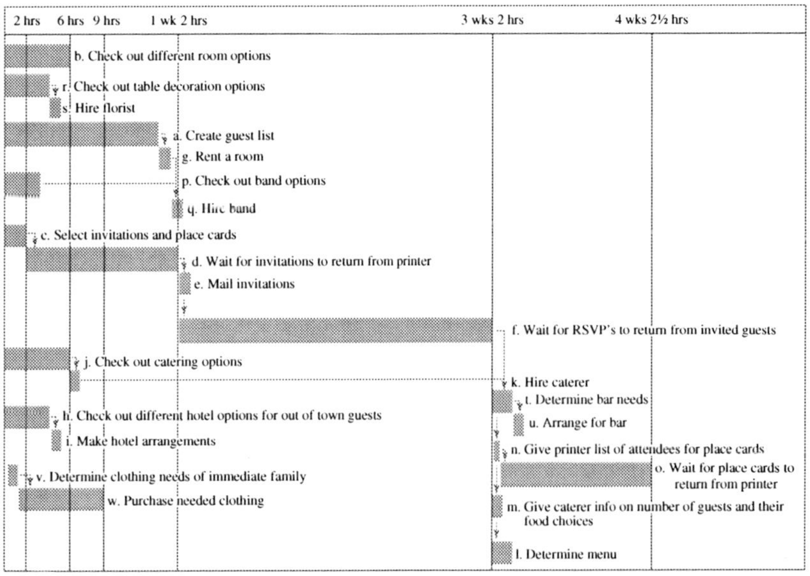

21. 3 weeks, 5 hours

23. 37 wk

25. Start: 24 weeks
Finish: 28 weeks

27. The inspection takes place at p. So start right before p, 35 weeks, and finish right after p, 37 weeks.

29. It was dependent on running closed ESS in the lab. The task's approximate length is 1 year. The task's completion date is end of the 1999.

31. It was dependent on install solar array, perform targeted lightweighting, improve reliability of motors, complete environmental control system installation, upgrade PMRF facilities, solar cell procurement, and Helios prototype functional test. The task's approximate length is 1 year. The task's completion date is the end of 2001.

33. 9 years
The approximate completion date is the end of 2003.

Chapter 9 Review

1. a. 3 vertices, 6 edges, 0 loops **b.** Answers will vary.
c. AB, BC **d.** A and B **e.** A-4, B-6, C-2
f. Yes; every pair of vertices is connected by a trail.

3. a. 5 vertices, 5 edges, 1 loop **b.** Answers will vary.
c. AB, BE **d.** A and B **e.** A-2, B-3, C-2, D-2, E-3
f. No; there is no trail connecting any of the 4 vertices on the left to vertex C.

5. a. 6 vertices, 12 edges, 0 loops **b.** Answers will vary.
c. AB, BC **d.** A and B
e. A-4, B-4, C-5, D-3, E-4, F-4
f. Yes; every pair of vertices is connected by a trail.

7.

9.
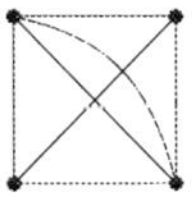

11. a.

b.
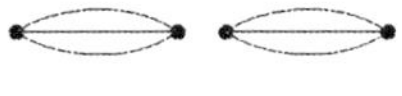

c.
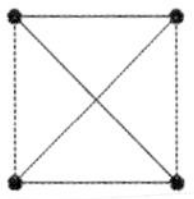

d.
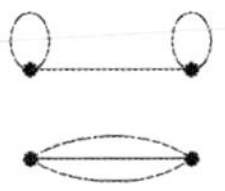

13. a. All even vertices means that there is an Euler circuit.
b. BABABCB

15. a. None because the graph is not connected.
b.
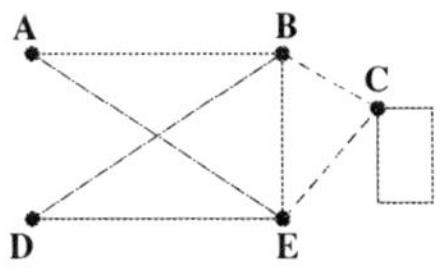

ABCCEBDEA

17. a. There are exactly two odd vertices. There is an Euler trail starting at C and ending at D.
b. CAFCEBFEDCBAD

19. a. There are more than two odd vertices. There is no Euler circuit. Eulerize.
b.
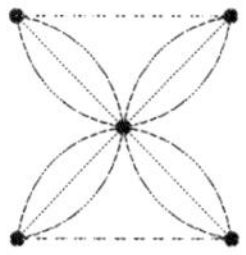
Add two edges, .one connecting the two top vertices and the other connecting the two bottom vertices.

21. a. No. There are odd vertices.
b.
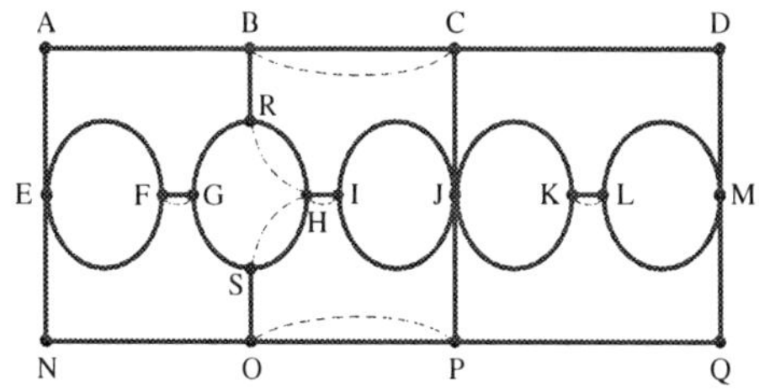

The route is E-F-G-R-B-C-J-P-O-S-H-I-J-K-L-M-L-K-J-I-H-R-H-S-G-F-E-A-B-C-D-M-Q-P-O-N-E.

23.
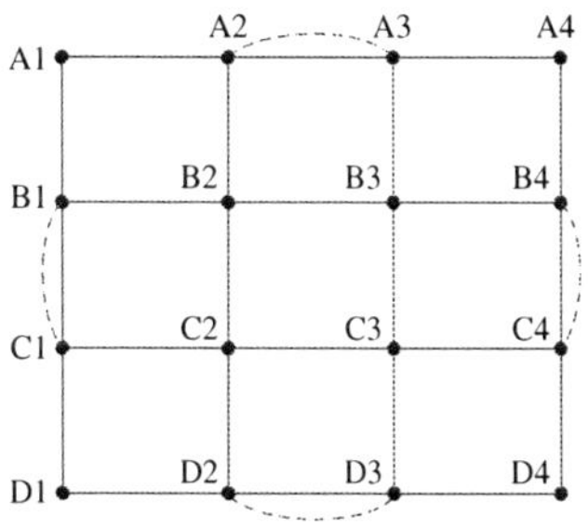

Start on A and 1st, go right on A one block, down to D, over to 3rd, back to A, loop to 2nd and back, over to 4th, down to D, over to 1st, up to C, over to 4th, up to B over to 1st, loop to C and back to A1.

25. a. 6 **b.** CMNLC: $571 **c.** CNLMC: $560
d. CMNLC:$571 **e.** CNLMC: $560

27. a. 24 **b.** CMNSLC: $650 **c.** CSLNMC: $490
d. CSLNMC: $490 **e.** Too many possibilities.

29. CMNSLHC: $638

31. CMNSLHC: $638

33. $(0, 0) \to (2, 4) \to (5, 7) \to (6, 3) \to (7, 2) \to (1, 1) \to (0, 0)$: 28 mm

35. Answers will vary. Possible answer:

37. Answers will vary. Possible answer:
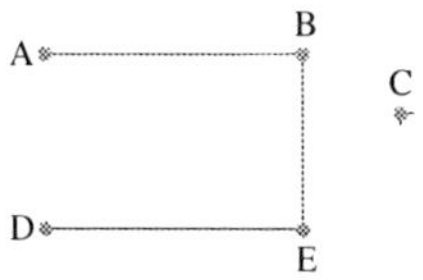

39. Answers will vary. Possible answer:
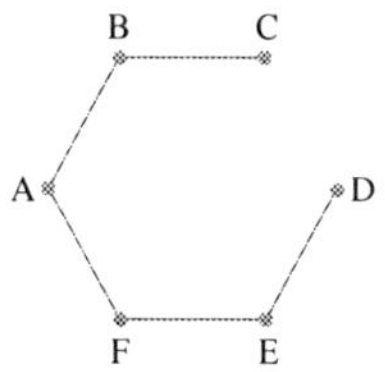

41. Answers will vary. Possible answer:

43. 60

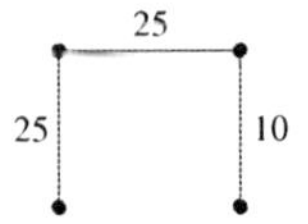

45. 200

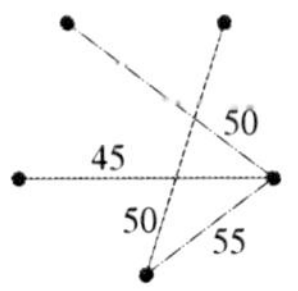

47. 1,000

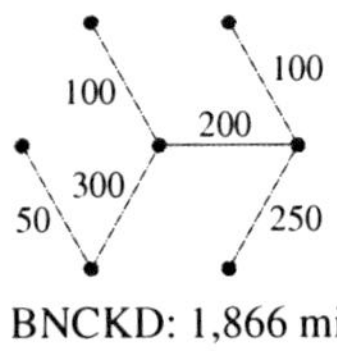

49. BNCKD: 1,866 mi **51.** BNPhCKDS: 2,849 mi

53. Since the three points form a triangle with angles that are less than 120°, find a Steiner point inside the triangle. Let y = perpendicular edge length and let h = length of each other edge. $y + 2h = 125(\tan 35° + \sqrt{3})$

55. Since the three points form a triangle with an angle that is 120° or more (Pleasant Hill, 124°), then the shortest network consists of the sum of the two shortest sides of the triangle. Let h = length of one short side.

$$2h = \frac{300}{\sin 62°} \text{ or } \frac{300}{\cos 28°}$$

57. Place two Steiner points, so the shorter sides are bases of isosceles triangles. $700\sqrt{3} + 900$

59.

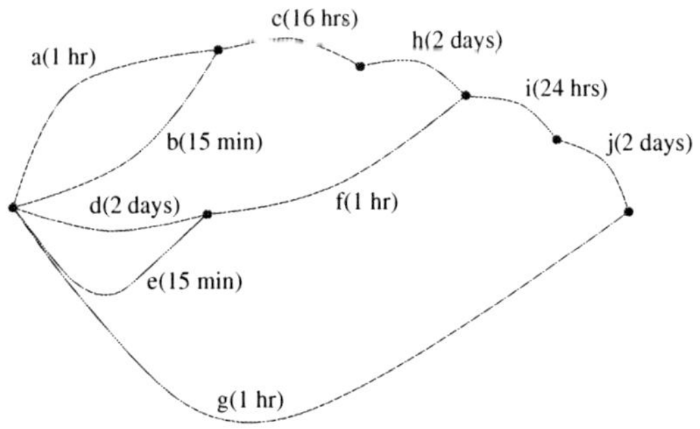

The critical path is a, c, h, i, j.

61. 2 workers

63. Create the Gantt chart from the PERT chart.

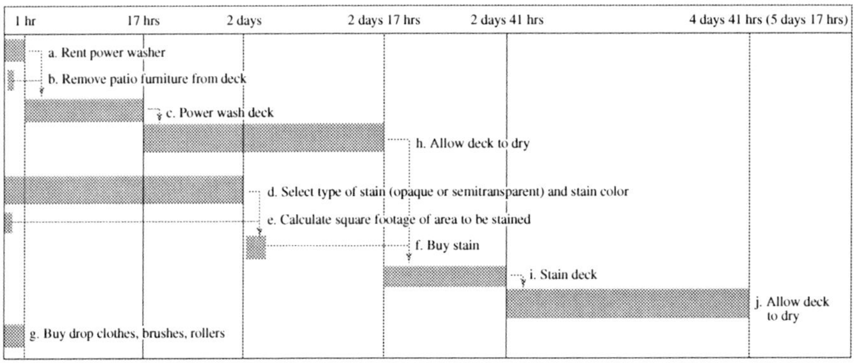

CHAPTER 10 Exponential and Logarithmic Functions

10.0A Review of Exponentials and Logarithms

1. $v = 2$ **3.** $v = -4$ **5.** $25 = u$

7. $1 = u$ **9.** $b = 4$ **11.** $b = \frac{1}{2}$

13. $b^P = Q$ **15.** $b^M = N + T$ **17.** $b^{M+R} = N + T$

19. $\log_b G = F$ **21.** $\log_b G = F + 2$

23. $\log_b (E - F) = CD$ **25.** $\log_b (Z + 3) = 2 - H$

27. a. 4.05519996684 **b.** 25.1188643151

29. a. 2.14501636251 **b.** 2.34979510988

31. a. 0.301194211912 **b.** 0.063095734448

33. a. 3.93313930533 **b.** 2.87719968669

35. a. 4.51250674972 **b.** 79.2446596231

37. a. 18.0500269989 **b.** 316.978638492

39. a. 5.67969701231 **b.** 0.176065729885

41. a. 0.982078472412 **b.** 0.426511261365

43. a. −0.162518929498 **b.** −0.070581074286

45. a. 4.1 **b.** 4.1

47. a. 4.79314718056 **b.** 4.40102999566

49. a. 2.3 **b.** 2.3

51. a. 8 **b.** 8

53. a. 2.30258509299 **b.** 4.60517018599 **c.** 6.90775527898

55. a. 0.434294481903 **b.** 0.868588963807 **c.** 1.30288344571

57. $\log 2.9, \ln 2.9, e^{2.9}, 10^{2.9}$

10.0B Review of Properties of Logarithms

1. $6x$ **3.** $-0.036x$ **5.** $2x + 5$

7. $1 - x$ **9. a.** x^2 **b.** x^2

11. a. $9x^2$ **b.** $9x^2$

13. $\log x - \log 4$ **15.** $\ln 1.8 + \ln x$

17. $x \log 1.225$ **19.** $\ln 3 + 4 \ln x$

21. $\log 5 + 2 \log x - \log 7$

23. $\ln 20x$ **25.** $\log 3x$

27. $\ln 2x$ **29.** $\log x^3$

31. 0

33. a. ln(0.35) **b.** log(0.35)

37. a. $250 \ln \frac{17}{4}$ **b.** $250 \log \frac{17}{4}$

41. a. $\frac{10{,}000}{73} \ln \frac{16}{13}$ **b.** $\frac{10{,}000}{73} \log \frac{16}{13}$

45. a. $e^{3.66}$ **b.** $10^{3.66}$

49. a. $6e^2$ **b.** 600

53. a. $\frac{5}{18} e^{3.1}$ **b.** $\frac{5}{18}(10)^{3.1}$

59. $\text{pH} \approx 3.5 < 7$ acid

61. $\text{pH} = 7.4 > 7$ base

63. $\text{pH} = -\log(1.3 \times 10^{-5}) = 4.9 < 7$ acid

65. $x = 10^{-7}$ mole per liter

67. a. Yes; plant because pH = 6.5 is acceptable.
b. No; do not plant because pH = 3.5 is not acceptable.

69. Paprika prefers soil that has a hydrogen ion concentration from $3.16 \times 10^{-8.5}$ to 10^{-7} mole per liter.

71. a. $t = 14.21$ years; $n = 14.21$ periods
b. $t = 13.95$ years
c. $t = 13.89$ years
d. $t = 13.86$ years
e. The larger n is (the shorter the compounding period), the shorter the doubling time.

73. To accumulate \$15,000:
$t = 4.99$ years
To accumulate \$100,000:
$t = 28.34$ years

75. To accumulate \$30,000: $t = 6.49$ years
To accumulate \$100,000: $t = 25.75$ years

77. a. $d = \frac{\log 3}{\log 2}$
b. (Note: The answers for part b for exercises 77–81 are not given per text reference.)

79. a. $d = \frac{\log 6}{\log 3}$

81. a. $d = \frac{\log 8}{\log 4}$

10.1 Exponential Growth

1. Population is 32,473. **3.** July 2012

5. a. (0,9392) and (4,9786) **b.** 4 years
c. 394 thousand people **d.** 98.5 thousand people/year
e. 1.049% per year

7. a. (0,18731) and (4,19007) **b.** 4 years
c. 276 thousand people **d.** 69 thousand people/year
e. 0.368%/year

9. a. $p = 9392e^{0.0102736324t}$ **b.** 10,197 thousand
c. 10,734 thousand **d.** In 2071

11. a. $P(t) = 18731e^{0.00365685681t}$
b. 19,147 thousand
c. 19,643 thousand
d. In 2115

13. a. $p(t) = 2{,}510e^{0.2541352t}$
b. 14,868
c. 2.7 days

15. a. $p(t) = 230{,}000e^{0.0497488314t}$ **b.** October 2010
c. \$417,826 **d.** \$15,652 per month

17. a. $p(t) = 1.6e^{0.1899057139t}$
b. $t = 9.65$ months from April 2009
February 2010

19. a. $p(t) = 109{,}478e^{0.1256876607t}$ **b.** 180,996 thousand
c. 232,723 thousand

21. a. $p(t) = 3{,}929{,}214e^{0.0300866701t}$
b. 1810. The exponential growth can't continue forever.
c. 1810: 7,171,916
2000: 2,179,040,956

23. Africa: 32.9 years
North America: 55,3 years
Europe: 19.4 years

25. 49 years

27. a. Each year the house is worth 10% more than its current value (not the original value)
b. 4.25 years

29. $P(t) = 18{,}731(1.0036837328)^t$

31. $P(t) = 9392(1.0104876491)^t$

35. 5060.3 days or 13.86 years

37. 1989 days

10.2 Exponential Decay

1. 3.2 g **3.** 7.4

5. a. $Q(t) = 50e^{-0.266595069t}$ **b.** 38.3 mg
c. 0.08 mg **d.** −11.7 mg per hour
e. −0.234 per hour or −23.4% per hour.
f. −2.1 mg per hour
g. −0.042 per hour or −4.2% per hour.
h. Radioactive substances decay faster when there is more substance present. The rate of decay is proportional to the amount present.

7. a. 21.0 hours **b.** 42.0 hours **c.** 63.0 hours

9. 30.2 years **11.** 50.3 years

13. 81,055 years **15.** 99.7 seconds

17. 53% of the original amount expected in a living organism

19. 63.5 days **21.** 1,441 years old

23. 14,648 years old **25.** 2,949 years old

27. Shroud made in 1350 A.D.: 92.6% of the original amount expected in a living organism
Shroud made in 33 A.D.: 78.9% of the original amount expected in a living organism

29. 50% of the original amount expected in a living organism

31. A 5,000 year old mummy should have about 55% of the original amount.
62% remaining carbon-14 would indicate approximately 3,950 years.
The museum's claim is not justified.

33. 8,300 years old

35.

t Hours After Injection (hr)	1	2	3	4	5
Q Portion Remaining (%)	0.891	0.794	0.708	0.631	0.562

37. 46.0 hours

39. a. From the article on pg. 777, there are 3.7×10^7 atomic disintegrations/second.
b. 3.7×10^8 atomic disintegrations/second
c. 7.4×10^8 atomic disintegrations/second

10.3 Logarithmic Scales

1. 7.6

3. Each recording yields a magnitude of 5.2

5. a. The 1906 earthquake's amplitude was almost 16 times that of the 1989 quake.
b. The 1906 quake released about 55 times as much energy as the 1989 quake.
7. a. The San Francisco earthquake's amplitude was about 32 times that of the LA quake.
b. The San Francisco quake released about 150 times as much energy as the LA quake.
9. a. The Indian Ocean earthquake's amplitude was about 126 times that of the 1989 San Francisco quake.
b. The Indian Ocean quake released about 1,109 times as much energy as the 1989 San Francisco quake.
11. a. The New Madrid earthquake's amplitude was more than 158 times that of the Coalinga quake.
b. The New Madrid quake released about 1,549 times as much energy as the Coalinga quake.
13. a. About a 26% increase.
b. About a 40% increase.
15. Energy released is magnified by a factor of 28.
17. 70 dB
19. 74 dB
21. 10 dB gain
23. 9.1 dB gain
25. It requires 5 singers to reach the higher dB level. Thus, 4 singers joined him.
27. It requires 2 singers to reach the higher dB level. Thus, 1 singer joined her.
29. It requires 6 players to reach the higher dB level. Thus, 5 players joined in.
31. 3 dB gain
33. a. The American iPod's impact is about 1,000 times that of the European iPod.
b. American: less than 2 minutes, European: 2 hours
35. $I_1 = 10^{-3.5} I_2$

Chapter 10 Review

1. $x = 4$
3. $x = 5^5 = 3{,}125$
5. $\ln(e^x) = x$ and $e^{\ln x} = x$
7. $\ln\frac{A}{B} = \ln A - \ln B$
9. $\ln(A^n) = n \cdot \ln A$
11. $\ln(A \cdot B) = \ln A + \ln B$
13. $\log(x + 2)$
15. $x = 11.30719075$
17. $x = 447{,}213.6$
19. a. $p(t) = 300e^{-0.1307824861t}$ **b.** They lose about 1.5 g
c. 263.2 g **d.** 17.6 years
21. 3.1 on the Richter scale
23. 102 dB
25. It requires 4 players to reach the highest dB level. Thus, 3 trumpet players have joined in.

CHAPTER 11 Matrices and Markov Chains

11.0 Review of Matrices

1. a. 3×2 **b.** none of these
3. a. 2×1 **b.** column matrix
5. a. 1×2 **b.** row matrix
7. a. 3×3 **b.** square matrix
9. a. 3×1 **b.** column matrix
11. $a_{21} = 22$ **13.** $c_{21} = 41$
15. $e_{11} = 3$ **17.** $g_{12} = -11$
19. $j_{21} = -3$

21. a. $AC = \begin{bmatrix} 5\cdot 23 + 0\cdot 41 \\ 22\cdot 23 - 3\cdot 41 \\ 18\cdot 23 + 9\cdot 41 \end{bmatrix} = \begin{bmatrix} 115 \\ 383 \\ 783 \end{bmatrix}$

b. CA does not exist.
23. a. AD does not exist. **b.** DA does not exist.
25. a. CG does not exist. **b.** GC does not exist.
27. a. JB does not exist.

b. $BJ = \begin{bmatrix} 2243 \\ 1056 \\ 52 \end{bmatrix}$

29. a. $AF = \begin{bmatrix} -10 & 50 \\ -56 & 229 \\ 0 & 153 \end{bmatrix}$

b. FA does not exist.
31. Sale: \$130, Regular: \$170
33. Blondie: \$3.45, Slice Man: \$3.70
35. Jim: 3.07, Eloise: 3.00, Sylvie: 3.50

37. a. $\begin{bmatrix} -0.3 \\ 4.4 \end{bmatrix}$

b. This represents the change in sales from 2002 to 2003 for hotels and restaurants.

39. a. $\begin{bmatrix} 6.25 & 8.97 & 4.97 & 24.85 & 6.98 & 3.88 \\ 6.10 & 8.75 & 5.25 & 22.12 & 6.98 & 3.75 \end{bmatrix} \begin{bmatrix} 2 \\ 1 \\ 2 \\ 1 \\ 9 \\ 4 \end{bmatrix}$

b. The cost is \$134.60 at Piedmont Lumber and \$131.39 at Truitt and White.

41. a. $[53{,}594 \quad 64{,}393 \quad 100{,}237 \quad 63{,}198]$

b. $\begin{bmatrix} 0.9885 & 0.0015 & 0.0076 & 0.0024 \\ 0.0011 & 0.9901 & 0.0057 & 0.0032 \\ 0.0018 & 0.0042 & 0.9897 & 0.0043 \\ 0.0017 & 0.0035 & 0.0077 & 0.9870 \end{bmatrix}$

c. $[53{,}336 \quad 64{,}478 \quad 100{,}466 \quad 63{,}142]$

This is the new population in 2001.

43. $BC = \begin{bmatrix} 1 \\ 22 \end{bmatrix}$, so $A(BC) = \begin{bmatrix} 106 \\ 68 \end{bmatrix}$

$AB = \begin{bmatrix} 17 & 20 & -6 \\ -3 & 12 & -8 \end{bmatrix}$, so $(AB)C = \begin{bmatrix} 106 \\ 68 \end{bmatrix}$

45. $\begin{bmatrix} 3 & -2 \\ 4 & 0 \end{bmatrix}$

47. Does not exist

49. $\begin{bmatrix} 19 & 7 & 34 \\ 74 & 0 & -11 \\ 13 & -2 & 44 \end{bmatrix}$

63. $AB = \begin{bmatrix} 62 & 32 \\ -40 & 56 \end{bmatrix}$

65. **a.** $EF = \begin{bmatrix} -1178 & -2101 & -2378 \\ 5970 & -3091 & 498 \\ 5580 & -660 & 2340 \end{bmatrix}$

b. $FE = \begin{bmatrix} 1892 & -2520 \\ 10{,}723 & -3821 \end{bmatrix}$

67. **a.** $(BF)C = \begin{bmatrix} 109{,}348 & 23{,}813 & -16{,}663 \\ -23{,}840 & 1688 & -7720 \end{bmatrix}$

b. $B(FC) = \begin{bmatrix} 109{,}348 & 23{,}813 & -16{,}663 \\ -23{,}840 & 1688 & -7720 \end{bmatrix}$

69. **a.** $C^2 = \begin{bmatrix} 376 & 64 & 281 \\ -932 & -1203 & 614 \\ 952 & -808 & 293 \end{bmatrix}$

b. $C^5 = \begin{bmatrix} 4{,}599{,}688 & -2{,}586{,}492 & 9{,}810{,}101 \\ -1{,}887{,}820 & -24{,}086{,}567 & 2{,}293{,}357 \\ 42{,}244{,}296 & -7{,}140{,}820 & -4{,}471{,}911 \end{bmatrix}$

11.1 Markov Chains

1. **a.** 0.9
b. 0.2 = 20%
c. [0.22 0.78]
d. [0.568 0.432]
e.

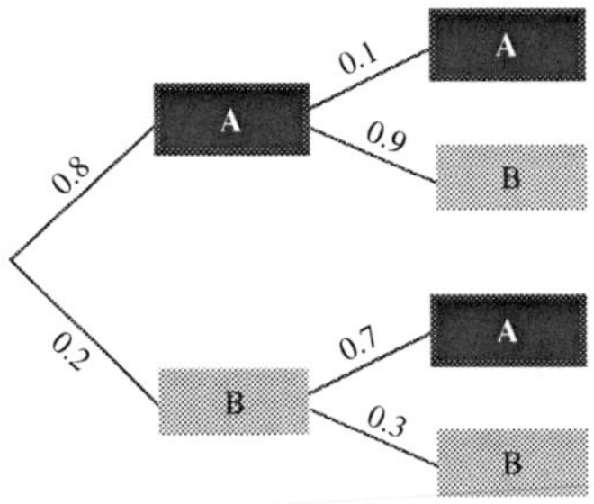

3. **a.** 0.8
b. 0.4 = 40%
c. [0.12 0.88]
d. [0.024 0.976]
e.

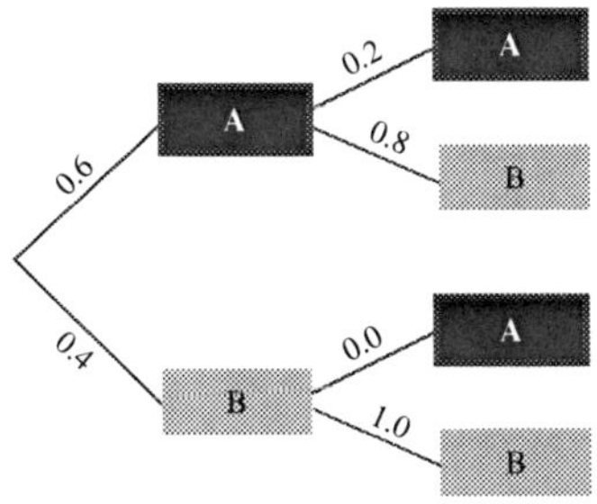

5. **a.** p(a cola drinker chooses KickKola) = 0.14
p(a cola drinker doesn't choose KickKola) = 0.86
b. [0.14 0.86]

7. **a.** p(health club user in Metropolis uses Silver's Gym) = 0.48
p(health club user in Metropolis uses Fitness Lab) = 0.37
p(health club user in Metropolis uses ThinNFit) = 0.15
b. [0.48 0.37 0.15]

9. **a.** p(next purchases is KickKola | current purchase is not KickKola) = 0.12
p(next purchase is not KickKola | current purchase is not KickKola) = 0.88
p(next purchase is KickKola | current purchase is KickKola) = 0.63
p(next purchase is not KickKola | current purchase is KickKola) = 0.37
b. The transition matrix is given by:

$$\begin{array}{cc} K & K' \end{array}$$
$$\begin{bmatrix} 0.63 & 0.37 \\ 0.12 & 0.88 \end{bmatrix} \begin{array}{c} K \\ K' \end{array}$$

11. **a.** p(go next to Silvers | go now to Silvers) = 0.71
p(go next to Fitness Lab | go now to Silvers) = 0.12
p(go next to ThinNFit | go now to Silvers) = 1 − 0.71 − 0.12 = 0.17
p(go next to Silvers | go now to Fitness Lab) = 0.32
p(go next to ThinNFit | go now to Fitness Lab) = 0.34
p(go next to Fitness lab | go now to Fitness Lab) = 1 − 0.32 − 0.34 = 0.34
p(go next to ThinNFit | go now to ThinNFit) = 0.96
0 p(go next to Silvers | go now to ThinNFit) = 0.02
p(go next to Fitness Lab | go now to ThinNFit) = 0.02
b. The transition matrix is given by:

$$\begin{array}{ccc} S & F & T \end{array}$$
$$\begin{bmatrix} 0.71 & 0.12 & 0.17 \\ 0.32 & 0.34 & 0.34 \\ 0.02 & 0.02 & 0.96 \end{bmatrix} \begin{array}{c} S \\ F \\ T \end{array}$$

13. **a.** Market share: 0.0882 + 0.1032 = 0.1914 ≈ 19%
b. Market share: 0.217614 ≈ 22%
c. Market share: 0.1914 ≈ 19%
d. Market share: 0.217614 ≈ 22%
e. Market share: 0.2431 ≈ 24%
f. Trees don't require any knowledge of matrices and are very visual. The matrices are quicker and don't require drawings.

15. **a.** Market share: Silver's: 0.4622 ≈ 46%
Fitness Lab: 0.1864 ≈ 19%
ThinNFit: 0.3514 ≈ 35%
b. Market share: Silver's = 0.4622 ≈ 46%, Fitness Lab = 0.1864 ≈ 19%, ThinNFit = 0.3514 ≈ 35%
c. Market share: Silver's = 0.3948 ≈ 39%, Fitness Lab = 0.1259 ≈ 13%, ThinNFit = 0.4793 ≈ 48%
d. Market share: Silver's = 0.3302 ≈ 33%, Fitness Lab = 0.0998 ≈ 10%, ThinNFit = 0.5700 ≈ 57%
e. Market share: Silver's = 0.2371 ≈ 24%, Fitness Lab = 0.0750 ≈ 8%, ThinNFit = 0.6879 ≈ 69% (More than 100% due to rounding.)

17. **a.** Homeowners: 39%; Renters: 61%
b. Homeowners: 45%; Renters: 55%
c. Homeowners: 51%; Renters: 49%

19. Market share: 0.2659 ≈ 27%

11.2 Systems of Linear Equations

1. Solution
3. Not a solution
5. Not a solution
7. a. $2y = -5x + 4$: slope $= -\frac{5}{2}$ $\quad y - \text{int} = 2$
   $-19y = -6x + 72$: slope $= \frac{6}{19}$ $\quad y - \text{int} = -\frac{72}{19}$
   b. One solution
9. a. $3y = -4x + 12$: slope $= -\frac{4}{3}$ $\quad y - \text{int} = 4$
   $6y = -8x + 24$: slope $= -\frac{4}{3}$ $\quad y - \text{int} = 4$
   b. Infinite number of solutions
11. a. $y = -x + 7$: slope $= -1$ $\quad y - \text{int} = 7$
    $2y = -3x + 8$: slope $= -\frac{3}{2}$ $\quad y - \text{int} = 4$
    $2y = -2x + 14$: slope $= -1$ $\quad y - \text{int} = 7$
    b. One solution
13. This system could have a single solution since it has 3 equations and 3 unknowns.
15. This system could not have a single solution because the first and second equations are equivalent.
17. This system could not have a single solution because there are less equations than unknowns.

11.3 Long-Range Predictions with Markov Chains

5. Market share: 24%
7. 20% will rent, 80% will own
   We assume that the trend won't change and that the residents' moving plans are realized.
9. Sierra Cruiser will eventually control 26% of the market.

11.4 Solving Larger Systems of Equations

1. (4, −2, 1)
3. (2, 5, 7)
5. (5, 10, −2)
7. (3, 5, 7)
9. (4, 0, −3)
11. (5, −6, 0)
13. (4, −2, 1)
15. (2, 5, 7)
17. (5, 10, −2)
19. (3, 5, 7)
21. (4, 0, −3)
23. (5, −6, 0)
25. (5, −4)
27. (2, 3)

11.5 More on Markov Chains

5. Silver's 10.8%, Fitness Lab 4.5%, ThinNFit 84.6%
7. Safe Shop 48.6%, PayNEat 37.6%, other markets 13.8%

Chapter 11 Review

1. Neither; the dimensions are 3 × 2.
3. Column; the dimensions are 4 × 1.
5. $\begin{bmatrix} -3 & -24 \\ -24 & 28 \end{bmatrix}$
7. Does not exist
9. $\begin{bmatrix} 35 & -19 & -72 & 24 \\ 12 & 34 & 51 & 24 \end{bmatrix}$
11. NYC: $43,000, DC: $41,300
13. a. (2, −4)
    b. Substituting $x = 2$ and $y = -4$ into the original equations yields true statements.
15. a. (−3, 2)
    b. Substituting $x = -3$ and $y = 2$ into the original equations yields true statements.
17. (1, 7, 2)
    b. Substituting $x = 1$, $y = 7$, and $z = 2$ into the original equations yields true statements.
19. a. 0.7 b. 0.37 c. [0.226 0.774]
    d. [0.1452 0.8548]
    e.

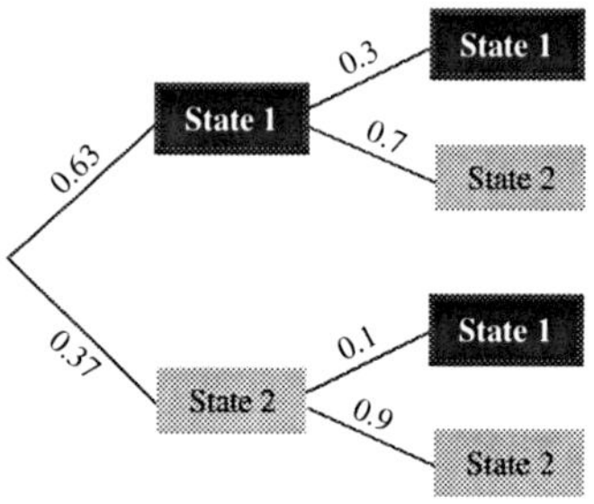

21. a. $L = [0.4 \quad 0.6]$
    b. $LT = [0.4 \quad 0.6]\begin{bmatrix} 0.7 & 0.3 \\ 0.2 & 0.8 \end{bmatrix} = [0.4 \quad 0.6]$
23. a. $L = [\frac{1}{7} \quad \frac{6}{7}]$
    b. $LT = [\frac{1}{7} \quad \frac{6}{7}]\begin{bmatrix} 0.4 & 0.6 \\ 0.1 & 0.9 \end{bmatrix} = [\frac{1}{7} \quad \frac{6}{7}]$
25. a. $L = [\frac{2}{5} \quad \frac{1}{5} \quad \frac{2}{5}]$
    b. $LT = [\frac{2}{5} \quad \frac{1}{5} \quad \frac{2}{5}]\begin{bmatrix} 0.5 & 0.1 & 0.4 \\ 0.2 & 0.2 & 0.6 \\ 0.4 & 0.3 & 0.3 \end{bmatrix} = [\frac{2}{5} \quad \frac{1}{5} \quad \frac{2}{5}]$
27. Market share: 21.5%
29. Long term prediction 29.6%.
31. Apartment: 22.1%, Condo/Townhouse: 19.6%, House: 58.3%
    Thus, more condominiums and townhouses will be needed.
33. Emigration and immigration, also affordability based or economy. We don't know if the prospective buyers can afford houses. It ignores economic conditions.

CHAPTER 12 Linear Programming

12.0 Review of Linear Inequalities

1.

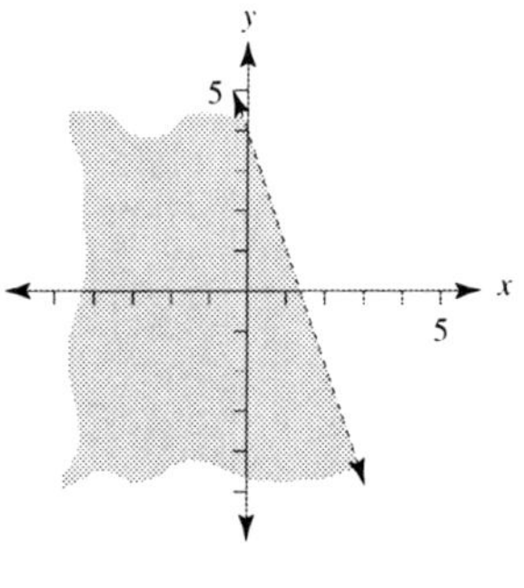

3.

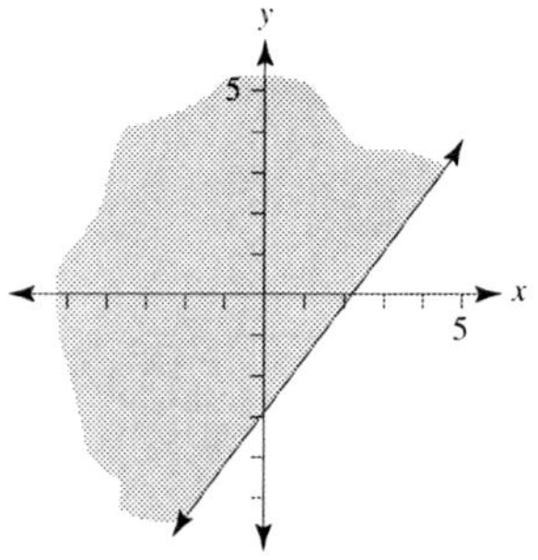

5.

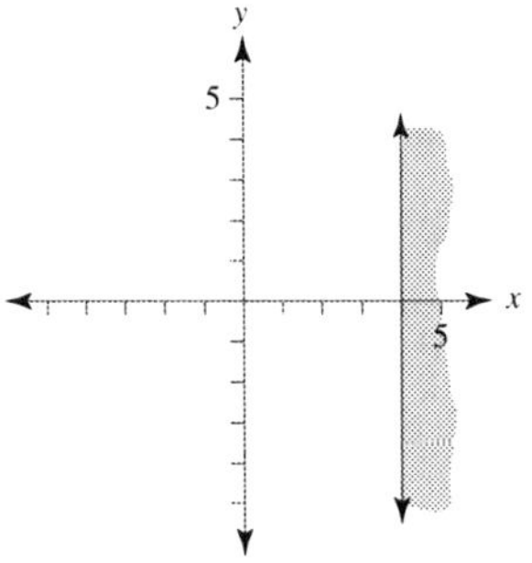

7.

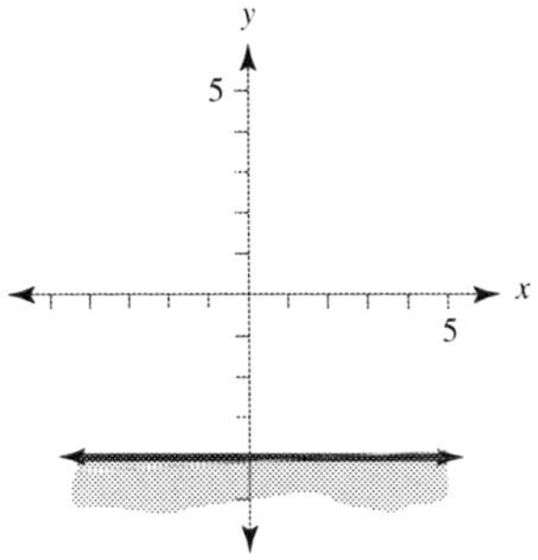

9. a.

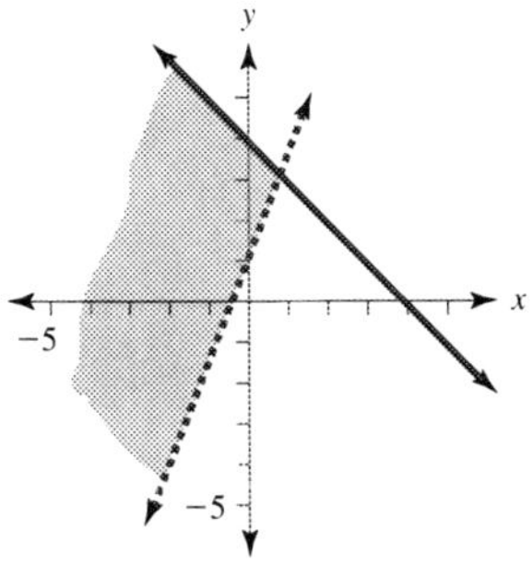

b. Unbounded **c.** (1, 3)

11. a.

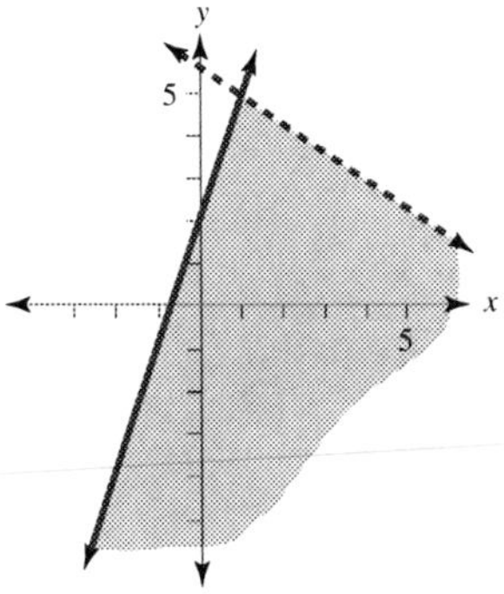

b. Unbounded **c.** (1, 5)

13. a.

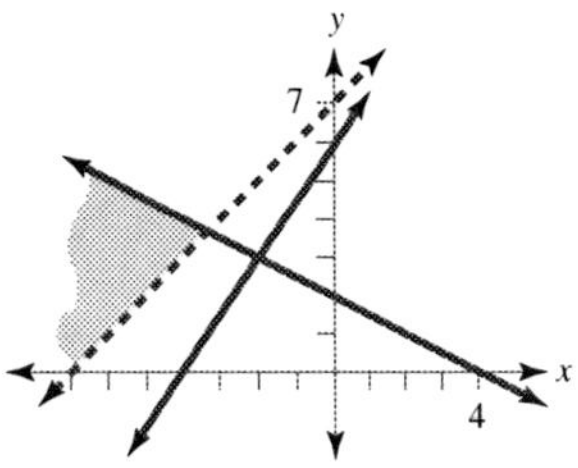

b. Unbounded

c. $(-3\frac{1}{3}, 3\frac{2}{3})$

15. a.

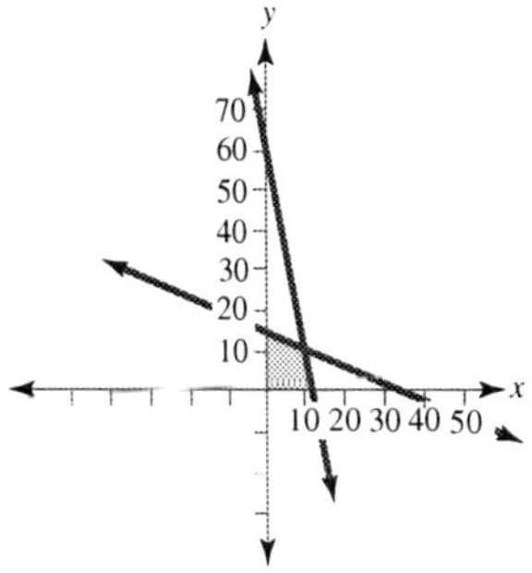

b. Bounded

c. Corner Points: (0, 14), (0, 0), (12, 0), (10, 10)

17. a.

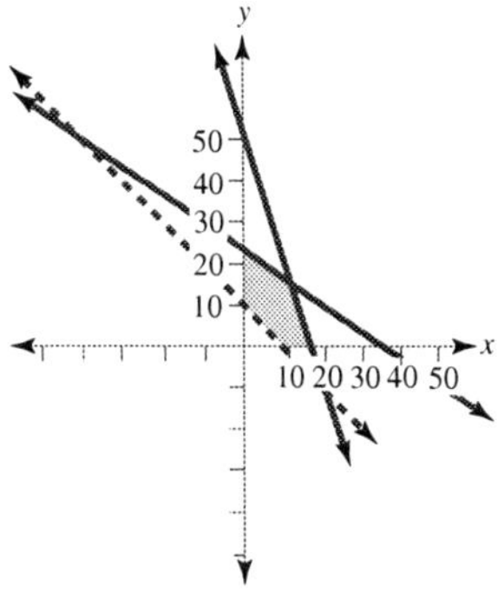

b. Bounded

c. Corner Points:

$(0, 10), (0, 23\frac{2}{11}), (17\frac{1}{7}, 0), (10, 0), (12, 15)$

19. a.

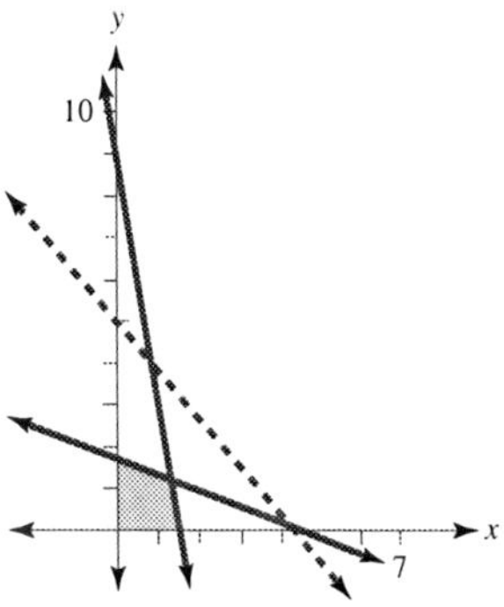

b. Bounded

c. Corner Points: (0, 0), (0, 1.7), (1.5, 0), (1.3, 1.2)

21. a.

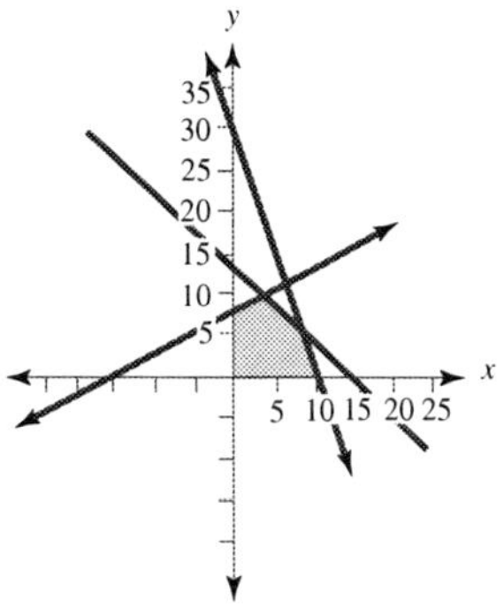

b. Bounded

c. Corner Points: (0, 0), (0, 8), (10, 0), (4, 10), (8, 6)

12.1 The Geometry of Linear Programming

1. Let x = number of shrubs
y = number of trees
$1x + 3y \leq 100$

3. Let x = number of hardbacks
y = number of paperbacks
$4.50x + 1.25y \geq 5{,}000$

5. Let x = number of refrigerators
y = number of dishwashers
$63x + 41y \leq 1{,}650$

7. She should make 45 table lamps and no floor lamps. The profit is \$1,485.

9. They should make 60 pounds of Morning Blend and 120 pounds of South American Blend. The profit is \$480.

11. They should send 15,000 loaves to Shopgood and 20,000 loaves to Rollie's. The cost is \$3,000.

13. Global has many choices, including the following:
15 Orvilles and 30 Wilbers
48 Orvilles and 8 Wilburs
Each of these generates a cost of at least \$720,000. Another way to express the answer would be any point on the line $y = -\frac{2}{3}x + 40$ produces a minimum of \$720,000.

15. **a.** He should order 20 large refrigerators and 30 smaller refrigerators. The profit is \$9,500.
b. He should order 40 large refrigerators and no small refrigerators. The profit is \$10,000.
c. 40

17. 0 Mexican; 0 Columbian

19. 150 minutes unused; \$0

21. **a.** There is no maximum.
b. No maximum
c. (6, 0)
d. $z = 12$

23. They should use 3 weeks of production in Detroit and 6 weeks of production in Los Angeles. The cost is \$3,780,000.

Chapter 12 Review

1.

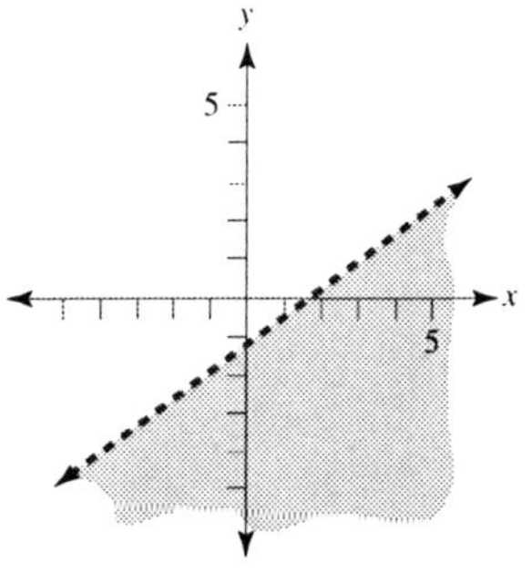

3.

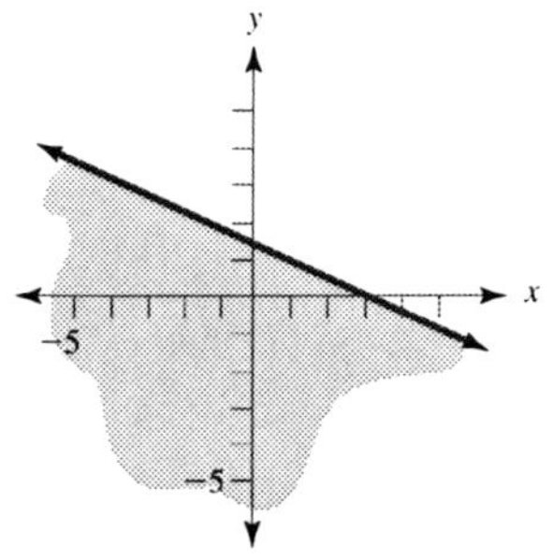

5. Corner Point: $(\frac{3}{2}, \frac{1}{2})$

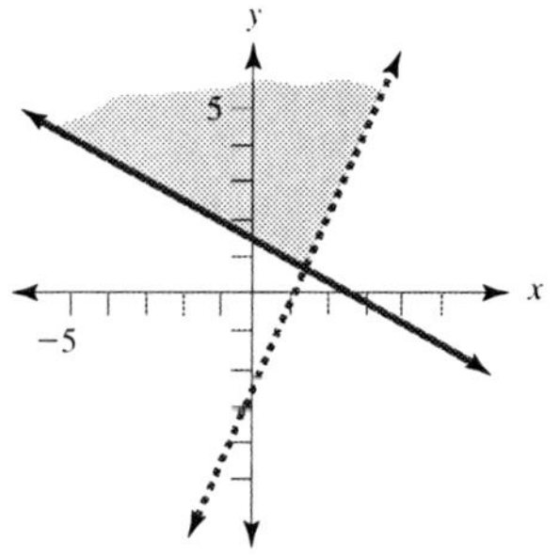

7. Corner Point: $(-3, -23)$

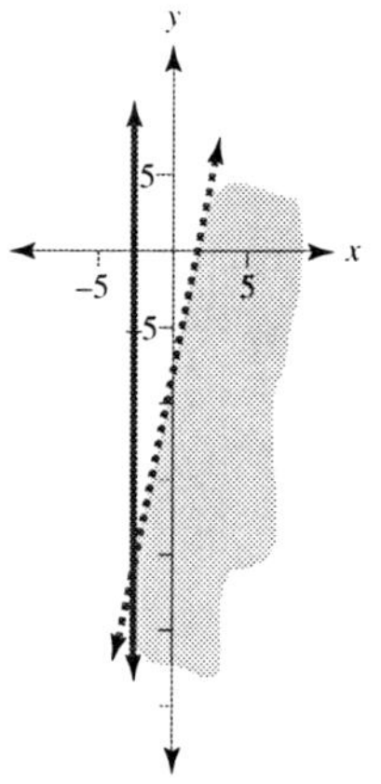

9. Corner Points: $(\frac{9}{5}, 0)$, $(0, 3)$

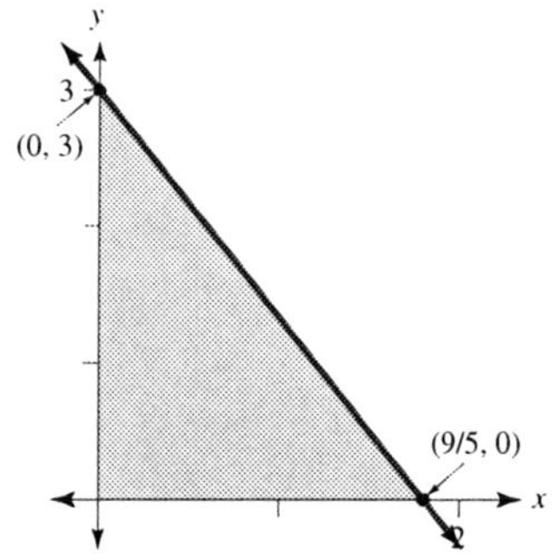

11. The minimum value of z is 0 at $(0, 0)$.

13. The minimum value of z is 25.12 at $(\frac{140}{41}, \frac{250}{41})$.

15. They should make 30 of each assembly for a maximum income of \$16,500.

17. Answers will vary. Possible answer: Think of the objective function as part of a 'profit line.' For instance, if the profit was to be 100, all points in the feasible region that satisfy the equation, objective function = 100, would make a profit line. Increasing the profit, would produce another profit line. The farther the profit lines move away from the origin, the higher the profit represented by the line. The highest profit line would be located at a corner point or along a line segment joining two corners as that is the farthest the line could go and still have points in the feasible region.

19. George Dantzig

INDEX

All hyphenated entries (12- and 13-) pertain to sections of the book found at academic.cengage.com/math/johnson.

H

I

Q

R